AF769327

Understanding Mathematical Analysis

Edited by **Lucas Lincoln**

NY RESEARCH PRESS
New York

Published by NY Research Press,
23 West, 55th Street, Suite 816,
New York, NY 10019, USA
www.nyresearchpress.com

Understanding Mathematical Analysis
Edited by Lucas Lincoln

© 2015 NY Research Press

International Standard Book Number: 978-1-63238-460-7 (Hardback)

This book contains information obtained from authentic and highly regarded sources. Copyright for all individual chapters remain with the respective authors as indicated. A wide variety of references are listed. Permission and sources are indicated; for detailed attributions, please refer to the permissions page. Reasonable efforts have been made to publish reliable data and information, but the authors, editors and publisher cannot assume any responsibility for the validity of all materials or the consequences of their use.

The publisher's policy is to use permanent paper from mills that operate a sustainable forestry policy. Furthermore, the publisher ensures that the text paper and cover boards used have met acceptable environmental accreditation standards.

Trademark Notice: Registered trademark of products or corporate names are used only for explanation and identification without intent to infringe.

Printed in China.

Contents

Preface

Mathematical analysis is a sub-field of mathematical science, which basically focuses on theories of differentiation, integration, measurements, limits, and infinite series like zeno's dichotomy. A major part of this branch is the study of analytical functions. All these theories are studied in terms of real and complex numbers. Functions also play an important part in the study of mathematical analysis. Analysis is a component of maths that was evolved from calculus based studies, which involves elementary concepts and techniques of analysis itself. Analysis and geometry are two completely different mathematical concepts, but the use of analysis in the study of space of mathematical objects, bridges the gap between these two branches. The two subjects where analysis is used in geometry are topological space and metric space.

If we turn the pages of history all the world over, mathematical analysis can be traced to the early days of ancient Greek mathematics. However, as a discipline, it was formally established during the 17th century. Mathematical analysis can be divided into various branches such as real, complex, functional, numerical, differential, measure theory among numerous others.

I hope that this book will contribute significantly to both basic and advanced studies of mathematical analysis. This book is a combined effort of authors from all over the world, who're experts in the field of Mathematical analysis. Without their hard work, support and timely submissions, this book would not have been possible. I wish to thank all the authors for their contributions to this book and it is my pleasure to acknowledge the assistance of the publishing team for their assistance during the compilation of this book into its final format. Lastly, I'd like to thank my family for their constant important support all through my life.

Editor

Existence Theorems for Vector Equilibrium Problems via Quasi-Relative Interior

Capătă Adela Elisabeta

Technical University of Cluj-Napoca, Department of Mathematics, Cluj-Napoca, G. Bariţiu 25-28, 400027 Jud. Cluj, Romania

Correspondence should be addressed to Capătă Adela Elisabeta; adela.capata@math.utcluj.ro

Academic Editors: L. B. Ciric, M. Hajarian, and Y. Xia

The aim of this paper is to present new existence theorems for solutions of vector equilibrium problems, by using weak interior type conditions and weak convexity assumptions.

1. Introduction

Starting with the paper of Blum and Oettli [1], the field of equilibrium problems is intensively studied by researchers, and many papers dealing with vector equilibrium problems were written. The interest of the researchers on this topic is due to the fact that equilibrium problems represent a natural and unified framework for other problems, such as optimization problems, variational inequality problems, and saddle-point problems, problems which until now were separately studied and have applications in physics, economics, and so forth, (see, for instance, [2]). Since then, a large variety of vector equilibrium problems were considered and the authors studied the existence of solutions (see, for instance, [3–10]), well posedness (see, for instance, [11, 12]), and sensitivity analysis (see, for instance, [13, 14]).

Most of the existence results are based on the hypothesis of nonemptyness of the ordering cone. In this paper, based on weak convexity assumptions defined by the means of quasi-relative interior of a convex set introduced in [15] and by using a quite recent separation theorem, whose statement was given in [16], we establish existence theorems for solutions of vector equilibrium problems. Then, the results are applied to vector optimization problems and to vector variational inequalities.

The paper is organized as follows. In Section 2 we recall some notions and auxiliary results that we need throughout this paper. Then, in Section 3 we present the main results of this paper, whose statements are given in the terms of quasi-relative interior and weak convexity assumptions. These results are applied to a vector optimization problem and to vector variational inequalities in Sections 4 and 5, respectively.

2. Preliminaries

Let Y be a separated locally convex space, and let $C \subseteq Y$ be a nontrivial convex cone. Having Y^*, the dual space of Y, the dual cone of C is

$$C^* := \{y^* \in Y^* \mid y^*(c) \geq 0, \forall c \in C\}. \quad (1)$$

We recall the definitions of quasi-interior and quasi-relative interior of a convex set and some useful properties of the quasi-relative interior notion.

Definition 1. Let M be a nonempty convex subset of Y. The quasi interior of M is the set

$$\text{qi } M := \{y \in M \mid \text{cl cone}(M - y) = Y\}. \quad (2)$$

Definition 2 (see [17]). Let M be a nonempty convex subset of Y. The quasi-relative interior of M is the set

$$\text{qri } M := \{y \in M \mid \text{cl cone}(M - y) \text{ is a linear subspace of } Y\}. \quad (3)$$

In a separable Banach space the quasi-relative interior of any nonempty closed convex set is nonempty (cf. [17]).

For the proof of the following properties and other useful properties of the quasi-relative interior of a convex set, we refer the reader to [17, 18], respectively.

Proposition 3. *Let M and N be nonempty convex subsets of Y, $y \in Y$, and $\alpha \in \mathbb{R}$. Then the following statements are true:*

(i) $\text{qri } M + \text{qri } N \subseteq \text{qri } (M+N)$;

(ii) $\text{qri } (\alpha M) = \alpha \text{ qri } M$;

(iii) $\text{qri } (\text{qri } M) = \text{qri } M$;

(iv) $\text{qri } (M-y) = \text{qri } M - y$;

If $\text{qri } M \neq \emptyset$, *then*

(v) $\text{cl qri } M = \text{cl } M$;

(vi) $\text{cl cone}(\text{qri } M) = \text{cl cone } M$.

The normal cone of a convex subset M of Y at $y_0 \in Y$ is defined as

$$N_M(y_0) := \{y^* \in Y^* \mid y^*(y-y_0) \le 0, \forall y \in M\}. \tag{4}$$

By means of the normal cone, the next characterizations of the abovementioned interior notions hold, and they can be found in [16, 17].

Theorem 4. *Let M be a nonempty convex subset of Y and $y \in M$. Then $y \in \text{qi } M$ if and only if $N_M(y) = \{0\}$.*

Theorem 5. *Let M be a nonempty convex subset of Y and $y \in M$. Then $y \in \text{qri } M$ if and only if $N_M(y)$ is a linear subspace of Y^*.*

For other generalizations of the classical interior we refer the reader to [19–22]. Whenever the interior of the set M is nonempty, then $\text{int } M = \text{qri } M$ (see [17, Corollary 2.14]). If Y is finite dimensional, then $\text{qri } M = \text{ri } M$ (see [17]), where by $\text{ri } M$ we understand the relative interior of M, that is, the interior with respect to the affine hull.

The following characterization for the quasi-interior of a convex cone holds, and it can be found in [23].

Proposition 6. *Let C be a convex cone in a locally convex space Y, and let $c \in C$. Then $c \in \text{qi } C$ if and only if*

$$y^*(c) > 0 \quad \forall y^* \in C^* \setminus \{0\}. \tag{5}$$

Lemma 7 (see [17]). *Let C be a convex cone in a separable locally convex space Y. Then $c \in \text{qi } C$ if and only if $c \in \text{qri } C$ and $\text{cl }(C-C) = Y$.*

Lemma 8 (see [17]). *Let C be a convex cone in a locally convex space Y. Then*

$$tC + (1-t) \text{ qri } C \subseteq \text{qri } C \tag{6}$$

for all $t \in [0,1)$.

The statement of the next theorem is due to [16], where it was proved for normed spaces, and, later on, it was proved for separated locally convex spaces by [24].

Theorem 9. *Let M be a nonempty convex subset of Y and $y_0 \in M \backslash \text{qri } M$. Then, there exists $y^* \in Y^* \setminus \{0\}$ such that $y^*(y) \ge y^*(y_0)$ for all $y \in M$.*

For other separation theorems which involve the quasi-relative interior we refer the reader to [25].

Definition 10. A function $f: A \to Y$ is said to be generalized C-subconvexlike on A if $\text{cone } f(A) + \text{qri } C$ is convex.

Proposition 11. *If $f(A)+C$ is convex, then* $\text{cone } f(A) + \text{qri } C$ *is also convex.*

Proof. Let $z_1, z_2 \in \text{cone } f(A) + \text{qri } C$ and let $t \in (0,1)$. So, there exist $t_1 \ge 0$, $t_2 \ge 0$, $a_1, a_2 \in A$, and $c_1, c_2 \in \text{qri } C$ such that

$$z_1 = t_1 f(a_1) + c_1, \qquad z_2 = t_2 f(a_2) + c_2. \tag{7}$$

Case 1. If $t_1 = 0$ and $t_2 = 0$, then $z_1 = c_1$, $z_2 = c_2$ and

$$\begin{aligned} &tz_1 + (1-t) z_2 \\ &\quad = tc_1 + (1-t) c_2 \in \text{qri } C \subseteq \text{cone } f(A) + \text{qri } C. \end{aligned} \tag{8}$$

Case 2. If $t_1 \neq 0$ or $t_2 \neq 0$, denote

$$\begin{gathered} t_1' = \frac{t_1}{t_1+t_2} \in [0,1], \qquad t_2' = \frac{t_2}{t_1+t_2} \in [0,1], \\ c_1' = \frac{c_1}{t_1+t_2} \in \text{qri } C, \qquad c_2' = \frac{c_2}{c_1+c_2} \in \text{qri } C, \\ z_1' = \frac{z_1}{t_1+t_2}, \qquad z_2' = \frac{z_2}{t_1+t_2}. \end{gathered} \tag{9}$$

So, we have

$$z_1' = t_1' f(a_1) + c_1', \quad z_2' = t_2' f(a_2) + c_2'. \tag{10}$$

In what follows we will prove that $tz_1' + (1-t)z_2' \in \text{cone } f(A) + \text{qri } C$. For this, we evaluate

$$tz_1' + (1-t) z_2' = tt_1' f(a_1) + tc_1' + (1-t) t_2' f(a_2) + (1-t) c_2'. \tag{11}$$

Dividing this equality by the positive number $tt_1' + (1-t)t_2'$ it yields

$$\begin{aligned} &\frac{tz_1' + (1-t) z_2'}{tt_1' + (1-t) t_2'} \\ &\quad = \frac{tt_1'}{tt_1' + (1-t) t_2'} f(a_1) + \frac{(1-t) t_2'}{tt_1' + (1-t) t_2'} f(a_2) \\ &\quad + \frac{t}{tt_1' + (1-t) t_2'} c_1' + \frac{1-t}{tt_1' + (1-t) t_2'} c_2' \\ &\quad = \frac{tt_1'}{tt_1' + (1-t) t_2'} \left[f(a_1) + c_1'\right] + \frac{(1-t) t_2'}{tt_1' + (1-t) t_2'} \left[f(a_2) + c_2'\right] \\ &\quad + \frac{t\left(1-t_1'\right)}{tt_1' + (1-t) t_2'} c_1' + \frac{(1-t)\left(1-t_2'\right)}{tt_1' + (1-t) t_2'} c_2'. \end{aligned} \tag{12}$$

By the above equalities and the convexity of $f(A)+C$, we have

$$\frac{tz_1' + (1-t)\,z_2'}{tt_1' + (1-t)\,t_2'} \in f(A) + C + t\,\mathrm{qri}\,C + (1-t)\,\mathrm{qri}\,C \subseteq f(A) + C + \mathrm{qri}\,C \subseteq f(A) + \mathrm{qri}\,C. \tag{13}$$

Thus,

$$tz_1' + (1-t)\,z_2' \in \mathrm{cone}\ f(A) + \mathrm{qri}\,C, \tag{14}$$

whence multiplying this relation by $t_1 + t_2$ we obtain the conclusion. □

3. Existence Results

Throughout this paper we study the following strong vector equilibrium problem:

$$\text{find } a \in A \text{ such that } f(a,b) \notin -C \setminus \{0\} \quad \forall b \in B, \tag{VEP}$$

where A and B are non-empty sets, $f : A \times B \to Y$, and C has a non-empty quasi-relative interior.

Definition 12. We say that a point $a_0 \in A$ is a quasi-relative solution of (VEP) if

$$f(a_0, b) \notin -\mathrm{qri}\,C \quad \forall b \in B. \tag{15}$$

Theorem 13. *Suppose that the following conditions are satisfied:*

(i) *for every $b \in B$, the function $b \mapsto f(\cdot, b)$ is generalized C-subconvexlike on A;*

(ii) $\mathrm{cl}\,(C - C) = Y$;

(iii) *for every $b \in B$, $\mathrm{cl\ cone(cone}\ f(A,b) + \mathrm{qri}\ C)$ is not a linear subspace of Y.*

Then, problem (VEP) admits a quasi-relative solution.

Proof. Suppose by contradiction that (VEP) has no solutions; that is, for any $a \in A$ there exists $b_0 \in B$ such that

$$f(a, b_0) \in -\mathrm{qri}\,C. \tag{16}$$

This leads us to

$$0 \in f(a, b_0) + \mathrm{qri}\,C, \tag{17}$$

which implies that

$$0 \in \mathrm{cone}\ f(A, b_0) + \mathrm{qri}\,C. \tag{18}$$

Assumption (i) assures the convexity of the set $\mathrm{cone}\ f(A, b_0) + \mathrm{qri}\,C$, and, together with the relation (18) and assumption (iii), it gives

$$0 \notin \mathrm{qri}\,(\mathrm{cone}\ f(A, b_0) + \mathrm{qri}\,C). \tag{19}$$

The separation Theorem 9 of convex sets assures the existence of a nonzero functional $y^* \in Y^*$ such that

$$y^*(y) \geq y^*(0) \quad \forall y \in \mathrm{cone}\ f(A, b_0) + \mathrm{qri}\,C. \tag{20}$$

So

$$y^*(tf(a, b_0) + c) \geq 0 \quad \forall t \geq 0,\ a \in A,\ c \in \mathrm{qri}\,C. \tag{21}$$

Taking $t = 0$, by $\mathrm{cl}\,(\mathrm{qri}\,C) = \mathrm{cl}\ C$ and (21) we have

$$y^*(c) \geq 0 \quad \forall c \in C, \tag{22}$$

which gives

$$y^* \in C^*. \tag{23}$$

Further, for any $t > 0$, dividing by t relation (21) it implies

$$y^*\left(f(a, b_0) + \frac{c}{t}\right) \geq 0 \quad \text{for any } a \in A,\ c \in \mathrm{qri}\,C. \tag{24}$$

Now letting $t \to \infty$ in the above relation we get

$$y^*(f(a, b_0)) \geq 0 \quad \forall a \in A. \tag{25}$$

Since $f(a, b_0) \in -\mathrm{qri}\,C$, by assumption (ii), Lemma 7, and Proposition 6, we get

$$y^*(f(a, b_0)) < 0, \tag{26}$$

which is a contradiction to (25), and this completes the proof. □

This theorem allows us to obtain existence results for important practical spaces, whose ordering cones have empty interiors, but nonempty quasi-relative interiors. This is the case of the Banach space $L^P(T, \mu)$ with the positive cone

$$L_+^p(T, \mu) = \{u \in L^p(T, \mu) \mid u(t) \geq 0 \text{ a.e. in } [0, T]\}, \tag{27}$$

where $T > 0$ is a real constant, $[0, T]$ is an interval, (T, μ) is a σ-finite measure space, and $p \in [1, \infty)$.

The next result deals with stronger assumptions than the ones presented in Theorem 13.

Corollary 14. *Suppose that the following conditions are satisfied:*

(i) *for every $b \in B$, the function $b \mapsto f(\cdot, b)$ is generalized C-subconvexlike on A;*

(ii) $\mathrm{cl}\,(C - C) = Y$;

(iii) *for every $b \in B$, $0 \notin \mathrm{qri}\ [\mathrm{cone\,(cone}\ f(A,b) + \mathrm{qri}\ C)]$.*

Then, problem (VEP) admits a quasi-relative solution.

Proof. Since $0 \notin \mathrm{qri}\ [\mathrm{cone\,(cone}\ f(A,b) + \mathrm{qri}\ C)]$ and

$$0 \in \mathrm{cone}(\mathrm{cone}\ f(A, b) + \mathrm{qri}\ C) \tag{28}$$

for every $b \in B$, by the definition of the quasi-relative interior, we get

$$\mathrm{cl\ cone}[\mathrm{cone}(\mathrm{cone}\ f(A, b) + \mathrm{qri}\ C)] = \mathrm{cl\ cone}(\mathrm{cone}\ f(A, b) + \mathrm{qri}\ C) \tag{29}$$

is not a linear subspace of Y. So all the assumption of Theorem 13 are satisfied, and the conclusion follows now by this theorem. □

Remark 15. In Theorem 13 and Corollary 14, according to Proposition 11, the generalized C-subconvexlike assumption of the function $b \mapsto f(\cdot, b)$ on A can be replaced by the convexity of the set $f(A, b) + C$.

4. Existence Results for Vector Optimization Problems

Let $B = A$, and let the function $F : A \to Y$. In this section we study the vector optimization problem,

$$\text{find } a \in A \text{ such that } F(b) - F(a) \notin -C \setminus \{0\} \quad \forall b \in A. \tag{VOP}$$

According to [26], the point $F(a)$ is called quasi-relative minimal point of the set $F(A)$, that is,

$$F(A) - F(a) \notin -\text{qri } C, \tag{30}$$

while a is a quasi-relative minimizer of (VOP), that is,

$$F(b) - F(a) \notin -\text{qri } C \quad \forall b \in A. \tag{31}$$

By Theorem 13 and Corollary 14 we have the following results.

Theorem 16. *Suppose that the following conditions are satisfied:*

(i) *for every $b \in A$, the function $F(b) - F(\cdot)$ is generalized C-subconvexlike on A;*

(ii) $\text{cl}(C - C) = Y$;

(iii) *for every $b \in A$, $\text{cl cone}(\text{cone}(F(b) - F(A)) + \text{qri } C)$ is not a linear subspace of Y.*

Then, problem (VOP) admits a quasi-relative solution.

Proof. Define the function $f : A \times A \to Y$ by

$$f(a, b) = F(b) - F(a) \quad \forall a, b \in A. \tag{32}$$

It is easy to see that all the assumptions of the Theorem 13 are satisfied by this function f. So, problem (VEP) admits a solution, which implies that problem (VOP) has a solution, and the proof is completed. □

Corollary 17. *Suppose that the following conditions are satisfied:*

(i) *for every $b \in A$, the function $b \mapsto F(b) - F(\cdot)$ is generalized C subconvexlike on A;*

(ii) $\text{cl}(C - C) = Y$;

(iii) *for every $b \in A$, $0 \notin \text{qri}[\text{cone}(\text{cone}(F(b) - F(A)) + \text{qri } C)]$.*

Then, problem (VOP) admits a quasi-relative solution.

To show that the set of functions which satisfies the assumptions of Corollary 17 is nonempty, we give the following example.

Example 18. Let $A = [0, 1]$, $Y = \mathbb{R}^2$, $C = \mathbb{R}^2_+$, and $F(a) = (a, 0)$. We have that

$$F(b) - F(a) = (b - a, 0) \quad \forall a, b \in [0, 1]. \tag{33}$$

Since $F(b) - F(A) + C = (b - A, 0) + C$ and A is a convex set, we deduce that

$$F(b) - F(A) + C \tag{34}$$

is also convex and that, by Proposition 11, the first assumption of Corollary 17 is verified. The second assumption is obviously satisfied, and we still have to verify its third assumption. Because $\text{int } C \neq \emptyset$, then it is equal to $\text{qri } C$. The set

$$\begin{aligned} \text{int}[\text{cone}(\text{cone}(F(b) - F(A)) + \text{int } C)] &= \text{int}[\mathbb{R} \times [0, \infty)] \\ &= \mathbb{R} \times (0, \infty), \end{aligned} \tag{35}$$

for all $b \in (0, 1]$, while for $b = 1$

$$\text{int}[\text{cone}(\text{cone}(F(b) - F(A)) + \text{int } C)] = (0, \infty) \times (0, \infty). \tag{36}$$

Thus,

$$(0, 0) \notin \text{int}[\text{cone}(\text{cone}(F(b) - F(A)) + \text{int } C)] \quad \forall b \in A, \tag{37}$$

and assumption (iii) is checked.

5. Existence Results for Vector Variational Inequalities

Let A be a non-empty convex subset of a vector space, let $B = A$, and let the operator $F : A \to \mathcal{L}(X, Y)$, where $\mathcal{L}(X, Y)$ denotes the set of all linear and continuous functions defined on X with values on Y. Throughout this section we study the Minty vector variational inequality:

$$\text{find } a \in A \text{ such that } \langle F(b), b - a \rangle \notin -C \setminus \{0\} \quad \forall b \in A. \tag{MVI}$$

Definition 19. We say that a point $a_0 \in A$ is a quasi-relative solution of (MVI) if

$$\langle F(b), b - a_0 \rangle \notin -\text{qri } C \quad \forall b \in A. \tag{38}$$

By Theorem 13 and Corollary 14 we have the following results.

Theorem 20. *Suppose that the following conditions are satisfied:*

(i) $\text{cl}(C - C) = Y$;

(ii) *for every $b \in A$, $\text{cl cone}(\text{cone}\langle F(b), b - A \rangle + \text{qri } C)$ is not a linear subspace of Y.*

Then, problem (MVI) admits a quasi-relative solution.

Proof. Let the function $f : A \times A \to Y$ be defined by

$$f(a, b) = \langle F(b), b - a \rangle \quad \forall a, b \in A. \tag{39}$$

Obviously f fulfills assumptions (ii) and (iii) of Theorem 13. It remains to show that the first assumption of this theorem is also verified by this function f.

Since A is a convex set, by the definition of the function f we deduce that $f(A, b)$ is a convex set for every $b \in A$. By this we get

$$f(A, b) + C \tag{40}$$

convex, which together with Proposition 11 gives the conclusion, and proof is completed. □

Corollary 21. *Suppose that the following conditions are satisfied:*

(i) $\operatorname{cl}(C - C) = Y$;

(ii) *for every* $b \in A$, $0 \notin \operatorname{qri}[\operatorname{cone}(\operatorname{cone}\langle F(b), b - A\rangle + \operatorname{qri} C)]$.

Then, problem (MVI) admits a quasi-relative solution.

In what follows we turn our attention to existence results for the Stampacchia vector variational inequality

$$\text{find } a_0 \in A \text{ such that } \langle F(a), b - a\rangle \notin -C \setminus \{0\} \quad \forall b \in A. \tag{SVI}$$

Definition 22. We say that a point $a_0 \in A$ is a quasi-relative solution of (SVI) if

$$\langle F(a_0), b - a_0\rangle \notin -\operatorname{qri} C \quad \forall b \in A. \tag{41}$$

Next we recall a definition concerning vector variational inequalities (see [27]).

Definition 23. Let $c^* \in C^* \setminus \{0\}$. The operator F is said to be c^*-upper hemicontinuous if A is a convex set, and, for all $a, b \in A$ the function

$$\forall \lambda \in [0, 1] \longmapsto c^*(\langle F(\lambda b + (1 - \lambda) a), b - a\rangle) \in \mathbb{R} \tag{42}$$

is upper semicontinuous at 0.

Theorem 24. *Let* $c^* \in C^* \setminus \{0\}$, *let* $a_0 \in A$, *and suppose that the following conditions are satisfied:*

(i) $c^*(\langle F(b), b - a_0\rangle + \operatorname{qri} C) \geq 0$ *for all* $b \in A$;

(ii) F *is* c^**-upper hemicontinuous.*

Then, problem (SVI) admits a quasi-relative solution.

Proof. By assumption (i)

$$c^*(\langle F(b), b - a_0\rangle + \operatorname{qri} C) \geq 0 \quad \forall b \in A, \tag{43}$$

whence, by using Lemma 8, we deduce that

$$\langle F(b), b - a_0\rangle \notin -C \setminus \{0\} - \operatorname{qri} C \subseteq -\operatorname{qri} C \quad \forall b \in A, \tag{44}$$

that is, a_0 is a quasi-relative solution of (MVI). In what follows we will prove that a_0 is also a quasi-relative solution of (SVI).

Because A is convex, then

$$\lambda b + (1 - \lambda) a_0 \in A, \tag{45}$$

for all $b \in A$ and all $\lambda \in [0, 1]$, and by (43), we get

$$c^*(\langle F(\lambda b + (1 - \lambda) a_0), b - a_0\rangle + \operatorname{qri} C) \geq 0 \quad \forall \lambda \in [0, 1]. \tag{46}$$

For each $\epsilon > 0$, there exists, according to (ii), a number $\lambda \in (0, 1]$ such that

$$\begin{aligned} c^*(\langle F(\lambda b + (1 - \lambda) a_0), b - a_0\rangle + \operatorname{qri} C) \\ < c^*(\langle F(a_0), b - a_0\rangle + \operatorname{qri} C) + \epsilon. \end{aligned} \tag{47}$$

This, together with (46), gives

$$0 < c^*(\langle F(a_0), b - a_0\rangle + \operatorname{qri} C) + \epsilon. \tag{48}$$

Since the inequality holds for each $\epsilon > 0$, we deduce that

$$c^*(\langle F(a_0), b - a_0\rangle + \operatorname{qri} C) \geq 0, \tag{49}$$

whence, by using Lemma 8, we get

$$\langle F(a_0), b - a_0\rangle \notin -C \setminus \{0\} - \operatorname{qri} C \subseteq -\operatorname{qri} C, \tag{50}$$

which completes the proof. □

References

[1] E. Blum and W. Oettli, "From optimization and variational inequalities to equilibrium problems," *The Mathematics Student*, vol. 63, no. 1–4, pp. 123–145, 1994.

[2] J. P. Aubin, *Optima and Equilibria, An Introduction to Nonlinear Analysis*, Springer, 1993.

[3] Q. H. Ansari, I. V. Konnov, and J. C. Yao, "On generalized vector equilibrium problems," *Nonlinear Analysis, Theory, Methods and Applications*, vol. 47, no. 1, pp. 543–554, 2001.

[4] Q. H. Ansari, W. Oettli, and D. Schläger, "A generalization of vectorial equilibria," *Mathematical Methods of Operations Research*, vol. 46, no. 2, pp. 147–152, 1997.

[5] Q. H. Ansari, X. Q. Yang, and J. C. Yao, "Existence and duality of implicit vector variational problems," *Numerical Functional Analysis and Optimization*, vol. 22, no. 7-8, pp. 815–829, 2001.

[6] M. Bianchi, N. Hadjisavvas, and S. Schaible, "Vector equilibrium problems with generalized monotone bifunctions," *Journal of Optimization Theory and Applications*, vol. 92, no. 3, pp. 527–542, 1997.

[7] M. Bianchi, G. Kassay, and R. Pini, "Ekeland's principle for vector equilibrium problems," *Nonlinear Analysis, Theory, Methods and Applications*, vol. 66, no. 7, pp. 1454–1464, 2007.

[8] G. Bigi, A. Capătă, and G. Kassay, "Existence results for strong vector equilibrium problems and their applications," *Optimization*, vol. 61, no. 5, pp. 567–583, 2012.

[9] A. Capătă, "Existence results for proper efficient solutions of vector equilibrium problems and applications," *Journal of Global Optimization*, vol. 51, no. 4, pp. 657–675, 2011.

[10] A. Capătă and G. Kassay, "On vector equilibrium problems and applications," *Taiwanese Journal of Mathematics*, vol. 15, no. 1, pp. 365–380, 2011.

[11] M. Bianchi, G. Kassay, and R. Pini, "Well-posedness for vector equilibrium problems," *Mathematical Methods of Operations Research*, vol. 70, no. 1, pp. 171–182, 2009.

[12] J. W. Chen, Y. J. Cho, and Z. Wan, "The existence of solutions and wellposedness for bilevel mixed equilibrium problems in Banach spaces," *Taiwanese Journal of Mathematics*, vol. 17, pp. 724–748, 2013.

[13] K. Kimura and J. C. Yao, "Sensitivity analysis of vector equilibrium problems," *Taiwanese Journal of Mathematics*, vol. 12, no. 3, pp. 649–669, 2008.

[14] K. Kimura and J. C. Yao, "Sensitivity analysis of solution mappings of parametric generalized quasi vector equilibrium problems," *Taiwanese Journal of Mathematics*, vol. 12, no. 9, pp. 2233–2268, 2008.

[15] Z. A. Zhou and X. M. Yang, "Optimality conditions of generalized subconvexlike set-valued optimization problems based on the quasi-relative interior," *Journal of Optimization Theory and Applications*, vol. 150, no. 2, pp. 327–340, 2011.

[16] P. Daniele, S. Giuffrè, G. Idone, and A. Maugeri, "Infinite dimensional duality and applications," *Mathematische Annalen*, vol. 339, no. 1, pp. 221–239, 2007.

[17] J. M. Borwein and A. S. Lewis, "Partially finite convex programming. Part I: quasi relative interiors and duality theory," *Mathematical Programming*, vol. 57, no. 1–3, pp. 15–48, 1992.

[18] J. Borwein and R. Goebel, "Notions of relative interior in Banach spaces," *Journal of Mathematical Sciences*, vol. 115, no. 4, pp. 2542–2553, 2003.

[19] R. B. Holmes, *Geometric Functional Analysis and Its Applications*, Springer, Berlin, Germany, 1975.

[20] M. A. Limber and R. K. Goodrich, "Quasi interiors, lagrange multipliers, and L^p spectral estimation with lattice bounds," *Journal of Optimization Theory and Applications*, vol. 78, no. 1, pp. 143–161, 1993.

[21] R. T. Rockafellar, *Conjugate Duality and Optimization*, vol. 16 of *CBMS Regional Conference Series*, Society for Industrial and Applied Mathematics, Philadelphia, Pa, USA, 1974.

[22] C. Zălinescu, *Convex Analysis in General Vector Spaces*, World Scientific Publishing, Singapore, 2002.

[23] X. H. Gong, "Optimality conditions for Henig and globally proper efficient solutions with ordering cone has empty interior," *Journal of Mathematical Analysis and Applications*, vol. 307, no. 1, pp. 12–31, 2005.

[24] R. I. Boţ, E. R. Csetnek, and G. Wanka, "Regularity conditions via quasi-relative interior in convex programming," *SIAM Journal on Optimization*, vol. 19, no. 1, pp. 217–233, 2008.

[25] F. Cammaroto and B. D. Bella, "Separation theorem based on the quasirelative interior and application to duality theory," *Journal of Optimization Theory and Applications*, vol. 125, no. 1, pp. 223–229, 2005.

[26] T. Q. Bao and B. S. Mordukhovich, "Relative Pareto minimizers for multiobjective problems: existence and optimality conditions," *Mathematical Programming A*, vol. 122, no. 2, pp. 301–347, 2010.

[27] X. H. Gong, "Efficiency and Henig efficiency for vector equilibrium problems," *Journal of Optimization Theory and Applications*, vol. 108, no. 1, pp. 139–154, 2001.

2

On a Family of Multivariate Modified Humbert Polynomials

Rabia Aktaş[1] **and Esra Erkuş-Duman**[2]

[1] *Department of Mathematics, Faculty of Science, Ankara University, Tandoğan, 06100 Ankara, Turkey*
[2] *Department of Mathematics, Faculty of Science, Gazi University, Teknikokullar, 06500 Ankara, Turkey*

Correspondence should be addressed to Rabia Aktaş; raktas@science.ankara.edu.tr

Academic Editors: F. J. Garcia-Pacheco and V. Privman

This paper attempts to present a multivariable extension of generalized Humbert polynomials. The results obtained here include various families of multilinear and multilateral generating functions, miscellaneous properties, and also some special cases for these multivariable polynomials.

1. Introduction

The generalized Humbert polynomials are generated by [1]

$$(C - mxt + yt^m)^p = \sum_{n=0}^{\infty} P_n(m, x, y, p, C)\, t^n, \qquad (1)$$

where m is a positive integer and other parameters are unrestricted in general (see also [2, pages 77, 86] and [3, 4]). This definition includes many well-known special polynomials such as Humbert, Louville, Gegenbauer, Legendre, Tchebycheff, Pincherle, and Kinney polynomials.

In this paper, we consider the following multivariable extension of the generalized Humbert polynomials which are completely different from the polynomials introduced in [5]. This class of polynomials is generated by

$$\left(1 - \sum_{i=1}^{r} (m_i x_i t_i - y_i t_i^{m_i})\right)^{-\alpha} = \sum_{n_1,\ldots,n_r=0}^{\infty} \Theta_{n_1,\ldots,n_r}^{(\alpha)}(\mathbf{x},\mathbf{y},\mathbf{m})\, t_1^{n_1}\cdots t_r^{n_r}, \qquad (2)$$

$$\left(\left|\sum_{i=1}^{r} (m_i x_i t_i - y_i t_i^{m_i})\right| < 1\right),$$

where $\mathbf{x} = (x_1,\ldots,x_r)$, $\mathbf{y} = (y_1,\ldots,y_r)$, $\mathbf{m} = (m_1,\ldots,m_r)$ and m_i $(i = 1,2,\ldots,r)$ is a positive integer. It follows from (2) that

$$\Theta_{n_1,\ldots,n_r}^{(\alpha)}(\mathbf{x},\mathbf{y},\mathbf{m}) = \sum_{k_1=0}^{[n_1/m_1]}\cdots\sum_{k_r=0}^{[n_r/m_r]} \frac{(\alpha)_{n_1+\cdots+n_r-(m_1-1)k_1-\cdots-(m_r-1)k_r}}{k_1!\cdots k_r!\,(n_1-m_1k_1)!\cdots(n_r-m_rk_r)!} \times (-1)^{k_1+\cdots+k_r}(m_1x_1)^{n_1-m_1k_1}\cdots(m_rx_r)^{n_r-m_rk_r} y_1^{k_1}\cdots y_r^{k_r}, \qquad (3)$$

where $(\lambda)_k := \lambda(\lambda+1)\cdots(\lambda+k-1)$, $(\lambda)_0 := 1$ is the Pochhammer symbol.

The aim of this paper is to derive various families of multilinear and multilateral generating functions and to give several recurrence relations and expansions in the series of orthogonal polynomials for the family of multivariable polynomials given explicitly by (3). We present some special cases of our results and also obtain some other properties for these special cases.

2. Bilinear and Bilateral Generating Functions

In this section, with the help of the similar method as considered in [5–9], we derive several families of bilinear and bilateral generating functions for the family of multivariable polynomials generated by (2) and given explicitly by (3).

We begin by stating the following theorem.

Theorem 1. *Corresponding to an identically nonvanishing function* $\Omega_\mu(\mathbf{z})$ *of s complex variables* $z_1,\ldots,z_s$ $(s\in\mathbb{N})$ *and of complex order* μ, *let*

$$\Lambda_{\mu,\nu}(\mathbf{z};w) := \sum_{k=0}^{\infty} a_k\Omega_{\mu+\nu k}(\mathbf{z})\,w^k, \tag{4}$$

where $(a_k\neq 0,\ \mu,\nu\in\mathbb{C})$; $\mathbf{z}=(z_1,\ldots,z_s)$. *Then, for* $p\in\mathbb{N}$; $\mathbf{x}=(x_1,\ldots,x_r)$; $\mathbf{y}=(y_1,\ldots,y_r)$; $\mathbf{m}=(m_1,\ldots,m_r)$, $m_i\in\mathbb{N}\,(i=1,2,\ldots,r)$, *one has*

$$\begin{aligned}\sum_{n_1,\ldots,n_r=0}^{\infty}\sum_{k=0}^{[n_1/p]} & a_k\Theta^{(\alpha)}_{n_1-pk,n_2,\ldots,n_r}(\mathbf{x},\mathbf{y},\mathbf{m})\\ &\times t_2^{n_2}\cdots t_r^{n_r}\Omega_{\mu+\nu k}(\mathbf{z})\,\eta^k t_1^{n_1-pk}\\ &=\left(1-\sum_{i=1}^{r}\left(m_ix_it_i-y_it_i^{m_i}\right)\right)^{-\alpha}\Lambda_{\mu,\nu}(\mathbf{z};\eta)\end{aligned}\tag{5}$$

provided that each member of (5) *exists.*

Proof. For convenience, let S denote the first member of the assertion (5) of Theorem 1. Replacing n_1 by n_1+pk, we may write that

$$\begin{aligned}S&=\sum_{n_1,\ldots,n_r=0}^{\infty}\sum_{k=0}^{\infty}a_k\,\Theta^{(\alpha)}_{n_1,n_2,\ldots,n_r}(\mathbf{x},\mathbf{y},\mathbf{m})\,\Omega_{\mu+\nu k}(\mathbf{z})\,\eta^k t_1^{n_1}\cdots t_r^{n_r}\\ &=\sum_{n_1,\ldots,n_r=0}^{\infty}\Theta^{(\alpha)}_{n_1,n_2,\ldots,n_r}(\mathbf{x},\mathbf{y},\mathbf{m})\,t_1^{n_1}\cdots t_r^{n_r}\sum_{k=0}^{\infty}a_k\Omega_{\mu+\nu k}(\mathbf{z})\,\eta^k\\ &=\left(1-\sum_{i=1}^{r}\left(m_ix_it_i-y_it_i^{m_i}\right)\right)^{-\alpha}\Lambda_{\mu,\nu}(\mathbf{z};\eta),\end{aligned}\tag{6}$$

which completes the proof. □

By using a similar idea, we also get the next result immediately.

Theorem 2. *For a nonvanishing function* $\varphi_{\lambda_1,\ldots,\lambda_r}(\mathbf{z})$ *of s complex variables* $z_1,\ldots,z_s$ $(s\in\mathbb{N})$ *and for* $p_i\in\mathbb{N}$, $\lambda_i,\eta_i\in\mathbb{C}$, $\mathbf{z}=(z_1,\ldots,z_s)$, $\mathbf{w}=(w_1,\ldots,w_r)$, $\lambda:=(\lambda_1,\ldots,\lambda_r)$, $\eta:=(\eta_1,\ldots,\eta_r)$ *let*

$$\begin{aligned}\Omega^{\mathbf{n},\mathbf{p}}_{\lambda,\eta,\alpha,\beta,\mathbf{m}}&(\mathbf{x},\mathbf{y};\mathbf{z};\mathbf{w})\\ &:=\sum_{k_1=0}^{[n_1/p_1]}\cdots\sum_{k_r=0}^{[n_r/p_r]}a_{k_1,\ldots,k_r}\Theta^{(\alpha+\beta)}_{n_1-p_1k_1,\ldots,n_r-p_rk_r}(\mathbf{x},\mathbf{y},\mathbf{m})\\ &\qquad\times\varphi_{\lambda_1+\eta_1k_1,\ldots,\lambda_r+\eta_rk_r}(\mathbf{z})\,w_1^{k_1}\cdots w_r^{k_r},\end{aligned}\tag{7}$$

where $a_{k_1,\ldots,k_r}\neq 0$; $n_i,(i=1,2,\ldots,r)\in\mathbb{N}_0$; $\mathbb{N}_0:=\mathbb{N}\cup\{0\}$. *Then, one has*

$$\begin{aligned}\sum_{l_1=0}^{n_1}\cdots\sum_{l_r=0}^{n_r}\sum_{k_1=0}^{[l_1/p_1]}\cdots\sum_{k_r=0}^{[l_r/p_r]}&a_{k_1,\ldots,k_r}\Theta^{(\alpha)}_{n_1-l_1,\ldots,n_r-l_r}(\mathbf{x},\mathbf{y},\mathbf{m})\\ &\times\Theta^{(\beta)}_{l_1-p_1k_1,\ldots,l_r-p_rk_r}(\mathbf{x},\mathbf{y},\mathbf{m})\\ &\times\varphi_{\lambda_1+\eta_1k_1,\ldots,\lambda_r+\eta_rk_r}(\mathbf{z})\\ &\times w_1^{k_1}\cdots w_r^{k_r}\\ =\Omega^{\mathbf{n},\mathbf{p}}_{\lambda,\eta,\alpha,\beta,\mathbf{m}}&(\mathbf{x},\mathbf{y};\mathbf{z};\mathbf{w})\end{aligned}\tag{8}$$

provided that each member of (8) *exists.*

3. Special Cases

As an application of the above theorems, when the multivariable function $\Omega_{\mu+\nu k}(\mathbf{z})$, $\mathbf{z}=(z_1,\ldots,z_s)$, $k\in\mathbb{N}_0$, $s\in\mathbb{N}$, is expressed in terms of simpler functions of one and more variables, then we can give further applications of the above theorems. We first set

$$s=r,\qquad \Omega_{\mu+\nu k}(\mathbf{z})=g^{(\beta_1,\ldots,\beta_r)}_{\mu+\nu k}(\mathbf{z})\tag{9}$$

in Theorem 1, where the Chan-Chyan-Srivastava polynomials $g^{(\beta_1,\ldots,\beta_r)}_{\mu+\nu k}(\mathbf{z})$ [10] are generated by

$$\begin{aligned}&\prod_{i=1}^{r}(1-x_it)^{-\alpha_i}=\sum_{n=0}^{\infty}g_n^{(\alpha_1,\ldots,\alpha_r)}(x_1,\ldots,x_r)\,t^n\\ &\left(\alpha_i\in\mathbb{C}\,(i=1,2,\ldots,r);|t|<\min\left\{|x_1|^{-1},\ldots,|x_r|^{-1}\right\}\right).\end{aligned}\tag{10}$$

We are thus led to the following result which provides a class of bilateral generating functions for the Chan-Chyan-Srivastava polynomials and the family of multivariable polynomials given explicitly by (3).

Corollary 3. *If* $\Lambda_{\mu,\nu}(\mathbf{z};w):=\sum_{k=0}^{\infty}a_k\,g^{(\beta_1,\ldots,\beta_r)}_{\mu+\nu k}(\mathbf{z})w^k$, $a_k\neq 0$, $\mu,\nu\in\mathbb{C}$, $\mathbf{z}=(z_1,\ldots,z_r)$ *then, one has*

$$\begin{aligned}\sum_{n_1,\ldots,n_r=0}^{\infty}\sum_{k=0}^{[n_1/p]}&a_k\Theta^{(\alpha)}_{n_1-pk,n_2,\ldots,n_r}(\mathbf{x},\mathbf{y},\mathbf{m})\\ &\times t_2^{n_2}\cdots t_r^{n_r}\,g^{(\beta_1,\ldots,\beta_r)}_{\mu+\nu k}(\mathbf{z})\,\eta^k t_1^{n_1-pk}\\ &=\left(1-\sum_{i=1}^{r}\left(m_ix_it_i-y_it_i^{m_i}\right)\right)^{-\alpha}\Lambda_{\mu,\nu}(\mathbf{z};\eta)\end{aligned}\tag{11}$$

provided that each member of (11) *exists.*

Remark 4. Using the generating relation (10) for the Chan-Chyan-Srivastava polynomials and getting $a_k = 1,\ \mu = 0,\ \nu = 1$, we find that

$$\sum_{n_1,\ldots,n_r=0}^{\infty} \sum_{k=0}^{[n_1/p]} \Theta^{(\alpha)}_{n_1-pk,n_2,\ldots,n_r}(\mathbf{x},\mathbf{y},\mathbf{m})\, t_2^{n_2}\cdots t_r^{n_r} g_k^{(\beta_1,\ldots,\beta_r)}(\mathbf{z}) \times \eta^k t_1^{n_1-pk} = \left(1-\sum_{i=1}^{r}\left(m_i x_i t_i - y_i t_i^{m_i}\right)\right)^{-\alpha} \prod_{i=1}^{r}(1-z_i\eta)^{-\beta_i},$$

$$|\eta| < \min\left\{|z_1|^{-1},\ldots,|z_r|^{-1}\right\}, \left|\sum_{i=1}^{r}\left(m_i x_i t_i - y_i t_i^{m_i}\right)\right| < 1. \tag{12}$$

On the other hand, choosing $s = 2r$ and

$$\varphi_{\lambda_1+\eta_1 k_1,\ldots,\lambda_r+\eta_r k_r}(\mathbf{z}) = \Theta^{(\gamma)}_{\lambda_1+\eta_1 k_1,\ldots,\lambda_r+\eta_r k_r}(\mathbf{u},\mathbf{v},\mathbf{m}), \tag{13}$$

$\lambda_i, \eta_i\ (i = 1,2,\ldots,r) \in \mathbb{N}_0$, $\mathbf{u} = (u_1,\ldots,u_r)$, $\mathbf{v} = (v_1,\ldots,v_r)$ in Theorem 2, we obtain the following class of bilinear generating functions for the family of multivariable polynomials given explicitly by (3).

Corollary 5. *If*

$$\Lambda^{\mathbf{n},\mathbf{p}}_{\lambda,\eta,\alpha,\beta,\gamma,\mathbf{m}}(\mathbf{x},\mathbf{y};\mathbf{u},\mathbf{v};\mathbf{w}) := \sum_{k_1=0}^{[n_1/p_1]}\cdots\sum_{k_r=0}^{[n_r/p_r]} a_{k_1,\ldots,k_r} \Theta^{(\alpha+\beta)}_{n_1-p_1k_1,\ldots,n_r-p_rk_r}(\mathbf{x},\mathbf{y},\mathbf{m}) \times \Theta^{(\gamma)}_{\lambda_1+\eta_1k_1,\ldots,\lambda_r+\eta_rk_r}(\mathbf{u},\mathbf{v},\mathbf{m})\, w_1^{k_1}\cdots w_r^{k_r}, \tag{14}$$

where $a_{k_1,\ldots,k_r} \neq 0$; $p_i \in \mathbb{N}$; $n_i, \lambda_i, \eta_i \in \mathbb{N}_0, i = 1,2,\ldots,r$, then, we get

$$\sum_{l_1=0}^{n_1}\cdots\sum_{l_r=0}^{n_r}\sum_{k_1=0}^{[l_1/p_1]}\cdots\sum_{k_r=0}^{[l_r/p_r]} a_{k_1,\ldots,k_r}\Theta^{(\alpha)}_{n_1-l_1,\ldots,n_r-l_r}(\mathbf{x},\mathbf{y},\mathbf{m}) \times \Theta^{(\beta)}_{l_1-p_1k_1,\ldots,l_r-p_rk_r}(\mathbf{x},\mathbf{y},\mathbf{m}) \times \Theta^{(\gamma)}_{\lambda_1+\eta_1k_1,\ldots,\lambda_r+\eta_rk_r}(\mathbf{u},\mathbf{v},\mathbf{m}) \times w_1^{k_1}\cdots w_r^{k_r} = \Lambda^{\mathbf{n},\mathbf{p}}_{\lambda,\eta,\alpha,\beta,\gamma,\mathbf{m}}(\mathbf{x},\mathbf{y};\mathbf{u},\mathbf{v};\mathbf{w}) \tag{15}$$

provided that each member of (15) *exists.*

Furthermore, for every suitable choice of the coefficients $a_k\ (k \in \mathbb{N}_0)$, if the multivariable functions $\Omega_{\mu+\nu k}(\mathbf{y})$ and $\varphi_{\lambda_1+\eta_1k_1,\ldots,\lambda_r+\eta_rk_r}(\mathbf{y})$, $\mathbf{y} = (y_1,\ldots,y_s)$, $(s \in \mathbb{N})$, are expressed as an appropriate product of several simpler functions, the assertions of Theorems 1 and 2 can be applied in order to derive various families of multilinear and multilateral generating functions for the family of multivariable polynomials given explicitly by (3).

4. Some Miscellaneous Properties

In this section, we now discuss some further properties of the family of multivariable polynomials given by (3). We start with the following theorems.

Theorem 6. *Let $\Psi_{n_1,\ldots,n_r}(\mathbf{x},\mathbf{y},\mathbf{m})$ be a family of functions generated by*

$$G\left(m_1x_1t_1 - y_1t_1^{m_1},\ldots,m_rx_rt_r - y_rt_r^{m_r}\right) = \sum_{n_1,\ldots,n_r=0}^{\infty} \Psi_{n_1,\ldots,n_r}(\mathbf{x},\mathbf{y},\mathbf{m})\, t_1^{n_1}\cdots t_r^{n_r}. \tag{16}$$

The following relations hold:

$$n_i\Psi_{n_1,\ldots,n_r}(\mathbf{x},\mathbf{y},\mathbf{m}) = x_i\frac{\partial}{\partial x_i}\Psi_{n_1,\ldots,n_r}(\mathbf{x},\mathbf{y},\mathbf{m}) - y_i\frac{\partial}{\partial x_i}\Psi_{n_1,\ldots,n_{i-1},n_i-m_i+1,n_{i+1},\ldots,n_r}(\mathbf{x},\mathbf{y},\mathbf{m}) \tag{17}$$

for $n_i \geq m_i - 1$, $i = 1,2,\ldots,r; n_j \geq 0,\ j \neq i$ and

$$n_i\Psi_{n_1,\ldots,n_r}(\mathbf{x},\mathbf{y},\mathbf{m}) = x_i\frac{\partial}{\partial x_i}\Psi_{n_1,\ldots,n_r}(\mathbf{x},\mathbf{y},\mathbf{m}) \tag{18}$$

for $n_i = 0,1,\ldots,m_i-2$, $i = 1,2,\ldots,r$; $n_j \geq 0,\ j \neq i$. Also, one finds that

$$m_ix_i\frac{\partial}{\partial y_i}\Psi_{n_1,\ldots,n_r}(\mathbf{x},\mathbf{y},\mathbf{m}) - m_iy_i\frac{\partial}{\partial y_i} \times \Psi_{n_1,\ldots,n_{i-1},n_i-m_i+1,n_{i+1},\ldots,n_r}(\mathbf{x},\mathbf{y},\mathbf{m}) = -\left(n_i - m_i + 1\right)\Psi_{n_1,\ldots,n_{i-1},n_i-m_i+1,n_{i+1},\ldots,n_r}(\mathbf{x},\mathbf{y},\mathbf{m}) \tag{19}$$

for $n_i \geq m_i - 1$, $i = 1,2,\ldots,r$; $n_j \geq 0,\ j \neq i$ and

$$\frac{\partial}{\partial y_i}\Psi_{n_1,\ldots,n_r}(\mathbf{x},\mathbf{y},\mathbf{m}) = 0 \tag{20}$$

for $n_i = 0,1,\ldots,m_i-2$, $i = 1,2,\ldots,r$; $n_j \geq 0,\ j \neq i$.

Proof. Fix $i = 1, 2, \ldots, r$. Then, by differentiating (16) with respect to x_i and t_i, after making necessary calculations we obtain that

$$\begin{aligned}
&\sum_{n_1,\ldots,n_r=0}^{\infty} n_i \Psi_{n_1,\ldots,n_r}(\mathbf{x},\mathbf{y},\mathbf{m}) t_1^{n_1}\cdots t_r^{n_r} \\
&= \left(x_i - y_i t_i^{m_i-1}\right) \sum_{n_1,\ldots,n_r=0}^{\infty} \frac{\partial}{\partial x_i} \Psi_{n_1,\ldots,n_r}(\mathbf{x},\mathbf{y},\mathbf{m}) t_1^{n_1}\cdots t_r^{n_r} \\
&= x_i \sum_{n_1,\ldots,n_r=0}^{\infty} \frac{\partial}{\partial x_i} \Psi_{n_1,\ldots,n_r}(\mathbf{x},\mathbf{y},\mathbf{m}) t_1^{n_1}\cdots t_r^{n_r} \\
&\quad - y_i \sum_{\substack{n_1,\ldots,n_{i-1},n_{i+1},\ldots,n_r=0 \\ (n_i \geq m_i-1)}}^{\infty} \frac{\partial}{\partial x_i} \Psi_{n_1,\ldots,n_{i-1},n_i-m_i+1,n_{i+1},\ldots,n_r}(\mathbf{x},\mathbf{y},\mathbf{m}) \\
&\qquad \times t_1^{n_1}\cdots t_r^{n_r}.
\end{aligned} \tag{21}$$

Comparing the coefficients of $t_1^{n_1}\cdots t_r^{n_r}$, we obtain (17) and (18) for the fixed i. Similarly, if we differentiate (16) with respect to y_i and t_i, we can find the relation (19). □

Theorem 7. *If $\Psi_{n_1,\ldots,n_r}(\mathbf{x},\mathbf{y},\mathbf{m})$ is a family of functions generated by (16), then it satisfies the relations*

$$\begin{aligned}
&- m_i \frac{\partial}{\partial y_i} \Psi_{n_1,\ldots,n_{i-1},n_i-1,n_{i+1},\ldots,n_r}(\mathbf{x},\mathbf{y},\mathbf{m}) \\
&\quad = \frac{\partial}{\partial x_i} \Psi_{n_1,\ldots,n_{i-1},n_i-m_i,n_{i+1},\ldots,n_r}(\mathbf{x},\mathbf{y},\mathbf{m})
\end{aligned} \tag{22}$$

for $n_i \geq m_i$, $i = 1, 2, \ldots, r$; $n_j \geq 0$, $j \neq i$ and

$$\frac{\partial}{\partial y_i} \Psi_{n_1,\ldots,n_{i-1},n_i-1,n_{i+1},\ldots,n_r}(\mathbf{x},\mathbf{y},\mathbf{m}) = 0 \tag{23}$$

for $n_i = 1, 2, \ldots, m_i - 1$, $i = 1, 2, \ldots, r$; $n_j \geq 0$, $j \neq i$.

Proof. By comparing the derivatives of (16) with respect to x_i and y_i, we have

$$\begin{aligned}
&- m_i t_i \sum_{n_1,\ldots,n_r=0}^{\infty} \frac{\partial}{\partial y_i} \Psi_{n_1,\ldots,n_r}(\mathbf{x},\mathbf{y},\mathbf{m}) t_1^{n_1}\cdots t_r^{n_r} \\
&\quad = t_i^{m_i} \sum_{n_1,\ldots,n_r=0}^{\infty} \frac{\partial}{\partial x_i} \Psi_{n_1,\ldots,n_r}(\mathbf{x},\mathbf{y},\mathbf{m}) t_1^{n_1}\cdots t_r^{n_r},
\end{aligned} \tag{24}$$

which implies that

$$\begin{aligned}
&- m_i \frac{\partial}{\partial y_i} \Psi_{n_1,\ldots,n_{i-1},n_i-1,n_{i+1},\ldots,n_r}(\mathbf{x},\mathbf{y},\mathbf{m}) \\
&\quad = \frac{\partial}{\partial x_i} \Psi_{n_1,\ldots,n_{i-1},n_i-m_i,n_{i+1},\ldots,n_r}(\mathbf{x},\mathbf{y},\mathbf{m})
\end{aligned} \tag{25}$$

for $n_i \geq m_i$, $i = 1, 2, \ldots, r$; $n_j \geq 0$, $j \neq i$ and

$$\frac{\partial}{\partial y_i} \Psi_{n_1,\ldots,n_{i-1},n_i-1,n_{i+1},\ldots,n_r}(\mathbf{x},\mathbf{y},\mathbf{m}) = 0 \tag{26}$$

for $n_i = 1, 2, \ldots, m_i - 1$, $i = 1, 2, \ldots, r$; $n_j \geq 0$, $j \neq i$. Thus, the proof is completed. □

As a result of these theorems, if we choose

$$\begin{aligned}
&G\left(m_1 x_1 t_1 - y_1 t_1^{m_1}, \ldots, m_r x_r t_r - y_r t_r^{m_r}\right) \\
&\quad = \left(1 - \sum_{i=1}^{r} \left(m_i x_i t_i - y_i t_i^{m_i}\right)\right)^{-\alpha},
\end{aligned} \tag{27}$$

then we can give some recurrence relations for the family of multivariable polynomials given explicitly by (3). In the view of Theorem 6, we get

Corollary 8. *For the family of multivariable polynomials generated by (2), the following relations hold*

$$\begin{aligned}
n_i \Theta_{n_1,\ldots,n_r}^{(\alpha)}(\mathbf{x},\mathbf{y},\mathbf{m}) &= x_i \frac{\partial}{\partial x_i} \Theta_{n_1,\ldots,n_r}^{(\alpha)}(\mathbf{x},\mathbf{y},\mathbf{m}) \\
&\quad - y_i \frac{\partial}{\partial x_i} \Theta_{n_1,\ldots,n_{i-1},n_i-m_i+1,n_{i+1},\ldots,n_r}^{(\alpha)}(\mathbf{x},\mathbf{y},\mathbf{m}),
\end{aligned} \tag{28}$$

$$\begin{aligned}
&m_i x_i \frac{\partial}{\partial y_i} \Theta_{n_1,\ldots,n_r}^{(\alpha)}(\mathbf{x},\mathbf{y},\mathbf{m}) - m_i y_i \frac{\partial}{\partial y_i} \\
&\quad \times \Theta_{n_1,\ldots,n_{i-1},n_i-m_i+1,n_{i+1},\ldots,n_r}^{(\alpha)}(\mathbf{x},\mathbf{y},\mathbf{m}) \\
&\quad = -\left(n_i - m_i + 1\right) \Theta_{n_1,\ldots,n_{i-1},n_i-m_i+1,n_{i+1},\ldots,n_r}^{(\alpha)}(\mathbf{x},\mathbf{y},\mathbf{m})
\end{aligned} \tag{29}$$

for $n_i \geq m_i - 1$, $i = 1, 2, \ldots, r$; $n_j \geq 0$, $j \neq i$. Also, for $n_i = 0, 1, \ldots, m_i - 2$, $i = 1, 2, \ldots, r$; $n_j \geq 0$, $j \neq i$, we have

$$n_i \Theta_{n_1,\ldots,n_r}^{(\alpha)}(\mathbf{x},\mathbf{y},\mathbf{m}) = x_i \frac{\partial}{\partial x_i} \Theta_{n_1,\ldots,n_r}^{(\alpha)}(\mathbf{x},\mathbf{y},\mathbf{m}), \tag{30}$$

$$\frac{\partial}{\partial y_i} \Theta_{n_1,\ldots,n_r}^{(\alpha)}(\mathbf{x},\mathbf{y},\mathbf{m}) = 0. \tag{31}$$

Corollary 9. *By summing the relations given by (28) and (29), respectively, for $i = 1, 2, \ldots, r$, we get*

$$\begin{aligned}
&\sum_{i=1}^{r} n_i \Theta_{n_1,\ldots,n_r}^{(\alpha)}(\mathbf{x},\mathbf{y},\mathbf{m}) \\
&\quad = \sum_{i=1}^{r} x_i \frac{\partial}{\partial x_i} \Theta_{n_1,\ldots,n_r}^{(\alpha)}(\mathbf{x},\mathbf{y},\mathbf{m}) \\
&\quad - \sum_{i=1}^{r} y_i \frac{\partial}{\partial x_i} \Theta_{n_1,\ldots,n_{i-1},n_i-m_i+1,n_{i+1},\ldots,n_r}^{(\alpha)}(\mathbf{x},\mathbf{y},\mathbf{m}),
\end{aligned}$$

$$\sum_{i=1}^{r} m_i x_i \frac{\partial}{\partial y_i} \Theta^{(\alpha)}_{n_1,\ldots,n_r}(\mathbf{x},\mathbf{y},\mathbf{m}) - \sum_{i=1}^{r} m_i y_i \frac{\partial}{\partial y_i} \Theta^{(\alpha)}_{n_1,\ldots,n_{i-1},n_i-m_i+1,n_{i+1},\ldots,n_r}(\mathbf{x},\mathbf{y},\mathbf{m}) = -\sum_{i=1}^{r}(n_i - m_i + 1) \times \Theta^{(\alpha)}_{n_1,\ldots,n_{i-1},n_i-m_i+1,n_{i+1},\ldots,n_r}(\mathbf{x},\mathbf{y},\mathbf{m}) \quad (32)$$

for $n_i \geq m_i - 1$, $i = 1,2,\ldots,r$; $n_j \geq 0$, $j \neq i$.

Similarly, as a consequence of Theorem 7, we can give the next result at once.

Corollary 10. *Other recurrence relations for the family of multivariable polynomials* $\Theta^{(\alpha)}_{n_1,\ldots,n_r}(\mathbf{x},\mathbf{y},\mathbf{m})$ *are*

$$-m_i \frac{\partial}{\partial y_i} \Theta^{(\alpha)}_{n_1,\ldots,n_{i-1},n_i-1,n_{i+1},\ldots,n_r}(\mathbf{x},\mathbf{y},\mathbf{m}) = \frac{\partial}{\partial x_i} \Theta^{(\alpha)}_{n_1,\ldots,n_{i-1},n_i-m_i,n_{i+1},\ldots,n_r}(\mathbf{x},\mathbf{y},\mathbf{m}) \quad (33)$$

for $n_i \geq m_i$, $i = 1,2,\ldots,r$; $n_j \geq 0$, $j \neq i$ *and*

$$\frac{\partial}{\partial y_i} \Theta^{(\alpha)}_{n_1,\ldots,n_{i-1},n_i-1,n_{i+1},\ldots,n_r}(\mathbf{x},\mathbf{y},\mathbf{m}) = 0 \quad (34)$$

for $n_i = 1,2,\ldots,m_i - 1$, $i = 1,2,\ldots,r$; $n_j \geq 0$, $j \neq i$.

Theorem 11. *The generating relation* (2) *yields the following addition formula for the family of multivariable polynomials given by* (3)

$$\Theta^{(\alpha+\beta)}_{n_1,\ldots,n_r}(\mathbf{x},\mathbf{y},\mathbf{m}) = \sum_{k_1=0}^{n_1} \cdots \sum_{k_r=0}^{n_r} \Theta^{(\alpha)}_{n_1-k_1,\ldots,n_r-k_r}(\mathbf{x},\mathbf{y},\mathbf{m}) \times \Theta^{(\beta)}_{k_1,\ldots,k_r}(\mathbf{x},\mathbf{y},\mathbf{m}). \quad (35)$$

Proof. From (2), we have

$$\sum_{n_1,\ldots,n_r=0}^{\infty} \Theta^{(\alpha+\beta)}_{n_1,\ldots,n_r}(\mathbf{x},\mathbf{y},\mathbf{m}) t_1^{n_1} \cdots t_r^{n_r} = \left(1 - \sum_{i=1}^{r}(m_i x_i t_i - y_i t_i^{m_i})\right)^{-(\alpha+\beta)} = \left(1 - \sum_{i=1}^{r}(m_i x_i t_i - y_i t_i^{m_i})\right)^{-\alpha} \times \left(1 - \sum_{i=1}^{r}(m_i x_i t_i - y_i t_i^{m_i})\right)^{-\beta}$$
$$= \sum_{n_1,\ldots,n_r=0}^{\infty} \Theta^{(\alpha)}_{n_1,\ldots,n_r}(\mathbf{x},\mathbf{y},\mathbf{m}) t_1^{n_1} \cdots t_r^{n_r} \times \sum_{k_1,\ldots,k_r=0}^{\infty} \Theta^{(\beta)}_{k_1,\ldots,k_r}(\mathbf{x},\mathbf{y},\mathbf{m}) t_1^{k_1} \cdots t_r^{k_r}. \quad (36)$$

Replacing n_i by $n_i - k_i$, $i = 1,2,\ldots,r$, the right-hand side of the last equality is

$$\sum_{n_1,\ldots,n_r=0}^{\infty} \sum_{k_1=0}^{n_1} \cdots \sum_{k_r=0}^{n_r} \Theta^{(\alpha)}_{n_1-k_1,\ldots,n_r-k_r}(\mathbf{x},\mathbf{y},\mathbf{m}) \times \Theta^{(\beta)}_{k_1,\ldots,k_r}(\mathbf{x},\mathbf{y},\mathbf{m}) t_1^{n_1} \cdots t_r^{n_r}. \quad (37)$$

Comparing the coefficients of $t_1^{n_1} \cdots t_r^{n_r}$ completes the proof. □

We now give expansions of the family of multivariable polynomials $\Theta^{(\alpha)}_{n_1,\ldots,n_r}(\mathbf{x},\mathbf{y},\mathbf{m})$ given explicitly by (3) in series of Legendre, Gegenbauer, Hermite, and Laguerre polynomials.

Theorem 12. *Expansions of* $\Theta^{(\alpha)}_{n_1,\ldots,n_r}(\mathbf{x},\mathbf{y},\mathbf{m})$ *in series of Legendre, Gegenbauer, Hermite, and Laguerre polynomials are as follows*

$$\Theta^{(\alpha)}_{n_1,\ldots,n_r}(\mathbf{x},\mathbf{y},\mathbf{m}) = \prod_{i=1}^{r} \left\{ \sum_{k_i=0}^{[n_i/m_i]} \sum_{s_i=0}^{[(n_i-m_ik_i)/2]} \frac{(2n_i - 2m_ik_i - 4s_i + 1)}{k_i! s_i! (3/2)_{n_i-m_ik_i-s_i}} \times (-1)^{k_i} y_i^{k_i} P_{n_i-m_ik_i-2s_i}\left(\frac{m_i x_i}{2}\right) \right\} \times (\alpha)_{n_1+\cdots+n_r-(m_1-1)k_1-\cdots-(m_r-1)k_r},$$

$$\Theta^{(\alpha)}_{n_1,\ldots,n_r}(\mathbf{x},\mathbf{y},\mathbf{m}) = \prod_{i=1}^{r} \left\{ \sum_{k_i=0}^{[n_i/m_i]} \sum_{s_i=0}^{[(n_i-m_ik_i)/2]} \frac{(\nu_i + n_i - m_ik_i - 2s_i)}{k_i! s_i! (\nu_i)_{n_i-m_ik_i-s_i+1}} \times (-y_i)^{k_i} C^{\nu_i}_{n_i-m_ik_i-2s_i}\left(\frac{m_i x_i}{2}\right) \right\} \times (\alpha)_{n_1+\cdots+n_r-(m_1-1)k_1-\cdots-(m_r-1)k_r},$$

$$\Theta^{(\alpha)}_{n_1,\ldots,n_r}(\mathbf{x},\mathbf{y},\mathbf{m}) = \prod_{i=1}^{r} \left\{ \sum_{k_i=0}^{[n_i/m_i]} \sum_{s_i=0}^{[(n_i-m_ik_i)/2]} \frac{(-y_i)^{k_i} H_{n_i-m_ik_i-2s_i}(m_i x_i/2)}{k_i! s_i! (n_i - m_ik_i - 2s_i)!} \right\} \times (\alpha)_{n_1+\cdots+n_r-(m_1-1)k_1-\cdots-(m_r-1)k_r},$$

$$\Theta^{(\alpha)}_{n_1,\ldots,n_r}(\mathbf{x},\mathbf{y},\mathbf{m}) = \prod_{i=1}^{r}\left\{\sum_{k_i=0}^{[n_i/m_i]}\sum_{s_i=0}^{n_i-m_ik_i}\frac{(\beta_i+1)_{n_i-m_ik_i}2^{n_i-m_ik_i}(-1)^{s_i}}{k_i!\,(n_i-m_ik_i-s_i)!(\beta_i+1)_{s_i}} \times(-y_i)^{k_i}L^{(\beta_i)}_{s_i}\left(\frac{m_ix_i}{2}\right)\right\} \times(\alpha)_{n_1+\cdots+n_r-(m_1-1)k_1-\cdots-(m_r-1)k_r}. \tag{38}$$

Proof. By (2), we get

$$\sum_{n_1,\ldots,n_r=0}^{\infty}\Theta^{(\alpha)}_{n_1,\ldots,n_r}(\mathbf{x},\mathbf{y},\mathbf{m})\,t_1^{n_1}\cdots t_r^{n_r} = \left(1-\sum_{i=1}^{r}\left(m_ix_it_i-y_it_i^{m_i}\right)\right)^{-\alpha}$$
$$= \sum_{n_1,\ldots,n_r=0}^{\infty}\sum_{k_1,\ldots,k_r=0}^{\infty}\frac{(\alpha)_{n_1+\cdots+n_r+k_1+\cdots+k_r}}{k_1!\cdots k_r!n_1!\cdots n_r!}(-1)^{k_1+\cdots+k_r} \times(m_1x_1)^{n_1}\cdots(m_rx_r)^{n_r}y_1^{k_1}\cdots y_r^{k_r} \times t_1^{n_1+m_1k_1}\cdots t_r^{n_r+m_rk_r}. \tag{39}$$

If we use the result in [11, page 181]

$$\frac{(mx)^n}{n!} = \sum_{s=0}^{[n/2]}\frac{(2n-4s+1)\,P_{n-2s}(mx/2)}{s!(3/2)_{n-s}}, \tag{40}$$

we can write that

$$\sum_{n_1,\ldots,n_r=0}^{\infty}\Theta^{(\alpha)}_{n_1,\ldots,n_r}(\mathbf{x},\mathbf{y},\mathbf{m})\,t_1^{n_1}\cdots t_r^{n_r} = \prod_{i=1}^{r}\left\{\sum_{n_i=0}^{\infty}\sum_{k_i=0}^{\infty}\sum_{s_i=0}^{[n_i/2]}\frac{(2n_i-4s_i+1)}{s_i!k_i!(3/2)_{n_i-s_i}}P_{n_i-2s_i}\left(\frac{m_ix_i}{2}\right) \times(-y_i)^{k_i}t_i^{n_i+m_ik_i}\right\}(\alpha)_{n_1+\cdots+n_r+k_1+\cdots+k_r}. \tag{41}$$

Getting $n_i-m_ik_i$ instead of n_i in the last equality, we have

$$\sum_{n_1,\ldots,n_r=0}^{\infty}\Theta^{(\alpha)}_{n_1,\ldots,n_r}(\mathbf{x},\mathbf{y},\mathbf{m})\,t_1^{n_1}\cdots t_r^{n_r}$$
$$= \sum_{n_1,\ldots,n_r=0}^{\infty}\prod_{i=1}^{r}\left\{\sum_{k_i=0}^{[n_i/m_i]}\sum_{s_i=0}^{[(n_i-m_ik_i)/2]}\frac{(2n_i-2m_ik_i-4s_i+1)}{s_i!k_i!(3/2)_{n_i-m_ik_i-s_i}} \times P_{n_i-m_ik_i-2s_i}\left(\frac{m_ix_i}{2}\right)(-y_i)^{k_i}t_i^{n_i}\right\} \times(\alpha)_{n_1+\cdots+n_r-(m_1-1)k_1-\cdots-(m_r-1)k_r}$$
$$= \sum_{n_1,\ldots,n_r=0}^{\infty}\left\{\prod_{i=1}^{r}\sum_{k_i=0}^{[n_i/m_i]}\sum_{s_i=0}^{[(n_i-m_ik_i)/2]}\frac{(2n_i-2m_ik_i-4s_i+1)}{s_i!k_i!(3/2)_{n_i-m_ik_i-s_i}} \times P_{n_i-m_ik_i-2s_i}\left(\frac{m_ix_i}{2}\right)(-y_i)^{k_i}\right\} \times(\alpha)_{n_1+\cdots+n_r-(m_1-1)k_1-\cdots-(m_r-1)k_r}t_1^{n_1}\cdots t_r^{n_r}. \tag{42}$$

Comparing the coefficients of $t_1^{n_1}\cdots t_r^{n_r}$ gives the desired relation.

In a similar manner, in (39), using the following results, respectively, [11, page 283 (36), page 194 (4), page 207 (2)]

$$\frac{(mx)^n}{n!} = \sum_{k=0}^{[n/2]}\frac{(\nu+n-2k)\,C^{\nu}_{n-2k}(mx/2)}{k!(\nu)_{n+1-k}},$$
$$\frac{(mx)^n}{n!} = \sum_{k=0}^{[n/2]}\frac{H_{n-2k}(mx/2)}{k!\,(n-2k)!}, \tag{43}$$
$$\frac{(mx)^n}{n!} = 2^n\sum_{k=0}^{n}\frac{(-1)^k(\alpha+1)_nL^{(\alpha)}_k(mx/2)}{(n-k)!(\alpha+1)_k},$$

one can easily obtain the other expansions of $\Theta^{(\alpha)}_{n_1,\ldots,n_r}(\mathbf{x},\mathbf{y},\mathbf{m})$ in series of Gegenbauer, Hermite, and Laguerre polynomials. □

5. The Special Cases of $\Theta^{(\alpha)}_n(\mathbf{x},\mathbf{y},\mathbf{m})$ and Some Properties

In this section, we discuss some special cases of the family of multivariable polynomials $\Theta^{(\alpha)}_n(\mathbf{x},\mathbf{y},\mathbf{m})$ and give their several properties.

5.1. The Case of $m_i = 2$, $y_i = 1$, $i = 1,2,\ldots,r$ in (2). This case gives a multivariable analogue of Gegenbauer polynomials

$$\left(1-2x_1t_1+t_1^2-\cdots-2x_rt_r+t_r^2\right)^{-\alpha} = \sum_{n_1,\ldots,n_r=0}^{\infty}C^{(\alpha)}_{n_1,\ldots,n_r}(x_1,\ldots,x_r)\,t_1^{n_1}\cdots t_r^{n_r}, \tag{44}$$

which reduces to two variables Gegenbauer polynomials given by [12] for $r = 2$. Equation (44) yields the following formula:

$$C^{(\alpha)}_{n_1,\ldots,n_r}(x_1,\ldots,x_r) = \sum_{k_1=0}^{[n_1/2]}\cdots\sum_{k_r=0}^{[n_r/2]}\left(\left((\alpha)_{n_1+\cdots+n_r-k_1-\cdots-k_r}\right) \times \left(k_1!\cdots k_r!\,(n_1-2k_1)!\cdots(n_r-2k_r)!\right)^{-1}\right) \times (-1)^{k_1+\cdots+k_r}(2x_1)^{n_1-2k_1}\cdots(2x_r)^{n_r-2k_r}. \quad (45)$$

It follows that $C^{(\alpha)}_{n_1,\ldots,n_r}(x_1,\ldots,x_r)$ is a polynomial of degree n_i with respect to the fixed variable x_i $(i = 1,2,\ldots,r)$. Thus, $C^{(\alpha)}_{n_1,\ldots,n_r}(x_1,\ldots,x_r)$ is a polynomial of total degree $(n_1+\cdots+n_r)$ with respect to the variables $x_1,\ldots,x_r$. Equation (45) also yields

$$C^{(\alpha)}_{n_1,\ldots,n_r}(x_1,\ldots,x_r) = \frac{2^{n_1+\cdots+n_r}(\alpha)_{n_1+\cdots+n_r}}{n_1!\cdots n_r!}x_1^{n_1}\cdots x_r^{n_r} + \Pi(x_1,\ldots,x_r), \quad (46)$$

where $\Pi(x_1,\ldots,x_r)$ is a polynomial of degree $(n_1+\cdots+n_r-2)$ with respect to the variables $x_1,\ldots,x_r$. In (44), by getting $x_1 \to -x_1$ and $t_1 \to -t_1$, we have

$$C^{(\alpha)}_{n_1,\ldots,n_r}(-x_1,x_2,\ldots,x_r) = (-1)^{n_1}C^{(\alpha)}_{n_1,\ldots,n_r}(x_1,\ldots,x_r). \quad (47)$$

Similarly, for $i = 1,2,\ldots,r,$ we get

$$C^{(\alpha)}_{n_1,\ldots,n_r}(x_1,\ldots,x_{i-1},-x_i,x_{i+1},\ldots,x_r) = (-1)^{n_i}C^{(\alpha)}_{n_1,\ldots,n_r}(x_1,\ldots,x_r). \quad (48)$$

Taking $x_i \to -x_i$ and $t_i \to -t_i$, $i = 1,2,\ldots,r$, in (44), we obtain

$$C^{(\alpha)}_{n_1,\ldots,n_r}(-x_1,\ldots,-x_r) = (-1)^{n_1+\cdots+n_r}C^{(\alpha)}_{n_1,\ldots,n_r}(x_1,\ldots,x_r). \quad (49)$$

Theorem 13. *For the polynomials $C^{(\alpha)}_{n_1,\ldots,n_r}(x_1,\ldots,x_r)$, one has*

$$C^{(\alpha)}_{2n_1,\ldots,2n_r}(0,0,\ldots,0) = \frac{(-1)^{n_1+\cdots+n_r}(\alpha)_{n_1+\cdots+n_r}}{n_1!\cdots n_r!} \quad (50)$$

and if at least one of n_i, $i = 1,2,\ldots,r$, is odd; then

$$C^{(\alpha)}_{n_1,\ldots,n_r}(0,0,\ldots,0) = 0. \quad (51)$$

Proof. If we set all $x_i = 0$, $i = 1,2,\ldots,r$ in (44), we have

$$\left(1+t_1^2+\cdots+t_r^2\right)^{-\alpha} = \sum_{n_1,\ldots,n_r=0}^{\infty} C^{(\alpha)}_{n_1,\ldots,n_r}(0,0,\ldots,0)\,t_1^{n_1}\cdots t_r^{n_r}. \quad (52)$$

On the other hand, we get

$$\left(1+t_1^2+\cdots+t_r^2\right)^{-\alpha} = \sum_{n_1,\ldots,n_r=0}^{\infty}\frac{(-1)^{n_1+\cdots+n_r}}{n_1!\cdots n_r!}(\alpha)_{n_1+\cdots+n_r}t_1^{2n_1}\cdots t_r^{2n_r}. \quad (53)$$

By comparing the coefficients of $t_1^{n_1}\cdots t_r^{n_r}$, we obtain the desired. □

From the theorems and corollaries given in Section 4, we can give some other properties of $C^{(\alpha)}_{n_1,\ldots,n_r}(x_1,\ldots,x_r)$.

Remark 14. By Corollaries 8 and 9, for the family of multivariable polynomials generated by (44), the following relations:

$$n_iC^{(\alpha)}_{n_1,\ldots,n_r}(x_1,\ldots,x_r) = x_i\frac{\partial}{\partial x_i}C^{(\alpha)}_{n_1,\ldots,n_r}(x_1,\ldots,x_r) - \frac{\partial}{\partial x_i}C^{(\alpha)}_{n_1,\ldots,n_{i-1},n_i-1,n_{i+1},\ldots,n_r}(x_1,\ldots,x_r),$$
$$\sum_{i=1}^{r}n_iC^{(\alpha)}_{n_1,\ldots,n_r}(x_1,\ldots,x_r) = \sum_{i=1}^{r}x_i\frac{\partial}{\partial x_i}C^{(\alpha)}_{n_1,\ldots,n_r}(x_1,\ldots,x_r) - \sum_{i=1}^{r}\frac{\partial}{\partial x_i}C^{(\alpha)}_{n_1,\ldots,n_{i-1},n_i-1,n_{i+1},\ldots,n_r}(x_1,\ldots,x_r) \quad (54)$$

hold for $n_i \geq 1$, $i = 1,2,\ldots,r$.

Remark 15. From Theorem 11, the multivariable polynomials $C^{(\alpha)}_{n_1,\ldots,n_r}(x_1,\ldots,x_r)$ satisfy the following addition formula:

$$C^{(\alpha+\beta)}_{n_1,\ldots,n_r}(x_1,\ldots,x_r) = \sum_{k_1=0}^{n_1}\cdots\sum_{k_r=0}^{n_r}C^{(\alpha)}_{n_1-k_1,\ldots,n_r-k_r}(x_1,\ldots,x_r) \times C^{(\beta)}_{k_1,\ldots,k_r}(x_1,\ldots,x_r). \quad (55)$$

Remark 16. As a result of Theorem 12, expansions of $C^{(\alpha)}_{n_1,\dots,n_r}(x_1,\dots,x_r)$ in series of Legendre, Gegenbauer, Hermite, and Laguerre polynomials are as follows:

$$\begin{aligned}
C^{(\alpha)}_{n_1,\dots,n_r}&(x_1,\dots,x_r)\\
&=\prod_{i=1}^{r}\left\{\sum_{k_i=0}^{[n_i/2]}\sum_{s_i=0}^{[(n_i-2k_i)/2]}\frac{(2n_i-4k_i-4s_i+1)}{k_i!s_i!(3/2)_{n_i-2k_i-s_i}}\times(-1)^{k_i}P_{n_i-2k_i-2s_i}(x_i)\right\}\\
&\quad\times(\alpha)_{n_1+\cdots+n_r-k_1-\cdots-k_r},\\
C^{(\alpha)}_{n_1,\dots,n_r}&(x_1,\dots,x_r)\\
&=\prod_{i=1}^{r}\left\{\sum_{k_i=0}^{[n_i/2]}\sum_{s_i=0}^{[(n_i-2k_i)/2]}\frac{(\nu_i+n_i-2k_i-2s_i)}{k_i!s_i!(\nu_i)_{n_i-2k_i-s_i+1}}\times(-1)^{k_i}C^{\nu_i}_{n_i-2k_i-2s_i}(x_i)\right\}\\
&\quad\times(\alpha)_{n_1+\cdots+n_r-k_1-\cdots-k_r},\\
C^{(\alpha)}_{n_1,\dots,n_r}&(x_1,\dots,x_r)\\
&=\prod_{i=1}^{r}\left\{\sum_{k_i=0}^{[n_i/2]}\sum_{s_i=0}^{[(n_i-2k_i)/2]}\frac{(-1)^{k_i}H_{n_i-2k_i-2s_i}(x_i)}{k_i!s_i!(n_i-2k_i-2s_i)!}\right\}\\
&\quad\times(\alpha)_{n_1+\cdots+n_r-k_1-\cdots-k_r},\\
C^{(\alpha)}_{n_1,\dots,n_r}&(x_1,\dots,x_r)\\
&=\prod_{i=1}^{r}\left\{\sum_{k_i=0}^{[n_i/2]}\sum_{s_i=0}^{n_i-2k_i}\frac{(\beta_i+1)_{n_i-2k_i}2^{n_i-2k_i}(-1)^{s_i}}{k_i!(n_i-2k_i-s_i)!(\beta_i+1)_{s_i}}\times(-1)^{k_i}L^{(\beta_i)}_{s_i}(x_i)\right\}\\
&\quad\times(\alpha)_{n_1+\cdots+n_r-k_1-\cdots-k_r}.
\end{aligned}\tag{56}$$

We now give a hypergeometric representation for the multivariable polynomials $C^{(\alpha)}_{n_1,\dots,n_r}(x_1,\dots,x_r)$ given by (44).

Theorem 17. *The multivariable polynomials $C^{(\alpha)}_{n_1,\dots,n_r}(x_1,\dots,x_r)$ have the following hypergeometric representation:*

$$\begin{aligned}
C^{(\alpha)}_{n_1,\dots,n_r}&(x_1,\dots,x_r)\\
&=\frac{(\alpha)_{n_1+\cdots+n_r}}{n_1!\cdots n_r!}(2x_1)^{n_1}\cdots(2x_r)^{n_r}\\
&\quad\times F^{(r)}_B\left[-\frac{n_1}{2},\dots,-\frac{n_r}{2},\frac{-n_1+1}{2},\dots,\frac{-n_r+1}{2};\right.\\
&\quad\left.1-\alpha-n_1-\cdots-n_r;\frac{1}{x_1^2},\dots,\frac{1}{x_r^2}\right],\\
&\quad\left(\max\left\{\frac{1}{x_1^2},\dots,\frac{1}{x_r^2}\right\}<1\right),
\end{aligned}\tag{57}$$

where $F^{(r)}_B$ is a Lauricella function of r-variable defined by [13] (see also [2])

$$\begin{aligned}
F^{(r)}_B&[a_1,\dots,a_r,b_1,\dots,b_r;c;x_1,\dots,x_r]\\
&=\sum_{n_1,\dots,n_r=0}^{\infty}\frac{(a_1)_{n_1}\cdots(a_r)_{n_r}(b_1)_{n_1}\cdots(b_r)_{n_r}}{(c)_{n_1+\cdots+n_r}}\frac{x_1^{n_1}}{n_1!}\cdots\frac{x_r^{n_r}}{n_r!},\\
&\quad(\max\{|x_1|,\dots,|x_r|\}<1).
\end{aligned}\tag{58}$$

Proof. We have the following results from [11]:

$$\begin{aligned}
(\alpha)_{n-k}&=\frac{(-1)^k(\alpha)_n}{(1-\alpha-n)_k},\qquad (-n)_{2k}=2^{2k}\left(-\frac{n}{2}\right)_k\left(\frac{-n+1}{2}\right)_k,\\
(n-2k)!&=\frac{n!}{(-n)_{2k}}.
\end{aligned}\tag{59}$$

In view of these relations, the equality (45) can be written as follows:

$$\begin{aligned}
C^{(\alpha)}_{n_1,\dots,n_r}&(x_1,\dots,x_r)\\
&=\sum_{k_1=0}^{[n_1/2]}\cdots\sum_{k_r=0}^{[n_r/2]}\left(\left((\alpha)_{n_1+\cdots+n_r}2^{2(k_1+\cdots+k_r)}\left(-\frac{n_1}{2}\right)_{k_1}\right.\right.\\
&\quad\times\left(\frac{-n_1+1}{2}\right)_{k_1}\cdots\left(-\frac{n_r}{2}\right)_{k_r}\\
&\quad\left.\times\left(\frac{-n_r+1}{2}\right)_{k_r}\right)\\
&\quad\times\left(k_1!\cdots k_r!n_1!\cdots n_r!\right.\\
&\quad\left.\left.\times(1-\alpha-n_1-\cdots-n_r)_{k_1+\cdots+k_r}\right)^{-1}\right)\\
&\quad\times(2x_1)^{n_1-2k_1}\cdots(2x_r)^{n_r-2k_r}\\
&=\frac{(\alpha)_{n_1+\cdots+n_r}}{n_1!\cdots n_r!}(2x_1)^{n_1}\cdots(2x_r)^{n_r}
\end{aligned}$$

$$\times \sum_{k_1=0}^{[n_1/2]} \cdots \sum_{k_r=0}^{[n_r/2]} \left(\left(\left(-\frac{n_1}{2}\right)_{k_1} \left(\frac{-n_1+1}{2}\right)_{k_1} \cdots \left(-\frac{n_r}{2}\right)_{k_r} \left(\frac{-n_r+1}{2}\right)_{k_r} \right) \times \left((1-\alpha-n_1-\cdots-n_r)_{k_1+\cdots+k_r} \times k_1! \cdots k_r! \right)^{-1} \right) \times \left(\frac{1}{x_1^2}\right)^{k_1} \cdots \left(\frac{1}{x_r^2}\right)^{k_r}$$
$$= F_B^{(r)} \left[-\frac{n_1}{2}, \ldots, -\frac{n_r}{2}, \frac{-n_1+1}{2}, \ldots, \frac{-n_r+1}{2}; 1-\alpha-n_1-\cdots-n_r; \frac{1}{x_1^2}, \ldots, \frac{1}{x_r^2} \right] \times \frac{(\alpha)_{n_1+\cdots+n_r}}{n_1! \cdots n_r!} (2x_1)^{n_1} \cdots (2x_r)^{n_r}, \quad (60)$$

where $\max\{1/x_1^2, \ldots, 1/x_r^2\} < 1$. The proof is completed. □

Remark 18. For $r = 2$, Theorem 17 reduces to the known result for two variable Gegenbauer polynomials given by [12].

5.2. The Case of $x_i = 0$, $y_i = -x_i$, $i = 1, 2, \ldots, r$ in (2). This case yields

$$(1 - x_1 t_1^{m_1} - x_2 t_2^{m_2} - \cdots - x_r t_r^{m_r})^{-\alpha} = \sum_{n_1,\ldots,n_r=0}^{\infty} U_{n_1,\ldots,n_r}^{(\alpha)}(x_1, \ldots, x_r)\, t_1^{n_1} \cdots t_r^{n_r}, \quad (61)$$

which is a different unification from that in [8].

Remark 19. From Corollaries 8 and 9, the multivariable polynomials $U_{n_1,\ldots,n_r}^{(\alpha)}(x_1, \ldots, x_r)$ satisfy

$$m_i x_i \frac{\partial}{\partial x_i} U_{n_1,\ldots,n_{i-1},n_i-m_i+1,n_{i+1},\ldots,n_r}^{(\alpha)}(x_1, \ldots, x_r) = (n_i - m_i + 1)\, U_{n_1,\ldots,n_{i-1},n_i-m_i+1,n_{i+1},\ldots,n_r}^{(\alpha)}(x_1, \ldots, x_r),$$

$$\sum_{i=1}^{r} m_i x_i \frac{\partial}{\partial x_i} \times U_{n_1,\ldots,n_{i-1},n_i-m_i+1,n_{i+1},\ldots,n_r}^{(\alpha)}(x_1, \ldots, x_r) = \sum_{i=1}^{r} (n_i - m_i + 1) \times U_{n_1,\ldots,n_{i-1},n_i-m_i+1,n_{i+1},\ldots,n_r}^{(\alpha)}(x_1, \ldots, x_r). \quad (62)$$

for $n_i \geq m_i - 1$, $i = 1, 2, \ldots, r$; $n_j \geq 0$, $j \neq i$.

Remark 20. As result of Theorem 11, the multivariable polynomials $U_{n_1,\ldots,n_r}^{(\alpha)}(x_1, \ldots, x_r)$ have the following addition formula:

$$U_{n_1,\ldots,n_r}^{(\alpha+\beta)}(x_1, \ldots, x_r) = \sum_{k_1=0}^{n_1} \cdots \sum_{k_r=0}^{n_r} U_{n_1-k_1,\ldots,n_r-k_r}^{(\alpha)}(x_1, \ldots, x_r) \times U_{k_1,\ldots,k_r}^{(\beta)}(x_1, \ldots, x_r). \quad (63)$$

5.2.1. The Case of $m_i = 1$, $i = 1, 2, \ldots, r$ in Section 5.2. In this case, we get a r-variable analogue of Lagrange polynomials (see [14]) which is different from that in [10]

$$(1 - x_1 t_1 - \cdots - x_r t_r)^{-\alpha} = \sum_{n_1,\ldots,n_r=0}^{\infty} G_{n_1,\ldots,n_r}^{(\alpha)}(x_1, \ldots, x_r)\, t_1^{n_1} \cdots t_r^{n_r}. \quad (64)$$

They are given explicitly by

$$G_{n_1,\ldots,n_r}^{(\alpha)}(x_1, \ldots, x_r) = \frac{(\alpha)_{n_1+\cdots+n_r}}{n_1! \cdots n_r!} x_1^{n_1} \cdots x_r^{n_r}. \quad (65)$$

Remark 21. From Corollaries 8 and 9, the multivariable polynomials $G_{n_1,\ldots,n_r}^{(\alpha)}(x_1, \ldots, x_r)$ satisfy

$$n_i G_{n_1,\ldots,n_r}^{(\alpha)}(x_1, \ldots, x_r) = x_i \frac{\partial}{\partial x_i} G_{n_1,\ldots,n_r}^{(\alpha)}(x_1, \ldots, x_r) \quad (66)$$

for $i = 1, 2, \ldots, r$ and

$$\sum_{i=1}^{r} n_i G_{n_1,\ldots,n_r}^{(\alpha)}(x_1, \ldots, x_r) = \sum_{i=1}^{r} x_i \frac{\partial}{\partial x_i} G_{n_1,\ldots,n_r}^{(\alpha)}(x_1, \ldots, x_r). \quad (67)$$

Remark 22. By Theorem 11, the multivariable polynomials $G_{n_1,\ldots,n_r}^{(\alpha)}(x_1, \ldots, x_r)$ have the following addition formula:

$$G_{n_1,\ldots,n_r}^{(\alpha+\beta)}(x_1, \ldots, x_r) = \sum_{k_1=0}^{n_1} \cdots \sum_{k_r=0}^{n_r} G_{n_1-k_1,\ldots,n_r-k_r}^{(\alpha)}(x_1, \ldots, x_r) \times G_{k_1,\ldots,k_r}^{(\beta)}(x_1, \ldots, x_r). \quad (68)$$

Remark 23. From Theorem 12, expansions of $G^{(\alpha)}_{n_1,\ldots,n_r}(x_1, \ldots, x_r)$ in series of Legendre, Gegenbauer, Hermite, and Laguerre polynomials are given by

$$
\begin{aligned}
G^{(\alpha)}_{n_1,\ldots,n_r}(x_1,\ldots,x_r) &= \prod_{i=1}^{r}\left\{\sum_{s_i=0}^{[n_i/2]}\frac{(2n_i-4s_i+1)}{s_i!(3/2)_{n_i-s_i}}P_{n_i-2s_i}\left(\frac{x_i}{2}\right)\right\} \\
&\quad\times(\alpha)_{n_1+\cdots+n_r}, \\
G^{(\alpha)}_{n_1,\ldots,n_r}(x_1,\ldots,x_r) &= \prod_{i=1}^{r}\left\{\sum_{s_i=0}^{[n_i/2]}\frac{(\nu_i+n_i-2s_i)}{s_i!(\nu_i)_{n_i-s_i+1}}C^{\nu_i}_{n_i-2s_i}\left(\frac{x_i}{2}\right)\right\} \\
&\quad\times(\alpha)_{n_1+\cdots+n_r}, \\
G^{(\alpha)}_{n_1,\ldots,n_r}(x_1,\ldots,x_r) &= \prod_{i=1}^{r}\left\{\sum_{s_i=0}^{[n_i/2]}\frac{H_{n_i-2s_i}(x_i/2)}{s_i!\,(n_i-2s_i)!}\right\}(\alpha)_{n_1+\cdots+n_r}, \\
G^{(\alpha)}_{n_1,\ldots,n_r}(x_1,\ldots,x_r) &= \prod_{i=1}^{r}\left\{\sum_{s_i=0}^{n_i}\frac{(\beta_i+1)_{n_i}2^{n_i}(-1)^{s_i}}{(n_i-s_i)!(\beta_i+1)_{s_i}}L^{(\beta_i)}_{s_i}\left(\frac{x_i}{2}\right)\right\} \\
&\quad\times(\alpha)_{n_1+\cdots+n_r}.
\end{aligned}
\tag{69}
$$

We can discuss some generating functions for the multivariable polynomials $G^{(\alpha)}_{n_1,\ldots,n_r}(x_1,\ldots,x_r)$.

Theorem 24. *For the polynomials $G^{(\alpha)}_{n_1,\ldots,n_r}(x_1,\ldots,x_r)$, the following generating function holds true for $k \in \mathbb{N}_0$:*

$$
\begin{aligned}
&\sum_{n_1,\ldots,n_r=0}^{\infty} G^{(\alpha)}_{n_1,\ldots,n_r}(x_1,\ldots,x_r)\,[k+n_i]_m z_1^{n_1}\cdots z_r^{n_r} \\
&\quad = m!(1-x_1z_1-\cdots-x_rz_r)^{-\alpha} \\
&\qquad\times \mathrm{g}_m^{(\alpha,-k)}\left(\frac{x_iz_i}{1-x_1z_1-\cdots-x_rz_r},-1\right),
\end{aligned}
\tag{70}
$$

for each $i = 1, 2, \ldots, r$, where

$$[k]_m := k(k-1)\cdots(k-m+1), \quad m \in \mathbb{N}, [k]_0 = 1 \tag{71}$$

and $\mathrm{g}_n^{(\alpha,\beta)}(x, y)$ denotes the classical Lagrange polynomials defined by [14]

$$
\begin{aligned}
&(1-xt)^{-\alpha}(1-yt)^{-\beta} = \sum_{n=0}^{\infty}\mathrm{g}_n^{(\alpha,\beta)}(x,y)\,t^n \\
&\left(\alpha,\beta\in\mathbb{C};|t|<\min\left\{|x|^{-1},|y|^{-1}\right\}\right).
\end{aligned}
\tag{72}
$$

Proof. Fix $i = 1, 2, \ldots, r$. Let $\mathscr{C}_{\xi_i}$ and $\mathscr{C}^*_{\xi_i}$ denote the circles of radius ε_i, centered $\xi_i = z_i$ and $\xi_i = 0$, respectively, where

$$0 < \varepsilon_i < |x_i|^{-1}\left|1-\sum_{k=1}^{r}x_kz_k\right|. \tag{73}$$

Here these circles are described in the positive direction. By Cauchy's Integral Formula, we have

$$
\begin{aligned}
&D^m_{z_i}\left\{(1-x_1z_1-\cdots-x_rz_r)^{-\alpha}z_i^k\right\} \\
&= \frac{m!}{2\pi i}\oint_{\mathscr{C}_{\xi_i}}\frac{\xi_i^k(1-x_1z_1-\cdots-x_i\xi_i-\cdots-x_rz_r)^{-\alpha}}{(\xi_i-z_i)^{m+1}}d\xi_i \\
&= \frac{m!}{2\pi i} \\
&\quad\times\oint_{\mathscr{C}^*_{\xi_i}}\frac{(\xi_i+z_i)^k(1-x_1z_1-\cdots-x_i(\xi_i+z_i)-\cdots-x_rz_r)^{-\alpha}}{\xi_i^{m+1}}d\xi_i \\
&= \frac{m!}{2\pi i}z_i^k(1-x_1z_1-\cdots-x_rz_r)^{-\alpha} \\
&\quad\times\oint_{\mathscr{C}^*_{\xi_i}}\frac{(1+\xi_i/z_i)^k}{\xi_i^{m+1}}\left(1-\frac{x_i\xi_i}{1-x_1z_1-\cdots-x_rz_r}\right)^{-\alpha}d\xi_i \\
&= m!z_i^k(1-x_1z_1-\cdots-x_rz_r)^{-\alpha} \\
&\quad\times\mathrm{g}_m^{(\alpha,-k)}\left(\frac{x_i}{1-x_1z_1-\cdots-x_rz_r},-\frac{1}{z_i}\right) \\
&= m!z_i^{k-m}(1-x_1z_1-\cdots-x_rz_r)^{-\alpha} \\
&\quad\times\mathrm{g}_m^{(\alpha,-k)}\left(\frac{x_iz_i}{1-x_1z_1-\cdots-x_rz_r},-1\right).
\end{aligned}
\tag{74}
$$

On the other hand, we can write that

$$
\begin{aligned}
&D^m_{z_i}\left\{(1-x_1z_1-\cdots-x_rz_r)^{-\alpha}z_i^k\right\} \\
&= D^m_{z_i}\left\{z_i^k\sum_{n_1,\ldots,n_r=0}^{\infty}G^{(\alpha)}_{n_1,\ldots,n_r}(x_1,\ldots,x_r)\,z_1^{n_1}\cdots z_r^{n_r}\right\} \\
&= \sum_{n_1,\ldots,n_r=0}^{\infty}G^{(\alpha)}_{n_1,\ldots,n_r}(x_1,\ldots,x_r)\frac{d^m}{dz_i^m}z_1^{n_1}\cdots z_i^{n_i+k}\cdots z_r^{n_r} \\
&= z_i^{k-m}\sum_{n_1,\ldots,n_r=0}^{\infty}G^{(\alpha)}_{n_1,\ldots,n_r}(x_1,\ldots,x_r)\,[k+n_i]_m z_1^{n_1}\cdots z_r^{n_r}.
\end{aligned}
\tag{75}
$$

From (74) and (75), we see that

$$
\begin{aligned}
&\sum_{n_1,\ldots,n_r=0}^{\infty}G^{(\alpha)}_{n_1,\ldots,n_r}(x_1,\ldots,x_r)\,[k+n_i]_m z_1^{n_1}\cdots z_r^{n_r} \\
&\quad= m!(1-x_1z_1-\cdots-x_rz_r)^{-\alpha} \\
&\qquad\times\mathrm{g}_m^{(\alpha,-k)}\left(\frac{x_iz_i}{1-x_1z_1-\cdots-x_rz_r},-1\right).
\end{aligned}
\tag{76}
$$

□

Corollary 25. *From (71), one observes that*

$$
\begin{aligned}
[n_i+m]_m &= (n_i+m)(n_i+m-1)\cdots(n_i+1) \\
&= (n_i+1)_m
\end{aligned}
\tag{77}
$$

for each $i = 1, 2, \ldots, r$. *By setting* $k = m$ *in* (70) *and then using* (77), *one has*

$$\sum_{n_1,\ldots,n_r=0}^{\infty} (n_i + 1)_m G^{(\alpha)}_{n_1,\ldots,n_r}(x_1,\ldots,x_r) z_1^{n_1}\cdots z_r^{n_r} = m!(1 - x_1z_1 - \cdots - x_rz_r)^{-\alpha} \times g_m^{(\alpha,-m)}\left(\frac{x_iz_i}{1 - x_1z_1 - \cdots - x_rz_r}, -1\right). \quad (78)$$

5.2.2. The Case of $m_i = i$, $i = 1, 2, \ldots, r$ *in Section 5.2.* This case reduces to a multivariable analogue of Lagrange-Hermite polynomials

$$\left(1 - x_1t_1 - x_2t_2^2 - \cdots - x_rt_r^r\right)^{-\alpha} = \sum_{n_1,\ldots,n_r=0}^{\infty} H^{(\alpha)}_{n_1,\ldots,n_r}(x_1,\ldots,x_r) t_1^{n_1}\cdots t_r^{n_r}, \quad (79)$$

which is different from that in [6].

Remark 26. By Corollaries 8 and 9, the multivariable polynomials $H^{(\alpha)}_{n_1,\ldots,n_r}(x_1,\ldots,x_r)$ satisfy

$$ix_i\frac{\partial}{\partial x_i}H^{(\alpha)}_{n_1,\ldots,n_{i-1},n_i-i+1,n_{i+1},\ldots,n_r}(x_1,\ldots,x_r) = (n_i - i + 1) H^{(\alpha)}_{n_1,\ldots,n_{i-1},n_i-i+1,n_{i+1},\ldots,n_r}(x_1,\ldots,x_r),$$

$$\sum_{i=1}^{r} ix_i\frac{\partial}{\partial x_i}H^{(\alpha)}_{n_1,\ldots,n_{i-1},n_i-i+1,n_{i+1},\ldots,n_r}(x_1,\ldots,x_r) = \sum_{i=1}^{r}(n_i - i + 1) \times H^{(\alpha)}_{n_1,\ldots,n_{i-1},n_i-i+1,n_{i+1},\ldots,n_r}(x_1,\ldots,x_r). \quad (80)$$

for $n_i \geq i - 1$, $i = 1, 2, \ldots, r$; $n_j \geq 0$, $j \neq i$.

Remark 27. From Theorem 11, the multivariable polynomials $H^{(\alpha)}_{n_1,\ldots,n_r}(x_1,\ldots,x_r)$ have the following addition formula:

$$H^{(\alpha+\beta)}_{n_1,\ldots,n_r}(x_1,\ldots,x_r) = \sum_{k_1=0}^{n_1}\cdots\sum_{k_r=0}^{n_r} H^{(\alpha)}_{n_1-k_1,\ldots,n_r-k_r}(x_1,\ldots,x_r) \times H^{(\beta)}_{k_1,\ldots,k_r}(x_1,\ldots,x_r). \quad (81)$$

Furthermore, setting $m_i = i$, $x_i = 0$, $y_i = -x_i$, $i = 1, 2, \ldots, r$ in Corollary 5, we obtain the following class of bilinear generating functions for the polynomials $H^{(\alpha)}_{n_1,\ldots,n_r}(x_1,\ldots,x_r)$.

Remark 28. If

$$\Xi^{\mathbf{n},\mathbf{p}}_{\lambda,\eta,\alpha,\beta,\gamma}(x_1,\ldots,x_r; u_1,\ldots,u_r; \mathbf{w}) := \sum_{k_1=0}^{[n_1/p_1]}\cdots\sum_{k_r=0}^{[n_r/p_r]} a_{k_1,\ldots,k_r} H^{(\alpha+\beta)}_{n_1-p_1k_1,\ldots,n_r-p_rk_r}(x_1,\ldots,x_r) \times H^{(\gamma)}_{\lambda_1+\eta_1k_1,\ldots,\lambda_r+\eta_rk_r}(u_1,\ldots,u_r) w_1^{k_1}\cdots w_r^{k_r}, \quad (82)$$

where $a_{k_1,\ldots,k_r} \neq 0$; $p_i \in \mathbb{N}$; $n_i, \lambda_i, \eta_i \in \mathbb{N}_0$, $i = 1, 2, \ldots, r$, and $\lambda = (\lambda_1,\ldots,\lambda_r)$, $\eta = (\eta_1,\ldots,\eta_r)$, $\mathbf{w} = (w_1,\ldots,w_r)$, then we have

$$\sum_{l_1=0}^{n_1}\cdots\sum_{l_r=0}^{n_r}\sum_{k_1=0}^{[l_1/p_1]}\cdots\sum_{k_r=0}^{[l_r/p_r]} a_{k_1,\ldots,k_r} H^{(\alpha)}_{n_1-l_1,\ldots,n_r-l_r}(x_1,\ldots,x_r) \times H^{(\beta)}_{l_1-p_1k_1,\ldots,l_r-p_rk_r}(x_1,\ldots,x_r) \times H^{(\gamma)}_{\lambda_1+\eta_1k_1,\ldots,\lambda_r+\eta_rk_r}(u_1,\ldots,u_r) \times w_1^{k_1}\cdots w_r^{k_r} = \Xi^{\mathbf{n},\mathbf{p}}_{\lambda,\eta,\alpha,\beta,\gamma}(x_1,\ldots,x_r; u_1,\ldots,u_r; \mathbf{w}). \quad (83)$$

References

[1] H. W. Gould, "Inverse series relation and other expansions involving Humbert polynomials," *Duke Mathematical Journal*, vol. 32, pp. 697–711, 1965.

[2] H. M. Srivastava and H. L. A. Manocha, *Treatise on Generating Functions*, Halsted Press, Ellis Horwood Limited, John Wiley and Sons, Chichester, UK, 1984.

[3] J. P. Singhal, "A note on generalized Humbert polynomials," *Glasnik Matematicki III*, vol. 5, no. 25, pp. 241–245, 1970.

[4] D. Zeitlin, "On a class of polynomials obtained from generalized Humbert polynomials," *Proceedings of the American Mathematical Society*, vol. 18, pp. 28–34, 1967.

[5] R. Aktaş, R. Şahin, and A. Altın, "On a multivariable extension of Humbert polynomials," *Applied Mathematics and Computation*, vol. 218, pp. 662–666, 2011.

[6] A. Altın and E. Erkuş, "On a multivariable extension of the Lagrange-Hermite polynomials," *Integral Transforms and Special Functions*, vol. 17, no. 4, pp. 239–244, 2006.

[7] A. Altın, E. Erkuş, and M. A. Özarslan, "Families of linear generating functions for polyno- mials in two variables," *Integral Transforms and Special Functions*, vol. 17, no. 5, pp. 315–320, 2006.

[8] E. Erkuş and H. M. Srivastava, "A unfied presentation of some families of multivariable polynomials," *Integral Transforms and Special Functions*, vol. 17, pp. 267–273, 2006.

[9] E. Özergin, M. A. Özarslan, and H. M. Srivastava, "Some families of generating functions for a class of bivariate polynomials," *Mathematical and Computer Modelling*, vol. 50, pp. 1113–1120, 2009.

[10] W.-C. C. Chan, C.-J. Chyan, and H. M. Srivastava, "The Lagrange polynomials in several variables," *Integral Transforms and Special Functions*, vol. 12, no. 2, pp. 139–148, 2001.

[11] E. D. Rainville, *Special Functions*, The Macmillan Company, New York, NY, USA, 1960.

[12] M. A. Khan and G. S. Kahmmash, "A study of a two variables Gegenbauer polynomials," *Applied Mathematical Sciences*, vol. 2, no. 13, pp. 639–657, 2008.

[13] G. Lauricella, "Sulle funzioni ipergeometriche a più variabili," *Rendiconti del Circolo Matematico di Palermo*, vol. 7, pp. 111–158, 1893.

[14] A. Erdélyi, W. Magnus, F. Oberhettinger, and F. G. Tricomi, *Higher Transcendental Functions*, vol. 3, McGraw-Hill Book Company, New York, NY, USA, 1955.

3

Fixed Point Results of Locally Contractive Mappings in Ordered Quasi-Partial Metric Spaces

Abdullah Shoaib,[1] Muhammad Arshad,[1] and Jamshaid Ahmad[2]

[1] *Department of Mathematics, International Islamic University, H-10, Islamabad 44000, Pakistan*
[2] *Department of Mathematics, COMSATS Institute of Information Technology, Chak Shahzad, Islamabad 44000, Pakistan*

Correspondence should be addressed to Muhammad Arshad; marshad_zia@yahoo.com

Academic Editors: A. Bekir, A. Ibeas, and A. M. Peralta

Fixed point results for a self-map satisfying locally contractive conditions on a closed ball in an ordered 0-complete quasi-partial metric space have been established. Instead of monotone mapping, the notion of dominated mappings is applied. We have used weaker metric, weaker contractive conditions, and weaker restrictions to obtain unique fixed points. An example is given which shows that how this result can be used when the corresponding results cannot. Our results generalize, extend, and improve several well-known conventional results.

1. Introduction and Preliminaries

Fixed points results of mappings satisfying certain contractive conditions on the entire domain have been at the centre of vigorous research activity, (see [1–3]) and it has a wide range of applications in different areas such as nonlinear and adaptive control systems, parameterize estimation problems, fractal image decoding, computing magnetostatic fields in a nonlinear medium, and convergence of recurrent networks (see [4–6]).

Recently, many results appeared related to fixed point theorem in complete metric spaces endowed with a partial ordering in the literature. Ran and Reurings [7] proved an analogue of Banach's fixed point theorem in metric space endowed with a partial order and gave applications to matrix equations. In this way, they weakened the usual contraction condition. Subsequently, Nieto and Rodríguez-López [8] extended the result in [7] for nondecreasing mappings and applied it to obtain a unique solution for a 1st-order ordinary differential equation with periodic boundary conditions. Thereafter, many works related to fixed point problems have also been considered in partially ordered metric spaces (see [7–11]).

On the other hand notion of a partial metric space was introduced by Matthews in [12]. In partial metric spaces, the distance of a point from itself may not be zero. Partial metric spaces have applications in theoretical computer science (see [13]). Altun and Erduran [14] and Paesano and Vetro [15] used the idea of partial metric space and partial order and gave some fixed point theorems for contractive condition on ordered partial metric spaces. Recently, Karapınar et al. [16] introduced the concept of quasi-partial metric space. Romaguera [17] has given the idea of 0-complete partial metric space. Nashine et al. [18] used this concept and proved some classical results.

From the application point of view the situation is not yet completely satisfactory because it frequently happens that a mapping T is a contraction not on the entire space X but merely on a subset Y of X. However, if Y is closed and a sequence $\{x_n\}$ in X converges to some x in X, then by imposing a subtle restriction on the choice of x_0, one may force the sequence to stay eventually in Y. In this case, one can establish the existence of a fixed point of T. Arshad et al. [19] proved a significant result concerning the existence of fixed points of a mapping satisfying a contractive conditions on closed ball in a complete dislocated metric space. Other results on closed ball can be seen in [20, 21]. In this paper we have obtained fixed point theorems for dominated self-mappings in a 0-complete ordered quasi-partial metric space on closed ball under several contractive conditions to generalize, extend and improve some classical fixed point results. We have used weaker contractive condition and weaker restrictions to obtain unique fixed point. Our results do not exist even yet in metric spaces. An example shows how this result can be used when the corresponding results cannot.

Consistent with [16, 22, 23] the following definitions and results will be needed in the sequel.

Definition 1 (see [16]). A quasi-partial metric is a function $q : X \times X \to R^+$ satisfying the following:

(i) if $0 \le q(x,x) = q(x,y) = q(y,y)$, then $x = y$ (equality);

(ii) $q(x,x) \le q(y,x)$ (small self-distances);

(iii) $q(x,x) \le q(x,y)$ (small self-distances);

(iv) $q(x,z) + q(y,y) \le q(x,y) + q(y,z)$ (triangle inequality), for all $x, y, z \in X$.

The pair (X,q) is called a quasi-partial metric space.

Note that if $q(x,y) = q(y,x)$ for all $x, y \in X$, then (X,q) becomes a partial metric space (X,p). Moreover if $q(x,x) = 0$ for all $x \in X$, then (X,q) and (X,p) become a quasimetric space and a metric space, respectively. Also $p_q(x,y) = (1/2)[q(x,y)+q(y,x)]$, $x, y \in X$ is a partial metric on X. The function $d_{p_q} : X \times X \to R^+$ defined by $d_{p_q}(x,y) = q(x,y)+q(y,x)-q(x,x)-q(y,y)$ is a (usual) metric on X. The ball $\overline{B(x,\varepsilon)}$, where $\overline{B(x,\varepsilon)} = \{y \in X : q(x,y) \le \varepsilon + q(x,x)\}$, is a closed ball in quasi-partial metric space, for some $x \in X$ and $\varepsilon > 0$.

Definition 2 (see [16]). Let (X,q) be a quasi-partial metric. Then, we have the following;

(a) A sequence $\{x_n\}$ in (X,q) converges to a point $x \in X$ if and only if $\lim_{n\to\infty} q(x,x_n) = \lim_{n\to\infty} q(x_n,x) = q(x,x)$.

(b) A sequence $\{x_n\}$ in (X,q) is called a Cauchy sequence if the limits $\lim_{n,m\to\infty} q(x_n,x_m)$ and $\lim_{n,m\to\infty} q(x_m,x_n)$ exist (and are finite).

(c) The space (X,q) is said to be complete if every Cauchy sequence $\{x_n\}$ in (X,q) converges to a point $x \in X$ such that $(x,x) = \lim_{n,m\to\infty} q(x_n,x_m) = \lim_{n,m\to\infty} q(x_m,x_n)$.

Lemma 3 (see [16]). *Let (X,q) be a quasi-partial metric space, let (X,p_q) be the corresponding partial metric space, and let (X,d_{p_q}) be the corresponding metric space. These statements are equivalent. (i) The sequence $\{x_n\}$ is Cauchy in (X,q). (ii) The sequence $\{x_n\}$ is Cauchy in (X,p_q). (iii) The sequence $\{x_n\}$ is Cauchy in (X,d_{p_q}). These statements are also equivalent. (i) (X,q) is complete. (ii) (X,p_q) is complete. (iii) (X,d_{p_q}) is complete.*

Definition 4. Let $(X,\preceq)$ be a partial ordered set. Then $x, y \in X$ are called comparable if $x \preceq y$ or $y \preceq x$ holds.

Definition 5 (see [22]). Let $(X,\preceq)$ be a partially ordered set. A self-mapping f on X is called dominated if $fx \preceq x$ for each x in X.

Example 6 (see [22]). Let $X = [0,1]$ be endowed with the usual ordering and $f : X \to X$ defined by $fx = x^n$ for some $n \in \mathbb{N}$. Since $fx = x^n \le x$ for all $x \in X$, therefore f is a dominated map.

Theorem 7 (see [23]). *Let (X,d) be a complete metric space, $S : X \to X$ a mapping, $r > 0$, and x_0 an arbitrary point in X. Suppose there exists $k \in [0,1)$ with*

$$d(Sx,Sy) \le kd(x,y), \quad \text{for all } x,y \in Y = \overline{B(x_0,r)} \tag{1}$$

and $d(x_0,Sx_0) < (1-k)r$. Then there exists a unique point x^* in $\overline{B(x_0,r)}$ such that $x^* = Sx^*$.

2. Fixed Points of Dominated Mapping

In this section we introduce the notion of a 0-complete quasi-partial metric spaces. We also prove some results in these spaces.

Definition 8. Let (X,q) be a quasi-partial metric space.

(a) A sequence $\{x_n\}$ in (X,q) is called 0-Cauchy if $\lim_{n,m\to\infty} q(x_n,x_m) = 0$ or $\lim_{n,m\to\infty} q(x_m,x_n) = 0$.

(b) The space (X,q) is called 0-complete if every 0-Cauchy sequence in X converges to a point $x \in X$ such that $q(x,x) = 0$.

It is easy to see that every 0-Cauchy sequence in (X,q) is Cauchy in (X,d_{p_q}) and if (X,q) is complete, then it is 0-complete but the converse assertions do not hold. For example, the space $X = [0,+\infty) \cap Q$ with $q(x,y) = |x-y| + |x|$ is a 0-complete quasi-partial metric space but it is not complete (since $d_{p_q}(x,y) = 2|x-y|$ and (X,d_{p_q}) is not complete).

Moreover, if (X,d) is complete quasimetric space which is also quasi-partial, then it is 0-complete, quasi-partial metric space.

Definition 9. Let X be a nonempty set. Then $(X,\preceq,q)$ is called an ordered quasi-partial metric space if (i) q is a quasi-partial metric on X and (ii) $\preceq$ is a partial order on X.

Theorem 10. *Let $(X,\preceq,q)$ be a 0-complete ordered quasi-partial metric space, $S : X \to X$ a dominated map, and x_0 an arbitrary point in X. Suppose that there exists $k \in [0,1)$ such that*

$$q(Sx,Sy) \le kq(x,y) \quad \text{for all comparable elements } x, y \text{ in } \overline{B(x_0,r)}. \tag{2}$$

And

$$q(x_0,Sx_0) \le (1-k)[r + q(x_0,x_0)]. \tag{3}$$

If, for a nonincreasing sequence $\{x_n\}$ in $\overline{B(x_0,r)}$, $\{x_n\} \to u$ implies that $u \preceq x_n$, then there exists a point x^* in $\overline{B(x_0,r)}$ such that $x^* = Sx^*$ and $q(x^*,x^*) = 0$. Moreover, x^* is unique, if for every pair of elements x, y in X there exists a point $z \in X$ such that $z \preceq x$ and $z \preceq y$.

Proof. Consider a Picard sequence $x_{n+1} = Sx_n$ with initial guess x_0, as $x_{n+1} = Sx_n \preceq x_n$ for all $n \in \{0\} \cup N$. We will prove that $x_n \in \overline{B(x_0, r)}$ for all $n \in N$ by mathematical induction. By using inequality (3), we have

$$\begin{aligned} q(x_0, x_1) &\leq (1-k)[r + q(x_0, x_0)] \\ &\leq r + q(x_0, x_0). \end{aligned} \tag{4}$$

Therefore $x_1 \in \overline{B(x_0, r)}$. Now let $x_2, \ldots, x_j \in \overline{B(x_0, r)}$ for some $j \in N$. As $x_{n+1} \preceq x_n$, so using inequality (2), we obtain

$$q(x_j, x_{j+1}) = q(Sx_{j-1}, Sx_j) \leq kq(x_{j-1}, x_j), \tag{5}$$

which implies that,

$$q(x_j, x_{j+1}) \leq k^2 q(x_{j-2}, x_{j-1}) \leq \cdots \leq k^j q(x_0, x_1). \tag{6}$$

Now

$$\begin{aligned} q(x_0, x_{j+1}) &\leq q(x_0, x_1) + \cdots + q(x_j, x_{j+1}) \\ &\quad - [q(x_1, x_1) + \cdots + q(x_j, x_j)] \\ &\leq q(x_0, x_1)[1 + \cdots + k^{j-1} + k^j] \\ &\qquad \text{(by inequality (6))}, \\ q(x_0, x_{j+1}) &\leq (1-k)[r + q(x_0, x_0)] \frac{(1-k^{j+1})}{1-k} \\ &\qquad \text{(by inequality (3))}. \end{aligned} \tag{7}$$

Thus $x_{j+1} \in \overline{B(x_0, r)}$. Hence $x_n \in \overline{B(x_0, r)}$ for all $n \in N$. Also $x_{n+1} \preceq x_n$ for all $n \in N$. It implies that

$$q(x_n, x_{n+1}) \leq k^n q(x_0, x_1) \quad \text{for all } n \in N. \tag{8}$$

It follows that

$$\begin{aligned} q(x_n, x_{n+i}) &\leq q(x_n, x_{n+1}) + \cdots + q(x_{n+i-1}, x_{n+i}) \\ &\leq k^n q(x_0, x_1)[1 + \cdots + k^{i-2} + k^{i-1}] \longrightarrow 0 \\ &\qquad \text{as } n \longrightarrow \infty. \end{aligned} \tag{9}$$

Notice that the sequence $\{x_n\}$ is a 0-Cauchy sequence in $(\overline{B(x_0, r)}, q)$. As $\overline{B(x_0, r)}$ is closed so is 0-complete. Therefore there exists a point $x^* \in \overline{B(x_0, r)}$ with

$$q(x^*, x^*) = \lim_{n \to \infty} q(x_n, x^*) = \lim_{n \to \infty} q(x^*, x_n) = 0. \tag{10}$$

Now,

$$q(x^*, Sx^*) \leq q(x^*, x_n) + q(Sx_{n-1}, Sx^*) - q(x_n, x_n). \tag{11}$$

On taking limit as $n \to \infty$ and using the fact that $x^* \preceq x_n \preceq x_{n-1}$, when $x_n \to x^*$, we have,

$$q(x^*, Sx^*) \leq \lim_{n \to \infty} [q(x^*, x_n) + kq(x_{n-1}, x^*)]. \tag{12}$$

Then by inequality (10), we have

$$q(x^*, Sx^*) \leq 0. \tag{13}$$

Similarly,

$$q(Sx^*, x^*) \leq 0, \tag{14}$$

and hence $x^* = Sx^*$. □

Uniqueness. Let y be another point in $\overline{B(x_0, r)}$ such that $y = Sy$. If x^* and y are comparable, then

$$q(x^*, y) = q(Sx^*, Sy) \leq kq(x^*, y). \tag{15}$$

This shows that $x^* = y$. Now if x^* and y are not comparable then there exists a point $z \in \overline{B(x_0, r)}$ which is lower bound of both x^* and y, that is, $z \preceq x^*$ and $z \preceq y$. Moreover by assumption $z \preceq x^* \preceq x_n \cdots \preceq x_0$. Now we will prove that $S^n z \in \overline{B(x_0, r)}$:

$$\begin{aligned} q(x_0, Sz) &\leq q(x_0, x_1) + q(x_1, Sz) - q(x_1, x_1) \\ &\leq (1-k)[r + q(x_0, x_0)] + kq(x_0, z) \\ &\qquad \text{(by inequality (2) and inequality (3))}, \\ q(x_0, Sz) &\leq (1-k)r + (1-k)q(x_0, x_0) + k[r + q(x_0, x_0)] \\ &= r - rk + q(x_0, x_0) - kq(x_0, x_0) + rk \\ &\quad + kq(x_0, x_0) \\ &= r + q(x_0, x_0). \end{aligned} \tag{16}$$

It follows that $Sz \in \overline{B(x_0, r)}$. Now let $S^j z \in \overline{B(x_0, r)}$

$$\begin{aligned} q(x_0, S^{j+1} z) &\leq q(x_0, x_1) + q(x_1, S^{j+1} z) - q(x_1, x_1) \\ &\leq (1-k)[r + q(x_0, x_0)] + kq(x_0, S^j z) \\ &\qquad \text{(by inequality (2) and inequality (3))}, \\ q(x_0, S^{j+1} z) &\leq (1-k)r + (1-k)q(x_0, x_0) \\ &\quad + k[r + q(x_0, x_0)] \\ &\leq r - rk + q(x_0, x_0) - kq(x_0, x_0) \\ &\quad + rk + kq(x_0, x_0) \\ &\leq r + q(x_0, x_0). \end{aligned} \tag{17}$$

It follows that $S^{j+1} z \in \overline{B(x_0, r)}$. Hence $S^n z \in \overline{B(x_0, r)}$ for all $n \in N$. Now as S is dominated, it follows that $S^{n-1} z \preceq S^{n-2} z \preceq \cdots \preceq z \preceq x^*$ and $S^{n-1} z \preceq y$ for all $n \in N$, which further implies $S^{n-1}z \preceq S^n x^*$ and $S^{n-1}z \preceq S^n y$ for all $n \in N$ as $S^n x^* = x^*$ and $S^n y = y$ for all $n \in N$:

$$\begin{aligned} q(x^*, y) &= q(S^n x^*, S^n y) \\ &\leq q(S^n x^*, S^{n-1}z) + q(S^{n-1}z, S^n y) \\ &\quad - q(S^{n-1}z, S^{n-1}z) \\ &\leq kq(S^{n-1}x^*, S^{n-2}z) + kq(S^{n-2}z, S^{n-1}y) \\ &\qquad (\because \text{ by inequality (2)}) \\ &\;\;\vdots \\ &\leq k^{n-2}q(x^*, Sz) + k^{n-2}q(Sz, y) \longrightarrow 0 \\ &\qquad \text{as } n \longrightarrow \infty. \end{aligned} \tag{18}$$

Hence $x^* = y$.

Example 11. Let $X = [0, +\infty) \cap Q$ be endowed with order $x \preceq y$ if $q(x, x) \leq q(y, y)$ and let $q : X \times X \to R^+$ be the 0-complete ordered quasi-partial metric on X defined by $q(x, y) = \max\{y - x, 0\} + x$. Define

$$Sx = \begin{cases} \frac{1}{10}x & \text{if } x \in [0, 1] \cap X \\ x - \frac{4}{9} & \text{if } x \in (1, \infty) \cap X. \end{cases} \tag{19}$$

Clearly, S is dominated mapping. Take $x_0 = 1/2$, $r = 1/2$; then $\overline{B(x_0, r)} = [0, 1] \cap X$; we have $q(x_0, x_0) = 1/2$, $k = 4/9$ with

$$\begin{aligned} (1 - k)[r + q(x_0, x_0)] &= \left(1 - \frac{4}{9}\right)\left[\frac{1}{2} + \frac{1}{2}\right] = \frac{5}{9}, \\ q(x_0, Sx_0) &= q\left(\frac{1}{2}, S\left(\frac{1}{2}\right)\right) \\ &= q\left(\frac{1}{2}, \frac{1}{20}\right) = \frac{1}{2} < \frac{5}{9}. \end{aligned} \tag{20}$$

Also if $x, y \in (1, \infty) \cap X$, then,

$$\begin{aligned} q(Sx, Sy) &= \max\left\{y - \frac{4}{9} - x + \frac{4}{9}, 0\right\} + x - \frac{4}{9} \\ &= \max\{y - x, 0\} + x - \frac{4}{9} \\ &\geq \frac{4}{9}[\max\{y - x, 0\} + x] = kq(x, y). \end{aligned} \tag{21}$$

So the contractive condition does not hold on X in each case.

Now if

$$x, y \in \overline{B(x_0, r)} \cap X, \tag{22}$$

then,

$$\begin{aligned} q(Sx, Sy) &= \max\left\{\frac{1}{10}y - \frac{1}{10}x, 0\right\} + \frac{1}{10}x \\ &= \frac{1}{10}q(x, y) < \frac{4}{9}q(x, y). \end{aligned} \tag{23}$$

Therefore, all the conditions of Theorem 13 are satisfied. Moreover, 0 is the unique fixed point of S and $q(0, 0) = 0$.

In Theorem 10, the condition (3) is imposed to restrict the condition (2) only for x, y in $\overline{B(x_0, r)}$ and Example 11 explains the utility of this restriction. However, the following result relaxes the condition (3) but imposes the condition (2) for all comparable elements in the whole space X.

Theorem 12. *Let $(X, \preceq, q)$ be a 0-complete ordered quasi-partial metric space, $S : X \to X$ a dominated map, and x_0 an arbitrary point in X. Suppose there exists $k \in [0, 1)$ with*

$$\begin{gathered} q(Sx, Sy) \leq kq(x, y), \\ \text{for all comparable elements } x, y \text{ in } X. \end{gathered} \tag{24}$$

If, for a nonincreasing sequence $\{x_n\}$ in X, $\{x_n\} \to u$ implies that $u \preceq x_n$, then there exists a point x^* in X such that $x^* = Sx^*$ and $q(x^*, x^*) = 0$. Moreover, x^* is unique, if for every pair of elements x, y in X, there exists a point $z \in X$ such that $z \preceq x$ and $z \preceq y$.

In Theorem 10, the existence of a lower bound and for a nonincreasing sequence $\{x_n\}$ in X, $\{x_n\} \to u$ implies that $u \preceq x_n$ are imposed to restrict the condition (2) only for comparable elements. However, the following result relaxes these conditions but imposes the condition (2) for all elements in $\overline{B(x_0, r)}$.

Theorem 13. *Let (X, q) be a 0-complete partial metric space, $S : X \to X$ a map, and x_0 an arbitrary point in X. Suppose there exists $k \in [0, 1)$ with*

$$\begin{gathered} q(Sx, Sy) \leq kq(x, y), \\ \text{for all elements } x, y \text{ in } \overline{B(x_0, r)}, \\ q(x_0, Sx_0) \leq (1 - k)[r + q(x_0, x_0)]. \end{gathered} \tag{25}$$

Then there exists a unique point x^* in $\overline{B(x_0, r)}$ such that $x^* = Sx^*$. Further $q(x^*, x^*) = 0$.

Metric version of Theorem 10 is given below.

Theorem 14. *Let $(X, \preceq, d)$ be a complete ordered metric space, $S : X \to X$ a dominated map, and x_0 an arbitrary point in X. Suppose that there exists $k \in [0, 1)$ such that*

$$\begin{gathered} d(Sx, Sy) \leq kd(x, y) \\ \text{for all comparable elements } x, y \text{ in } \overline{B(x_0, r)}, \\ d(x_0, Sx_0) \leq (1 - k)r. \end{gathered} \tag{26}$$

If, for a nonincreasing sequence $\{x_n\}$ *in* $\overline{B(x_0,r)}$, $\underline{\{x_n\}} \to u$ *implies that* $u \preceq x_n$, *then there exists a point* x^* *in* $\overline{B(x_0,r)}$ *such that* $x^* = Sx^*$ *and* $q(x^*,x^*) = 0$. *Moreover,* x^* *is unique, if for every pair of elements* x, y *in* X *there exists a point* $z \in X$ *such that* $z \preceq x$ *and* $z \preceq y$.

Theorem 15. *Let* $(X, \preceq, q)$ *be a 0-complete ordered quasi-partial metric space,* $S : X \to X$ *a dominated map, and* x_0 *an arbitrary point in* X. *Suppose that there exists* $b \in [0, 1/2)$ *such that*

$$q(Sx,Sy) \le b\left[q(x,Sx) + q(y,Sy)\right] \tag{27}$$

for all comparable elements x, y *in* $\overline{B(x_0,r)}$.

And

$$q(x_0,Sx_0) \le (1-k)\left[r + q(x_0,x_0)\right], \tag{28}$$

where $k = b/(1-b)$. *If, for a nonincreasing sequence* $\{x_n\}$ *in* $\overline{B(x_0,r)}$, $\underline{\{x_n\}} \to u$ *implies that* $u \preceq x_n$, *then there exists a point* x^* *in* $\overline{B(x_0,r)}$ *such that* $x^* = Sx^*$ *and* $q(x^*,x^*) = 0$. *Moreover,* x^* *is unique, if for any two points* x, y *in* $\overline{B(x_0,r)}$ *there exists a point* $z \in \overline{B(x_0,r)}$ *such that* $z \preceq x$ *and* $z \preceq y$, *and*

$$q(x_0,Sx_0)+q(z,Sz) \le q(x_0,z)+q(Sx_0,Sz) \quad \textit{for all } z \preceq Sx_0. \tag{29}$$

Proof. Consider a Picard sequence $x_{n+1} = Sx_n$ with initial guess x_0, as $x_{n+1} = Sx_n \preceq x_n$ for all $n \in \{0\} \cup N$. We will prove that $x_n \in \overline{B(x_0,r)}$ for all $n \in N$ by mathematical induction. By using inequality (28), we have

$$\begin{aligned} q(x_0,x_1) &\le (1-k)\left[r + q(x_0,x_0)\right] \\ &\le r + q(x_0,x_0). \end{aligned} \tag{30}$$

Therefore $x_1 \in \overline{B(x_0,r)}$. Now let $x_2, \ldots, x_j \in \overline{B(x_0,r)}$ for some $j \in N$. As $x_{j+1} \preceq x_j$, so using inequality (27), we obtain

$$\begin{aligned} q(x_j,x_{j+1}) &= q(Sx_{j-1},Sx_j) \\ &\le b\left[q(x_{j-1},x_j) + q(x_j,x_{j+1})\right] \\ &\le kq(x_{j-1},x_j), \end{aligned} \tag{31}$$

which implies that

$$q(x_j,x_{j+1}) \le k^2 q(x_{j-2},x_{j-1}) \le \cdots \le k^j q(x_0,x_1). \tag{32}$$

Now,

$$\begin{aligned} q(x_0,x_{j+1}) &\le q(x_0,x_1) + \cdots + q(x_j,x_{j+1}) \\ &\quad - \left[q(x_1,x_1) + \cdots + q(x_j,x_j)\right] \\ &\le q(x_0,x_1)\left[1 + \cdots + k^{j-1} + k^j\right] \\ &\qquad \text{(by inquality (32))}, \\ q(x_0,x_{j+1}) &\le (1-k)\left[r + q(x_0,x_0)\right]\frac{(1-k^{j+1})}{1-k} \\ &\qquad \text{(by inquality (28))}. \end{aligned} \tag{33}$$

Thus $x_{j+1} \in \overline{B(x_0,r)}$. Hence $x_n \in \overline{B(x_0,r)}$ for all $n \in N$. Also $x_{n+1} \preceq x_n$ for all $n \in N$. It implies that

$$q(x_n,x_{n+1}) \le k^n q(x_0,x_1) \quad \text{for all } n \in N. \tag{34}$$

It follows that

$$\begin{aligned} q(x_n,x_{n+i}) &\le q(x_n,x_{n+1}) + \cdots + q(x_{n+i-1},x_{n+i}) \\ q(x_n,x_{n+i}) &\le k^n q(x_0,x_1)\left[1 + \cdots + k^{i-2} + k^{i-1}\right] \to 0 \\ &\qquad \text{as } n \to \infty. \end{aligned} \tag{35}$$

Notice that the sequence $\{x_n\}$ is a 0-Cauchy sequence in $(\overline{B(x_0,r)}, q)$. As $\overline{B(x_0,r)}$ is closed, so is 0-complete. Therefore there exists a point $x^* \in \overline{B(x_0,r)}$ with

$$q(x^*,x^*) = \lim_{n\to\infty} q(x_n,x^*) = \lim_{n\to\infty} q(x^*,x_n) = 0. \tag{36}$$

Now,

$$q(x^*,Sx^*) \le q(x^*,x_n) + q(Sx_{n-1},Sx^*) - q(x_n,x_n). \tag{37}$$

On taking limit as $n \to \infty$ and using the fact that $x^* \preceq x_n \preceq x_{n-1}$, when $x_n \to x^*$, we have,

$$\begin{aligned} q(x^*,Sx^*) &\le \lim_{n\to\infty} [q(x^*,x_n) \\ &\qquad + b\{q(x_{n-1},Sx_{n-1}) + q(x^*,Sx^*)\}] \\ &\le \lim_{n\to\infty} b\left\{k^{n-1}q(x_0,x_1) + q(x^*,Sx^*)\right\} \\ &\le \lim_{n\to\infty} k^n q(x_0,x_1) \le 0. \end{aligned} \tag{38}$$

Similarly,

$$q(Sx^*,x^*) \le 0, \tag{39}$$

and hence $x^* = Sx^*$. Now,

$$\begin{aligned} q(x^*,x^*) &= q(Sx^*,Sx^*) \\ &\le b\{q(x^*,Sx^*) + q(x^*,Sx^*)\} \end{aligned} \tag{40}$$

$$(1-2b)\,q(x^*,x^*) \le 0.$$

This implies that

$$q(x^*,x^*) = 0. \tag{41}$$

□

Uniqueness. Let y be another point in $\overline{B(x_0, r)}$ such that $y = Sy$. If x^* and y are comparable, then

$$\begin{aligned} q(x^*, y) &= q(Sx^*, Sy) \\ &\le b[q(x^*, Sx^*) + q(y, Sy)] \\ &= b[q(x^*, x^*) + q(y, y)] \\ &= bq(y, y) \le q(y, y). \end{aligned} \tag{42}$$

Using the fact that $q(y, y) \le q(x^*, y)$, we have $q(x^*, y) = 0$. Similarly, $q(y, x^*) = 0$. Now if x^* and y are not comparable to then there exists a point $z \in \overline{B(x_0, r)}$ which is a lower bound of both x^* and y. Moreover by assumption $Sz \preceq z \preceq x^* \preceq x_n \cdots \preceq x_0$. Now by using inequality (27), we have

$$\begin{aligned} q(Sx_0, Sz) &\le b[q(x_0, x_1) + q(z, Sz)] \\ &\le b[q(x_0, z) + q(x_1, Sz)] \\ &\quad \text{(by inequality (29))}, \end{aligned} \tag{43}$$

$$q(x_1, Sz) \le kq(x_0, z).$$

$$\begin{aligned} q(x_0, Sz) &\le q(x_0, x_1) + q(x_1, Sz) - q(x_1, x_1) \\ &\le q(x_0, x_1) + kq(x_0, z) \quad \text{(by inequality (43))}, \end{aligned}$$

$$\begin{aligned} q(x_0, Sz) &\le (1 - k)[r + q(x_0, x_0)] + k[r + q(x_0, x_0)] \\ &= r + q(x_0, x_0). \end{aligned} \tag{44}$$

It follows that $Sz \in \overline{B(x_0, r)}$. Now, we will prove that $S^n z \in \overline{B(x_0, r)}$, by using mathematical induction to apply inequality (27). Let $S^2 z, \ldots, S^j z \in \overline{B(x_0, r)}$ for some $j \in N$. As $S^j z \preceq S^{j-1} z \preceq \cdots \preceq z \preceq x^* \preceq x_n \cdots \preceq x_0$; then,

$$\begin{aligned} q(S^j z, S^{j+1} z) &= q(S(S^{j-1} z), S(S^j z)) \\ &\le b[q(S^{j-1} z, S^j z) + q(S^j z, S^{j+1} z)], \end{aligned} \tag{45}$$

which implies that

$$\begin{aligned} q(S^j z, S^{j+1} z) &\le kq(S^{j-1} z, S^j z) \\ &\le k^2 q(S^{j-2} z, S^{j-1} z) \le \cdots \le k^j q(z, Sz). \end{aligned} \tag{46}$$

Now,

$$\begin{aligned} q(x_{j+1}, S^{j+1} z) &= q(Sx_j, S(S^j z)) \\ &\le b[q(x_j, Sx_j) + q(S^j z, S^{j+1} z)] \\ &\le b[k^j q(x_0, x_1) + k^j q(z, Sz)] \\ &\quad \text{(by inequality (32) and inequality (46))} \\ &\le bk^j [q(x_0, z) + q(x_1, Sz)] \\ &\quad \text{(by inequality (29))} \\ &\le bk^j [q(x_0, z) + kq(x_0, z)] = k^{j+1} q(x_0, z_0). \end{aligned} \tag{47}$$

Now,

$$\begin{aligned} q(x_0, S^{j+1} z) &\le q(x_0, x_1) + q(x_1, x_2) + \cdots + q(x_j, x_{j+1}) \\ &\quad + q(x_{j+1}, S^{j+1} z) \\ &\le q(x_0, x_1) + kq(x_0, x_1) + \cdots + k^{j+1} q(x_0, z), \\ &\quad \text{(by inequality (32) and inequality (47))} \end{aligned}$$

$$q(x_0, S^{j+1} z) \le q(x_0, x_1)[1 + k + \cdots k^j] + k^{j+1} r, \quad \left(\text{as } z \in \overline{B(x_0, r)}\right)$$

$$q(x_0, S^{j+1} z) \le (1 - k) r \frac{(1 - k^{j+1})}{1 - k} + k^{j+1} r = r. \tag{48}$$

It follows that $S^{j+1} z \in \overline{B(x_0, r)}$ and hence $S^n z \in \overline{B(x_0, r)}$. Now inequality (46) can be written as

$$q(S^n z, S^{n+1} z) \le k^n q(z, Sz) \to 0 \quad \text{as } n \to \infty. \tag{49}$$

Now,

$$\begin{aligned} q(x^*, y) &= q(Sx^*, Sy) \\ &\le q(Sx^*, S^{n+1} z) + q(S^{n+1} z, Sy) \\ &\le b[q(x^*, Sx^*) + q(S^n z, S^{n+1} z)] \\ &\quad + b[q(S^n z, S^{n+1} z) + q(y, Sy)] \\ &\le bq(x^*, x^*) + 2bq(S^n z, S^{n+1} z) + bq(y, y) \\ &\le bq(y, y) \le q(y, y) \\ &\quad \text{(by inequality (41) and inequality (49)).} \end{aligned} \tag{50}$$

Using the fact that $q(y, y) \le q(x^*, y)$, we have $q(x^*, y) = 0$. Similarly, $q(y, x^*) = 0$. Hence $x^* = y$.

In Theorem 15, the conditions (29), the existence of a lower bound and for a nonincreasing sequence $\{x_n\}$ in X, $\{x_n\} \to u$ implies that $u \preceq x_n$ are imposed to restrict the condition (27) only for comparable elements. However, the following result relaxes these conditions but imposes the condition (27) for all elements in $\overline{B(x_0, r)}$.

Theorem 16. *Let (X, d_q) be a complete dislocated quasimetric space, $S : X \to X$ a map, and x_0 an arbitrary point in X. Suppose there exists $k \in [0, 1/2)$ with*

$$d_q(Sx, Sy) \le k[d_q(x, Sx) + d_q(y, Sy)], \tag{51}$$

for all elements x, y in $\overline{B(x_0, r)}$ and

$$d_q(x_0, Sx_0) \le (1 - \theta) r, \tag{52}$$

where $\theta = k/(1 - k)$. Then there exists a unique point x^* in $\overline{B(x_0, r)}$ such that $x^* = Sx^*$ and $d_q(x^*, x^*) = 0$.

In Theorem 10, the conditions (28) and (29) are imposed to restrict the condition (27) only for comparable elements. However, the following result relaxes these conditions but imposes the condition (27) for all elements in X.

Theorem 17. *Let $(X, \preceq, q)$ be a 0-complete ordered quasi-partial metric space and let $S : X \to X$ be a dominated map. Suppose that there exists $b \in [0, 1/2)$ such that*

$$q(Sx, Sy) \leq b[q(x, Sx) + q(y, Sy)] \tag{53}$$

for all comparable elements x, y in X. If, for a nonincreasing sequence $\{x_n\}$ in X, $\{x_n\} \to u$ implies that $u \preceq x_n$, then there exists a point x^* in X such that $x^* = Sx^*$ and $q(x^*, x^*) = 0$. Moreover, x^* is unique, if for any two points x, y in $\overline{B(x_0, r)}$ there exists a point $z \in \overline{B(x_0, r)}$ such that $z \preceq x$ and $z \preceq y$.

Theorem 18. *Let $(X, \preceq, q)$ be a 0-complete ordered quasi-partial metric space, $S : X \to X$ a dominated map, and x_0 an arbitrary point in X. Suppose that there exists $c \in [0, 1/2)$ such that*

$$q(Sx, Sy) \leq c[q(x, Sy) + q(Sx, y)] \tag{54}$$

for all comparable elements x, y in $\overline{B(x_0, r)}$.

And

$$q(x_0, Sx_0) \leq (1 - k)[r + q(x_0, x_0)], \tag{55}$$

where $k = c/(1 - c)$. If, for a nonincreasing sequence $\{x_n\}$ in $\overline{B(x_0, r)}$, $\{x_n\} \to u$ implies that $u \preceq x_n$, then there exists a point x^* in $\overline{B(x_0, r)}$ such that $x^* = Sx^*$. Further $q(x^*, x^*) = 0$.

Proof. Consider a Picard sequence $x_{n+1} = Sx_n$ with initial guess x_0. As $x_{n+1} = Sx_n \preceq x_n$ for all $n \in \{0\} \cup N$. We will prove that $x_n \in \overline{B(x_0, r)}$ for all $n \in N$ by mathematical induction. By using inequality (55), we have

$$q(x_0, x_1) \leq r + q(x_0, x_0). \tag{56}$$

Therefore $x_1 \in \overline{B(x_0, r)}$. Now let $x_2, \ldots, x_j \in \overline{B(x_0, r)}$ for some $j \in N$. As $x_{n+1} \preceq x_n$, so using inequality (54), we obtain

$$\begin{aligned} q(x_j, x_{j+1}) &\leq c[q(x_{j-1}, x_j) + q(x_j, x_{j+1}) \\ &\quad - q(x_j, x_j) + q(x_j, x_j)] \\ &\leq kq(x_{j-1}, x_j), \end{aligned} \tag{57}$$

which implies that

$$q(x_j, x_{j+1}) \leq k^2 q(x_{j-2}, x_{j-1}) \leq \cdots \leq k^j q(x_0, x_1). \tag{58}$$

Now

$$\begin{aligned} q(x_0, x_{j+1}) &\leq q(x_0, x_1) + \cdots + q(x_j, x_{j+1}) \\ &\quad - [q(x_1, x_1) + \cdots + q(x_j, x_j)] \\ q(x_0, x_{j+1}) &\leq (1 - k)[r + q(x_0, x_0)] \frac{(1 - k^{j+1})}{1 - k} \\ &\quad (\text{by inequality (55)}). \end{aligned} \tag{59}$$

Thus $x_{j+1} \in \overline{B(x_0, r)}$. Hence $x_n \in \overline{B(x_0, r)}$ for all $n \in N$. Also $x_{n+1} \preceq x_n$ for all $n \in N$. It implies that

$$q(x_n, x_{n+1}) \leq k^n q(x_0, x_1) \quad \text{for all } n \in N. \tag{60}$$

It follows that

$$\begin{aligned} q(x_n, x_{n+i}) &\leq q(x_n, x_{n+1}) + \cdots + q(x_{n+i-1}, x_{n+i}) \\ q(x_n, x_{n+i}) &\leq k^n q(x_0, x_1)[1 + \cdots + k^{i-2} + k^{i-1}] \to 0 \\ &\quad \text{as } n \to \infty. \end{aligned} \tag{61}$$

Notice that the sequence $\{x_n\}$ is a 0-Cauchy sequence in $(\overline{B(x_0, r)}, q)$. As $\overline{B(x_0, r)}$ is closed so is 0-complete. Therefore there exists a point $x^* \in \overline{B(x_0, r)}$ with

$$q(x^*, x^*) = \lim_{n\to\infty} q(x_n, x^*) = \lim_{n\to\infty} q(x^*, x_n) = 0. \tag{62}$$

Now,

$$q(x^*, Sx^*) \leq q(x^*, x_n) + q(Sx_{n-1}, Sx^*) - q(x_n, x_n). \tag{63}$$

On taking limit as $n \to \infty$ and using the fact that $x^* \preceq x_n \preceq x_{n-1}$, when $x_n \to x^*$, we have

$$\begin{aligned} q(x^*, Sx^*) &\leq \lim_{n\to\infty} [q(x^*, x_n) \\ &\quad + c\{q(x_{n-1}, Sx^*) + q(x_n, x^*)\} \\ &\leq \lim_{n\to\infty} [c\{q(x_{n-1}, x^*) + q(x^*, Sx^*) \\ &\quad - q(x^*, x^*) + q(x_n, x^*)\}] \\ (1 - c)\,q(x^*, Sx^*) &\leq 0 \quad (\text{by inequality (62)}). \end{aligned} \tag{64}$$

Similarly,

$$q(Sx^*, x^*) \leq 0 \tag{65}$$

and hence $x^* = Sx^*$. □

Remark 19. We can obtain the partial metric, quasi-metric, and metric version of all theorems which are still not present in the literature.

Conflict of Interests

The authors declare that they have no competing interests.

References

[1] M. Abbas, M. Arshad, and A. Azam, "Fixed points of asymptotically regular mappings in complex-valued metric spaces," *Georgian Mathematical Journal*, vol. 20, no. 2, pp. 213–221, 2013.

[2] J. Ahmad, M. Arshad, and C. Vetro, "On a theorem of Khan in a generalized metric space," *International Journal of Analysis*, vol. 2013, Article ID 852727, 6 pages, 2013.

[3] M. Arshad, J. Ahmad, and E. Karapinar, "Some common fixed point results in rectangular metric spaces," *International Journal of Analysis*, vol. 2013, Article ID 307234, 7 pages, 2013.

[4] K. Leibovic, "The principle of contration mapping in nonlinear and adoptive controle systems," *IEEE Transactions on Automatic Control*, vol. 9, pp. 393–398, 1964.

[5] G. A. Medrano-Cerda, "A fixed point formulation to parameter estimation problems," in *Proceedings of the 26th IEEE Conference on Decision and Control*, pp. 1468–1476, 1987.

[6] J. E. Steck, "Convergence of recurrent networks as contraction mappings," *Neural Networks*, vol. 3, pp. 7–11, 1992.

[7] A. C. M. Ran and M. C. B. Reurings, "A fixed point theorem in partially ordered sets and some applications to matrix equations," *Proceedings of the American Mathematical Society*, vol. 132, no. 5, pp. 1435–1443, 2004.

[8] J. J. Nieto and R. Rodríguez-López, "Contractive mapping theorems in partially ordered sets and applications to ordinary differential equations," *Order*, vol. 22, no. 3, pp. 223–239, 2005.

[9] I. Altun and H. Simsek, "Some fixed point theorems on ordered metric spaces and application," *Fixed Point Theory and Applications*, vol. 2010, Article ID 621492, 17 pages, 2010.

[10] T. G. Bhaskar and V. Lakshmikantham, "Fixed point theorems in partially ordered metric spaces and applications," *Nonlinear Analysis: Theory, Methods & Applications*, vol. 65, no. 7, pp. 1379–1393, 2006.

[11] H. K. Nashine, B. Samet, and C. Vetro, "Monotone generalized nonlinear contractions and fixed point theorems in ordered metric spaces," *Mathematical and Computer Modelling*, vol. 54, no. 1-2, pp. 712–720, 2011.

[12] S. G. Matthews, "Partial metric topology," *Annals of the New York Academy of Sciences*, vol. 728, pp. 183–197, 1994.

[13] M. A. Bukatin and S. Yu. Shorina, "Partial metrics and co-continuous valuations," in *Foundations of Software Science and Computation Structure*, M. Nivat et al., Ed., vol. 1378 of *Lecture Notes in Computer Science*, pp. 125–139, Springer, 1998.

[14] I. Altun and A. Erduran, "Fixed point theorems for monotone mappings on partial metric spaces," *Fixed Point Theory and Applications*, vol. 2011, Article ID 508730, 10 pages, 2011.

[15] D. Paesano and P. Vetro, "Suzuki's type characterizations of completeness for partial metric spaces and fixed points for partially ordered metric spaces," *Topology and Its Applications*, vol. 159, no. 3, pp. 911–920, 2012.

[16] E. Karapınar, İ. M. Erhan, and A. Öztürk, "Fixed point theorems on quasi-partial metric spaces," *Mathematical and Computer Modelling*, vol. 57, no. 9-10, pp. 2442–2448, 2013.

[17] S. Romaguera, "A Kirk type characterization of completeness for partial metric spaces," *Fixed Point Theory and Applications*, vol. 2010, Article ID 493298, 6 pages, 2010.

[18] H. K. Nashine, Z. Kadelburg, S. Radenovic, and J. K. Kim, "Fixed point theorems under Hardy-Rogers contractive conditions on 0-complete ordered partial metric spaces," *Fixed Point Theory and Applications*, vol. 2012, article 180, 2012.

[19] M. Arshad, A. Shoaib, and I. Beg, "Fixed point of a pair of contractive dominated mappings on a closed ball in an ordered dislocated metric space," *Fixed Point Theory and Applications*, vol. 2013, article 115, 2013.

[20] A. Azam, S. Hussain, and M. Arshad, "Common fixed points of Chatterjea type fuzzy mappings on closed balls," *Neural Computing and Applications*, vol. 21, no. 1, supplement, pp. 313–317, 2012.

[21] A. Azam, M. Waseem, and M. Rashid, "Fixed point theorems for fuzzy contractive mappings in quasi-pseudo-metric spaces," *Fixed Point Theory and Applications*, vol. 2013, article 27, 2013.

[22] M. Abbas and S. Z. Németh, "Finding solutions of implicit complementarity problems by isotonicity of the metric projection," *Nonlinear Analysis: Theory, Methods & Applications*, vol. 75, no. 4, pp. 2349–2361, 2012.

[23] E. Kryeyszig, *Introductory Functional Analysis with Applications*, John Wiley & Sons, New York, NY, USA, 1989.

4

Some Analogies of the Banach Contraction Principle in Fuzzy Modular Spaces

Kittipong Wongkum, Parin Chaipunya, and Poom Kumam

Department of Mathematics, Faculty of Science, King Mongkut's University of Technology Thonburi (KMUTT), 126 Pracha Uthit Road, Bang Mod, Thung Khru, Bangkok 10140, Thailand

Correspondence should be addressed to Poom Kumam; poom.kum@kmutt.ac.th

Academic Editors: J. Fernández, J. Liu, K. Sadarangani, and H. Xu

We established some theorems under the aim of deriving variants of the Banach contraction principle, using the classes of inner contractions and outer contractions, on the structure of fuzzy modular spaces.

1. Introduction and Preliminaries

The concept of a modular space was introduced by Nakano [1]. Soon after, Musielak and Orlicz [2] redefined and generalized the notation of modular space. After that, Kumam et al. [3–7] studied fixed points and some properties in modular spaces. In 2007, Nourouzi [8] proposed probabilistic modular spaces based on the theory of modular spaces and some researches on the Menger's probabilistic metric spaces. A pair (X, ρ) is called a *probabilistic modular space* if X is a real vector space and ρ is a mapping from X into the set of all distribution functions (for $x \in X$, the distribution function $\rho(x)$ is denoted by ρ_x, and $\rho_x(t)$ is the value ρ_x at $t \in \mathbb{R}$) satisfying the following conditions:

(PM1) $\rho_x(0) = 0$;

(PM2) $\rho_x(t) = 1$ for all $t > 0$ if and only if $x = \theta$;

(PM3) $\rho_{-x}(t) = \rho_x(t)$;

(PM4) $\rho_{\alpha x+\beta y}(s + t) \geq \rho_x(s) \wedge \rho_y(t)$ for all $x, y \in X$ and $\alpha, \beta, s, t \in \mathbb{R}_0^+$, $\alpha + \beta = 1$.

For every $x \in X$, $t > 0$ and $\alpha \in \mathbb{R} \setminus \{0\}$, if

$$\rho_{\alpha x}(t) = \rho_x\left(\frac{t}{|\alpha|^\beta}\right), \tag{1}$$

where $\beta \in (0, 1]$, then we say that (X, ρ) is β-*homogeneous.*

Recently, further studies have been made on the probabilistic modular spaces. Nourouzi [8] extended the well-known Baire's Theorem to probabilistic modular spaces by using a special condition. Nourouzi [8] investigated the continuity and boundedness of linear operators defined between probabilistic modular spaces in the probabilistic sense. After that, Shen and Chen [9] following the idea of probabilistic modular space and the definition of fuzzy metric space in the sense of George and Veeramani [10], applied fuzzy concept to the classical notions of modular and modular spaces, and in 2013, Shen and Chen [9] introduced the concept of a fuzzy modular space.

Definition 1 (Vasuki [11]). A *fuzzy set* A in X is a function with domain X and value in $[0, 1]$.

Definition 2 (Arul Selvaraj and Sivakumar [12]). A *triangular norm* is a function $* : [0, 1] \times [0, 1] \to [0, 1]$ satisfying, for each $a, b, c, d \in [0, 1]$, the following conditions:

(1) $a * 1 = a$;

(2) $a * b \leq c * d$ whenever $a \leq c, b \leq d$;

(3) $a * b = b * a$ and $(a * b) * c = a * (b * c)$.

Definition 3 (Shen and Chen [9]). Let V be a real or complex vector space with a zero θ, $*$ a continuous triangular norm, and μ a fuzzy set on the product $V \times \mathbb{R}^+$. Suppose that the following properties hold for $x, y \in V$ and $s, t > 0$:

(FM1) $\mu(x,t) > 0$;

(FM2) $\mu(x,t) = 1$ for all $t > 0$ if and only if $x = \theta$;

(FM3) $\mu(x,t) = \mu(-x,t)$;

(FM4) $\mu(z,s+t) \geq \mu(x,s) * \mu(y,t)$ whenever z is the convex combination between x and y;

(FM5) the mapping $t \mapsto \mu(x,t)$ is continuous at each fixed $x \in V$.

Then, we write $(V,\mu,*)$ to represent the space with the predefined properties. In particular, we call μ a *fuzzy modular* and the triple $(V,\mu,*)$ a *fuzzy modular space.*

It is worth noting that every fuzzy modular is non-decreasing with respect to $t > 0$.

Example 4. Let X be a real or complex vector space and ρ be a modular on X. Take the t-norm $a * b = \min\{a,b\}$. For every $t \in (0,\infty)$, define $\mu(x,t) = t/(t+\rho(x))$ for all $x \in X$. Then $(X,\mu,*)$ is a F-modular space.

Remark 5. Note that the above conclusion still holds even if the t-norm is replaced by $a * b = a \cdot b$ and $a * b = \max\{a+b-1,0\}$, respectively.

Shen and Chen [9] also studied the topological properties of a fuzzy modular space with a special property that for every $x \in V$ and a non-zero real λ, the equality

$$\mu(\lambda x,t) = \mu\left(x, \frac{t}{|\lambda|^{\beta}}\right) \tag{2}$$

holds for some fixed $\beta \in (0,1]$. If the fuzzy modular μ has this property, we shall say that it is *β-homogeneous.*

The *μ-ball* in $(V,\mu,*)$ is the set of the form

$$B(x,r,t) := \{y \in V; \mu(x-y,t) > 1-r\}, \tag{3}$$

where $r \in (0,1)$ and $t > 0$. Now, suppose that μ is β-homogeneous for some $\beta \in (0,1]$. According to Shen and Chen [9], the family $\mathfrak{B}$ of all μ-balls forms a base for a first-countable Hausdorff topology, written as $\mathcal{T}_{\mu}$. With the notion of the μ-balls, it is easy to see that a sequence (x_n) in V *μ-converges* (i.e., it converges in the topology $\mathcal{T}_{\mu}$) to its *μ-limit* $x \in V$ if and only if $\mu(x-x_n,t) \to 1$ as $n \to \infty$ for all $t > 0$. Note here that the μ-limit is unique if it does exists after all. It is then natural to say that (x_n) is *μ-Cauchy* if for any given $\varepsilon \in (0,1)$ and $t > 0$, there exists $N \in \mathbb{N}$ with $\mu(x_m - x_n,t) > 1-\varepsilon$ whenever $m,n > N$.

At this point, let us turn to a typical example of a triangular norm which is defined by $(a * b) = \min\{a,b\}$. This triangular norm has a very special property that if $*'$ is an arbitrary triangular norm, then $(a*'b) \leq (a * b)$ for all $a,b \in [0,1]$. With this property, it is suitable to call this $*$ a *strongest triangular norm.* As is claimed by Shen and Chen [9], if V is a real vector space equipped with a β-homogeneous fuzzy modular μ and a strongest triangular norm $*$, then a μ-convergent sequence is μ-Cauchy. The authors also mentioned that if $*$ is not the strongest one, such implementation is not always true.

The space $(V,\mu,*)$ is said to be *μ-complete* if μ-Cauchy sequences actually μ-converges. We note that it makes more sense if we require a complete fuzzy modular space to be equipped with the strongest triangular norm.

A mapping f from V into itself is said to be an *inner contraction* if there exists a positive constant $q < 1$ such that

$$\mu(fx-fy,qt) \geq \mu(x-y,t), \tag{4}$$

for all $x,y \in V$ and $t > 0$.

On the other hand, f is said to be an *outer contraction* if there exists a positive constant $q < 1$ such that

$$\mu(fx-fy,t) - 1 \geq q(\mu(x-y,t)-1), \tag{5}$$

for all $x,y \in V$ and $t > 0$.

In this paper, we shall be working under the aim of deriving some variants of the Banach contraction principle, by using the classes of mappings defined above, on the structure of fuzzy modular spaces.

2. Main Results

We divide this section into two parts, discussing independently about the two main categories of our contractions predefined in the previous section. Notice that it is clear from the definition that every fuzzy metric space is in turn a fuzzy modular space. Hence, our results are also supplied with corollaries in fuzzy metric spaces. We shall, however, omit such consequences since they are obvious.

2.1. Fixed-Point Theorem for an Inner Contraction

Theorem 6. *Let V be a real vector space equipped with a β-homogeneous fuzzy modular μ and the strongest triangular norm $*$ such that $(V,\mu,*)$ is μ-complete. Suppose that at each $x \in V$, $\mu(x,t) \to 1$ as $t \to \infty$. If $f : V \to V$ is an inner contraction with constant $q \in (0,1)$, then f has a unique fixed point.*

Proof. Given a point $x_0 \in V$, we suppose that $f^n x_0 \neq f^{n+1} x_0$ for all $n \in \mathbb{N}$. Let $t > 0$, observe that

$$\begin{aligned}
\mu(f^n x_0 - f^{n+1} x_0, t) &\geq \mu\left(f^{n-1} x_0 - f^n x_0, \frac{t}{q}\right) \\
&\geq \mu\left(f^{n-2} x_0 - f^{n-1} x_0, \frac{t}{q^2}\right) \\
&\;\;\vdots \\
&\geq \mu\left(x_0 - f x_0, \frac{t}{q^n}\right).
\end{aligned} \tag{6}$$

As $n \to \infty$, we have $f^n x_0 - f^{n+1} x_0 \to \theta$ for every $t > 0$. That is, for any given $t > 0$ and $\varepsilon \in (0,1)$, there exists $N \in \mathbb{N}$ such that

$$\mu(f^N x_0 - f^{N+1} x_0, t) \geq \mu\left(f^N x_0 - f^{N+1} x_0, \frac{t}{2^{\beta+1}}\right) > 1-\varepsilon. \tag{7}$$

We now claim to show by induction that $\mu(f^N x_0 - f^{N+p}x_0, t/2^{\beta+1}) > 1-\varepsilon$ for all $p \in \mathbb{N}$. Let us assume first that $\mu(f^N x_0 - f^{N+j}x_0, t) > 1-\varepsilon$ holds at some $j \in \mathbb{N}$. Observe that

$$\begin{aligned}
&\mu\left(f^N x_0 - f^{N+j+1}x_0, t\right) \\
&= \mu\left(\frac{1}{2}\left(f^N x_0 - f^{N+1}x_0\right) + \frac{1}{2}\left(f^{N+1}x_0 - f^{N+j+1}x_0\right), \frac{t}{2^{\beta}}\right) \\
&\geq \mu\left(f^N x_0 - f^{N+1}x_0, \frac{t}{2^{\beta+1}}\right) * \mu\left(f^{N+1}x_0 - f^{N+j+1}x_0, \frac{t}{2^{\beta+1}}\right) \\
&\geq \mu\left(f^N x_0 - f^{N+1}x_0, \frac{t}{2^{\beta+1}}\right) * \mu\left(f^{N}x_0 - f^{N+j}x_0, \frac{t}{2^{\beta+1}\cdot q}\right) \\
&\geq \mu\left(f^N x_0 - f^{N+1}x_0, \frac{t}{2^{\beta+1}}\right) * \mu\left(f^{N}x_0 - f^{N+j}x_0, \frac{t}{2^{\beta+1}}\right) \\
&> (1-\varepsilon) * (1-\varepsilon) \\
&= 1-\varepsilon.
\end{aligned} \tag{8}$$

We have thus proved our claim.

Next, we shall show that $(f^n x_0)$ is Cauchy. Let $t > 0$ and $\varepsilon \in (0,1)$ be arbitrary, and we choose $N \in \mathbb{N}$ according to the claim given above. For $n > m > N$, we may write $m = N + s$ and $n = N + s + t$, for some $s, t \in \mathbb{N}$. Now, consider that

$$\begin{aligned}
\mu\left(f^m x_0 - f^n x_0, t\right) &= \mu\left(f^{N+s}x_0 - f^{N+s+t}x_0, t\right) \\
&\geq \mu\left(f^N x_0 - f^{N+t}x_0, \frac{t}{q^s}\right) \\
&\geq \mu\left(f^N x_0 - f^{N+t}x_0, t\right) \\
&> 1-\varepsilon.
\end{aligned} \tag{9}$$

Thus, $(f^n x_0)$ is Cauchy, and so the μ-completeness yields that $f^n x_0 \longrightarrow x_*$ for some $x_* \in V$. It follows that

$$\begin{aligned}
\lim_{n\to\infty}\mu\left(f^{n+1}x_0 - fx_*, t\right) &\geq \lim_{n\to\infty}\mu\left(f^n x_0 - x_*, \frac{t}{q}\right) \\
&\geq \lim_{n\to\infty}\mu\left(f^n x_0 - x_*, t\right) = 1, \\
&\qquad \forall t > 0.
\end{aligned} \tag{10}$$

This means $fx_* = x_*$, since $\mathcal{T}_\mu$ is Hausdorff. To show that the fixed point of f is unique, assume that $y_* \in V$ is a fixed point of f as well. Finally, we obtain that

$$\begin{aligned}
\mu\left(x_* - y_*, t\right) &= \mu\left(f^n x_* - f^n y_*, t\right) \\
&\geq \mu\left(x_* - y_*, \frac{t}{q^n}\right) \longrightarrow 1, \quad \forall t > 0.
\end{aligned} \tag{11}$$

Therefore, it must be the case that $x_* = y_*$, and so the conclusion is fulfilled. □

2.2. Fixed-Point Theorem for an Outer Contraction. For this part, we consider a weaker form of a μ-Cauchy sequence, namely, a *μ-G-Cauchy sequence*. This concept has been used all along together with the notion of fuzzy spaces. For a fuzzy modular space V, (x_n) is called a μ-G-Cauchy sequence if for each fixed $p \in \mathbb{N}$ and $t > 0$, we have $\lim_{n\to\infty}(x_{n+p}, x_n, t) = 1$. If every μ-G-Cauchy sequence μ-converges, V is said to be *μ-G-complete*. It is to be noted that the notion of μ-G-completeness is slightly stronger than the ordinary completeness. It is enough to see that every μ-Cauchy sequence is also a μ-G-Cauchy sequence. For our result, it is still a question whether or not the μ-G-completeness assumption can be weakened.

Theorem 7. *Let V be a real vector space equipped with a β-homogeneous fuzzy modular μ and the strongest triangular norm $*$ such that $(V, \mu, *)$ is μ-G-complete. If $f : V \longrightarrow V$ is an outer contraction with constant $q \in (0,1)$, then f has a unique fixed point.*

Proof. Given a point $x_0 \in V$, we suppose that $f^n x_0 \neq f^{n+1}x_0$ for all $n \in \mathbb{N}$. By the definition of an outer contraction, we can rewrite this notion in the following:

$$\mu\left(fx - fy, t\right) \geq q\mu\left(x - y, t\right) + \left(1 - q\right), \tag{12}$$

for all $x, y \in V$ and $t > 0$.

Let $t > 0$, observe that

$$\begin{aligned}
\mu\left(f^n x_0 - f^{n+1}x_0, t\right) &\geq q\mu\left(f^{n-1}x_0 - f^n x_0, t\right) + (1-q) \\
&\geq q^2\mu\left(f^{n-2}x_0 - f^{n-1}x_0, t\right) \\
&\quad + q(1-q) + (1-q) \\
&\ \ \vdots \\
&\geq q^n\mu\left(x_0 - fx_0, t\right) + \sum_{k=0}^{n-1} q^k(1-q).
\end{aligned} \tag{13}$$

As $n \longrightarrow \infty$, we have

$$\mu\left(f^n x_0 - f^{n+1}x_0, t\right) > 1-\varepsilon, \quad \text{for } n \text{ being sufficiently large.} \tag{14}$$

Next, we shall show that $(f^n x_0)$ is Cauchy. Let $t > 0$ and $\varepsilon \in (0,1)$ be arbitrary, and we choose $N \in \mathbb{N}$. For $n > N$, $n \in \mathbb{N}$ and for each $p > 0$. Now, consider that

$$\begin{aligned}
&\mu\left(f^n x_0 - f^{n+p}x_0, t\right) \\
&\geq \mu\left(f^n x_0 - f^{n+1}x_0, \frac{t}{2^{\beta+1}}\right) * \mu\left(f^{n+1}x_0 - f^{n+p}x_0, \frac{t}{2^{\beta+1}}\right) \\
&\geq \mu\left(f^n x_0 - f^{n+1}x_0, \frac{t}{2^{\beta+1}}\right) * \mu\left(f^{n+1}x_0 - f^{n+2}x_0, \frac{t}{2^{2(\beta+1)}}\right)
\end{aligned}$$

$$
\begin{aligned}
&* \mu\left(f^{n+2}x_0 - f^{n+p}x_0, \frac{t}{2^{2(\beta+1)}}\right) \\
&\vdots \\
&\geq \mu\left(f^{n}x_0 - f^{n+1}x_0, \frac{t}{2^{\beta+1}}\right) * \mu\left(f^{n+1}x_0 - f^{n+2}x_0, \frac{t}{2^{2(\beta+1)}}\right) \\
&\quad * \mu\left(f^{n+2}x_0 - f^{n+3}x_0, \frac{t}{2^{3(\beta+1)}}\right) \\
&\quad * \cdots * \mu\left(f^{n+p-1}x_0 - f^{n+p}x_0, \frac{t}{2^{(p-1)(\beta+1)}}\right) \\
&> (1-\varepsilon) * (1-\varepsilon) * \cdots * (1-\varepsilon) \\
&= 1-\varepsilon.
\end{aligned}
\tag{15}
$$

Thus, the sequence $(f^n x_0)$ is Cauchy, and so the μ-completeness yields that $f^n x_0 \to x_*$ for some $x_* \in V$. It follows that

$$
\mu\left(f^{n+1}x_0 - fx_*, t\right) \geq q\mu\left(f^n x_0 - x_*, t\right) + (1-q), \quad \forall t > 0. \tag{16}
$$

Taking $n \to \infty$, we have

$$
\mu\left(f^{n+1}x_0 - fx_*, t\right) \longrightarrow 1. \tag{17}
$$

This means $fx_* = x_*$, since $\mathcal{T}_\mu$ is Hausdorff. To show that the fixed point of f is unique, assume that $y_* \in V$ is a fixed point of f as well. Finally, we obtain that

$$
\begin{aligned}
\mu(x_* - y_*, t) &= \mu(fx_* - fy_*, t) \\
&\geq q\mu(x_* - y_*, t) + (1-q).
\end{aligned}
\tag{18}
$$

Hence, we have $\mu(x_* - y_*, t)(1-q) \geq (1-q)$ which implies that $\mu(x_* - y_*, t) = 1$. Therefore, it must be the case that $x_* = y_*$, and so the conclusion is fulfilled. □

Open Question 1. Is Theorem 7 true under the assumption that V is μ-complete as in Theorem 6?

Acknowledgments

The authors would like to thank the Higher Education Research Promotion and National Research University Project of Thailand's Office of the Higher Education Commission for financial support (under NRU-CSEC Project no. 56000508).

References

[1] H. Nakano, *Modular Semi-Ordered Spaces*, 1956.

[2] J. Musielak and W. Orlicz, "On modular spaces," *Studia Mathematica*, vol. 18, pp. 49–65, 1959.

[3] K. Kuaket and P. Kumam, "Fixed points of asymptotic pointwise contractions in modular spaces," *Applied Mathematics Letters*, vol. 24, no. 11, pp. 1795–1798, 2011.

[4] P. Kumam, "Some geometric properties and fixed point theorem in modular spaces," in *Fixed Point Theorem and Its Applications*, J. Garcia Falset, L. Fuster, and B. Sims, Eds., pp. 173–188, Yokohama Publishers, 2004.

[5] P. Kumam, "Fixed point theorems for nonexpansive mappings in modular spaces," *Archivum Mathematicum*, vol. 40, no. 4, pp. 345–353, 2004.

[6] C. Mongkolkeha and P. Kumam, "Fixed point and common fixed point theorems for generalized weak contraction mappings of integral type in modular spaces," *International Journal of Mathematics and Mathematical Sciences*, vol. 2011, Article ID 705943, 12 pages, 2011.

[7] C. Mongkolkeha and P. Kumam, "Some fixed point results for generalized weak contraction mappings in Modular spaces," *International Journal of Analysis*, vol. 2013, Article ID 247378, 6 pages, 2013.

[8] K. Nourouzi, "Probabilistic modular spaces," in *Proceedings of the 6th International ISAAC Congress*, Ankara, Turkey, 2007.

[9] Y. Shen and W. Chen, "On fuzzy modular spaces," *Journal of Applied Mathematics*, vol. 2013, Article ID 576237, 8 pages, 2013.

[10] A. George and P. Veeramani, "On some results in fuzzy metric spaces," *Fuzzy Sets and Systems*, vol. 64, no. 3, pp. 395–399, 1994.

[11] R. Vasuki, "A common fixed point theorem in a fuzzy metric space," *Fuzzy Sets and Systems*, vol. 97, no. 3, pp. 395–397, 1998.

[12] X. Arul Selvaraj and D. Sivakumar, "t-Norm (λ, μ)-fuzzy quotient near-rings and t-Norm (λ, μ)-Fuzzy quasi-ideals," *International Mathematical Forum*, vol. 6, no. 5, pp. 203–209, 2011.

A Liouville Property of Holomorphic Maps

Chengjie Yu

Department of Mathematics, Shantou University, Shantou, Guangdong 515063, China

Correspondence should be addressed to Chengjie Yu; cjyu@stu.edu.cn

Academic Editors: M. M. Cavalcanti and J.-S. Chen

We prove a Liouville property of holomorphic maps from a complete Kähler manifold with nonnegative holomorphic bisectional curvature to a complete simply connected Kähler manifold with a certain assumption on the sectional curvature.

1. Introduction

Liouville property is an interesting topic in analysis since Liouville found that a bounded holomorphic function on the complex plane must be a constant function. It has been studied by many geometers. For example, Yau [1] studied Liouville property of harmonic functions on complete Riemannian manifold with nonnegative Ricci curvature and Tam [2] studied Liouville property of harmonic maps. By Yau's Schwartz lemma [3], every holomorphic map from a complete Kähler manifold with nonnegative Ricci curvature, to a complete Kähler manifold with holomorphic bisectional curvature not greater than a negative constant is a constant map. This is a Liouville property for holomorphic maps.

In this paper, we prove the following Liouville property of holomorphic maps. This is, in some sense, a generalization of Yau's Liouville property for holomorphic maps.

Theorem 1. *Let M be a complete Kähler manifold with nonnegative holomorphic bisectional curvature and N a complete simply connected Kähler manifold with sectional curvature $\leq -c/(1+r)^2$, where c is some positive constant and r is the distance function of N to a fixed point o. Then, any holomorphic map from M to N must be constant.*

Our strategy to prove this result is by first showing that there are plenty of bounded pluri-subharmonic functions on the target N which is guaranteed by the classical Hessian comparison. Then, by using Ni-Tam's [4] Liouville type theorem for plurisubharmonic functions on complete Kähler manifolds with nonnegative holomorphic bisectional curvature and noting that the pull back of a pluri-subharmonic function along a holomorphic map is still pluri-subharmonic, we obtain the conclusion.

2. Proof of the Main Result

In order to show the existence of enough bounded plurisubharmonic functions on complete simply connected Kähler manifolds with negative quadratically decayed sectional curvature, we need the following technical lemma.

Lemma 2. *For any positive constant c, there is a bounded function $f \in C^2([0,\infty))$ such that*

(1) $f(0) = 0$ *and* $f(r) > 0$ *for* $r > 0$;

(2) $f'(r) > 0$ *for* $r \geq 0$;

(3) $f''(r) + c/(1+r)\coth(cr/(1+r))f'(r) \geq 0$ *for* $r \geq 0$.

Proof. Note that

$$c\coth\left(\frac{cr}{1+r}\right) \longrightarrow c\coth c \tag{1}$$

as $r \to \infty$ and $c\coth c > 1$. There are two positive numbers R and δ, such that

$$\frac{c}{1+r}\coth\left(\frac{cr}{1+r}\right) > \frac{1+\delta}{1+r} \tag{2}$$

for any $r \geq R$. Let

$$h(r) = \min\left\{\frac{1+\delta}{1+r}, \frac{c}{1+r}\coth\left(\frac{cr}{1+r}\right)\right\}. \quad (3)$$

Then, it is clear that $h \in C([0,\infty))$. Moreover,

$$h(r) = \frac{1+\delta}{1+r} \quad (4)$$

when $r \geq R$. Let

$$f(r) = \int_0^r e^{-\int_0^s h(t)dt} ds. \quad (5)$$

Then, it is clear that $f \in C^2([0,\infty))$ satisfying (1) and (2). Moreover, when $r \geq R$,

$$\begin{aligned} f(r) &= \int_0^r e^{-\int_0^R h(t)dt - \int_R^s h(t)dt} ds \\ &= \int_0^r e^{-\int_0^R h(t)dt - \int_R^s ((1+\delta)/(1+t))dt} ds \\ &= (1+R)^{1+\delta} e^{-\int_0^R h(t)dt} \int_0^r \frac{1}{(1+s)^{1+\delta}} ds \\ &\leq (1+R)^{1+\delta} e^{-\int_0^R h(t)dt} \int_0^\infty \frac{1}{(1+s)^{1+\delta}} ds. \end{aligned} \quad (6)$$

So, f is bounded.

Finally,

$$\left(\log f'\right)'(r) = -h \geq -\frac{c}{1+r}\coth\left(\frac{cr}{1+r}\right), \quad (7)$$

which is equivalent to (3). □

Lemma 3. *Let (M^n, g) be a complete simply connected Kähler manifold with sectional curvature $\leq -c/((1+r)^2)$, where c is some positive constant and r is the distance function of M with respect to a fixed point $o \in M$. Then, for any point $p \in M$, there is a bounded continuous plurisubharmonic function ϕ such that $\phi > 0$ on $M \setminus \{p\}$ and $\phi(p) = 0$.*

Proof. Let r_p be the distance function to p. Then, it is clear that the sectional curvature of M is not greater than $-c_1/(1+r_p)^2$, where c_1 is some positive constant. By Hessian comparison, we have

$$D^2 r_p \geq \frac{\sqrt{c_1}}{1+r_p}\coth\left(\frac{\sqrt{c_1}r_p}{1+r_p}\right)\left(g - dr_p \otimes dr_p\right). \quad (8)$$

Let $e_1 = (1/\sqrt{2})(\nabla r_p - J\nabla r_p)$ and $e_1, e_2, \ldots, e_n$ be a parallel unitary frame along geodesic rays emanating from p. Then

$$\left(r_p\right)_1 = \frac{1}{\sqrt{2}} \quad (9)$$

and $(r_p)_\alpha = 0$ for any $\alpha > 1$.

By Lemma 2, let f be a bounded function in $C^2([0,\infty))$ such that

(1) $f(0) = 0$ and $f(r) > 0$ for $r > 0$;

(2) $f'(r) > 0$ for $r \geq 0$;

(3) $f''(r) + (\sqrt{c_1}/(1+r))\coth(\sqrt{c_1}r/(1+r))f'(r) \geq 0$ for $r \geq 0$.

Let $\phi(x) = f(r_p(x))$. Then, by (8), we have

$$\begin{aligned} \left(\phi_{\alpha\overline{\beta}}\right) &= \left(f'\left(r_p\right)\left(r_p\right)_{\alpha\overline{\beta}} + f''\left(r_p\right)\left(r_p\right)_\alpha\left(r_p\right)_{\overline{\beta}}\right) \\ &\geq \left(f'\left(r_p\right)\frac{\sqrt{c_1}}{1+r_p}\coth\left(\frac{\sqrt{c_1}r_p}{1+r_p}\right)\left(g_{\alpha\overline{\beta}} - \left(r_p\right)_\alpha\left(r_p\right)_{\overline{\beta}}\right) + f''\left(r_p\right)\left(r_p\right)_\alpha\left(r_p\right)_{\overline{\beta}}\right) \\ &= \begin{pmatrix} \frac{1}{2}\left[\frac{\sqrt{c_1}}{1+r_p}\coth\left(\frac{\sqrt{c_1}r_p}{1+r_p}\right)f'\left(r_p\right) + f''\left(r_p\right)\right] & 0 \\ 0 & \frac{\sqrt{c_1}}{1+r_p}\coth\left(\frac{\sqrt{c_1}r_p}{1+r_p}\right)f'\left(r_p\right)I_{n-1} \end{pmatrix}, \end{aligned} \quad (10)$$

which is nonnegative. Hence, ϕ is a bounded continuous pluri-subharmonic function on M with $\phi > 0$ on $M \setminus \{p\}$ and $\phi(p) = 0$. □

We are now ready to prove Theorem 1.

Proof of Theorem 1. We proceed by contradiction. Let $f : M \to N$ be a nonconstant holomorphic map. Then, there are two points p, q in M, such that $f(p) \neq f(q)$. By Lemma 3, let ϕ be a bounded continuous pluri-subharmonic function on N such that $\phi(f(p)) = 0$, $\phi(f(q)) > 0$. Then, $\phi(f)$ is a bounded continuous pluri-subharmonic function on M that is nonconstant. This contradicts Theorem 3.2 in [4]. □

Remark 4. The unitary invariant metric on $\mathbb{C}^n$ constructed by Seshadri [5] has negative sectional curvature and sectional curvature $\leq -1/(r^2 \log r)$ outside a compact subset. This means that the curvature decayed rate cannot be raised to be greater than 2 in Theorem 1.

Acknowledgments

This research was partially supported by NSFC 11001161 and GDNSF S2012010010038.

References

[1] S. T. Yau, "Harmonic functions on complete Riemannian manifolds," *Communications on Pure and Applied Mathematics*, vol. 28, pp. 201–228, 1975.

[2] L. F. Tam, "Liouville properties of harmonic maps," *Mathematical Research Letters*, vol. 2, no. 6, pp. 719–735, 1995.

[3] S. T. Yau, "A general Schwarz lemma for Kähler manifolds," *The American Journal of Mathematics*, vol. 100, no. 1, pp. 197–203, 1978.

[4] L. Ni and L. F. Tam, "Plurisubharmonic functions and the structure of complete Kähler manifolds with nonnegative curvature," *Journal of Differential Geometry*, vol. 64, no. 3, pp. 457–524, 2003.

[5] H. Seshadri, "Negative sectional curvature and the product complex structure," *Mathematical Research Letters*, vol. 13, no. 2-3, pp. 495–500, 2006.

6

Weighted Moments Estimators of the Parameters for the Extreme Value Distribution Based on the Multiply Type II Censored Sample

Jong-Wuu Wu,[1] Sheau-Chiann Chen,[1] Wen-Chuan Lee,[2] and Heng-Yi Lai[1]

[1] *Department of Applied Mathematics, National Chiayi University, Chiayi, Taiwan*
[2] *Department of International Business, Chang Jung Christian University, Tainan 71101, Taiwan*

Correspondence should be addressed to Wen-Chuan Lee; chuanlee@mail.cjcu.edu.tw

Academic Editors: C. P. Bhunu, S. Hristova, and L. Jódar

We propose the weighted moments estimators (WMEs) of the location and scale parameters for the extreme value distribution based on the multiply type II censored sample. Simulated mean squared errors (MSEs) of best linear unbiased estimator (BLUE) and exact MSEs of WMEs are compared to study the behavior of different estimation methods. The results show the best estimator among the WMEs and BLUE under different combinations of censoring schemes.

1. Introduction

The lifetime Y of a special component has an extreme value distribution with the location parameter θ and scale parameter σ if its cumulative distribution function (c.d.f.) is given by

$$F_Y(y) = \exp\left[-e^{-(y-\theta)/\sigma}\right], \quad -\infty < y < \infty, \qquad (1)$$

where $-\infty < \theta < \infty, \sigma > 0$. If $(Y - \theta)/\sigma = X$, then the c.d.f. of X can be written as $F_X(x) = \exp[-e^{-x}]$, $-\infty < x < \infty$. The distribution has been applied in the fields of life testing and reliability [1, 2].

Many researches in estimators of location and scale parameters for the extreme value distribution have been discussed. Balakrishnan and Chan [3] presented tables of the means, variances, and covariances of order statistics from extreme value distribution based on the complete sample. In life testing experiments, the experimenter might not always be in a position to observe the lifetimes of all the items put on test due to time limitations and/or other restrictions (such as dearth of bankroll or material resources). Therefore, multiplying type II censored sample may arise in practice. Let us suppose, for instance, that out of n items put on life test, the first r lifetimes $Y_{(1)} < Y_{(2)} < \cdots < Y_{(r)}$ have only been observed and the lifetimes for the rest $(N - r)$ components remain unobserved or missing. This type of censoring is known as right type II censoring. Another way to get censored data is to observe the largest s lifetimes $Y_{(N-s+1)} < Y_{(N-s+2)} < \cdots < Y_{(N)}$. The lifetimes of first $(N - s)$ components are missing, called left type II censoring scheme. Moreover, if the left and right censoring situations arise together, this is known as doubly type II censoring scheme [4–9]. In addition, a reverse situation to doubly type II censoring is midcensoring where the data on two extremes are available, but some middle observations are censored [10]. However, these censoring schemes are special cases of multiply type II censoring scheme. Furthermore, if the left and right censoring cases arise together, this is known as doubly type II censoring scheme. In addition, a reverse situation to doubly type II censoring is middle censoring where the data on two extremes are available, but some middle observations are censored. However, these censoring schemes are special cases of multiply type II censoring scheme [11, 12].

In this paper, we propose the WMEs of the location parameter θ and the scale parameter σ from extreme value distribution based on the multiply type II censored sample. The WMEs are derived in Section 2. We also extend the methods of Balakrishnan and Cohen [13] and Balakrishnan and Chan [4] to find BLUEs of the location and scale parameters under multiply type II censored sample. To study the behavior of different estimation methods, simulated MSEs of BLUE and exact MSEs of WMEs are compared and discussed in Section 3. We obtain the best estimator of θ and σ among the WMEs and BLUE, respectively. Finally, conclusions are given in Section 4.

2. The Weighted Moments Estimators and the Best Linear Unbiased Estimator

Suppose that the experimenter fails to observe the first r, the last s, and the middle l observations. Observations $Y_{(r+1)} < Y_{(r+2)} < \cdots < Y_{(r+k)} < Y_{(r+k+l+1)} < Y_{(r+k+l+2)} < \cdots < Y_{(N-s)}$ are the available multiply type II censored sample from (1). To eliminate the effect of the location parameter, we subtract $Y_{(r+1)}$ from each statistic, and a new available multiply type II censored sample can be written a $0 < Y_{(r+2)} - Y_{(r+1)} < \cdots < Y_{(r+k)} - Y_{(r+1)} < Y_{(r+k+l+1)} - Y_{(r+1)} < Y_{(r+k+l+2)} - Y_{(r+1)} < \cdots < Y_{(N-s)} - Y_{(r+1)}$. First, by using this censored data, we propose the following WMEs to estimate the scale parameter σ:

$$\begin{aligned}\hat{\sigma}_1 &= w_{11}\left[\sum_{i=r+2}^{r+k}\left(Y_{(i)} - Y_{(r+1)}\right)\right] + w_{12}\left[\sum_{i=r+k+l+1}^{N-s}\left(Y_{(i)} - Y_{(r+1)}\right)\right] \\ &= w_{11}C_{11} + w_{12}C_{12},\end{aligned} \tag{2}$$

$$\begin{aligned}\hat{\sigma}_2 &= w_{21}\left[\sum_{i=r+2}^{r+k}\left(Y_{(i)} - Y_{(r+1)}\right) + l\left(Y_{(r+k)} - Y_{(r+1)}\right)\right] \\ &\quad + w_{22}\left[\sum_{i=r+k+l+1}^{N-s}\left(Y_{(i)} - Y_{(r+1)}\right) + s\left(Y_{(N-s)} - Y_{(r+1)}\right)\right] \\ &= w_{21}C_{21} + w_{22}C_{22},\end{aligned} \tag{3}$$

$$\begin{aligned}\hat{\sigma}_3 &= w_{31}\left[\sum_{i=r+2}^{r+k}\left(Y_{(i)} - Y_{(r+1)}\right)\right] \\ &\quad + w_{32}\left[\sum_{i=r+k+l+1}^{N-s}\left(Y_{(i)} - Y_{(r+1)}\right) + s\left(Y_{(N-s)} - Y_{(r+1)}\right)\right] \\ &= w_{31}C_{31} + w_{32}C_{32},\end{aligned} \tag{4}$$

$$\begin{aligned}\hat{\sigma}_4 &= w_{41}\left[\sum_{i=r+2}^{r+k}\left(Y_{(i)} - Y_{(r+1)}\right) + l\left(Y_{(r+k)} - Y_{(r+1)}\right)\right] \\ &\quad + w_{42}\left[\sum_{i=r+k+l+1}^{N-s}\left(Y_{(i)} - Y_{(r+1)}\right)\right] \\ &= w_{41}C_{41} + w_{42}C_{42}.\end{aligned} \tag{5}$$

We apply (2) to (5) to compute their weights by minimizing their MSEs $\min_{w_{i1},w_{i2}} \mathrm{MSE}(\hat{\sigma}_i)$, $i = 1, 2, 3, 4$, or equivalently

$$\begin{aligned}&\min_{w_{i1},w_{i2}} w_{i1}^2 b_{i1} + w_{i2}^2 b_{i2} + 2w_{i1}w_{i2}b_{i3} \\ &\quad + \left(w_{i1}a_{i1} + w_{i2}a_{i2} - 1\right)^2, \quad i = 1, 2, 3, 4,\end{aligned} \tag{6}$$

where $a_{ij} = E(C_{ij})/\sigma$, $b_{ij} = \mathrm{Var}(C_{ij})/\sigma^2$, $i = 1, 2, 3, 4$, $j = 1, 2$ and $b_{i3} = \mathrm{Cov}(C_{i1}, C_{i2})/\sigma^2$, $i = 1, 2, 3, 4$. By using the results of Lieblein [14] and $Y_{(i)} = \theta + \sigma X_{(i)}$, taking $i = 1$ as example, a_{11}, a_{12}, b_{11}, b_{12}, and b_{13} are obtained as follows:

$$\begin{aligned}a_{11} &= \sum_{i=r+2}^{r+k} E\left(X_{(i)}\right) - (k-1)\,E\left(X_{(r+1)}\right), \\ a_{12} &= \sum_{i=r+k+l+1}^{N-s} E\left(X_{(i)}\right) - (N-s-r-k-l)\,E\left(X_{(r+1)}\right), \\ b_{11} &= \sum_{i=r+2}^{r+k} \mathrm{Var}\left(X_{(i)}\right) + 2\sum_{i=r+2}^{r+k-1}\sum_{j=i+1}^{r+k} \mathrm{Cov}\left(X_{(i)}, X_{(j)}\right) \\ &\quad + (k-1)^2\,\mathrm{Var}\left(X_{(r+1)}\right) \\ &\quad - 2(k-1)\sum_{i=r+2}^{r+k} \mathrm{Cov}\left(X_{(i)}, X_{(r+1)}\right), \\ b_{12} &= \sum_{i=r+k+l+1}^{N-s} \mathrm{Var}\left(X_{(i)}\right) \\ &\quad + 2\sum_{i=r+k+l+1}^{N-s-1}\sum_{j=i+1}^{N-s} \mathrm{Cov}\left(X_{(i)}, X_{(j)}\right) \\ &\quad + (N-s-r-k-l)^2\,\mathrm{Var}\left(X_{(r+1)}\right) \\ &\quad - 2(N-s-r-k-l)\sum_{i=r+k+l+1}^{N-s} \mathrm{Cov}\left(X_{(i)}, X_{(r+1)}\right), \\ b_{13} &= \sum_{i=r+2}^{r+k}\sum_{j=r+k+l+1}^{N-s} \mathrm{Cov}\left(X_{(i)}, X_{(j)}\right) \\ &\quad - (N-s-r-k-l)\sum_{i=r+2}^{r+k} \mathrm{Cov}\left(X_{(i)}, X_{(r+1)}\right) \\ &\quad - (k-1)\sum_{j=r+k+l+1}^{N-s} \mathrm{Cov}\left(X_{(r+1)}, X_{(j)}\right) \\ &\quad + (k-1)(N-s-r-k-l)\,\mathrm{Var}\left(X_{(r+1)}\right),\end{aligned} \tag{7}$$

where

$$\begin{aligned}E\left(X_{(i)}\right) &= NC_{i-1}^{N-1}\sum_{p=1}^{N-i+1}(-1)^{p-1} \\ &\quad \times C_{p-1}^{N-i}\frac{1}{i+p-1}\left[\gamma + \log(i+p-1)\right],\end{aligned}$$

$$\begin{aligned}
&\operatorname{Cov}\left(X_{(i)}, X_{(j)}\right) \\
&= ijC_j^N C_i^{j-1} \sum_{p=1}^{j-i} \sum_{q=1}^{N-j+1} (-1)^{p+q-2} \\
&\quad \times C_{p-1}^{j-i-1} C_{q-1}^{N-j} \phi\left(i+p-1, j-i-p+q\right) \\
&\quad - \left\{ NC_{i-1}^{N-1} \sum_{p=1}^{N-i+1} (-1)^{p-1} C_{p-1}^{N-i} \right. \\
&\quad \left. \times \frac{1}{i+p-1} \left[\gamma + \log\left(i+p-1\right)\right] \right\} \\
&\quad \times \left\{ NC_{j-1}^{N-1} \sum_{q=1}^{N-j+1} (-1)^{q-1} C_{q-1}^{N-j} \right. \\
&\quad \left. \times \frac{1}{j+q-1} \left[\gamma + \log\left(j+q-1\right)\right] \right\},
\end{aligned} \tag{8}$$

$\gamma = -\Gamma'(1)$ is Euler's constant and $\phi(t,u) = \int_{-\infty}^{\infty}\int_{-\infty}^{y} xye^{-x-te^{-x}}e^{-y-ue^{-y}}dx\,dy$ [14].

Finally, by differentiating (6) with respect to w_{i1} and w_{i2}, we obtain

$$\begin{aligned}
w_{i1} &= \frac{b_{i2}a_{i1} - b_{i3}a_{i2}}{\left(b_{i1} + a_{i1}^2\right)\left(b_{i2} + a_{i2}^2\right) - \left(b_{i3} + a_{i1}a_{i2}\right)^2}, \\
w_{i2} &= \frac{b_{i1}a_{i2} - b_{i3}a_{i1}}{\left(b_{i1} + a_{i1}^2\right)\left(b_{i2} + a_{i2}^2\right) - \left(b_{i3} + a_{i1}a_{i2}\right)^2},
\end{aligned} \tag{9}$$

where $a_{ij}, b_{ij}, j = 1,2$ and b_{i3} are as above definitions, $i = 1,2,3,4$. Hence, by applying the result in (9) to (6), a minimum value of $\operatorname{MSE}(\hat{\sigma}_i)$ can be obtained as follows:

$$\operatorname{MSE}\left(\hat{\sigma}_i\right) = \frac{b_{i1}b_{i2} - b_{i3}^2}{\left(b_{i1} + a_{i1}^2\right)\left(b_{i2} + a_{i2}^2\right) - \left(b_{i3} + a_{i1}a_{i2}\right)^2}, \quad i = 1,2,3,4, \tag{10}$$

where $a_{ij}, b_{ij}, j = 1,2$ and b_{i3} are as above definitions, $i = 1,2,3,4$ [12].

Based on (2) to (5), the estimators of location parameter θ for WMEs are computed by using moments method, and a weight is assigned in each equation from (11) to (14) as well. Thus, the estimators of location for WMEs are shown as follows:

$$\begin{aligned}
\hat{\theta}_1 &= w_1 \left[\frac{\sum_{i=r+1}^{r+k} Y_{(i)}}{k} - \frac{E_{11}}{k}\hat{\sigma}_1 \right] + \left(1 - w_1\right) \\
&\quad \times \left[\frac{\sum_{i=r+k+l+1}^{N-s} Y_{(i)}}{N-s-r-k-l} - \frac{E_{12}}{N-s-r-k-l}\hat{\sigma}_1 \right] \\
&= \theta + \sigma\left[w_1F_{11} + \left(1 - w_1\right)F_{12}\right],
\end{aligned} \tag{11}$$

$$\begin{aligned}
\hat{\theta}_2 &= w_2 \left[\frac{\sum_{i=r+1}^{r+k} Y_{(i)} + lY_{(r+k)}}{k+l} - \frac{E_{21}}{k+l}\hat{\sigma}_2 \right] + \left(1 - w_2\right) \\
&\quad \times \left[\frac{\sum_{i=r+k+l+1}^{N-s} Y_{(i)} + sY_{(N-s)}}{N-r-k-l} - \frac{E_{22}}{N-r-k-l}\hat{\sigma}_2 \right] \\
&= \theta + \sigma\left[w_2F_{21} + \left(1 - w_2\right)F_{22}\right],
\end{aligned} \tag{12}$$

$$\begin{aligned}
\hat{\theta}_3 &= w_3 \left[\frac{\sum_{i=r+1}^{r+k} Y_{(i)}}{k} - \frac{E_{11}}{k}\hat{\sigma}_3 \right] + \left(1 - w_3\right) \\
&\quad \times \left[\frac{\sum_{i=r+k+l+1}^{N-s} Y_{(i)} + sY_{(N-s)}}{N-r-k-l} - \frac{E_{22}}{N-r-k-l}\hat{\sigma}_3 \right] \\
&= \theta + \sigma\left[w_{33}F_{31} + \left(1 - w_3\right)F_{32}\right],
\end{aligned} \tag{13}$$

$$\begin{aligned}
\hat{\theta}_4 &= w_4 \left[\frac{\sum_{i=r+1}^{r+k} Y_{(i)} + lY_{(r+k)}}{k+l} - \frac{E_{21}}{k+l}\hat{\sigma}_4 \right] + \left(1 - w_4\right) \\
&\quad \times \left[\frac{\sum_{i=r+k+l+1}^{N-s} Y_{(i)}}{N-s-r-k-l} - \frac{E_{12}}{N-s-r-k-l}\hat{\sigma}_4 \right] \\
&= \theta + \sigma\left[w_4F_{41} + \left(1 - w_4\right)F_{42}\right],
\end{aligned} \tag{14}$$

where $\hat{\sigma}_1$, $\hat{\sigma}_2$, $\hat{\sigma}_3$, and $\hat{\sigma}_4$ as (2) to (5), respectively,

$$E_{11} = E\left[\sum_{i=r+1}^{r+k} X_{(i)}\right], \qquad E_{12} = E\left[\sum_{i=r+k+l+1}^{N-s} X_{(i)}\right],$$

$$E_{21} = E\left[\sum_{i=r+1}^{r+k} X_{(i)} + lX_{(r+k)}\right],$$

$$E_{22} = E\left[\sum_{i=r+k+l+1}^{N-s} X_{(i)} + sX_{(N-s)}\right],$$

$$F_{11} = \frac{\sum_{i=r+1}^{r+k} X_{(i)}}{k} - \frac{E_{11}}{k}\frac{\hat{\sigma}_1}{\sigma},$$

$$F_{12} = \frac{\sum_{i=r+k+l+1}^{N-s} X_{(i)}}{N-s-r-k-l} - \frac{E_{12}}{N-s-r-k-l}\frac{\hat{\sigma}_1}{\sigma},$$

$$F_{21} = \frac{\sum_{i=r+1}^{r+k} X_{(i)} + lX_{(r+k)}}{k+l} - \frac{E_{21}}{k+l}\frac{\hat{\sigma}_2}{\sigma},$$

$$F_{22} = \frac{\sum_{i=r+k+l+1}^{N-s} X_{(i)} + sX_{(N-s)}}{N-r-k-l} - \frac{E_{22}}{N-r-k-l}\frac{\hat{\sigma}_2}{\sigma},$$

$$F_{31} = \frac{\sum_{i=r+1}^{r+k} X_{(i)}}{k} - \frac{E_{11}}{k}\frac{\hat{\sigma}_3}{\sigma},$$

$$F_{32} = \frac{\sum_{i=r+k+l+1}^{N-s} X_{(i)} + sX_{(N-s)}}{N-r-k-l} - \frac{E_{22}}{N-r-k-l}\frac{\hat{\sigma}_3}{\sigma},$$

$$F_{41}=\frac{\sum_{i=r+1}^{r+k}X_{(i)}+lX_{(r+k)}}{k+l}-\frac{E_{21}}{k+l}\frac{\widehat{\sigma}_4}{\sigma},$$
$$F_{42}=\frac{\sum_{i=r+k+l+1}^{N-s}X_{(i)}}{N-s-r-k-l}-\frac{E_{12}}{N-s-r-k-l}\frac{\widehat{\sigma}_4}{\sigma}. \quad (15)$$

Similarly, the weights are computed by minimizing MSEs of $\widehat{\theta}_i$ and the optimal weights in MSEs of $\widehat{\theta}_i$ are obtained as

$$w_i=\left(\operatorname{Var}(F_{i2})-\operatorname{Cov}(F_{i1},F_{i2})-E(F_{i1})E(F_{i2})+[E(F_{i2})]^2\right)\times\left(\operatorname{Var}(F_{i1})+\operatorname{Var}(F_{i2})-2\operatorname{Cov}(F_{i1},F_{i2})+[E(F_{i1})]^2-2E(F_{i1})E(F_{i2})+[E(F_{i2})]^2\right)^{-1}. \quad (16)$$

The minimum value of MSEs of $\widehat{\theta}_i$ can then be written as

$$\begin{aligned}\operatorname{MSE}(\widehat{\theta}_i)=&\,w_i^2\{\operatorname{Var}(F_{i1})+\operatorname{Var}(F_{i2})-2\operatorname{Cov}(F_{i1},F_{i2})+[E(F_{i1})]^2-2E(F_{i1})E(F_{i2})+[E(F_{i2})]^2\}\\&-2w_i\{\operatorname{Var}(F_{i2})-\operatorname{Cov}(F_{i1},F_{i2})-E(F_{i1})E(F_{i2})+[E(F_{i2})]^2\}\\&+\operatorname{Var}(F_{i2})+[E(F_{i2})]^2,\end{aligned} \quad (17)$$

where $E(F_{i1})$, $E(F_{i2})$, $\operatorname{Var}(F_{i1})$, $\operatorname{Var}(F_{i2})$, and $\operatorname{Cov}(F_{i1},F_{i2})$ can be obtained based on (8) for $i=1,2,3,4$.

For the case of $l=0$, the WMEs of the scale parameter σ are proposed as (18) since it is reasonable with consecutive data. Consider the following:

$$\widehat{\sigma}_5=w_5\left[\sum_{i=r+2}^{N-s}\left(Y_{(i)}-Y_{(r+1)}\right)\right]=w_5C_5,$$
$$\widehat{\sigma}_6=w_6\left[\sum_{i=r+2}^{N-s}\left(Y_{(i)}-Y_{(r+1)}\right)+s\left(Y_{(N-s)}-Y_{(r+1)}\right)\right]=w_6C_6,$$
$$\widehat{\sigma}_7=w_7\left[r\left(Y_{(r+2)}-Y_{(r+1)}\right)+\sum_{i=r+2}^{N-s}\left(Y_{(i)}-Y_{(r+1)}\right)\right]=w_7C_7,$$
$$\widehat{\sigma}_8=w_8\left[r\left(Y_{(r+2)}-Y_{(r+1)}\right)+\sum_{i=r+2}^{N-s}\left(Y_{(i)}-Y_{(r+1)}\right)+s\left(Y_{(N-s)}-Y_{(r+1)}\right)\right]=w_8C_8. \quad (18)$$

In the same way, the minimum values of MSE of $\widehat{\sigma}_i$, $i=5,6,7,8$ can be obtained as

$$\operatorname{MSE}(\widehat{\sigma}_i)=\frac{a_i^2b_i+a_i^4}{(b_i+a_i^2)^2}-\frac{2a_i^2}{b_i+a_i^2}+1, \quad (19)$$

where $a_i=E(C_i)/\sigma$ and $b_i=\operatorname{Var}(C_i)/\sigma^2$, $i=5,6,7,8$. Thus, the WMEs $\widehat{\theta}_i$ of the location parameter θ for the case of $l=0$ can also be provided as follows:

$$\widehat{\theta}_5=\frac{\sum_{i=r+1}^{N-s}Y_{(i)}}{N-s-r}-\frac{E_5}{N-s-r}\widehat{\sigma}_5,$$
$$\widehat{\theta}_6=\frac{\sum_{i=r+1}^{N-s}Y_{(i)}+sY_{(N-s)}}{N-r}-\frac{E_6}{N-r}\widehat{\sigma}_6,$$
$$\widehat{\theta}_7=\frac{rY_{(r+1)}+\sum_{i=r+1}^{N-s}Y_{(i)}}{N-s}-\frac{E_7}{N-s}\widehat{\sigma}_7,$$
$$\widehat{\theta}_8=\frac{rY_{(r+1)}+\sum_{i=r+1}^{N-s}Y_{(i)}+sY_{(N-s)}}{N}-\frac{E_8}{N}\widehat{\sigma}_8, \quad (20)$$

where $E_5=E[\sum_{i=r+1}^{N-s}X_{(i)}]$, $E_6=E[\sum_{i=r+1}^{N-s}X_{(i)}+sX_{(N-s)}]$, $E_7=E[rX_{(r+1)}+\sum_{i=r+1}^{N-s}X_{(i)}]$, $E_8=E[rX_{(r+1)}+\sum_{i=r+1}^{N-s}X_{(i)}+sX_{(N-s)}]$, and $\widehat{\sigma}_5,\widehat{\sigma}_6,\widehat{\sigma}_7,\widehat{\sigma}_8$ as (18). The MSEs of $\widehat{\theta}_i$ are obtained by using $\operatorname{MSE}(\widehat{\theta}_i)=\operatorname{Var}(\widehat{\theta}_i)+[E(\widehat{\theta}_i)-\theta]^2$ for $i=5,6,7,8$.

Next, we extend the method of Balakrishnan and Cohen [13] and Balakrishnan and Chan [4] to derive the BLUE of the location and scale parameters in an explicit form under a multiply type II censored sample for the general case. BLUE $\widehat{\theta}_9$ and $\widehat{\sigma}_9$ are obtained as follows:

$$\widehat{\theta}_9=\left\{\frac{\alpha^T\beta^{-1}\alpha1^T\beta^{-1}-\alpha^T\beta^{-1}1\alpha^T\beta^{-1}}{(\alpha^T\beta^{-1}\alpha)(1^T\beta^{-1}1)-(\alpha^T\beta^{-1}1)^2}\right\}Y,$$
$$\widehat{\sigma}_9=\left\{\frac{1^T\beta^{-1}1\alpha^T\beta^{-1}-1^T\beta^{-1}\alpha1^T\beta^{-1}}{(\alpha^T\beta^{-1}\alpha)(1^T\beta^{-1}1)-(\alpha^T\beta^{-1}1)^2}\right\}Y, \quad (21)$$

where

$$1=(1,1,\ldots,1,1,\ldots,1)^T_{(N-s-r-l)\times1}$$
$$Y=\left(Y_{(r+1)},Y_{(r+2)},\ldots,Y_{(r+k)},Y_{(r+k+l+1)},Y_{(r+k+l+2)},\ldots,Y_{(N-s)}\right)^T$$
$$\alpha=\left(\alpha_{(r+1)},\alpha_{(r+2)},\ldots,\alpha_{(r+k)},\alpha_{(r+k+l+1)},\alpha_{(r+k+l+2)},\ldots,\alpha_{(N-s)}\right),$$

$$\beta = \begin{bmatrix} \beta_{(r+1,r+1)} & \beta_{(r+1,r+2)} & \cdots & \beta_{(r+1,r+k)} & \beta_{(r+1,r+k+l+1)} & \cdots & \beta_{(r+1,N-s)} \\ \beta_{(r+1,r+2)} & \beta_{(r+2,r+2)} & \cdots & \beta_{(r+2,r+k)} & \beta_{(r+1,r+k+l+1)} & \cdots & \beta_{(r+2,N-s)} \\ \vdots & \vdots & \ddots & \vdots & \vdots & \ddots & \vdots \\ \beta_{(r+1,r+k)} & \beta_{(r+2,r+k)} & \cdots & \beta_{(r+k,r+k)} & \beta_{(r+1,r+k+l+1)} & \cdots & \beta_{(r+k,N-s)} \\ \beta_{(r+1,r+k+l+1)} & \beta_{(r+2,r+k+l+1)} & \cdots & \beta_{(r+k,r+k+l+1)} & \beta_{(r+1,r+k+l+1)} & \cdots & \beta_{(r+k+l+1,N-s)} \\ \beta_{(r+1,r+k+l+2)} & \beta_{(r+2,r+k+l+2)} & \cdots & \beta_{(r+k,r+k+l+2)} & \beta_{(r+1,r+k+l+1)} & \cdots & \beta_{(r+k+l+2,N-s)} \\ \vdots & \vdots & \ddots & \vdots & \vdots & \ddots & \vdots \\ \beta_{(r+1,N-s)} & \beta_{(r+2,N-s)} & \cdots & \beta_{(r+k,N-s)} & \beta_{(r+1,r+k+l+1)} & \cdots & \beta_{(N-s,N-s)} \end{bmatrix},$$

$$\alpha_i = E\left(\frac{Y_{(i)} - \theta}{\sigma}\right) = E\left(X_{(i)}\right),$$

$$\beta_{(i,j)} = \text{Cov}\left(\frac{Y_{(i)} - \theta}{\sigma}, \frac{Y_{(j)} - \theta}{\sigma}\right) = \text{Cov}\left(X_{(i)}, X_{(j)}\right). \quad (22)$$

$E(X_{(i)})$ and $\text{Cov}(X_{(i)}, X_{(j)})$ are as above definition.

Further, the MSEs of BLUE $\hat{\theta}_9$ and $\hat{\sigma}_9$ are given by

$$\text{MSE}\left(\hat{\theta}_9\right) = \text{Var}\left(\hat{\theta}_9\right) = \sigma^2 \left\{\frac{\alpha^T \beta^{-1} \alpha}{(\alpha^T \beta^{-1} \alpha)(1^T \beta^{-1} 1) - (\alpha^T \beta^{-1} 1)^2}\right\},$$
$$\text{MSE}\left(\hat{\sigma}_9\right) = \text{Var}\left(\hat{\sigma}_9\right) = \sigma^2 \left\{\frac{1^T \beta^{-1} 1}{(\alpha^T \beta^{-1} \alpha)(1^T \beta^{-1} 1) - (\alpha^T \beta^{-1} 1)^2}\right\}, \quad (23)$$

where 1, α and β are as above definition. In addition, we find the best estimators based on the smallest MSEs among the WMEs and BLUE, respectively, as follows:

$$\underset{1 \le i \le 9}{\text{MSE}}\left(\hat{\theta}_i\right), \qquad \underset{1 \le i \le 9}{\text{MSE}}\left(\hat{\sigma}_i\right). \quad (24)$$

3. The Comparative Numerical Results

To study the behavior of our proposed WMEs and BLUE, the simulated MSEs of BLUE and exact MSEs of WMEs are compared. The numerical results were obtained to cover different censoring schemes assigning appropriate values to r, k, l, and s for sample size $N = 15, 25$, $\sigma = 1$, and $\theta = 1$. The study results are summarized in Tables 1 to 4.

For comparison results of scale estimators for σ, Tables 1 and 2 indicate that (i) WME2 is the best estimator for $N = 15$, $l \neq 0$, and all different censoring schemes except $(r,k,l,s) = (2,5,3,1), (1,5,2,3), (1,3,7,1), (1,5,1,4)$; (ii) WME3 is the best estimator for $N = 25$, $l \neq 0$, and all different censoring schemes except $(r,k,l,s) = (3,5,2,1), (3,5,1,2), (1,3,7,1)$; (iii) WME8 is the best estimator for $N = 15, 25$ and right censoring scheme; (iv) WME5(=WME6) is the best estimator for $N = 15, 25$ and left censoring scheme; (v) WME6 is the best estimator for $N = 15, 25$ and left censoring scheme. The magnitude of MSE decreases as sample size N increases. The value of the smallest MSE divided by BLUE is denoted as WME/BLUE, and its range is between 0.821384 and 0.996789 for $N = 15, 25$ and $l \neq 0$. It means that our smallest MSE of WMEs is smaller than that of BLUE. The range of WME/BLUE is between 0.794971 and 0.987798 for $N = 15, 25$ and $l = 0$. It means that our smallest MSE of WMEs is smaller than that of BLUE.

For comparison results of location estimators for θ, Tables 3 and 4 show that (i) WME1 is the best estimator for $N = 15$, $(r,k,l,s) = (1,5,3,2), (1,5,2,3), (1,3,7,1), (1,5,1,4)$, and middle censoring scheme; (ii) WME1 is the best estimator for $N = 25$ and all different censoring schemes except right censoring scheme; (iii) WME8 is the best estimator for $N = 15$ and right censoring scheme; (iv) WME5 is the best estimator for $N = 25$ and left and doubly censoring schemes. The magnitude of MSE decreases as sample size N increases. The range of WME/BLUE is between 0.861855 and 1.048036 for $N = 15, 25$ and $l \neq 0$. The range of WME/BLUE is between 0.843873 and 1.02405 for $N = 15, 25$ and $l = 0$.

4. Conclusion

The object of this paper is to propose WMEs of location parameter θ and scale parameter σ of the extreme value distribution based on the multiply type II censored sample. The comparison results of scale estimators find that our proposed WMEs are, in general, superior to BLUE under different combinations of censoring schemes and sample size. On the other hand, the comparison results of location estimators present that for $N = 15$, our proposed WMEs are, in general, superior to BLUE and different censoring schemes except $(r,k,l,s) = (2,5,2,2), (3,5,2,1), (3,5,1,2), (2,5,3,1), (2,5,1,3), (4,5,1,1)$, left and doubly censoring

TABLE 1: MSEs of BLUE and WMEs of the scale parameter σ at $N = 15$, $\sigma = 1.0$, and $\theta = 1$.

Censoring scheme type II	r	k	l	s	The simulated MSEs of BLUE and exact MSEs of WMEs					
					BLUE	WME1	WME2	WME3	WME4	WME/BLUE
Multiply	2	5	2	2	0.070396	0.067245	0.066049*	0.066182	0.067211	0.938249
	3	5	2	1	0.074345	0.069957	0.069488*	0.069732	0.069821	0.934669
	3	5	1	2	0.081776	0.077573	0.075709*	0.075784	0.077564	0.925810
	2	5	3	1	0.064955	0.061733	0.061551	0.061859	0.061543*	0.947471
	2	5	1	3	0.077774	0.074938	0.072323*	0.072350	0.074945	0.929912
	1	5	3	2	0.060664	0.058404	0.057795*	0.057900	0.058413	0.952707
	1	5	2	3	0.066040	0.064033	0.062353	0.062348*	0.064107	0.944094
	4	5	1	1	0.085554	0.079741	0.078920*	0.079045	0.079696	0.922458
	1	3	7	1	0.059028	0.056334	0.057378	0.057511	0.056253*	0.952984
	1	5	1	4	0.073128	0.071480	0.068433	0.068391*	0.071521	0.935223
Middle	0	3	6	0	0.045595	0.044507*	0.044635	0.044507*	0.044635	0.976138
					BLUE	WME5	WME6	WME7	WME8	WME/BLUE
Right	0	0	0	8	0.101814	0.132434	0.102793	0.132434	0.084350*	0.828472
Left	8	3	0	0	0.145212	0.127135*	0.127135*	0.349345	0.349345	0.875513
Doubly	7	3	0	2	0.169211	0.155749	0.145217*	0.477876	0.300771	0.858200

*The smallest MSE among the proposed WMEs and BLUE.

TABLE 2: MSEs of BLUE and WMEs of the scale parameter σ at $N = 25$, $\sigma = 1.0$, and $\theta = 1$.

Censoring scheme type II	r	k	l	s	The simulated MSEs of BLUE and exact MSEs of WMEs					
					BLUE	WME1	WME2	WME3	WME4	WME/BLUE
Multiply	2	5	2	2	0.033049	0.033973	0.032609	0.032605*	0.033950	0.986562
	3	5	2	1	0.034131	0.034709	0.034021*	0.034042	0.034667	0.996789
	3	5	1	2	0.036422	0.037066	0.035446*	0.035476	0.036991	0.973205
	2	5	3	1	0.038106	0.031840	0.031332	0.031300*	0.031862	0.821384
	2	5	1	3	0.034793	0.036526	0.034272	0.034250*	0.036474	0.984403
	1	5	3	2	0.030754	0.031121	0.030042	0.029959*	0.031178	0.974163
	1	5	2	3	0.031932	0.033203	0.031209	0.031211*	0.033194	0.977406
	4	5	1	1	0.041570	0.037365	0.036608	0.036605*	0.037358	0.880563
	1	3	7	1	0.030205	0.029124	0.028682*	0.028727	0.029052	0.949578
	1	5	1	4	0.033436	0.035586	0.032923	0.032814*	0.035537	0.981404
Middle	0	3	6	0	0.026008	0.025694*	0.025757	0.025694*	0.025757	0.987927
					BLUE	WME5	WME6	WME7	WME8	WME/BLUE
Right	0	0	0	8	0.038026	0.058844	0.046303	0.058844	0.037562*	0.987798
Left	8	3	0	0	0.060523	0.048114*	0.048114*	0.069671	0.069671	0.794971
Doubly	7	3	0	2	0.056175	0.053227	0.049471*	0.09685	0.064317	0.880659

*The smallest MSE among the proposed WMEs and BLUE.

schemes; for $N = 25$ our proposed WMEs are, in general, superior to BLUE and all different censoring schemes except right censoring scheme.

WMEs and BLUE can further be used to predict the jth $(N - s < j \leq N)$ ordered observation $Y_{(j)}$ in a sample of size n from the extreme value distribution based on multiply type II censored sample. Moreover, if observations $T = \exp(-Y)$ are from a Weibull distribution with a scale parameter α and a shape parameter β, then the best estimators of Weibull distribution can be obtained through proposing the best estimators of the extreme value distribution, where $\alpha = \exp(-\theta)$ and $\beta = 1/\sigma$. Moreover, by replacing Y by $-Y$, the corresponding distribution of $-Y$ is also called extreme value distribution with c.d.f. as given by

$$F_{-Y}(y) = 1 - \exp\left[-e^{(y-\mu)/\sigma}\right], \quad -\infty < y < \infty, \tag{25}$$

where $-\infty < \mu = -\theta < \infty$, $\sigma > 0$. Our method and the best linear unbiased method can be used to find the best estimator of μ and σ and predict the jth $(N - s < j \leq N)$ ordered observation $Y_{(j)}$ for extreme value distribution with c.d.f. as (25) based on negative transformation and the multiply type II censored sample.

TABLE 3: MSEs of BLUE and WMEs of the location parameter θ at $N = 15$, $\sigma = 1.0$, and $\theta = 1$.

Censoring scheme type II	r	k	l	s	The simulated MSEs of BLUE and exact MSEs of WMEs					
					BLUE	WME1	WME2	WME3	WME4	WME/BLUE
Multiply	2	5	2	2	0.081471*	0.084486	0.088129	0.087748	0.089356	1.037007
	3	5	2	1	0.087399*	0.093767	0.097813	0.091590	0.098849	1.047952
	3	5	1	2	0.087310*	0.093792	0.094834	0.091504	0.096280	1.048036
	2	5	3	1	0.081571*	0.084650	0.090575	0.085397	0.091483	1.037746
	2	5	1	3	0.081668*	0.084633	0.086168	0.090419	0.087180	1.036306
	1	5	3	2	0.078327	0.076445*	0.083378	0.085178	0.083145	0.975973
	1	5	2	3	0.078619	0.076260*	0.081964	0.089104	0.081176	0.969995
	4	5	1	1	0.097020*	0.105156	0.107617	0.100723	0.107746	1.038167
	1	3	7	1	0.083615	0.076417*	0.089073	0.089757	0.086154	0.913925
	1	5	1	4	0.079463	0.076634*	0.081216	0.093448	0.079339	0.964399
Middle	0	3	6	0	0.075444	0.071020*	0.079400	0.084294	0.079400	0.941360
					BLUE	WME5	WME6	WME7	WME8	WME/BLUE
Right	0	0	0	8	0.095452	0.105177	0.094909	0.105177	0.094913*	0.99431
Left	8	3	0	0	0.228809*	0.234311	0.234311	0.463530	0.463530	1.02405
Doubly	7	3	0	2	0.182548*	0.184806	0.186032	0.310332	0.268930	1.01237

*The smallest MSE among the proposed WMEs and BLUE.

TABLE 4: MSEs of BLUE and WMEs of the location parameter θ at $N = 25$, $\sigma = 1.0$, and $\theta = 1$.

Censoring scheme type II	r	k	l	s	The simulated MSEs of BLUE and exact MSEs of WMEs					
					BLUE	WME1	WME2	WME3	WME4	WME/BLUE
Multiply	2	5	2	2	0.047455	0.043392*	0.047048	0.046523	0.045351	0.914382
	3	5	2	1	0.047910	0.046756*	0.049381	0.047447	0.048703	0.975913
	3	5	1	2	0.048516	0.046415*	0.048644	0.047332	0.047459	0.956695
	2	5	3	1	0.048501	0.043917*	0.047599	0.046681	0.046598	0.905486
	2	5	1	3	0.046475	0.043009*	0.046365	0.046338	0.044072	0.925422
	1	5	3	2	0.046091	0.040401*	0.045429	0.046284	0.043109	0.876549
	1	5	2	3	0.045630	0.039876*	0.045084	0.045986	0.041846	0.873899
	4	5	1	1	0.051705	0.050038	0.051508	0.048829*	0.051088	0.944386
	1	3	7	1	0.047423	0.041238*	0.047731	0.050815	0.045522	0.861855
	1	5	1	4	0.045836	0.039504*	0.044303	0.045540	0.040560	0.861855
Middle	0	3	6	0	0.043759	0.039707*	0.043286	0.052436	0.043286	0.907402
					BLUE	WME5	WME6	WME7	WME8	WME/BLUE
Right	0	0	0	8	0.047960*	0.048812	0.049585	0.048812	0.04907	1.017765
Left	8	3	0	0	0.078955	0.066628*	0.066628*	0.076515	0.076515	0.843873
Doubly	7	3	0	2	0.062959	0.060392*	0.061361	0.068265	0.064084	0.959227

*The smallest MSE among the proposed WMEs and BLUE.

Acknowledgments

The authors thank the referees for their suggestions and helpful comments in revising the paper. This research was partially supported by the National Science Council, Taiwan (Plans nos. NSC 100-2221-E-309-002, NSC 101-2118-M-415-001, and NSC 101-2221-E-309-004).

References

[1] J. F. Lawless, *Statistical Models and Methods For Lifetime Data*, John Wiley & Sons, New York, NY, USA, 1982.

[2] N. R. Mann, R. E. Schafer, and N. D. Singpurwalla, *Methods For Statistical Analysis of Reliability and Life Data*, John Wiley & Sons, New York, NY, USA, 1974.

[3] N. Balakrishnan and P. Chan, "Order statistics from extreme value distribution, I: tables of means, variances and covariances," *Communications in Statistics-Simulation and Computation*, vol. 21, no. 4, pp. 1199–1217, 1992.

[4] N. Balakrishnan and P. Chan, "Order statistics from extreme value distribution, II: best linear unbiased estimates and some other uses," *Communications in Statistics-Simulation and Computation*, vol. 21, no. 4, pp. 1219–1246, 1992.

[5] A. J. Fernández, "On maximum likelihood prediction based on Type II doubly censored exponential data," *Metrika*, vol. 50, no. 3, pp. 211–220, 2000.

[6] L. M. Leemis and L. H. Shih, "Exponential parameter estimation for data sets containing left and right censored observations," *Communications in Statistics-Simulation and Computation*, vol. 18, no. 3, pp. 1077–1085, 1989.

[7] C. T. Lin and N. Balakrishnan, "Exact linear inference for scaled exponential distribution based on doubly type-II censored samples," *Journal of Statistical Computation and Simulation*, vol. 71, no. 3, pp. 183–199, 2001.

[8] M. L. Tiku, "A note on estimating the location and scale parameters of the exponential distribution from a censored sample," *Australian Journal of Statistics*, vol. 9, no. 2, pp. 49–54, 1967.

[9] S. K. Upadhyay, U. Singh, and V. Shastri, "Estimation of exponential parameters under multiply type II censoring," *Communications in Statistics B*, vol. 25, no. 3, pp. 801–815, 1996.

[10] A. E. Sarhan and B. G. Greenberg, *Contributions to Order Statistics*, John Wiley & Sons, New York, NY, USA, 1962.

[11] J. W. Wu, W. C. Lee, and S.-C. Chen, "Computational comparison for weighted moments estimators and BLUE of the scale parameter of a Pareto distribution with known shape parameter under type II multiply censored sample," *Applied Mathematics and Computation*, vol. 181, no. 2, pp. 1462–1470, 2006.

[12] J. W. Wu and C. C. Yang, "Weighted moments estimation of the scale parameter of the exponential distribution based on a multiply type II censored sample," *Quality and Reliability Engineering International*, vol. 18, no. 2, pp. 149–154, 2002.

[13] N. Balakrishnan and A. C. Cohen, *Order Statistics and Inference*, Academic Press, Boston, Mass, USA, 1991.

[14] J. Lieblein, "On the exact evaluation of the variances and covariances of order statistics in samples from the extreme-value distribution," *The Annals of Mathematical Statistics*, vol. 24, no. 2, pp. 282–287, 1953.

7

On Weak Exponential Expansiveness of Evolution Families in Banach Spaces

Tian Yue, Xiao-qiu Song, and Dong-qing Li

College of Science, China University of Mining and Technology, Xuzhou, Jiangsu 221008, China

Correspondence should be addressed to Xiao-qiu Song; sxqcumt@163.com

Academic Editors: G. Bonanno, F. J. Garcia-Pacheco, and F. Minhós

The aim of this paper is to give several characterizations for the property of weak exponential expansiveness for evolution families in Banach spaces. Variants for weak exponential expansiveness of some well-known results in stability theory (Datko (1973), Rolewicz (1986), Ichikawa (1984), and Megan et al. (2003)) are obtained.

1. Introduction

In recent years, the exponential stability theory of one parameter semigroups of operators and evolution families has witnessed significant development. A number of long-standing open problems have been solved, and the theory seems to have obtained a certain degree of maturity. One of the most important results of the stability theory is due to Datko, who proved in 1970 in [1] that a strongly continuous semigroup of operators $\{T(t)\}_{t\geq 0}$ is uniformly exponentially stable if and only if for each vector x from the Banach space X the function $t \to \|T(t)x\|$ lies in $L^2(\mathbb{R}_+)$. Later, Pazy generalizes the result in [2] for $L^p(\mathbb{R}_+)$, $p \geq 1$. In 1973, Dakto [3] generalized the results above and proved that an evolutionary process $\mathcal{U} = \{U(t,s)\}_{t\geq s\geq 0}$ with uniform exponential growth is uniformly exponentially stable if and only if there exists an exponent $p \geq 1$ such that $\sup_{s\geq 0}\int_s^\infty \|U(t,s)x\|^p dt < \infty$, for each $x \in X$. This result was improved by Rolewicz in 1986 (see [4]). In [5, 6], the authors generalized the results above in the case of C_0-semigroups and evolutionary process, respectively, and presented a unified treatment in terms of Banach function spaces.

In the last few years, new concepts of exponential expansiveness and in particular, of exponential instability, have been introduced and characterized (see[7–14]). The cases of uniform exponential instability have been considered in [8] for evolution families and in [10] for linear skew-product flows.

In the present paper, we introduce the concept of weak exponential expansiveness for evolution families which is an extension of classical concept of exponential expansiveness. Our main objective is to give some characterizations for weak exponential expansiveness properties of evolution families in Banach spaces, and variants for weak exponential expansiveness of some well-known results in stability theory (Datko [3], Rolewicz [4], Ichikawa [15], and Megan et al. [8]) are obtained.

2. Preliminaries

Let X be a real or complex Banach space. The norm on X and on the space $B(X)$ of all bounded linear operators on X will be denoted by $\|\cdot\|$.

Definition 1. A family $\mathcal{U} = \{U(t,s)\}_{t\geq s\geq 0}$ of bounded linear operators is called an evolution family if the following conditions are satisfied:

(i) $U(t,t) = I$, the identity operator on X, for all $t \geq 0$;

(ii) $U(t,r)U(r,s) = U(t,s)$ for all $t \geq r \geq s \geq 0$;

(iii) there exist $M \geq 1$ and $\omega > 0$ such that $\|U(t,s)x_0\| \leq Me^{\omega(t-s)}\|x_0\|$ for all $t \geq s \geq 0$ and $x_0 \in X$;

(iv) for every $x_0 \in X$ and every $t_0 \geq 0$, the mapping $r \mapsto \|U(r,t_0)x_0\|$ is continuous on $[t_0,\infty)$;

(v) for every $t \geq s \geq 0$, the operator $U(t,s)$ is injective.

Definition 2. An evolution family $\mathcal{U} = \{U(t,s)\}_{t\geq s\geq 0}$ is called uniformly expansive if there exists a constant $N > 0$ such that

$$\|U(t,t_0)x_0\| \geq N \|U(r,t_0)x_0\|, \quad (1)$$

for all $t \geq r \geq t_0 \geq 0$ and $x_0 \in X$.

Definition 3. An evolution family $\mathcal{U} = \{U(t,s)\}_{t\geq s\geq 0}$ is said to be uniformly exponentially expansive if there are $N, \nu > 0$ such that

$$\|U(t,t_0)x_0\| \geq N e^{\nu(t-r)} \|U(r,t_0)x_0\|, \quad (2)$$

for all $t \geq r \geq t_0 \geq 0$ and $x_0 \in X$.

Remark 4. It is obvious that an evolution family $\mathcal{U}$ is uniformly exponentially expansive if and only if there are $N, \nu > 0$ such that

$$\|U(t,t_0)x_0\| \geq N e^{\nu(t-t_0)} \|x_0\|, \quad (3)$$

for all $t \geq t_0 \geq 0$ and $x_0 \in X$.

Remark 5. If the evolution family $\mathcal{U}$ is uniformly exponentially expansive, then it is uniformly expansive. The converse is not necessarily valid. To show this, we consider the following example.

Example 6. Let $f : \mathbb{R}_+ \to \mathbb{R}_+$ be a monotone increasing and bounded continuous function, $X = \mathbb{R}_+$. The evolution family $\mathcal{U}$ defined by

$$U(t,t_0)x_0 = e^{f(t)-f(t_0)}x_0, \quad (4)$$

for all $t \geq t_0 \geq 0$ and $x_0 \in X$.

Proof. As a first step, we prove that $\mathcal{U}$ is uniformly expansive. The evolution family $\mathcal{U}$ satisfies the inequality

$$\begin{aligned}\|U(t,t_0)x_0\| &= e^{f(t)-f(t_0)}|x_0| \geq e^{f(r)-f(t_0)}|x_0| \\ &= N\|U(r,t_0)x_0\|,\end{aligned} \quad (5)$$

for all $t \geq r \geq t_0 \geq 0$ and $x_0 \in X$, where $N = 1$. Hence, $\mathcal{U}$ is uniformly expansive.

As a second step, we prove that $\mathcal{U}$ is not uniformly exponentially expansive. If we suppose that $\mathcal{U}$ is uniformly exponentially expansive, then, by Remark 4, there exist some constants $N, \nu > 0$ such that

$$e^{f(t)-f(t_0)} \geq N e^{\nu(t-t_0)}, \quad (6)$$

for all $t \geq t_0 \geq 0$.

In particular, for $t_0 = 0$, we obtain $e^{f(t)-f(0)} \geq Ne^{\nu t}$, which is absurd for $t \to \infty$. Hence, $\mathcal{U}$ is not uniformly exponentially expansive. □

Definition 7. An evolution family $\mathcal{U} = \{U(t,s)\}_{t\geq s\geq 0}$ is called weakly exponentially expansive if there are $N, \nu > 0$ such that for all $x_0 \in X$ there exists $t_0 \geq 0$ with

$$\|U(t,t_0)x_0\| \geq N e^{\nu(t-r)} \|U(r,t_0)x_0\|, \quad (7)$$

for all $t \geq r \geq t_0$.

Remark 8. If the evolution family $\mathcal{U}$ is uniformly exponentially expansive, then it is weakly exponentially expansive.

The following example shows that the converse is not valid.

Example 9. Let $X = \mathbb{R}^2$ with the Euclidean norm. Consider the evolution family generated by the matrix $U(t,t_0) = P(t, t_0)Q(t_0)$, where

$$\begin{aligned} P(t,t_0) &= \begin{pmatrix} e^{t-t_0}\sin t & e^{-(t-t_0)}\cos t \\ -e^{t-t_0}\cos t & e^{-(t-t_0)}\sin t \end{pmatrix}, \\ Q(t_0) &= \begin{pmatrix} \cos t_0 & \sin t_0 \\ \sin t_0 & -\cos t_0 \end{pmatrix}. \end{aligned} \quad (8)$$

Proof. We divide the proof into two steps.

As a first step, we prove that $\mathcal{U}$ is weakly exponentially expansive. For every $x_0 \in \mathbb{R}^2$, there exist $\rho \geq 0$ and $t_0 \in [0, 2\pi)$ such that $x_0 = (\rho\cos t_0, \rho\sin t_0)^T$.

It is easy to see that

$$\begin{aligned} U(t,t_0)x_0 &= P(t,t_0)Q(t_0)x_0 = P(t,t_0)(\rho,0)^T \\ &= \left(\rho e^{t-t_0}\sin t, -\rho e^{t-t_0}\cos t\right)^T, \end{aligned} \quad (9)$$

and hence

$$\|U(t,t_0)x_0\| = \rho e^{t-t_0} \geq N e^{t-r} \|U(r,t_0)x_0\|, \quad (10)$$

for all $t \geq r \geq t_0$ with $N = 1$, which shows that $\mathcal{U}$ is weakly exponentially expansive.

As a second step, we prove that $\mathcal{U}$ is not uniformly exponentially expansive. If we assume that $\mathcal{U}$ is uniformly exponentially expansive, then there exist some constants N, $\nu > 0$ such that

$$\|U(t,t_0)y_0\| \geq N e^{\nu(t-r)} \|U(r,t_0)y_0\|, \quad (11)$$

for all $t \geq r \geq t_0 \geq 0$ and $y_0 \in X$.

In particular, for $y_0 = (\sin t_0, -\cos t_0)^T$, we obtain

$$\begin{aligned} U(t,t_0)y_0 &= P(t,t_0)(0,1)^T \\ &= \left(e^{-(t-t_0)}\cos t, e^{-(t-t_0)}\sin t\right)^T, \end{aligned} \quad (12)$$

and hence

$$\|U(t,t_0)y_0\| = e^{-(t-t_0)} = e^{-(t-r)} \|U(r,t_0)y_0\|, \quad (13)$$

which shows that $\mathcal{U}$ is not uniformly exponentially expansive. □

3. The Main Results

Theorem 10. *The following assertions are equivalent:*

(i) *$\mathcal{U}$ is weakly exponentially expansive;*

(ii) *there are $\delta > 0$ and $c > 1$ such that for every $x_0 \in X$ there exist $t_0 \geq 0$ and $h_0 \in (0, \delta]$ with*

$$\|U(h_0 + r, t_0)x_0\| \geq c\|U(r,t_0)x_0\|, \quad (14)$$

for all $r \geq t_0$;

(iii) *there exist* $\delta > 0$ *and* $c > 1$ *such that for each* $x_0 \in X$ *there exists* $t_0 \geq 0$ *with the property that for every* $r \geq t_0$ *there is* $h_0 \in (0, \delta]$ *with*

$$\|U(h_0 + r, t_0) x_0\| \geq c \|U(r, t_0) x_0\|. \quad (15)$$

Proof. (i)⇒(ii) If $\mathcal{U}$ is weakly exponentially expansive, then by Definition 7, there exist $N, \nu > 0$ such that for all $x_0 \in X$ there exists $t_0 \geq 0$ with the property

$$\|U(t, t_0) x_0\| \geq N e^{\nu(t-r)} \|U(r, t_0) x_0\|, \quad (16)$$

for all $t \geq r \geq t_0$. Let $\delta > 0$ satisfy that $N e^{\nu\delta} > 1$. Then, for $h_0 = \delta$, we have

$$\|U(r + h_0, t_0) x_0\| \geq N e^{\nu\delta} \|U(r, t_0) x_0\| = c \|U(r, t_0) x_0\|, \quad (17)$$

for all $r \geq t_0$.

(ii)⇒(iii) It is obvious.

(iii)⇒(i) We define $N = 1/M e^{\omega\delta}$ and $\nu = \ln c/\delta$, where $\delta > 0$ and $c > 1$ are given by (iii).

From (iii), it results that for each $x_0 \in X$, there exists $t_0 \geq 0$ with the property that for every $r \geq t_0$ there is $h_0 \in (0, \delta]$ such that

$$\|U(h_0 + r, t_0) x_0\| \geq c \|U(r, t_0) x_0\|. \quad (18)$$

Let $r \geq t_0$, and we have that there is $h_1 \in (0, \delta]$ with

$$\begin{aligned} \|U(h_1 + h_0 + r, t_0) x_0\| &\geq c \|U(h_0 + r, t_0) x_0\| \\ &\geq c^2 \|U(r, t_0) x_0\|. \end{aligned} \quad (19)$$

By induction, we have that

$$\|U(r_n + r, t_0) x_0\| \geq c^n \|U(r, t_0) x_0\|, \quad \forall n \in \mathbb{N}, \quad (20)$$

where

$$r_n = \begin{cases} 0, & n = 0, \\ \sum_{i=0}^{n-1} h_i, & n \in \mathbb{N}^*, \end{cases} \quad h_i \in (0, \delta]. \quad (21)$$

It is easy to see that (r_n) is unbounded. In fact, if (r_n) is bounded, then there exists $r^* \in \mathbb{R}$ with $r_n \to r^*$ $(n \to \infty)$. From the relation (20) and $c > 1$, it follows that

$$\|U(r + r^*, t_0) x_0\| \geq \lim_{n \to \infty} c^n \|U(r, t_0) x_0\| \longrightarrow \infty, \quad (22)$$

which is a contradiction because $\{U(t, s)\}_{t \geq s \geq 0} \subseteq B(X)$.

So, (r_n) is unbounded, and then for $t \geq r$, there is $n \in \mathbb{N}$ such that

$$r_n \leq t - r \leq r_{n+1} \leq (n+1)\delta. \quad (23)$$

Then,

$$\begin{aligned} \|U(r_{n+1} + r, t_0) x_0\| &\leq M e^{\omega(r + r_{n+1} - t)} \|U(t, t_0) x_0\| \\ &\leq M e^{\omega(r_{n+1} - r_n)} \|U(t, t_0) x_0\| \\ &\leq M e^{\omega\delta} \|U(t, t_0) x_0\|, \end{aligned} \quad (24)$$

and hence

$$\begin{aligned} \|U(t, t_0) x_0\| &\geq \frac{1}{M e^{\omega\delta}} \|U(r + r_{n+1}, t_0) x_0\| \\ &\geq \frac{1}{M e^{\omega\delta}} c^{n+1} \|U(r, t_0) x_0\| \\ &= \frac{1}{M e^{\omega\delta}} e^{\nu(n+1)\delta} \|U(r, t_0) x_0\| \\ &\geq N e^{\nu(t-r)} \|U(r, t_0) x_0\|. \end{aligned} \quad (25)$$

□

Remark 11. Theorem 10 can be considered a generalization of some results from uniform exponential instability proved in [8].

An important set in what follows is $\mathcal{F}_1$, the set of all nondecreasing functions $F : \mathbb{R}_+ \to \mathbb{R}_+$ with the properties:

(f1) $F(tr) \leq F(t)F(r)$, for all $(t, r) \in \mathbb{R}_+^2$;

(f2) $F(t) > 0$, for every $t > 0$.

Theorem 12. *An evolution family* $\mathcal{U}$ *is weakly exponentially expansive if and only if there are* $F \in \mathcal{F}_1$ *and* $K > 0$ *such that for every* $x_0 \in X \setminus \{0\}$ *there is* $t_0 \geq 0$ *with*

$$\int_r^\infty F\left(\frac{1}{\|U(\tau, t_0) x_0\|}\right) d\tau \leq K F\left(\frac{1}{\|U(r, t_0) x_0\|}\right), \quad (26)$$

for all $r \geq t_0$.

Proof. Necessity. If $\mathcal{U}$ is weakly exponentially expansive, then by Definition 7, there are $N, \nu > 0$ such that for all $x_0 \in X \setminus \{0\}$ there exists $t_0 \geq 0$ with

$$\begin{aligned} \int_r^\infty \frac{1}{\|U(\tau, t_0) x_0\|} d\tau &\leq \int_r^\infty \frac{1}{N e^{\nu(\tau - r)} \|U(r, t_0) x_0\|} d\tau \\ &= \frac{1}{N\nu \|U(r, t_0) x_0\|}, \end{aligned} \quad (27)$$

for all $r \geq t_0$.

Thus, the inequality (26) is satisfied for $F(t) = t$ and $K = 1/N\nu$.

Sufficiency. We assume for a contradiction that for all $\delta > 0$ and $c > 1$ there exists $x_0 \in X$ such that for every $t_0 \geq 0$ there is $r \geq t_0$ with

$$\|U(t + r, t_0) x_0\| < c \|U(r, t_0) x_0\|, \quad (28)$$

for all $t \in (0, \delta]$.

In particular, for $\delta = KF(3)$ and $c = 3$, the inequality (28) implies

$$\begin{aligned} F(3)\int_r^{\infty} F\left(\frac{1}{\|U(t,t_0)x_0\|}\right)dt &\geq \int_r^{\infty} F\left(\frac{3}{\|U(t,t_0)x_0\|}\right)dt \\ &= \int_0^{\infty} F\left(\frac{3}{\|U(\tau+r,t_0)x_0\|}\right)d\tau \\ &\geq \int_0^{\delta} F\left(\frac{3}{\|U(\tau+r,t_0)x_0\|}\right)d\tau \\ &> \int_0^{\delta} F\left(\frac{1}{\|U(r,t_0)x_0\|}\right)d\tau \\ &= \delta F\left(\frac{1}{\|U(r,t_0)x_0\|}\right), \end{aligned} \tag{29}$$

which contradicts the inequality (26). This contradiction proves that $\mathscr{U}$ is weakly exponentially expansive. □

It makes sense to consider also the set $\mathscr{F}_2$ all non-decreasing functions $F : \mathbb{R}_+ \to \mathbb{R}_+$ with the properties:

(g1) $F(tr) \geq F(t)F(r)$, for all $(t,r) \in \mathbb{R}_+^2$;

(g2) $F(t) > 0$, for every $t > 0$.

Theorem 13. *An evolution family $\mathscr{U}$ is weakly exponentially expansive if and only if there are $F \in \mathscr{F}_2$ and $K > 0$ such that for every $x_0 \in X \setminus \{0\}$ there is $t_0 \geq 0$ with the relation (26).*

Proof. Necessity. This is a simple verification for $F(t) = t$.

Sufficiency. It is similar to the proof of Theorem 12. Indeed, from (28) for $c \in (1,+\infty)$ and $\delta = 2K/F(1/c)$, we have

$$\begin{aligned} \int_r^{\infty} F\left(\frac{1}{\|U(t,t_0)x_0\|}\right)dt &= \int_0^{\infty} F\left(\frac{1}{\|U(\tau+r,t_0)x_0\|}\right)d\tau \\ &\geq \int_0^{\delta} F\left(\frac{1}{\|U(\tau+r,t_0)x_0\|}\right)d\tau \\ &\geq \int_0^{\delta} F\left(\frac{1}{c\|U(r,t_0)x_0\|}\right)d\tau \\ &\geq \delta F\left(\frac{1}{c}\right)F\left(\frac{1}{\|U(r,t_0)x_0\|}\right) \\ &> KF\left(\frac{1}{\|U(r,t_0)x_0\|}\right), \end{aligned} \tag{30}$$

which contradicts the inequality (26). □

Remark 14. The preceding theorems are variants for the case of weak exponential expansiveness property of a well-known theorem due to Rolewicz [4].

Corollary 15. *An evolution family $\mathscr{U}$ is weakly exponentially expansive if and only if there are $p > 0$ and $K > 0$ such that for all $x_0 \in X \setminus \{0\}$ there exists $t_0 \geq 0$ with*

$$\int_r^{\infty} \frac{1}{\|U(t,t_0)x_0\|^p}dt \leq K\frac{1}{\|U(r,t_0)x_0\|^p}, \tag{31}$$

for all $r \geq t_0$.

Proof. It is immediate from Theorem 12 for $F(t) = t^p$. □

Remark 16. Corollary 15 is the version of a well-known theorem due to Datko [3], for the case of weak exponential expansiveness of evolution families.

In the following corollary, we give a discrete version of Theorems 12 and 13.

Corollary 17. *An evolution family $\mathscr{U}$ is weakly exponentially expansive if and only if there are $F \in \mathscr{F}_1 \cup \mathscr{F}_2$ and $K > 0$ such that for all $x_0 \in X \setminus \{0\}$ there exists $t_0 \geq 0$ with*

$$\sum_{n=0}^{\infty} F\left(\frac{1}{\|U(r+n,t_0)x_0\|}\right) \leq KF\left(\frac{1}{\|U(r,t_0)x_0\|}\right), \tag{32}$$

for all $r \geq t_0$.

Proof. Necessity. This is a simple verification for $F(t) = t$.

Sufficiency. By Definition 1, we know that $\|U(r+\tau,r+n)\| \leq Me^{\omega}$, $\tau \in [n,n+1]$ and $\tau \mapsto \|U(r+\tau,r+n)\|$ is continuous on $[n,n+1]$ for all $n \in \mathbb{N}$, so there exist $L_n > 0$ such that

$$L_n = \min_{\tau\in[n,n+1]} \|U(r+\tau,r+n)\|, \quad n \in \mathbb{N}. \tag{33}$$

Let $L = \inf_n L_n$. We suppose that $F \in \mathscr{F}_1$, and from (32), it results that

$$\begin{aligned} \int_r^{\infty} F\left(\frac{1}{\|U(t,t_0)x_0\|}\right)dt &= \sum_{n=0}^{\infty}\int_n^{n+1} F\left(\frac{1}{\|U(r+\tau,r+n)U(r+n,t_0)x_0\|}\right)d\tau \\ &\leq \sum_{n=0}^{\infty}\int_n^{n+1} F\left(\frac{1}{L\|U(r+n,t_0)x_0\|}\right)d\tau \\ &\leq F\left(\frac{1}{L}\right)\sum_{n=0}^{\infty} F\left(\frac{1}{\|U(r+n,t_0)x_0\|}\right) \\ &\leq KF\left(\frac{1}{L}\right)F\left(\frac{1}{\|U(r,t_0)x_0\|}\right), \end{aligned} \tag{34}$$

for all $r \geq t_0$.

If $F \in \mathcal{F}_2$, in a similar way we have

$$\begin{aligned} F(L) & \int_r^\infty F\left(\frac{1}{\|U(t,t_0)x_0\|}\right) dt \\ & \le \int_0^\infty F\left(\frac{L}{\|U(r+\tau,t_0)x_0\|}\right) d\tau \\ & = \sum_{n=0}^\infty \int_n^{n+1} F\left(\frac{L}{\|U(r+\tau,r+n)U(r+n,t_0)x_0\|}\right) d\tau \\ & \le \sum_{n=0}^\infty \int_n^{n+1} F\left(\frac{1}{\|U(r+n,t_0)x_0\|}\right) d\tau \\ & \le KF\left(\frac{1}{\|U(r,t_0)x_0\|}\right), \end{aligned} \tag{35}$$

for all $r \ge t_0$.

Applying Theorems 12 and 13, we conclude that $\mathcal{U}$ is weakly exponentially expansive. □

Corollary 18. *An evolution family $\mathcal{U}$ is weakly exponentially expansive if and only if there is $K > 0$ such that for all $x_0 \in X \setminus \{0\}$ there exists $t_0 \ge 0$ with*

$$\sum_{n=0}^\infty \frac{1}{\|U(r+n,t_0)x_0\|} \le K \frac{1}{\|U(r,t_0)x_0\|}, \tag{36}$$

for all $r \ge t_0$.

Another characterization of the weak exponential expansiveness is given by the following.

Theorem 19. *An evolution family $\mathcal{U}$ is weakly exponentially expansive if and only if there are $K, \alpha > 0$ such that for every $x_0 \in X \setminus \{0\}$ there is $t_0 \ge 0$ with*

$$\frac{1}{t-r}\int_r^t e^{\alpha(\tau-r)} \frac{1}{\|U(\tau,t_0)x_0\|} d\tau \le K \frac{1}{\|U(r,t_0)x_0\|}, \tag{37}$$

for all $t > r \ge t_0$.

Proof. Necessity. If $\mathcal{U}$ is weakly exponentially expansive then by Definition 7, there are $N, \nu > 0$ such that for all $x_0 \in X \setminus \{0\}$ there is $t_0 > 0$ with the property that for $\alpha = \nu/2$ we have

$$\begin{aligned} \frac{1}{t-r} & \int_r^t e^{\alpha(\tau-r)} \frac{1}{\|U(\tau,t_0)x_0\|} d\tau \\ & \le \frac{1}{N(t-r)} \int_r^t e^{-\alpha(\tau-r)} d\tau \frac{1}{\|U(r,t_0)x_0\|} \\ & = \frac{1-e^{-\alpha(t-r)}}{N\alpha(t-r)} \frac{1}{\|U(r,t_0)x_0\|} \\ & \le K \frac{1}{\|U(r,t_0)x_0\|}, \end{aligned} \tag{38}$$

for all $t > r \ge t_0$, where $K = (1/N)\sup_{\lambda>0}((1-e^{-\lambda})/\lambda) < \infty$.

Sufficiency. Let $\delta > 0$ be such that $e^{\alpha\delta} > 1 + 3\alpha\delta K$, where K and α are given by (37). We suppose that $\mathcal{U}$ is not weakly exponentially expansive. Then, by Theorem 10, for $c = 3$, there exists $x_0 \in X$ such that for all $t_0 \ge 0$ and all $u \in (0, \delta]$ there is $r \ge t_0$ with

$$\|U(u+r,t_0)x_0\| < 3\|U(r,t_0)x_0\|. \tag{39}$$

Then, for $t = r + \delta$, we have

$$\begin{aligned} \frac{1}{t-r} & \int_r^t e^{\alpha(\tau-r)} \frac{1}{\|U(\tau,t_0)x_0\|} d\tau \\ & = \frac{1}{\delta}\int_0^\delta e^{\alpha u} \frac{1}{\|U(u+r,t_0)x_0\|} du \\ & \ge \frac{1}{3\delta}\int_0^\delta e^{\alpha u} du \frac{1}{\|U(r,t_0)x_0\|} \\ & = \frac{e^{\alpha\delta}-1}{3\alpha\delta} \frac{1}{\|U(r,t_0)x_0\|} \\ & > K \frac{1}{\|U(r,t_0)x_0\|}, \end{aligned} \tag{40}$$

which contradicts the inequality (37), the proof is completed. □

Acknowledgments

The authors would like to thank the referee for helpful suggestions and comments. This work was supported by the "Fundamental Research Funds for the Central Universities" (no. 2012LWB53).

References

[1] R. Datko, "Extending a theorem of A. M. Liapunov to Hilbert space," *Journal of Mathematical Analysis and Applications*, vol. 32, no. 3, pp. 610–616, 1970.

[2] A. Pazy, *Semigroups of Linear Operators and Applications to Partial Differential Equations*, Springer, New York, NY, USA, 1983.

[3] R. Datko, "Uniform asymptotic stability of evolutionary processes in a Banach space," *SIAM Journal on Mathematical Analysis*, vol. 3, pp. 428–445, 1973.

[4] S. Rolewicz, "On uniform N-equistability," *Journal of Mathematical Analysis and Applications*, vol. 115, no. 2, pp. 434–441, 1986.

[5] J. M. A. M. Vanneerven, "Exponential stability of operators and operator semigroups," *Journal of Functional Analysis*, vol. 130, no. 2, pp. 293–309, 1995.

[6] P. Preda, A. Pogan, and C. Preda, "Functionals on function and sequence spaces connected with the exponential stability of evolutionary processes," *Czechoslovak Mathematical Journal*, vol. 56, no. 2, pp. 425–435, 2006.

[7] M. Megan and A. Pogan, "On exponential h-expansiveness of semigroups of operators in Banach spaces," *Nonlinear Analysis: Theory, Methods and Applications*, vol. 52, no. 2, pp. 545–556, 2003.

[8] M. Megan, A. L. Sasu, and B. Sasu, "Banach function spaces and exponential instability of evolution families," *Archivum Mathematicum*, vol. 39, pp. 227–286, 2003.

[9] M. Megan, B. Sasu, and A. L. Sasu, "Exponential expansiveness and complete admissibility for evolution families," *Czechoslovak Mathematical Journal*, vol. 54, no. 3, pp. 1485–1493, 2004.

[10] M. Megan, A. L. Sasu, and B. Sasu, "Exponential stability and exponential instability for linear skew-product ows," *Mathematica Bohemica*, vol. 129, pp. 225–243, 2004.

[11] M. Megan, B. Sasu, and A. L. Sasu, "Nonuniform exponential unstability of evolution operators in Banach spaces," *Glasnik Matematicki Series III*, vol. 36, pp. 287–295, 2001.

[12] M. Megan and C. Stoica, "Exponential instability of skew-evolution semi ows in Banach spaces," *Studia Universitatis "Babes-Bolyai" Mathematica*, vol. 53, pp. 17–24, 2008.

[13] N. van Minh, F. Räbiger, and R. Schnaubelt, "Exponential stability, exponential expansiveness, and exponential dichotomy of evolution equations on the half-line," *Integral Equations and Operator Theory*, vol. 32, no. 3, pp. 332–353, 1998.

[14] K. J. Palmer, "Exponential dichotomy and expansivity," *Annali di Matematica Pura ed Applicata*, vol. 185, no. 5, pp. S171–S185, 2006.

[15] A. Ichikawa, "Equivalence of L^p stability and exponential stability for a class of nonlinear semigroups," *Nonlinear Analysis: Theory, Methods and Applications*, vol. 8, no. 7, pp. 805–815, 1984.

8

On Generalized Carleson Operators of Periodic Wavelet Packet Expansions

Shyam Lal and Manoj Kumar

Department of Mathematics, Faculty of Science, Banaras Hindu University, Varanasi 221005, India

Correspondence should be addressed to Shyam Lal; shyam_lal@rediffmail.com

Academic Editors: M. Mansour and E. Momoniat

Three new theorems based on the generalized Carleson operators for the periodic Walsh-type wavelet packets have been established. An application of these theorems as convergence a.e. for the periodic Walsh-type wavelet packet expansion of block function with the help of summation by arithmetic means has been studied.

1. Introduction

Wavelet packet expansions have wide applications in engineering and technology. The Walsh-type wavelet packet expansions play an important role in signal processing, numerical analysis, and quantum mechanics. A family of nonstationary wavelet packets considered the smooth generalization of the Walsh functions having some of the same nice convergence properties for expansion of L^p-function, $1 < p < \infty$, as the Walsh-Fourier series. Walsh-type wavelet packet expansion has been studied by the researchers Billard [1], Nielsen [2], Sjölin [3] and others. In 1966, at first, Carleson operator has been introduced by Lennart Carleson (Carleson [4]). Several important properties of this operator has been studied by researcher Nielsen [2]. In this paper, the pointwise convergence almost everywhere by arithmetic means or $(C,1)$ summability method of the partial sum operator for Walsh-type wavelet packet expansion of functions from the block space, $\mathbb{B}_q$, $1 < q \le \infty$, $p^{-1} + q^{-1} = 1$ has been studied. Generalized Carleson operators are introduced and some new properties of generalized Carleson operators are investigated. Specific convergence properties of Walsh-type wavelet packet expansions of block functions using $(C,1)$ method and generalized Carleson operator have been obtained.

2. Definitions and Preliminaries

Walsh-Type Wavelet Packets. To every multiresolution analysis $\{V_j\}_{j\in\mathbb{Z}}$ for $L^2(\mathbb{R})$, an associated scaling function φ and a wavelet ψ are given with the properties that

$$V_j = \overline{\text{span}}\left\{2^{j/2}\varphi\left(2^j\cdot -k\right) : k\in\mathbb{Z}\right\},\quad j\in\mathbb{Z},$$
$$\left\{\psi_{j,k} \equiv 2^{j/2}\psi\left(2^j\cdot -k\right) : j,k\in\mathbb{Z}\right\} \tag{1}$$

is an orthonormal basis for $L^2(\mathbb{R})$.

We write

$$W_j = \overline{\text{span}}\left\{2^{j/2}\psi\left(2^j\cdot -k\right) : k\in\mathbb{Z}\right\},\quad j\in\mathbb{Z}. \tag{2}$$

Let $\mathbb{N}$ be the set of natural numbers. Let $(F_0^{(p)}, F_1^{(p)})$, $p\in\mathbb{N}$, be a family of bounded operators on $l^2(\mathbb{Z})$ of the form

$$\left(F_\epsilon^{(p)}a\right)_k = \sum_{n\in\mathbb{Z}} a_n h_\epsilon^{(p)}(n-2k),\quad \epsilon = 0,1 \tag{3}$$

with $h_1^{(p)}(n) = (-1)^n h_0^{(p)}(1-n)$ a real-valued sequence in $l^1(\mathbb{Z})$ such that

$$F_0^{(p)*}F_0^{(p)} + F_1^{(p)*}F_1^{(p)} = 1,$$
$$F_0^{(p)}F_1^{(p)*} = 0. \tag{4}$$

Define the family of functions $\{w_n\}_{n=0}^{\infty}$ recursively by letting $w_0 = \varphi$, $w_1 = \psi$ and then for $n \in \mathbb{N}$,

$$w_{2n}(x) = \sqrt{2}\sum_{l\in\mathbb{Z}} h_0^{(p)}(l)\, w_n(2x-l),$$
$$w_{2n+1}(x) = \sqrt{2}\sum_{l\in\mathbb{Z}} h_1^{(p)}(l)\, w_n(2x-l), \tag{5}$$

where $2^p \le n < 2^{p+1}$.

The family $\{w_n\}_{n=0}^{\infty}$ is basic non stationary wavelet packets. $\{w_n(\cdot - k) : n \ge 0,\ k \in \mathbb{Z}\}$ is an orthonormal basis for $L^2(\mathbb{R})$.

Moreover,

$$\left\{w_n(\cdot - k) : 2^j \le n < 2^{j+1}, k \in \mathbb{Z}\right\} \tag{6}$$

is an orthonormal basis for $W_j = \overline{\operatorname{span}}\{2^{j/2}\psi(2^j\cdot - k) : k \in \mathbb{Z}\}$.

Each pair $(F_0^{(p)}, F_1^{(p)})$ can be chosen as a pair of quadrature mirror filters associated with a multiresolution analysis, but this is not necessary.

The trigonometric polynomials given by

$$m_0^{(p)}(\xi) = \frac{1}{\sqrt{2}}\sum_{k\in\mathbb{Z}} h_0^{(p)}(k)\, e^{-ik\xi},$$
$$m_1^{(p)}(\xi) = \frac{1}{\sqrt{2}}\sum_{k\in\mathbb{Z}} h_1^{(p)}(k)\, e^{-ik\xi} \tag{7}$$

are called the symbols of the filters.

The Fourier transforms of (5) are given by

$$\widehat{w}_{2n}(\xi) = m_0^{(p)}\left(\frac{\xi}{2}\right)\widehat{w}_n\left(\frac{\xi}{2}\right),$$
$$\widehat{w}_{2n+1}(\xi) = m_1^{(p)}\left(\frac{\xi}{2}\right)\widehat{w}_n\left(\frac{\xi}{2}\right). \tag{8}$$

The Haar low-pass quadrature mirror filter $\{h_0(k)\}_k$ is given by $h_0(0) = h_0(1) = 1/\sqrt{2}$, $h_0(k) = 0$ otherwise, and the associated high-pass filter $\{h_1(k)\}_k$ is given by

$$h_1(k) = (-1)^k h_0(1-k). \tag{9}$$

Definition 1. Let $\{w_n\}_{n\ge 0, k\in\mathbb{Z}}$ be a family of non-stationary wavelet packets constructed by using a family $\{h_0^{(p)}(n)\}_{p=1}^{\infty}$ of finite filters for which there is a constant, $K \in \mathbb{Z}$ such that $h_0^{(p)}(n)$ is the Haar filter for every $p \ge K$. If $w_1 \in C^1(\mathbb{R})$ is compactly supported then $\{w_n\}_{n\ge 0}$ is called a family of Walsh-type wavelet packets.

Definition 2. Let $\{w_n\}_{n=0}^{\infty}$ be a family of Walsh-type basic wavelet packets. For $n \in \mathbb{N}_0$, define the corresponding periodic Walsh-type wavelet packets $\widetilde{w}_n$ by

$$\widetilde{w}_n(x) = \sum_{k\in\mathbb{Z}} w_n(x-k). \tag{10}$$

From Fubini's theorem, it follows that $\{\widetilde{w}_n\}_{n=0}^{\infty}$ is an orthonormal basis for $L^2[0,1)$.

Block Spaces. A dyadic q-block is a function $\beta \in L^q[0,1)$ which is supported on some dyadic interval I such that $\|\beta\|_q \le |I|^{1/q-1}$, where $\|\beta\|_q = \left[\int_0^1 |\beta(t)|^q dt\right]^{1/q}$, $1 < q < \infty$. Let $\mathbb{B}_q$ denote the space of measurable functions f on $[0,1)$ which has an expansion

$$f = \sum_{k=1}^{\infty} c_k \beta_k, \tag{11}$$

where each β_k is a q-block and the coefficients c_k, $k \in \mathbb{Z}$ satisfy

$$\||\{c_k\}\|| = \sum_{k: c_k \ne 0} |c_k| \left[1 + \log \frac{\sum_{j=1}^{\infty} |c_j|}{|c_k|}\right] < \infty. \tag{12}$$

The quasi norm of $f \in \mathbb{B}_q$ is given as the infimum of $\||\cdot\||$ over all possible decompositions of f into blocks

$$\|f\|_{\mathbb{B}_q} = \inf_{f=\sum c_k\beta_k} \||\{c_k\}\||. \tag{13}$$

Let $f \in \mathbb{B}_q$; then

$$\|f\|_1 \le \sum_{k=1}^{\infty} |c_k| \|\beta_k\|_1 \le \sum_{k=1}^{\infty} |c_k| < \infty, \tag{14}$$

using (12) and the fact that for each k, $\|\beta_k\|_q \le |I|^{1/q-1}$ which implies that $\|\beta_k\|_1 \le 1$; that is, $\mathbb{B}_q \subset L^1[0,1)$. Moreover, for

$$f \in L^q[0,1), \quad 1 < q < \infty, \quad \beta = \|f\|_q^{-1} f \tag{15}$$

is a q-block supported on $I = [0,1)$ so $L^q[0,1) \subset \mathbb{B}_q$.

The classical example to show that for each $q > 1$ there exists $f \in \mathbb{B}_q$ which belongs to none of the $L^p[0,1)$-space is the following.

Let

$$\beta_k(x) = \begin{cases} 2^k, & \dfrac{1}{2^k} < x \le \dfrac{3}{2^{(k+1)}}, \\ 0, & \text{otherwise}. \end{cases} \tag{16}$$

Then $f = \sum_{k=1}^{\infty} k^{-2}\beta_k \in \mathbb{B}_q$, but $\|f\|_p^p = \sum_{k=1}^{\infty}(1/2)k^{-2p}2^{k(p-1)} = \infty$ for every $p > 1$.

Summation of Series by Arithmetic Means. If a series $u_0 + u_1 + u_2 + \cdots$ is not convergent, that is, if $s_n = u_0 + u_1 + u_2 + \cdots + u_n$ does not tend to a limit, it is some time possible to associate with the series a "sum" in a less direct way. The simplest such method is "summation by arithmetic means". Let

$$\sigma_n = \frac{s_0 + s_1 + s_2 + \cdots + s_n}{n+1} \tag{17}$$

be the arithmetic mean of the partial sums of the given series.

If $s_n \to s$, then also $\sigma_n \to s$; for if $s_n = s + \delta_n$, then

$$\sigma_n = s + \frac{\delta_0 + \delta_1 + \delta_2 + \delta_2 + \cdots + \delta_n}{n+1}, \tag{18}$$

and the last term tends to zero if $\delta_n \to 0$. Consider

$$\begin{aligned}\sigma_n &= \frac{s_0 + s_1 + s_2 + \cdots + s_n}{n+1} \\ &= (u_0 + (u_0 + u_1) + \cdots + (u_0 + u_1 + \cdots + u_k) \\ &\quad + \cdots + (u_0 + u_1 + \cdots + u_n)) \times (n+1)^{-1} \\ &= \sum_{k=0}^{n} \left(1 - \frac{k}{n+1}\right) u_k.\end{aligned} \tag{19}$$

If $\sigma_n \to s$ as $n \to \infty$, $\sum_{n=0}^{\infty} u_n$ is said to be summable to s by Cesàro's means of order 1. We write

$$\sum_{n=0}^{\infty} u_n = s(C, 1). \tag{20}$$

But σ_n may tend to a limit even though s_n does not, for example, the series

$$1 - 1 + 1 - 1 + \cdots. \tag{21}$$

Here the partial sums s_n are alternately 1 and 0, and it is easily seen that $\sigma_n \to 1/2$.

2.1. Generalized Carleson Operators. Let $\{\widetilde{w}_n\}$ be a periodic Walsh-type wavelet packet basis. For any function $f \in L^1[0,1)$, define

$$(S_N f)(x) = \sum_{n=0}^{N} \langle f, \widetilde{w}_n \rangle \widetilde{w}_n(x). \tag{22}$$

The Carleson operator $\mathbb{G}$ is defined by

$$\begin{aligned}\mathbb{G} f(x) &= \sup_{N\ge 0} \left| \sum_{n=0}^{N} \langle f, \widetilde{w}_n \rangle \widetilde{w}_n(x) \right| \\ &= \sup_{N\ge 0} \left| (S_N f)(x) \right|.\end{aligned} \tag{23}$$

The generalized Carleson operator $\mathbb{G}_c$ is defined by

$$\begin{aligned}\mathbb{G}_c f(x) &= \sup_{N\ge 0} \left| \frac{(S_0 f)(x) + (S_1 f)(x) + \cdots + (S_N f)(x)}{N+1} \right| \\ &= \sup_{N\ge 0} \left| \frac{1}{N+1} \sum_{\nu=0}^{N} \sum_{n=0}^{\nu} \langle f, \widetilde{w}_n \rangle \widetilde{w}_n(x) \right| \\ &= \sup_{N\ge 0} \left| \sum_{n=0}^{N} \left(1 - \frac{n}{N+1}\right) \langle f, \widetilde{w}_n \rangle \widetilde{w}_n(x) \right|.\end{aligned} \tag{24}$$

The weak Carleson operator G is defined by

$$\begin{aligned}Gf(x) &= \limsup_{N\ge 0} \left| \sum_{n=0}^{N} \langle f, \widetilde{w}_n \rangle \widetilde{w}_n(x) \right|. \\ &= \limsup_{N\ge 0} \left| (S_N f)(x) \right|.\end{aligned} \tag{25}$$

The generalized weak Carleson operator G_c is define by

$$\begin{aligned}G_c f(x) &= \limsup_{N\ge 0} \left| \frac{(S_0 f)(x) + (S_1 f)(x) + \cdots + (S_N f)(x)}{N+1} \right| \\ &= \limsup_{N\ge 0} \left| \frac{1}{N+1} \sum_{\nu=0}^{N} \sum_{n=0}^{\nu} \langle f, \widetilde{w}_n \rangle \widetilde{w}_n(x) \right| \\ &= \limsup_{N\ge 0} \left| \sum_{n=0}^{N} \left(1 - \frac{n}{N+1}\right) \langle f, \widetilde{w}_n \rangle \widetilde{w}_n(x) \right|.\end{aligned} \tag{26}$$

The dyadic Carleson operator $\mathbb{G}^d$ is defined by

$$\begin{aligned}\mathbb{G}^d f(x) &= \sup_{N\ge 0} \left| \sum_{n=0}^{2^N-1} \langle f, \widetilde{w}_n \rangle \widetilde{w}_n(x) \right| \\ &= \sup_{N\ge 0} \left| (S_{2^N} f)(x) \right|.\end{aligned} \tag{27}$$

The generalized dyadic Carleson operator $\mathbb{G}_c^d$ is define by

$$\begin{aligned}\mathbb{G}_c^d f(x) &= \sup_{N\ge 0} \left| \frac{(S_0 f)(x) + (S_1 f)(x) + \cdots + (S_{2^N-1} f)(x)}{2^N} \right| \\ &= \sup_{N\ge 0} \left| \frac{1}{2^N} \sum_{\nu=0}^{2^N-1} \sum_{n=0}^{\nu} \langle f, \widetilde{w}_n \rangle \widetilde{w}_n(x) \right| \\ &= \sup_{N\ge 0} \left| \sum_{n=0}^{2^N-1} \left(1 - \frac{n}{2^N}\right) \langle f, \widetilde{w}_n \rangle \widetilde{w}_n(x) \right|.\end{aligned} \tag{28}$$

It is easy to prove that $\mathbb{G}_c$, G_c and $\mathbb{G}_c^d$ are sublinear operators.

Walsh Functions and Their Properties. The Walsh system $\{W_n\}_{n=0}^{\infty}$ is defined recursively on $[0,1)$ by letting

$$\begin{aligned}W_0(x) &= \begin{cases} 1, & 0 \le x < 1; \\ 0, & \text{otherwise}, \end{cases} \\ W_{2n}(x) &= W_n(2x) + W_n(2x-1), \\ W_{2n+1}(x) &= W_n(2x) - W_n(2x-1).\end{aligned} \tag{29}$$

Observe that the Walsh system is the family of wavelet packets obtained by considering $\varphi = W_0$,

$$\psi(x) = \begin{cases} 1, & 0 \le x < \frac{1}{2}; \\ -1, & \frac{1}{2} \le x < 1; \\ 0, & \text{otherwise} \end{cases} \tag{30}$$

and using the Haar filters in the definition of the nonstationary wavelet packets.

The Walsh system is closed under pointwise multiplication. Define the binary operator $\oplus : \mathbb{N}_0 \times \mathbb{N}_0 \to \mathbb{N}_0$ by

$$m \oplus n = \sum_{i=0}^{\infty} \left| m_i - n_i \right| 2^i, \tag{31}$$

where $m = \sum_{i=0}^{\infty} m_i 2^i$ and $n = \sum_{i=0}^{\infty} n_i 2^i$. Then

$$W_m(x) W_n(x) = W_{m \oplus n}(x), \quad (32)$$

(see Schipp et al. [5]).

We can carry over the operator $\oplus$ to the interval $[0,1]$ by identifying those $x \in [0,1]$ with a unique expansion $x = \sum_{j=0}^{\infty} x_j 2^{-j-1}$ (almost all $x \in [0,1]$ has such a unique expansion) by their associated binary sequence $\{x_j\}$. For two such points $x, y \in [0,1]$, define

$$x \oplus y = \sum_{j=0}^{\infty} \left| x_j - y_j \right| 2^{-j-1}. \quad (33)$$

The operation $\oplus$ is defined for almost all $x, y \in [0,1]$. With this definition, we have

$$W_n(x \oplus y) = W_n(x) W_n(y) \quad (34)$$

for every pair x, y for which $x \oplus y$ is defined, (Golubov et al. [6], page 11).

3. Main Results

In this paper, three new theorems for the generalized Carleson operators on the periodic Walsh-type wavelet packets have been determined in the following form.

Theorem 3. *Let $\{\widetilde{w}_n\}$ be a periodic Walsh-type wavelet packet basis. Then for every q-block β, $1 < q \le \infty$,*

$$\left|\left\{\mathbb{G}_c^d \beta > \alpha\right\}\right| \le \frac{C_q}{\alpha}, \quad \alpha > 0, \quad (35)$$

where $\mathbb{G}_c^d$ is the generalized dyadic Carleson operator defined by (28) and C_q is a positive finite constant.

Theorem 4. *Let $\{\widetilde{w}_n\}$ be a periodic Walsh-type wavelet packet basis. Then for every q-block β, $1 < q \le \infty$,*

$$\left|\left\{G_c \beta > \alpha\right\}\right| \le \frac{C_q}{\alpha}, \quad \alpha > 0, \quad (36)$$

where G_c is the generalized weak Carleson operator defined by (26) and C_q is a positive finite constant.

Theorem 5. *If a function f belongs to $\mathbb{B}_q$-class, $1 < q \le \infty$, then*

$$\left|\left\{G_c f > \alpha\right\}\right| = O\left(\|f\|_{\mathbb{B}_q}\right), \quad \text{for } \alpha > 0, \quad (37)$$

where G_c is the generalized weak Carleson operator.

4. Lemmas

For the proof of our theorems, the following lemmas are required.

Lemma 6 (Nielsen [7]). *Let $f_1 \in L^2(\mathbb{R})$, and define $\{f_n\}_{n\ge 2}$ recursively by*

$$\begin{aligned} f_{2n}(x) &= f_n(2x) + f_n(2x-1), \\ f_{2n+1}(x) &= f_n(2x) - f_n(2x-1). \end{aligned} \quad (38)$$

Then

$$f_n(x) = \sum_{s=0}^{2^J-1} W_{n-2^J}\left(s 2^{-J}\right) f_1\left(2^J x - s\right), \quad (39)$$

where $n, J \in \mathbb{N}$, $2^J \le n < 2^{J+1}$.

Lemma 7 (Zygmund [8], page 3). *Consider*

$$\sum_{\nu=1}^{n} u_\nu v_\nu = \sum_{\nu=1}^{n-1} \left(v_\nu - v_{\nu+1}\right) U_\nu + U_n v_n, \quad (40)$$

where $U_k = u_1 + u_2 + \cdots + u_k$ for $k = 1, 2, \ldots, n$; it is also called Abel's transformation.

Lemma 8. *Let $\{W_n\}_{n=0}^{\infty}$ be the Walsh system. Then*

$$\begin{aligned} &\left| \sum_{n=2^K}^{m} \left(1 - \frac{n}{m - 2^k + 1}\right) W_{n-2^K}\left(\left[2^K x\right] 2^{-K}\right) \right. \\ &\quad \left. \times W_{n-2^K}\left(\left[2^K y\right] 2^{-K}\right) \right| \\ &\le \frac{C}{x \oplus y}, \end{aligned} \quad (41)$$

where C is a finite positive constant, $K \ge 1$, $2^K \le n < 2^{K+1}$, and for all pairs $x, y \in [0,1)$ for which $x \oplus y$ is defined.

Proof. The Dirichlet kernel, $D_n(x) = \sum_{k=0}^{n-1} W_k(x)$, for the Walsh system satisfies

$$\left|D_n(x \oplus y)\right| \le \frac{1}{x \oplus y} \quad (42)$$

(see Golubov et al. [6], page 21).

Hence,

$$\begin{aligned} &\left| \sum_{n=2^K}^{m} \left(1 - \frac{n}{m - 2^k + 1}\right) W_{n-2^K}\left(\left[2^K x\right] 2^{-K}\right) \right. \\ &\quad \left. \times W_{n-2^K}\left(\left[2^K y\right] 2^{-K}\right) \right| \\ &= \left| \sum_{n=2^K}^{m} \left(1 - \frac{n}{m - 2^k + 1}\right) W_{n-2^K} \right. \\ &\quad \left. \times \left(\left[2^K x\right] 2^{-K} \oplus \left[2^K y\right] 2^{-K}\right) \right| \\ &= \left| \sum_{n=0}^{m-2^K} \left(1 - \frac{n}{m - 2^k + 1}\right) W_n \right. \\ &\quad \left. \times \left(\left[2^K x\right] 2^{-K} \oplus \left[2^K y\right] 2^{-K}\right) \right| \end{aligned}$$

$$= \left| \sum_{n=0}^{m-2^K-1} \left\{ \left(1 - \frac{n}{m-2^K+1}\right) - \left(1 - \frac{n+1}{m-2^K+1}\right) \right\} \right.$$

$$\times \sum_{r=0}^{n} W_r\left(\left[2^K x\right] 2^{-K} \oplus \left[2^K y\right] 2^{-K}\right)$$

$$+ \left(1 - \frac{m-2^K}{m-2^K+1}\right)$$

$$\left. \times \sum_{n=0}^{m-2^K} W_n\left(\left[2^K x\right] 2^{-K} \oplus \left[2^K y\right] 2^{-K}\right) \right|,$$

by Lemma 7,

$$= \left| \sum_{n=0}^{m-2^K-1} \frac{1}{m-2^K+1} \right.$$

$$\times \sum_{r=0}^{n} W_r\left(\left[2^K x\right] 2^{-K} \oplus \left[2^K y\right] 2^{-K}\right) + \frac{1}{m-2^K+1}$$

$$\left. \times \sum_{n=0}^{m-2^K} W_n\left(\left[2^K x\right] 2^{-K} \oplus \left[2^K y\right] 2^{-K}\right) \right|$$

$$\leq \left| \sum_{n=0}^{m-2^K-1} W_n\left(\left[2^K x\right] 2^{-K} \oplus \left[2^K y\right] 2^{-K}\right) + \frac{1}{m-2^K+1} \right.$$

$$\left. \times \sum_{n=0}^{m-2^K} W_n\left(\left[2^K x\right] 2^{-K} \oplus \left[2^K y\right] 2^{-K}\right) \right|$$

$$= \left| \sum_{n=2^K}^{m-1} W_{n-2^K}\left(\left[2^K x\right] 2^{-K} \oplus \left[2^K y\right] 2^{-K}\right) + \frac{1}{m-2^K+1} \right.$$

$$\left. \times \sum_{n=2^K}^{m} W_{n-2^K}\left(\left[2^K x\right] 2^{-K} \oplus \left[2^K y\right] 2^{-K}\right) \right|$$

$$\leq \left| \sum_{n=2^K}^{m-1} W_{n-2^K}\left(\left[2^K x\right] 2^{-K} \oplus \left[2^K y\right] 2^{-K}\right) \right| + \frac{1}{m-2^K+1}$$

$$\times \left| \sum_{n=2^K}^{m} W_{n-2^K}\left(\left[2^K x\right] 2^{-K} \oplus \left[2^K y\right] 2^{-K}\right) \right|$$

$$= \left| W_{2^K}\left(\left[2^K x\right] 2^{-K} \oplus \left[2^K y\right] 2^{-K}\right) D_{m-2^K} \right.$$

$$\left. \times \left(\left[2^K x\right] 2^{-K} \oplus \left[2^K y\right] 2^{-K}\right) \right| + \frac{1}{m-2^K+1}$$

$$\times \left| W_{2^K}\left(\left[2^K x\right] 2^{-K} \oplus \left[2^K y\right] 2^{-K}\right) D_{m-2^K+1} \right.$$

$$\left. \times \left(\left[2^K x\right] 2^{-K} \oplus \left[2^K y\right] 2^{-K}\right) \right|$$

$$= \left| D_{m-2^K}\left(x \oplus y\right) \right|$$

$$+ \frac{1}{m-2^K+1} \left| D_{m-2^K+1}\left(x \oplus y\right) \right|$$

$$\leq \frac{1}{(x \oplus y)} + \frac{1}{m-2^k+1} \frac{1}{(x \oplus y)}, \quad x \oplus y \neq 0$$

$$= \left(1 + \frac{1}{m-2^k+1}\right) \frac{1}{(x \oplus y)}$$

$$\leq \frac{C}{(x \oplus y)}, \tag{43}$$

where (32), (34), and the fact that $D_{\nu+1-2^K}$ is a constant on dyadic intervals of the form $[l2^{-K}, (l+1)2^{-K})$ are used. This completes the proof of Lemma 8. □

Lemma 9. *If*

$$K_{J,m}^{(\sigma)}(x,y) = \sum_{n=2^J}^{m} \left(1 - \frac{n}{m-2^J+1}\right) w_n(x)\, w_n(y),$$

$$\textit{for } 2^J \leq m < 2^{J+1}, \tag{44}$$

then

$$\left|K_{J,m}^{(\sigma)}(x,y)\right| \leq \sum_{l=-2N}^{2N} \frac{C}{\left|x-y+2^{K-J}l\right|}, \tag{45}$$

where C is an arbitrary constant.

Proof. The kernel can be expanded as

$$K_{J,m}^{(\sigma)}(x,y)$$

$$= \sum_{n=2^J}^{m} \left(1 - \frac{n}{m-2^J+1}\right) w_n(x)\, w_n(y)$$

$$= \sum_{n=2^J}^{m} \left(1 - \frac{n}{m-2^J+1}\right)$$

$$\times \left(\sum_{l=0}^{2^{J-K}-1} W_{n-2^{J-K}}\left(l 2^{-(J-K)}\right) w_{2^K}\left(2^{J-K}x - l\right) \right.$$

$$\times \sum_{k=0}^{2^{J-K}-1} W_{n-2^{J-K}}\left(k 2^{-(J-K)}\right)$$

$$\left. \times\, w_{2^K}\left(2^{J-K}y - k\right) \right),$$

by Lemma 6,

$$= \sum_{l=0}^{2^{J-K}-1} \sum_{k=0}^{2^{J-K}-1} \left\{ \sum_{n=2^J}^{m} \left(1 - \frac{n}{m-2^J+1}\right) \times \left(W_{n-2^{J-K}}\left(l2^{-(J-K)}\right) \times W_{n-2^{J-K}}\left(k2^{-(J-K)}\right)\right) \times w_{2^K}\left(2^{J-K}x - l\right) \times w_{2^K}\left(2^{J-K}y - k\right) \right\}. \quad (46)$$

Therefore, using Lemma 8,

$$\begin{aligned} \left|K_{J,m}^{(\sigma)}(x,y)\right| &\le \sum_{l=-N}^{N}{}' \sum_{k=-N}^{N}{}' \left| \sum_{n=2^J}^{m} \left(1 - \frac{n}{m-2^J+1}\right) \times W_{n-2^{J-K}}\left(\left[2^{J-K}\left(x + 2^{K-J}l\right)\right]2^{-(J-K)}\right) \times W_{n-2^{J-K}}\left(\left[2^{J-K}\left(y + 2^{K-J}k\right)\right]2^{-(J-K)}\right) \right| \times \left\|w_{2^K}\right\|_\infty^2 \\ &\le \sum_{l=-N}^{N}{}' \sum_{k=-N}^{N}{}' \frac{C}{\left(x + 2^{K-J}l\right) \oplus \left(y + 2^{K-J}k\right)}, \end{aligned} \quad (47)$$

where $\sum'$ indicates that only the terms for which $x + 2^{K-J}l \in [0,1)$ and $y + 2^{K-J}k \in [0,1)$, respectively, should be included in the sum. This implies the estimate

$$\left|K_{J,m}^{(\sigma)}(x,y)\right| \le \sum_{l=-N}^{N} \sum_{k=-N}^{N} \frac{\widetilde{C}}{\left|x - y + 2^{K-J}(l-k)\right|}, \quad (48)$$

since $a \oplus b \ge 2^{-\log_2[|a-b|]} \ge |a-b|/2$. This completes the proof of Lemma 9. □

5. Proof of Theorem 3

The dyadic arithmetic mean of partial sums for the expansion of a measurable (integrable) function f in the periodic Walsh-type wavelet packets,

$$\begin{aligned} \left(\sigma_{2^N} f\right)(x) &= \frac{1}{2^N} \sum_{n=0}^{2^N-1} \left(S_n f\right)(x) \\ &= \frac{1}{2^N} \sum_{n=0}^{2^N-1} \left(\sum_{k=0}^{n} \langle f, \widetilde{w}_k \rangle \widetilde{w}_k(x) \right), \quad \text{by (22)}, \\ &= \sum_{n=0}^{2^N-1} \left(1 - \frac{n}{2^N}\right) \langle f, \widetilde{w}_n \rangle \widetilde{w}_n(x), \end{aligned} \quad (49)$$

holds everywhere with the arithmetic mean of the projection onto the (periodized) scaling space $\widetilde{V}_N$ associated with the underlying multiresolution analysis (Hess-Nielsen and Wickerhauser [9]). Therefore, it suffices to consider the arithmetic mean of the projection operators $P_{\widetilde{V}_N}$ on to the space $\widetilde{V}_N$.

Suppose that the q-block β is associated with the dyadic interval $I \subset [0,1)$. If $1 < \alpha|I|$, then $|I|^{1-q}/\alpha^q \le 1/\alpha$, and using the fact that the operator $f \to \sup_N \sum_{n=0}^{N}(1 - n/(N+1))P_{\widetilde{V}_n} f(x)$ (and thus $f \to \mathbb{G}_c^d f(x)$) is of strong type (q,q). We have

$$\left|\left\{\mathbb{G}_c^d f(x) > \alpha\right\}\right| \le C_q \left(\frac{\|\beta\|_q}{\alpha}\right)^q \le C_q \frac{|I|^{1-q}}{\alpha^q} \le \frac{C_q}{\alpha}. \quad (50)$$

Now suppose that $1 \ge \alpha|I|$ with $I = [a,b)$. Put $\widetilde{I} = [(3a-b)/2, (3b-a)/2] \cap [0,1)$, and define $\overline{I} = [0,1) - \widetilde{I}$. We have

$$\begin{aligned} \left|\left\{\mathbb{G}_c^d f(x) > \alpha\right\}\right| &\le 2|I| + \left|\overline{I} \cap \left\{\mathbb{G}_c^d f(x) > \alpha\right\}\right| \\ &\le \frac{2}{\alpha} + \left|\overline{I} \cap \left\{\mathbb{G}_c^d f(x) > \alpha\right\}\right|. \end{aligned} \quad (51)$$

Fix $x \in \overline{I}$, and let $K_N(x,y)$ denote the operator kernel associated with the projection operators $P_{\widetilde{V}_N}$. Then there exists a finite constant C (independent of N) such that

$$\left|K_N(x,y)\right| \le \frac{C}{|x-y|} \quad (52)$$

(see Terence [10]).

Using the estimate (52) on the kernel K_N, we obtain

$$\begin{aligned} \left|\left(\sigma_{2^N}\beta\right)(x)\right| &= \left| \sum_{n=0}^{2^N-1} \left(1 - \frac{n}{2^N}\right) \langle \beta, \widetilde{w}_n \rangle \widetilde{w}_n(x) \right| \\ &= \left| \sum_{n=0}^{2^N-2} \left\{ \left(1 - \frac{n}{2^N}\right) - \left(1 - \frac{n+1}{2^N}\right) \right\} \times \sum_{r=0}^{n} \langle \beta, \widetilde{w}_r \rangle \widetilde{w}_r(x) + \left(1 - \frac{2^N-1}{2^N}\right) \sum_{n=0}^{2^N-1} \langle \beta, \widetilde{w}_n \rangle \widetilde{w}_n(x) \right|, \end{aligned}$$

by Lemma 7,

$$= \left| \sum_{n=0}^{2^N-2} \frac{1}{2^N} \sum_{r=0}^{n} \langle \beta, \widetilde{w}_r \rangle \widetilde{w}_r(x) + \frac{1}{2^N} \sum_{r=0}^{2^N-1} \langle \beta, \widetilde{w}_n \rangle \widetilde{w}_n(x) \right|$$

$$= \left| \sum_{n=0}^{2^N-2} \frac{1}{2^N} (S_n\beta)(x) + \frac{1}{2^N} (S_{2^N}\beta)(x) \right|$$

$$\le \sum_{n=0}^{2^N-2} \frac{1}{2^N} \left|(S_n\beta)(x)\right| + \frac{1}{2^N} \left|(S_{2^N}\beta)(x)\right|$$

$$\le \sum_{n=0}^{2^N-2} \frac{1}{2^N} \left| \int_I K_n(x,y)\beta(y)\,dy \right| + \frac{1}{2^N} \left| \int_I K_{2^N}(x,y)\beta(y)\,dy \right|$$

$$\le \sum_{n=0}^{2^N-2} \frac{1}{2^N} \left(\frac{C}{|x-a|} + \frac{C}{|x-b|} \right) \|\beta\|_1 + \frac{1}{2^N} \left(\frac{C}{|x-a|} + \frac{C}{|x-b|} \right) \|\beta\|_1$$

$$= \left(\frac{2^N-1}{2^N} + \frac{1}{2^N} \right) \left(\frac{C}{|x-a|} + \frac{C}{|x-b|} \right) \|\beta\|_1$$

$$= \left(\frac{C}{|x-a|} + \frac{C}{|x-b|} \right) \|\beta\|_1. \tag{53}$$

Since $\|\beta\|_1 \le 1$ and $x \in \overline{I}$ implies that $|x-a|, |x-b| \ge |I|/2$, therefore,

$$\left|(\sigma_{2^N}\beta)(x)\right| \le \left\{ \frac{2C}{|I|} + \frac{2C}{|I|} \right\} = \frac{4C}{|I|} \le \frac{\widetilde{C}}{\alpha}. \tag{54}$$

Finally we obtain

$$\left| \left\{ x \in \overline{I} : \sup_N \left|(\sigma_{2^N}\beta)(x)\right| > \alpha \right\} \right| \le \frac{\widetilde{C}}{\alpha}, \tag{55}$$

where $\widetilde{C}$ is independent of I and β and hence Theorem 3 follows.

6. Proof of Theorem 4

Fix $\alpha > 0$ and a q-block β supported on the dyadic interval $I \subset [0,1)$; two cases are considered.

Case I. If $1 < \alpha|I|$, then $|I|^{1-q}/\alpha^q \le 1/\alpha$. Therefore, using Theorem 5.1. [7], page 275, we have

$$\begin{aligned} \left|\{G_c\beta > \alpha\}\right| &\le C_q \left(\frac{\|\beta\|_q}{\alpha} \right)^q \\ &\le C_q \frac{\left(|I|^{1/q-1}\right)^q}{\alpha^q} \\ &= C_q \frac{|I|^{1-q}}{\alpha^q} \\ &\le \frac{C_q}{\alpha}. \end{aligned} \tag{56}$$

Case II. Let $1 \ge \alpha|I|$ with $I = [a,b)$. Let

$$\widetilde{I} = \left(\cup_{j=-1}^{1} \left(j + \left[\frac{3a-b}{2}, \frac{3b-a}{2} \right) \right) \right) \cap [0,1), \tag{57}$$

and define $\overline{I} = [0,1) \setminus \widetilde{I}$. Then

$$\begin{aligned} \left|\{G_c\beta > \alpha\}\right| &\le \left|\widetilde{I}\right| + \left|\overline{I} \cap \{G_c\beta > \alpha\}\right| \\ &\le 3|I| + \left|\overline{I} \cap \{G_c\beta > \alpha\}\right| \\ &\le \frac{6}{\alpha} + \left|\overline{I} \cap \{G_c\beta > \alpha\}\right|. \end{aligned} \tag{58}$$

Notice that

$$\begin{aligned} \left|\overline{I} \cap \{G_c\beta > \alpha\}\right| &\le \left|\overline{I} \cap \left\{G_c^d\beta > \frac{\alpha}{2}\right\}\right| \\ &+ \left|\overline{I} \cap \left\{\limsup_J M_J\beta > \frac{\alpha}{2}\right\}\right|, \end{aligned} \tag{59}$$

with

$$\begin{aligned} M_J\beta(x) &= \max_{2^J \le m < 2^{J+1}-1} M_J^m\beta(x), \\ M_J^m\beta(x) &= \left| \sum_{n=2^J}^{m} \left(1 - \frac{n}{m-2^J+1} \right) \langle \beta, \widetilde{w}_n \rangle \widetilde{w}_n(x) \right|. \end{aligned} \tag{60}$$

For $x \in [0,1)$, we have

$$\begin{aligned}
&\limsup_{J,m} M_J^m \beta(x) \\
&= \limsup_{J,m} \left| \sum_{n=2^J}^{m} \left(1 - \frac{n}{m-2^J+1}\right) \langle \beta, \widetilde{w}_n \rangle \widetilde{w}_n(x) \right| \\
&= \limsup_{J,m} \left| \sum_{l_1=-N}^{N} \sum_{l_2=-N}^{N} \sum_{n=2^J}^{m} \left(1 - \frac{n}{m-2^J+1}\right) \right. \\
&\qquad \left. \times \langle \beta, w_n(\cdot - l_1) \rangle w(x - l_2) \right| \\
&\le \sum_{l_1=-N}^{N} \sum_{l_2=-N}^{N} \limsup_{J,m} \left| \sum_{n=2^J}^{m} \left(1 - \frac{n}{m-2^J+1}\right) \right. \\
&\qquad \left. \times \langle \beta, w_n(\cdot - l_1) \rangle w(x - l_2) \right|.
\end{aligned} \tag{61}$$

Hence, it suffices to estimate $|E_\alpha^{l_1,l_2}|$ with

$$\begin{aligned}
&E_\alpha^{l_1,l_2} \\
&= \left\{ x \in \overline{I} : \limsup_{J,m} \left| \sum_{n=2^J}^{m} \left(1 - \frac{n}{m-2^J+1}\right) \right. \right. \\
&\qquad \left. \left. \times \langle \beta, w_n(\cdot - l_1) \rangle w(x - l_2) \right| > \alpha \right\}.
\end{aligned} \tag{62}$$

Fix $x \in \mathbb{R} \setminus I$; then

$$\begin{aligned}
&\left| \int_{-\infty}^{\infty} K_{J,m}^{\sigma}(x - l_1, y - l_2) \beta(y)\, dy \right| \\
&\quad \le \widetilde{C} \sum_{l=-2N}^{2N} \int_{-\infty}^{\infty} \frac{\beta(y)\, dy}{|x - y + l_2 - l_1 + 2^{K-J} l|},
\end{aligned} \tag{63}$$

which implies that whenever $x \in E_\alpha^{l_1,l_2}$, there is an increasing sequence $J_k \to \infty$ for which

$$\begin{aligned}
&\left(\frac{1}{|x - a + l_2 - l_1 + 2^{K-J_k} l|} \right. \\
&\qquad \left. + \frac{1}{|x - b + l_2 - l_1 + 2^{K-J_k} l|} \right) > C\alpha,
\end{aligned} \tag{64}$$

for some fixed $C > 0$ and for $k = 1, 2, \ldots$. Since $J_k \to \infty$, therefore

$$\left(\frac{1}{|x - a + l_2 - l_1|} + \frac{1}{|x - b + l_2 - l_1|} \right) > C\alpha. \tag{65}$$

Using that $\overline{I} = [0,1) \setminus \widetilde{I}$ and the same technique as in the proof of Lemma 9, we complete the proof to conclude that $|E_\alpha^{l_1,l_2}| \le 1/\alpha$ and consequently

$$\left| \overline{I} \cap \left\{ \limsup_{J} M_J \beta > \frac{\alpha}{2} \right\} \right| \le \frac{\widetilde{C}}{\alpha}, \tag{66}$$

which completes the proof of Theorem 4.

7. Proof of Theorem 5

Let $f = \sum_{k=1}^{\infty} c_k \beta_k$ be a function of $\mathbb{B}_q$. Then

$$\begin{aligned}
\sigma_N f &= \sum_{n=0}^{N} \left(1 - \frac{n}{N+1}\right) \langle f, \widetilde{w}_n \rangle \widetilde{w}_n \\
&= \sum_{n=0}^{N} \left(1 - \frac{n}{N+1}\right) \left\langle \sum_{k=1}^{\infty} c_k \beta_k, \widetilde{w}_n \right\rangle \widetilde{w}_n \\
&= \sum_{n=0}^{N} \left(1 - \frac{n}{N+1}\right) \sum_{k=1}^{\infty} c_k \langle \beta_k, \widetilde{w}_n \rangle \widetilde{w}_n \\
&= \sum_{k=1}^{\infty} c_k \left(\sum_{n=0}^{N} \left(1 - \frac{n}{N+1}\right) \langle \beta_k, \widetilde{w}_n \rangle \widetilde{w}_n \right) \\
&= \sum_{k=1}^{\infty} c_k (\sigma_N \beta_k),
\end{aligned} \tag{67}$$

due to the L^1 convergence of the average sum defining f. Since

$$\begin{aligned}
&\sum_{n=0}^{N} \left(1 - \frac{n}{N+1}\right) \langle f, \widetilde{w}_n \rangle \widetilde{w}_n \\
&\qquad = \sum_{k=1}^{\infty} c_k \sum_{n=0}^{N} \left(1 - \frac{n}{N+1}\right) \langle \beta_k, \widetilde{w}_n \rangle \widetilde{w}_n, \\
&\limsup_{N} \left| \sum_{n=0}^{N} \left(1 - \frac{n}{N+1}\right) \langle f, \widetilde{w}_n \rangle \widetilde{w}_n \right| \\
&\quad \le \sum_{k=1}^{\infty} |c_k| \limsup_{N} \left| \sum_{n=0}^{N} \left(1 - \frac{n}{N+1}\right) \langle \beta_k, \widetilde{w}_n \rangle \widetilde{w}_n \right| \\
&\qquad\qquad G_c f \le \sum_{k=1}^{\infty} |c_k| G_c \beta_k
\end{aligned} \tag{68}$$

therefore

$$\begin{aligned}|\{G_c f > \alpha\}| &\le \left|\left\{\sum_{k=1}^{\infty} |c_k| G_c\beta_k > \alpha\right\}\right| \\ &\le \frac{C_q}{\alpha}\sum_{k=1}^{\infty} |c_k|, \quad \text{by Theorem 4,} \\ &\le \frac{C_q}{\alpha}\|f\|_{\mathbb{B}_q}, \quad \text{by (12)}, \\ &= O\left(\|f\|_{\mathbb{B}_q}\right).\end{aligned} \tag{69}$$

This completes the proof of Theorem 5.

8. Applications

Following corollary can be deduced from our theorems.

Corollary 10. *Let $\{\widetilde{\omega}_n\}$ be a periodic Walsh-type wavelet packet basis. Then the Fourier expansion of any function $f \in \mathbb{B}_q$, $1 < q < \infty$, in $\{\widetilde{\omega}_n\}$ is summable by arithmetic means pointwise a.e.*

Proof. Let us write $(S_N f)(x) = \sum_{n=0}^{N}\langle f, \widetilde{\omega}_n\rangle \widetilde{\omega}_n(x)$ and

$$\begin{aligned}(\sigma_N f)(x) &= \frac{1}{N+1}\sum_{n=0}^{N}(S_n f)(x) \\ &= \sum_{n=0}^{N}\left(1 - \frac{n}{N+1}\right)\langle f, \widetilde{\omega}_n\rangle \widetilde{\omega}_n(x).\end{aligned} \tag{70}$$

With $f = \sum_{k=1}^{\infty} c_k\beta_k \in \mathbb{B}_q$, let $g_K = \sum_{k=1}^{K} c_k\beta_k$, and observe that $\|f - g_K\|_{\mathbb{B}_q} \to 0$. For each $x \in [0, 1)$, write

$$\begin{aligned}f - \sigma_N f = {}& (f - g_K) \\ &+ (g_K - \sigma_N g_K) + (\sigma_N g_K - \sigma_N f).\end{aligned} \tag{71}$$

Thus

$$\begin{aligned}&\left|\left\{x : \limsup_{N\to\infty} |f(x) - (\sigma_N f)(x)| > \alpha\right\}\right| \\ &\le \left|\left\{x : \limsup_{N\to\infty} |f(x) - g_K(x)| > \frac{\alpha}{3}\right\}\right| \\ &\quad + \left|\left\{x : \limsup_{N\to\infty} |g_K(x) - (\sigma_N g_K)(x)| > \frac{\alpha}{3}\right\}\right| \\ &\quad + \left|\left\{x : \limsup_{n\to\infty} |(\sigma_N g_K)(x) - (\sigma_N f)(x)| > \frac{\alpha}{3}\right\}\right| \\ &\le \left|\left\{x : \limsup_{N\to\infty} |f(x) - g_K(x)| > \frac{\alpha}{3}\right\}\right| \\ &\quad + \left|\left\{x : \limsup_{N\to\infty} \left|g_K(x) - \sum_{n=0}^{N}\left(1 - \frac{n}{N+1}\right) \times \langle g_K, \widetilde{\omega}_n\rangle \widetilde{\omega}_n(x)\right| > \frac{\alpha}{3}\right\}\right| \\ &\quad + \left|\left\{x : \limsup_{N\to\infty} \left|\sum_{n=0}^{N}\left(1 - \frac{n}{N+1}\right) \times \langle g_K - f, \widetilde{\omega}_n\rangle \widetilde{\omega}_n(x)\right| > \frac{\alpha}{3}\right\}\right| \\ &\le \left|\left\{x : \limsup_{N\to\infty} |f(x) - g_K(x)| > \frac{\alpha}{3}\right\}\right| \\ &\quad + \left|\left\{x : \limsup_{N\to\infty} \left|g_K(x) - \sum_{n=0}^{N}\langle g_K, \widetilde{\omega}_n\rangle \widetilde{\omega}_n(x)\right| > \frac{\alpha}{3}\right\}\right| \\ &\quad + \left|\left\{x : \limsup_{N\to\infty} \left|\sum_{n=0}^{N}\langle g_K - f, \widetilde{\omega}_n\rangle \widetilde{\omega}_n(x)\right| > \frac{\alpha}{3}\right\}\right| \\ &\le \left|\left\{x : \limsup_{N\to\infty} |f(x) - g_K(x)| > \frac{\alpha}{3}\right\}\right| \\ &\quad + \left|\left\{x : \limsup_{N\to\infty} |g_K(x) - (S_N g_K)(x)| > \frac{\alpha}{3}\right\}\right| \\ &\quad + \left|\left\{x : \limsup_{N\to\infty} |(S_N g_K)(x) - (S_N f)(x)| > \frac{\alpha}{3}\right\}\right| \\ &\le \frac{3}{\alpha}\|f - g_K\|_{\mathbb{B}_q} + 0 + \frac{3}{\alpha}C_q\|f - g_K\|_{\mathbb{B}_q}, \quad \text{by Theorem 5.}\end{aligned} \tag{72}$$

From this it follows that

$$\left|\left\{x : \limsup_{n\to\infty} |f(x) - (\sigma_N f)(x)| > \alpha\right\}\right| = 0. \tag{73}$$

This completes the proof of the corollary. □

Acknowledgments

Shyam Lal, one of the authors, is thankful to DST-CIMS for encouragement to this work. Manoj Kumar is grateful to CSIR, India in the form of Junior Research Fellowship vide Reference no. 17-06/2012 (i)EU-V dated 28-09-2012 for this research work.

References

[1] P. Billard, "Sur la convergence presque partout des séries de Fourier-Walsh des fontions de l'espace $L^2(0, 1)$," *Studia Mathematica*, vol. 28, pp. 363–388, 1967.

[2] M. Nielsen, *Size Properties of Wavelet Packets [Ph.D. thesis]*, Washington University, St. Louis, Mo, USA, 1999.

[3] P. Sjölin, "An inequality of paley and convergence a.e. of Walsh-Fourier series," in *Arkiv för Matematik*, vol. 7, pp. 551–570, 1969.

[4] L. Carleson, "On convergence and growth of partial sums of Fourier series," *Acta Mathematica*, vol. 116, no. 1, pp. 135–157, 1966.

[5] F. Schipp, W. R. Wade, and P. Simon, *Walsh Series, Introduction to Dyadic Harmonic Analysis*, Adam Hilger Ltd, Bristol, UK, 1990.

[6] B. Golubov, A. Efimov, and V. Skvortson, *Walsh Series and Transforms: Theory and Application*, Kluwer Academic, Dordrecht, The Netherlands, 1991, Translated From the 1987 Russian Original by W.R. Wade.

[7] M. Nielsen, "Walsh-type wavelet packet expansions," *Applied and Computational Harmonic Analysis*, vol. 9, no. 3, pp. 265–285, 2000.

[8] A. Zygmund, *Trigonometric Series Volume I*, Cambridge University Press, 1959.

[9] N. Hess-Nielsen and M. V. Wickerhauser, "Wavelets and time-frequency analysis," *Proceedings of the IEEE*, vol. 84, no. 4, pp. 523–540, 1996.

[10] T. Terence, "On the almost everywhere convergence of wavelet summation methods," *Applied and Computational Harmonic Analysis*, vol. 3, no. 4, pp. 384–387, 1996.

Solving Fuzzy Fractional Differential Equations Using Zadeh's Extension Principle

M. Z. Ahmad,[1] M. K. Hasan,[2] and S. Abbasbandy[3]

[1] *Institute of Engineering Mathematics, Universiti Malaysia Perlis, Kampus Tetap Pauh Putra, 02600 Arau, Perlis, Malaysia*
[2] *School of Information Technology, Faculty of Information Science and Technology, Universiti Kebangsaan Malaysia, 43600 Bangi, Selangor, Malaysia*
[3] *Department of Mathematics, Imam Khomeini International University, Ghazvin 34149-16818, Iran*

Correspondence should be addressed to M. Z. Ahmad; mzaini@unimap.edu.my

Academic Editors: M. Bruzón and S. Momani

We study a fuzzy fractional differential equation (FFDE) and present its solution using Zadeh's extension principle. The proposed study extends the case of fuzzy differential equations of integer order. We also propose a numerical method to approximate the solution of FFDEs. To solve nonlinear problems, the proposed numerical method is then incorporated into an unconstrained optimisation technique. Several numerical examples are provided.

1. Introduction

Fractional calculus is an important branch in mathematical analysis. It is a generalisation of ordinary calculus that allows noninteger order. At the beginning, it was slowly established. However, after Leibniz and Newton invented differential calculus, it has been a subject of interest among mathematicians, physicists, and engineers. Consequently, the theory of fractional calculus has been extensively developed and influenced in many areas of discipline. The fractional integral and fractional derivative of Riemann-Liouville, for example, have been applied to solve many mathematical problems [1–6]. One of the particular interests is the case of solving fractional differential equations [7–9]. The fractional derivative of Riemann-Liouville, however, has a common characteristic. It requires a quantity of fractional derivative of unknown solution at initial point. In practice, we do not clearly know what the meaning of the fractional derivative at that point is. In other words, the required quantity cannot be measured and perhaps may not be available [10, 11].

The well-known and popularly used method in solving fractional differential equations is the Caputo fractional derivative. It allows to specify a quantity of integer order derivatives at the initial point. This quantity typically is available and can be measured. It is therefore not surprising that there is a vast literature dealing with fractional differential equations involving the Caputo fractional derivative [12–16]. The theory and application of fractional differential equations under both types of fractional derivatives have been discussed by many authors [11, 17–25]. Some potential applications have been studied in [26–28].

In the context of mathematical modelling, developing an accurate fractional differential equation is not a simple task. It requires an understanding of real physical phenomena involved. The real physical phenomena, however, are always pervaded with uncertainty. This is obvious when dealing with "living" materials such as soil, water, and microbial populations [29]. When a real physical phenomenon is modelled by a fractional differential equation, namely,

$$D_a^{\beta} x(t) = f(t, x(t)), \quad 0 < \beta \leq 1, \ t > a, \qquad x(t_0) = x_0, \tag{1}$$

we cannot usually be sure that the model is perfect. For example, the initial value in (1) may not be known precisely. It may take any value in the form of "less than x_0," "about x_0," or "more than x_0." Classical mathematics, however, fail to cope with this situation. Therefore, it is necessary to have other theories in order to handle this issue. Various theories exist for describing this situation and the most popular one is the fuzzy set theory [30].

In order to obtain a more realistic model than (1), Agarwal et al. [31] have taken an initiative to introduce the concept of solution for fuzzy fractional differential equations. This contribution has motivated several authors to establish some results on the existence and uniqueness of solution (see [32]). In [33], the authors derived the explicit solution of fuzzy fractional differential equations using the Riemann-Liouville H-derivative. Recently, Salahshour et al. [34] applied fuzzy Laplace transforms [35] to solve fuzzy fractional differential equations. Basically, the proposed ideas are a generalisation of the theory and solution of fuzzy differential equations [36–41]. However, the authors considered fuzzy fractional differential equations under the Riemann-Liouville H-derivative. Again, it requires a quantity of fractional H-derivative of an unknown solution at the fuzzy initial point. In this paper, we propose a new interpretation of fuzzy fractional differential equations and present their solutions analytically and numerically. The proposed idea is a generalisation of the interpretation given in [42–49], where the authors used Zadeh's extension principle to interpret fuzzy differential equations. According to Mizukoshi et al. [43], this interpretation requires neither the concept of derivatives of a fuzzy function nor the use of selection theory to obtain a solution to the fuzzy differential equation.

This paper is organised in the following sequence. In Section 2, we recall some basic definitions and theoretical backgrounds needed in this paper. In Section 3, we present the solution of fuzzy fractional differential equations analytically and numerically. In Section 4, some numerical examples are given. Finally in Section 5, we give conclusions.

2. Basic Concepts

In this section, we briefly elaborate some definitions and important concepts of fractional calculus and fuzzy set.

2.1. The Fractional Integral and Fractional Derivative. The following definitions of fractional integral and fractional derivative are adopted from [11].

Definition 1. A real function $x(t)$, $t > 0$, is said to be in the space C_μ, $\mu \in \mathbb{R}$, if there exists a real number $p > \mu$, such that $x(t) = t^p x_1(t)$, where $x_1(t) \in C(0, \infty)$ and it is said to be in the space C_μ^n if and only if $x^n \in C_\mu$, $n \in \mathbb{N}$.

Definition 2. The Riemann-Liouville fractional integral of x of order $\beta > 0$ with $a \geq 0$ is defined as

$$I_a^\beta x(t) = \frac{1}{\Gamma(\beta)} \int_a^t (t-s)^{\beta-1} x(s)\, ds, \quad t > a, \tag{2}$$

and for $\beta = 0$, the Riemann-Liouville fractional integral of x is defined as

$$I_a^0 x(t) = x(t). \tag{3}$$

Here, $\Gamma(\beta)$ is the well-known gamma function defined as

$$\Gamma(\beta) = \int_0^\infty t^{\beta-1} e^{-t} dt. \tag{4}$$

Definition 3. The Caputo fractional derivative of x of order $\beta > 0$ with $a \geq 0$ is defined as

$${}^cD_a^\beta x(t) = \frac{1}{\Gamma(n-\beta)} \int_a^t (t-s)^{n-\beta-1} x^{(n)}(s)\, ds \tag{5}$$

for $n - 1 < \beta \leq n$, $n \in \mathbb{N}$, $t \geq a$, $x \in C_{-1}^n$.

The following are two basic properties of the Caputo fractional derivative [50].

(1) Let $x \in C_{-1}^n$, $n \in \mathbb{N}$. Then ${}^cD_a^\beta x$, $0 \leq \beta \leq n$, is well defined and ${}^cD_a^\beta x \in C_{-1}$.

(2) Let $n - 1 < \beta \leq n$, $n \in \mathbb{N}$, and $x \in C_\mu^n$, $\mu \geq -1$. Then

$$I_a^\beta \left({}^cD_a^\beta\right) x(t) = x(t) - \sum_{k=0}^{n-1} x^{(k)}(a) \frac{(t-a)^k}{k!}. \tag{6}$$

The Laplace transform of the Caputo fractional derivative is given by [51]

$$\mathscr{L}\left\{{}^cD^\beta x(t)\right\} = s^\beta X(s) - \sum_{k=0}^{n-1} s^{\beta-k-1} x^{(k)}(0); \quad n-1 < \beta \leq n. \tag{7}$$

2.2. Fuzzy Set Theory. According to Zadeh [30], a fuzzy set is a generalisation of a classical set that allows a membership function to take any value in the unit interval [0, 1].

Definition 4 (see [30]). Let U be a universal set. A fuzzy set A in U is defined by a membership function $A(x)$ that maps every element in U to the unit interval [0, 1].

Definition 5 (see [38]). Let A be a fuzzy set defined in $\mathbb{R}$. A is called a fuzzy number if

(i) A is normal: there exists $x_0 \in \mathbb{R}$ such that $A(x_0) = 1$;

(ii) A is convex: for all $x, y \in \mathbb{R}$ and $0 \leq \lambda \leq 1$, it holds that

$$A(\lambda x + (1-\lambda) y) \geq \min(A(x), A(y)); \tag{8}$$

(iii) A is upper semicontinuous: for any $x_0 \in \mathbb{R}$, it holds that

$$A(x_0) \geq \lim_{x \to x_0^\pm} A(x); \tag{9}$$

(iv) $[A]^0 = \overline{\{x \in \mathbb{R} \mid A(x) > 0\}}$ is a compact subset of $\mathbb{R}$.

Definition 6 (see [38]). Let A be a fuzzy number defined in $\mathbb{R}$. The α-cut of A is the crisp set $[A]^\alpha$ that contains all elements in $\mathbb{R}$ such that the membership value of A is greater than or equal to α; that is,

$$[A]^\alpha = \{x \in U \mid A(x) \geq \alpha\}, \quad \alpha \in (0, 1]. \tag{10}$$

For a fuzzy number A, its α-cuts are closed intervals in $\mathbb{R}$ and we denote them by

$$[A]^{\alpha} = [a_1^{\alpha}, a_2^{\alpha}]. \quad (11)$$

Definition 7 (see [52]). A fuzzy number A is called a triangular fuzzy number if its membership function has the following form:

$$A(x) = \begin{cases} 0, & \text{if } x < a, \\ \dfrac{x-a}{b-a}, & \text{if } a \le x < b, \\ \dfrac{c-x}{c-b}, & \text{if } b \le x \le c, \\ 0, & \text{if } x > c, \end{cases} \quad (12)$$

and its α-cuts are simply $[A]^{\alpha} = [a + \alpha(b-a), c - \alpha(c-b)]$, $\alpha \in (0, 1]$.

In this paper, we denote $A = (a, b, c)$ as the triangular fuzzy number and $\mathscr{F}(\mathbb{R})$ as the set of all triangular fuzzy numbers.

Any crisp function can be extended to take fuzzy set as arguments by applying Zadeh's extension principle [30]. Let f be a function from X to Y. Given a fuzzy set A in X, we want to find a fuzzy set $B = f(A)$ in Y that is induced by f. If f is a strictly monotone, then we can extend f to fuzzy set as follows:

$$f(A)(y) = \begin{cases} A\left(f^{-1}(y)\right), & \text{if } y \in \operatorname{range}(f), \\ 0, & \text{if } y \notin \operatorname{range}(f). \end{cases} \quad (13)$$

It is clear that (13) can be easily calculated by determining the membership at the endpoints of the α-cuts of A. However, in general, the process of finding the fuzzy set $B = f(A)$ is more complicated and cannot be gathered easily. For example, if f is nonmonotone, then the problem can arise when two or more distinct points in X are mapped to the same point in Y. If this is the case, then the above equation may take two or more different values. This requires a new extension of (13) as shown below:

$$f(A)(y) = \begin{cases} \sup\limits_{x \in f^{-1}(y)} A(x), & \text{if } y \in \operatorname{range}(f), \\ 0, & \text{if } y \notin \operatorname{range}(f), \end{cases} \quad (14)$$

where

$$f^{-1}(y) = \{x \in X \mid f(x) = y\}. \quad (15)$$

Some computational methods to compute (14) can be found in [53, 54].

Theorem 8 (see [55]). *If $f : \mathbb{R} \to \mathbb{R}$ is continuous, then $f : \mathscr{F}(\mathbb{R}) \to \mathscr{F}(\mathbb{R})$ is well defined and*

$$[f(A)]^{\alpha} = f([A]^{\alpha}), \quad \forall \alpha \in [0, 1], \ \forall A \in \mathscr{F}(\mathbb{R}), \quad (16)$$

where $f(A) = \{f(u) \mid u \in [A]^{\alpha}\}$.

For $A, B \in \mathscr{F}(\mathbb{R})$ and $\lambda \in \mathbb{R}$, the sum $A + B$ and the product λA are defined as follows, respectively:

$$[A+B]^{\alpha} = [A]^{\alpha} + [B]^{\alpha}, \\ [\lambda B]^{\alpha} = \lambda [A]^{\alpha} \quad (17)$$

for each $\alpha \in [0, 1]$.

Definition 9 (see [56]). If A and B are two fuzzy numbers, then the distance D between A and B is defined as

$$D(A, B) = \sup_{\alpha \in [0,1]} \max \left\{ \left| a_1^{\alpha} - b_1^{\alpha} \right|, \left| a_2^{\alpha} - b_2^{\alpha} \right| \right\}. \quad (18)$$

In [57], the authors have shown that $(\mathscr{F}(\mathbb{R}), D)$ is a complete metric space and the following properties are well known:

(i) $D(A + C, B + C) = D(A, B), \forall A, B, C \in \mathscr{F}(\mathbb{R})$,

(ii) $D(\lambda A, \lambda B) = |\lambda| D(A, B), \forall A, B \in \mathscr{F}(\mathbb{R})$ and $\lambda \in \mathbb{R}$,

(iii) $D(A + B, C + D) \le D(A, C) + D(B, D), \forall A, B, C, D \in \mathscr{F}(\mathbb{R})$.

3. Fuzzy Fractional Differential Equations

In this section, we present analytical and numerical solutions of fuzzy fractional differential equations.

3.1. Analytical Solution of Fuzzy Fractional Differential Equations. First, let us consider the following fractional differential equation:

$${}^{c}D_a^{\beta} x(t) = f(t, x(t)), \\ x(t_0) = x_0, \quad (19)$$

where $f : [t_0, T] \times \mathbb{R} \to \mathbb{R}$ is a real-valued function, $x_0 \in \mathbb{R}$, and $\beta \in (0, 1]$. If $\beta = 1$, then (19) becomes an ordinary differential equation.

Assume that the initial value is replaced by a fuzzy number; then we have the following fuzzy fractional differential equation:

$${}^{c}D_a^{\beta} X(t) = f(t, X(t)), \\ X(t_0) = X_0, \quad (20)$$

where $X_0 \in \mathscr{F}(\mathbb{R})$ and $\beta \in (0, 1]$. If $\beta = 1$, then (20) becomes a fuzzy differential equation.

In order to find the solution of (20), we first find the solution of (19). By taking the Laplace transform on both sides of (19), we get

$$\mathscr{L}\left[{}^{c}D_a^{\beta} x(t)\right] = \mathscr{L}\left[f(t, x(t))\right]. \quad (21)$$

It follows that

$$s^{\beta} \mathscr{L}\{x(t)\} - x(t_0) s^{\beta-1} = \mathscr{L}\left[f(t, x(t))\right]. \quad (22)$$

Assume that after simplifying (22), we get

$$\mathscr{L}[x(t)] = m(s). \tag{23}$$

Then by taking the inverse Laplace transform to (23), we have

$$x(t) = \mathscr{L}^{-1}[m(s)] = g(t, \beta, x_0), \tag{24}$$

for $t \in [t_0, T]$ and $x_0 \in \mathbb{R}$. In order to find the solution of (20), we fuzzify (24) using Zadeh's extension principle. Hence, we have

$$X(t) = \widehat{g}(t, \beta, X_0), \tag{25}$$

which is the solution of (20).

This procedure can be shown precisely in the following theorem.

Theorem 10. *Let G be an open set in* $\mathbb{R}$ *and* $[X_0]^\alpha \in \mathscr{F}(\mathbb{R}) \subset G$. *Suppose that f is continuous and that for each* $\beta \in (0, 1]$ *and each* $x_0 \in G$ *there exists a unique solution* $g(\cdot, \beta, x_0)$ *of the problem* (19) *and that* $g(t, \beta, \cdot)$ *is continuous in G for each* $t \in [t_0, T]$ *fixed. Then, there exists a unique fuzzy solution* $X(t) = \widehat{g}(t, \beta, X_0)$ *of the problem* (20).

Proof. The proof of this theorem uses the basic idea found in [46]. Since f is continuous, there exists a unique solution $g(t, \beta, x_0)$. This solution is well defined and continuous in G, for each $t \in [t_0, T]$ fixed. Then, from Theorem 8, we have $\widehat{g}(t, \beta, \cdot) : \mathscr{F}(G) \rightarrow \mathscr{F}(\mathbb{R})$, which is continuous and well defined. Therefore, there exists a unique fuzzy solution of the form $X(t) = \widehat{g}(t, \beta, X_0)$ for the problem (20). □

Remark 11. The existence of $X(t)$ is guaranteed by Theorem 8 since $x(t)$ is continuous.

Moreover, in order to have valid level sets, $X(t)$ should satisfy the following Staking Theorem [58].

Theorem 12. *If* $X : [t_0, T] \rightarrow \mathscr{F}(\mathbb{R})$ *is a fuzzy solution of* (20) *and denoting* $[X(t)]^\alpha = [x_1^\alpha(t), x_2^\alpha(t)]$ *for* $\alpha \in [0, 1]$, *then*

(i) $[X(t)]^\alpha$ *is nonempty compact subset of* $\mathbb{R}$;

(ii) $[X(t)]^{\alpha_2} \subseteq [X(t)]^{\alpha_1}$ *for* $0 \leq \alpha_1 \leq \alpha_2 \leq 1$;

(iii) $[X(t)]^\alpha = \bigcap_{n=1}^{\infty}[X(t)]^{\alpha_n}$ *for any nondecreasing sequence* $\alpha_n \rightarrow \alpha$ *in* $[0, 1]$.

In the following result, we will show that $x_1^\alpha(t)$ and $x_2^\alpha(t)$ do not interchange at all $t \in [t_0, \infty)$.

Theorem 13. *If* $X(t) = g(t, \beta, X_0)$ *is obtained by using Theorem 10 and* $[X(t)]^\alpha = [x_1^\alpha(t), x_2^\alpha(t)]$ *for* $\alpha \in [0, 1]$, *then* $x_1^\alpha(t)$ *and* $x_2^\alpha(t)$ *do not interchange at all* $t \in [t_0, \infty)$.

Proof. We know that $X(t)$ is obtained by Zadeh's extension principle through Theorem 10; then its membership function has the following form:

$$X(t)(y) = \begin{cases} \sup\limits_{x \in g^{-1}(t, \beta, y)} X_0(x), & \text{if } y \in \operatorname{range}(g), \\ 0, & \text{if } y \notin \operatorname{range}(g). \end{cases} \tag{26}$$

It follows that

$$\begin{aligned} x_1^\alpha(t) &= \min\left\{g(t, \beta, u) \mid u \in \left[x_{0,1}^\alpha, x_{0,2}^\alpha\right]\right\}, \\ x_2^\alpha(t) &= \max\left\{g(t, \beta, u) \mid u \in \left[x_{0,1}^\alpha, x_{0,2}^\alpha\right]\right\} \end{aligned} \tag{27}$$

for $\alpha \in [0, 1]$. It is obvious that

$$x_1^\alpha(t) \leq x_2^\alpha(t). \tag{28}$$

This holds for all $t \in [t_0, \infty)$. This completes the proof. □

In general, the solution of (20) may not be found analytically. Therefore, a numerical method must be proposed.

3.2. Numerical Solution of Fuzzy Fractional Differential Equations. Fractional Euler method under the Caputo fractional derivative has been proposed in [59]. However, in order to approximate the solution of fuzzy fractional differential equations, the fractional Euler method has to be extended in the fuzzy setting. In this case, Zadeh's extension principle plays an important role.

Let $x(t)$ be the solution of (19). The first two terms of fractional Taylor series for $x(t)$ at t_i can be written as [59]

$$x(t_{i+1}) \approx x(t_i) + {}^cD^\beta x(t_i)\frac{h^\beta}{\Gamma(\beta+1)}. \tag{29}$$

From (19), we have

$$x(t_{i+1}) \approx x(t_i) + \frac{h^\beta}{\Gamma(\beta+1)} f(t_i, x(t_i)). \tag{30}$$

Let $w_{i+1} \approx x(t_{i+1})$; then we have the following fractional Euler method [59]:

$$w_{i+1} = w_i + \frac{h^\beta}{\Gamma(\beta+1)} f(t_i, w_i), \tag{31}$$

for $i = 0, 1, 2, \ldots, N$. Let $g(\beta, h, t_i, w_i) = w_i + (h^\beta/\Gamma(\beta + 1)) f(t_i, w_i)$; then (31) becomes

$$w_{i+1} = g(\beta, h, t_i, w_i). \tag{32}$$

To approximate the solution of (20), (32) will be fuzzified using Zadeh's extension principle. We then obtain the following fuzzy fractional Euler method:

$$W_{i+1} = g(\beta, h, t_i, W_i) \tag{33}$$

for $i = 0, 1, \ldots, N$, and $W_i \in \mathscr{F}(\mathbb{R})$. Please note that if $\beta = 1$, then this fuzzy fractional Euler method becomes fuzzy Euler method as proposed in [47]. The membership function of g in (33) can be defined as

$$g(\beta, h, t_i, W_i)(y) = \begin{cases} \sup\limits_{x \in g^{-1}(\beta, h, t_i, y)} W_i(x), & \text{if } y \in \operatorname{range}(g), \\ 0, & \text{if } y \notin \operatorname{range}(g). \end{cases} \tag{34}$$

Input: Fractional Euler Equation, Fuzzy Initial Value, β, t_0, t_N and N
Descritise $\alpha \leftarrow \{\alpha_0, \alpha_1, \dots, \alpha_n\}$;
$h \leftarrow \frac{t_N - t_0}{N}$;
for $j = \#(\alpha)$ **do**
  $w_1(1, j) \leftarrow a + \alpha(j) * (b - a)$;
  $w_2(1, j) \leftarrow c - \alpha(j) * (c - b)$;
end
for $i = 1 : N$ **do**
  for $j = 1 : \#(\alpha)$ **do**
    $w_1(i+1, j) \leftarrow \min(g(\beta, h, t_i, w_1(i, j)), g(\beta, h, t_i, w_2(i, j)))$;
    $w_2(i+1, j) \leftarrow \max(g(\beta, h, t_i, w_1(i, j)), g(\beta, h, t_i, w_2(i, j)))$;
  end
end
Output: w_1 and w_2

ALGORITHM 1: Fuzzy fractional Euler method for a linear problem.

In general, the computation of (34) is not an easy task. By using the concept of α-cuts, (34) can be calculated as follows:

$$w^{\alpha}_{i+1,1} = \min\left\{\left(u + \frac{h^{\beta}}{\Gamma(\beta+1)} f(t_i, u)\right) \mid u \in \left[w^{\alpha}_{i,1}, w^{\alpha}_{i,2}\right]\right\},$$
$$w^{\alpha}_{i+1,2} = \max\left\{\left(u + \frac{h^{\beta}}{\Gamma(\beta+1)} f(t_i, u)\right) \mid u \in \left[w^{\alpha}_{i,1}, w^{\alpha}_{i,2}\right]\right\}. \quad (35)$$

The optimisation problems in (35) are performed as follows.

(a) If $g(\beta, h, t_i, u)$ is increasing or decreasing on the interval $[w^{\alpha}_{i,1}, w^{\alpha}_{i,2}]$, then the optimal solutions are obtained at the endpoints of that interval (see Algorithm 1).

(b) If $g(\beta, h, t_i, u)$ is nonmonotone, we first split the interval $[w^{\alpha}_{i,1}, w^{\alpha}_{i,2}]$ into several subintervals and then solve the optimisation problems on the subintervals. By taking the minimum and maximum of all the results, we obtain the optimal solutions on the interval $[w^{\alpha}_{i,1}, w^{\alpha}_{i,2}]$. These procedures are given in Algorithm 2.

4. Numerical Examples

In this section, we present three examples for solving fuzzy fractional differential equations.

Example 1. Consider the following linear fuzzy fractional differential equation:

$$^{c}D^{\beta}_{0} X(t) = X(t),$$
$$X(0) = X_0, \quad (36)$$

where $\beta \in (0, 1]$, $t > 0$, and X_0 is any triangular fuzzy number.

This problem is a generalisation of the following fractional differential equation:

$$^{c}D^{\beta}_{0} x(t) = x(t),$$
$$x(t) = x_0, \quad (37)$$

where $\beta \in (0, 1]$, $t > 0$, and x_0 is a real number.

In order to find the solution of (36), we first find the solution of (37). By taking the Laplace transform on both sides of (37), we have

$$\mathscr{L}\left[{}^{c}D^{\beta}_{0} x(t)\right] = \mathscr{L}[x(t)]. \quad (38)$$

We then obtain

$$s^{\beta}\mathscr{L}\{x(t)\} - x(t_0)s^{\beta-1} = \mathscr{L}\{x(t)\}. \quad (39)$$

Simplifying (39), we get

$$\mathscr{L}\{x(t)\} = \frac{x_0 s^{\beta-1}}{s^{\beta} - 1}. \quad (40)$$

By taking the inverse Laplace transform to (40), we obtain

$$x(t) = x_0 \mathscr{L}^{-1}\left\{\frac{s^{\beta-1}}{s^{\beta} - 1}\right\}, \quad (41)$$

which finally has the following solution:

$$x(t) = x_0 E_{\beta}(t^{\beta}), \quad (42)$$

where $E_{\beta}(*)$ is the Mittag-Leffler function defined as

$$E_{\beta}(z) = \sum_{k=0}^{\infty} \frac{z^k}{\Gamma(\beta k + 1)}, \quad \beta > 0. \quad (43)$$

By using Zadeh's extension principle to (42) in relation to x_0, we obtain

$$X(t) = X_0 E_{\beta}(t^{\beta}), \quad (44)$$

which is the solution of (36).

```
Input: Fractional Euler Equation, Fuzzy Initial Value, β, t_0, t_N and N
Descritise α ← {α_n, α_{n-1}, ..., α_0};
h ← (t_N − t_0)/N;
for j = #(α) do
    x0_1(1, j) ← a + α(j) * (b − a);
    x0_2(1, j) ← c − α(j) * (c − b);
end
w_1(1, :) ← fliplr(x0_1(1, :));
w_2(1, :) ← fliplr(x0_2(1, :));
for i = 1 : N do
  for j = 1 : #(α) do
    w_1(i + 1, j) ← min {g(β, h, t_i, u)|u ∈ [w_1(i, 1), w_2(i, 1)]};
    w_2(i + 1, j) ← max {g(β, h, t_i, u)|u ∈ [w_1(i, 1), w_2(i, 1)]};
  end
  for i = 1 : N do
    for j = 2 : #(α) do
      p_1(i, j) ← min {g(β, h, t_i, u)|u ∈ [w_1(i, j), w_2(i, j − 1)]};
      q_1(i, j) ← w_1(i, j − 1);
      r_1(i, j) ← min {g(β, h, t_i, u)|u ∈ [w_2(i, j − 1), w_2(i, j)]};
      w_1(i + 1, j) ← min([p_1(i, j), q_1(i, j), r_1(i, j)]);
      p_2(i, j) ← max {g(β, h, t_i, u)|u ∈ [w_1(i, j), w_1(i, j − 1)]};
      q_2(i, j) ← w_2(i, j − 1);
      r_2(i, j) ← max {g(β, h, t_i, u)|u ∈ [w_2(i, j − 1), w_2(i, j)]};
      w_2(i + 1, j) ← max([p_2(i, j), q_2(i, j), r_2(i, j)]);
    end
  end
end
Output: w_1 and w_2
```

Algorithm 2: Fuzzy fractional Euler method for a nonlinear problem.

Let $[X(t)]^{\alpha} = [x_1^{\alpha}(t), x_2^{\alpha}(t)]$ and $X_0 = (1, 2, 3)$, where $[X_0]^{\alpha} = [1 + \alpha, 3 - \alpha]$ for $\alpha \in [0, 1]$; then the first ten terms of (44) can be expressed as follows:

$$\begin{aligned} x_1^{\alpha}(t) = (1+\alpha)\Bigg(1 &+ \frac{t^{\beta}}{\Gamma(\beta+1)} + \frac{t^{2\beta}}{\Gamma(2\beta+1)} + \frac{t^{3\beta}}{\Gamma(3\beta+1)} + \frac{t^{4\beta}}{\Gamma(4\beta+1)} \\ &+ \frac{t^{5\beta}}{\Gamma(5\beta+1)} + \frac{t^{6\beta}}{\Gamma(6\beta+1)} + \frac{t^{7\beta}}{\Gamma(7\beta+1)} + \frac{t^{8\beta}}{\Gamma(8\beta+1)} + \frac{t^{9\beta}}{\Gamma(9\beta+1)}\Bigg), \\ x_2^{\alpha}(t) = (3-\alpha)\Bigg(1 &+ \frac{t^{\beta}}{\Gamma(\beta+1)} + \frac{t^{2\beta}}{\Gamma(2\beta+1)} + \frac{t^{3\beta}}{\Gamma(3\beta+1)} + \frac{t^{4\beta}}{\Gamma(4\beta+1)} \\ &+ \frac{t^{5\beta}}{\Gamma(5\beta+1)} + \frac{t^{6\beta}}{\Gamma(6\beta+1)} + \frac{t^{7\beta}}{\Gamma(7\beta+1)} + \frac{t^{8\beta}}{\Gamma(8\beta+1)} + \frac{t^{9\beta}}{\Gamma(9\beta+1)}\Bigg). \end{aligned} \tag{45}$$

Clearly, (45) are the valid α-cuts of the solution of (36). For numerical approximation, we set $t \in [0, 2]$ and $N = 100$. By using Algorithm 1, the results for different values of β are plotted in Figure 1. From the graphs, we can see that if β approaches 1, the approximate solutions will approach the approximate solution of fuzzy differential equation. Numerical values at $t = 2$ for different values of β are listed in Table 1.

Example 2. Consider the following linear fuzzy fractional differential equation:

$$\begin{aligned} {}^{c}D_0^{\beta} X(t) &= -X(t), \\ X(0) &= X_0. \end{aligned} \tag{46}$$

The nonfuzzy problem associated with (46) is

$$\begin{aligned} {}^{c}D_0^{\beta} x(t) &= -x(t), \\ x(0) &= x_0. \end{aligned} \tag{47}$$

TABLE 1: Numerical solutions of Example 1 with different values of β.

α	$\beta = 0.6$		$\beta = 0.8$		$\beta = 1$	
	$x_1^\beta(t)$	$x_2^\beta(t)$	$x_1^\beta(t)$	$x_2^\beta(t)$	$x_1^\beta(t)$	$x_2^\beta(t)$
0.0	26063.941926	78191.825779	98.369851	295.109553	7.244646	21.733938
0.1	28670.336119	75585.431586	108.206836	285.272568	7.969110	21.009473
0.2	31276.730311	72979.037394	118.043821	275.435583	8.693575	20.285009
0.3	33883.124504	70372.643201	127.880806	265.598598	9.418039	19.560544
0.4	36489.518697	67766.249008	137.717791	255.761613	10.142504	18.836079
0.5	39095.912889	65159.854816	147.554776	245.924627	10.866969	18.111615
0.6	41702.307082	62553.460623	157.391761	236.087642	11.591433	17.387150
0.7	44308.701274	59947.066430	167.228746	226.250657	12.315898	16.662686
0.8	46915.095467	57340.672238	177.065732	216.413672	13.040363	15.938221
0.9	49521.489660	54734.278045	186.902717	206.576687	13.764827	15.213756
1.0	52127.883852	52127.883852	196.739702	196.739702	14.489292	14.489292

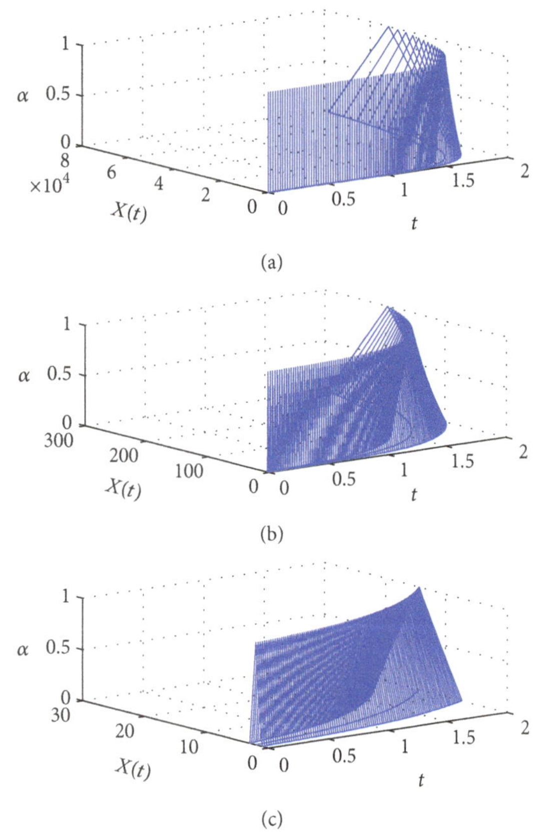

FIGURE 1: The numerical solution of (36) for (a) $\beta = 0.6$, (b) $\beta = 0.8$, and (c) $\beta = 1$.

In order to find the solution of (46), we first find the solution of (47). By taking the Laplace transform on both sides of (47), we have

$$\mathscr{L}\left\{{}^{c}D_0^\beta x(t)\right\} = -\mathscr{L}\{x(t)\}. \quad (48)$$

It follows that

$$s^\beta \mathscr{L}\{x(t)\} - x(t_0)s^{\beta-1} = -\mathscr{L}\{x(t)\}. \quad (49)$$

After simplifying, we get

$$\mathscr{L}\{x(t)\} = \frac{x_0 s^{\beta-1}}{s^\beta + 1}. \quad (50)$$

By taking the inverse Laplace transform to (50), we obtain

$$x(t) = x_0 \mathscr{L}^{-1}\left\{\frac{s^{\beta-1}}{s^\beta + 1}\right\}, \quad (51)$$

which finally has the following solution:

$$x(t) = x_0 E_\beta\left(-t^\beta\right), \quad (52)$$

where $E_\beta(*)$ is the Mittag-Leffler function. Using Zadeh's extension principle to (52) in relation to x_0, we obtain the solution of (46) as follows:

$$X(t) = X_0 E_\beta\left(-t^\beta\right). \quad (53)$$

Let $[X(t)]^\alpha = [x_1^\alpha(t), x_2^\alpha(t)]$ and $X_0 = (2, 3, 4)$ where $[X_0]^\alpha = [\alpha + 2, 4 - \alpha]$ for $\alpha \in [0, 1]$; then the first ten terms of (52) can be expressed as follows:

$$x_1^\alpha(t) = (\alpha + 2)\left(-1 - \frac{t^\beta}{\Gamma(\beta+1)} - \frac{t^{2\beta}}{\Gamma(2\beta+1)} - \frac{t^{3\beta}}{\Gamma(3\beta+1)} - \frac{t^{4\beta}}{\Gamma(4\beta+1)} - \frac{t^{5\beta}}{\Gamma(5\beta+1)} - \frac{t^{6\beta}}{\Gamma(6\beta+1)} - \frac{t^{7\beta}}{\Gamma(7\beta+1)} - \frac{t^{8\beta}}{\Gamma(8\beta+1)} - \frac{t^{9\beta}}{\Gamma(9\beta+1)}\right),$$

TABLE 2: Numerical solutions of Example 2 with different values of β.

α	$\beta = 0.6$ $x_1^\beta(t)$	$\beta = 0.6$ $x_2^\beta(t)$	$\beta = 0.8$ $x_1^\beta(t)$	$\beta = 0.8$ $x_2^\beta(t)$	$\beta = 1$ $x_1^\beta(t)$	$\beta = 1$ $x_2^\beta(t)$
0.0	0.000024	0.000048	0.016304	0.032608	0.265239	0.530478
0.1	0.000025	0.000047	0.017119	0.031793	0.278501	0.517216
0.2	0.000026	0.000046	0.017934	0.030978	0.291763	0.503954
0.3	0.000027	0.000044	0.018750	0.030163	0.305024	0.490692
0.4	0.000029	0.000043	0.019565	0.029347	0.318286	0.477430
0.5	0.000030	0.000042	0.020380	0.028532	0.331548	0.464168
0.6	0.000031	0.000041	0.021195	0.027717	0.344810	0.450906
0.7	0.000032	0.000040	0.022010	0.026902	0.358072	0.437644
0.8	0.000033	0.000038	0.022826	0.026087	0.371334	0.424382
0.9	0.000035	0.000037	0.023641	0.025271	0.384596	0.411120
1.0	0.000036	0.000036	0.024456	0.024456	0.397858	0.397858

$$\begin{aligned} x_2^\alpha(t) = (4-\alpha)\Bigg(&-1 - \frac{t^\beta}{\Gamma(\beta+1)} - \frac{t^{2\beta}}{\Gamma(2\beta+1)} \\ &- \frac{t^{3\beta}}{\Gamma(3\beta+1)} - \frac{t^{4\beta}}{\Gamma(4\beta+1)} - \frac{t^{5\beta}}{\Gamma(5\beta+1)} \\ &- \frac{t^{6\beta}}{\Gamma(6\beta+1)} - \frac{t^{7\beta}}{\Gamma(7\beta+1)} \\ &- \frac{t^{8\beta}}{\Gamma(8\beta+1)} - \frac{t^{9\beta}}{\Gamma(9\beta+1)}\Bigg). \end{aligned} \tag{54}$$

Clearly, (54) are the valid α-cuts of the solution of (46). By using Algorithm 1 with the same interval t and interval N as in Example 1, the numerical solutions of (46) for different values of β are plotted in Figure 2. Again, we can see that the numerical solutions will approach the numerical solution of fuzzy differential equation as β increases to 1. Numerical solutions at $t = 2$ for different values of β are listed in Table 2.

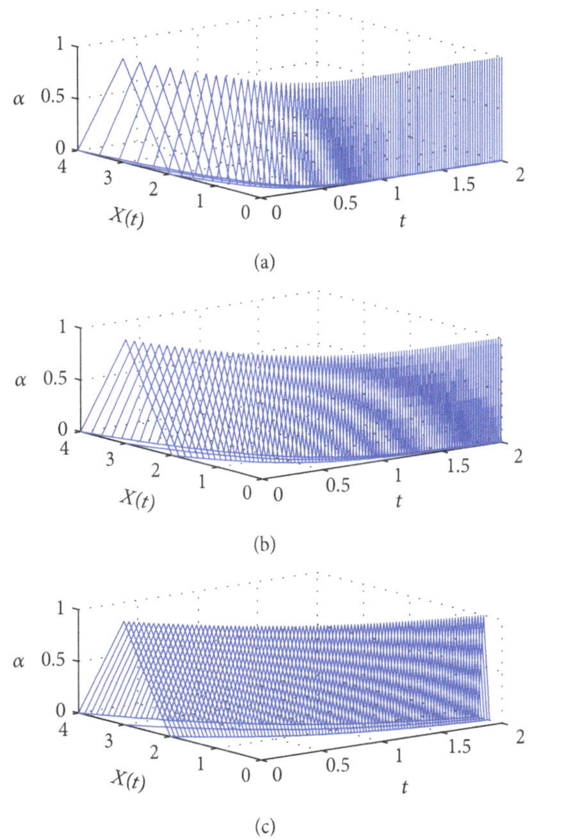

FIGURE 2: The numerical solutions of (46) for (a) $\beta = 0.6$, (b) $\beta = 0.8$, and (c) $\beta = 1$.

In the following example, we provide detailed procedures for solving a nonlinear fuzzy fractional differential equation.

Example 3. Let us consider the following problem:

$$\begin{aligned} {}^cD_0^\beta X(t) &= \cos(tX), \\ X(0) &= X_0, \end{aligned} \tag{55}$$

where $\beta \in (0,1]$, $t = [0,5]$, and X_0 is any triangular fuzzy number.

To solve this problem, we use Algorithm 2. First, let $[X_0]^\alpha = [x_{0,1}^\alpha, x_{0,2}^\alpha]$. We discretise α up to 11 points, which are $\alpha_0 = 0 < \alpha_1 = 0.1 < \cdots < \alpha_{10} = 1$. Let $X_0 = W_0$; then we have

$$\begin{aligned} w_{1,1}^{\alpha_{10}} &= g\left(\beta, h, t_0, w_{0,1}^{\alpha_{10}}\right) = g\left(\beta, h, t_0, w_{0,2}^{\alpha_{10}}\right) = w_{1,2}^{\alpha_{10}}, \\ w_{2,1}^{\alpha_9} &= \min\left[\min_{u\in[w_{1,1}^{\alpha_9}, w_{1,1}^{\alpha_{10}}]} g(\beta,h,t_1,u), w_{1,1}^{\alpha_{10}}, \min_{u\in[w_{1,2}^{\alpha_{10}}, w_{1,2}^{\alpha_9}]} g(\beta,h,t_1,u)\right], \\ w_{2,2}^{\alpha_9} &= \max\left[\max_{u\in[w_{1,1}^{\alpha_9}, w_{1,1}^{\alpha_{10}}]} g(\beta,h,t_1,u), w_{1,2}^{\alpha_{10}}, \max_{u\in[w_{1,2}^{\alpha_{10}}, w_{1,2}^{\alpha_9}]} g(\beta,h,t_1,u)\right], \\ &\vdots \\ w_{N+1,1}^{\alpha_0} &= \min\left[\min_{u\in[w_{N,1}^{\alpha_0}, w_{N,1}^{\alpha_1}]} g(\beta,h,t_N,u), w_{N,1}^{\alpha_1}, \min_{u\in[w_{N,2}^{\alpha_1}, w_{N,2}^{\alpha_0}]} g(\beta,h,t_N,u)\right], \\ w_{N+1,2}^{\alpha_0} &= \max\left[\max_{u\in[w_{N,1}^{\alpha_0}, w_{N,1}^{\alpha_1}]} g(\beta,h,t_N,u), w_{N,2}^{\alpha_1}, \max_{u\in[w_{N,2}^{\alpha_1}, w_{N,2}^{\alpha_0}]} g(\beta,h,t_N,u)\right], \end{aligned} \tag{56}$$

TABLE 3: Numerical solutions of Example 3 with different values of β.

α	$\beta = 0.6$		$\beta = 0.8$		$\beta = 1$	
	$x_1^{\beta}(t)$	$x_2^{\beta}(t)$	$x_1^{\beta}(t)$	$x_2^{\beta}(t)$	$x_1^{\beta}(t)$	$x_2^{\beta}(t)$
0.0	0.316778	0.316778	0.320052	0.320052	0.328592	1.645658
0.1	0.316778	0.316778	0.320052	0.320052	0.328592	1.645476
0.2	0.316778	0.316778	0.320052	0.320052	0.328593	1.643608
0.3	0.316778	0.316778	0.320052	0.320052	0.328594	0.328664
0.4	0.316778	0.316778	0.320052	0.320052	0.328594	0.328629
0.5	0.316778	0.316778	0.320052	0.320052	0.328595	0.328617
0.6	0.316778	0.316778	0.320052	0.320052	0.328596	0.328610
0.7	0.316778	0.316778	0.320052	0.320052	0.328597	0.328606
0.8	0.316778	0.316778	0.320052	0.320052	0.328598	0.328604
0.9	0.316778	0.316778	0.320052	0.320052	0.328599	0.328602
1.0	0.316778	0.316778	0.320052	0.320052	0.328600	0.328600

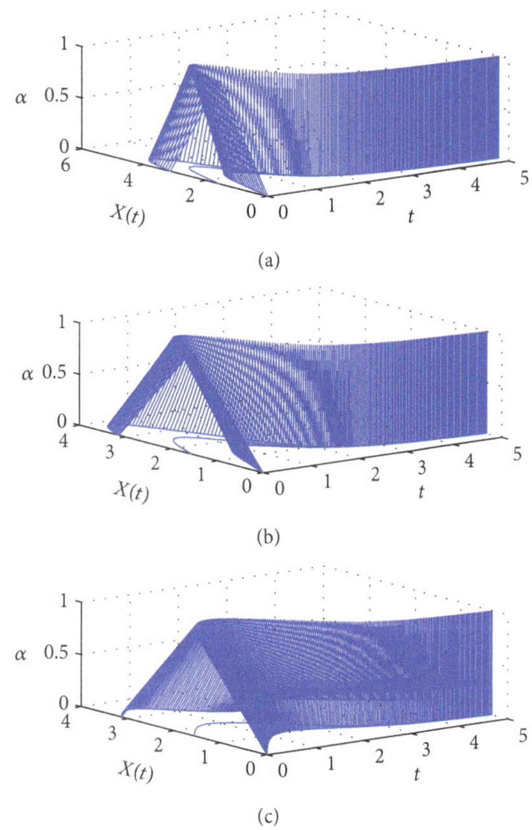

FIGURE 3: The approximate solutions of (55) for (a) $\beta = 0.6$, (b) $\beta = 0.8$, and (c) $\beta = 1$.

where

$$g(\beta, h, t_i, u) = u + \frac{h^{\beta}}{\Gamma(\beta+1)} \cos(t_i u), \quad u \in \left[w_{i,1}^{\alpha_j}, w_{i,2}^{\alpha_j}\right] \tag{57}$$

for $i = 0, 1, \ldots, N$ and $j = 0, 1, \ldots, 10$.

Let $X_0 = (0, \pi/2, \pi)$ and $N = 200$; then these procedures will result in the approximate solutions of (55) at different values of β as plotted in Figure 3. From the graphs, we can see that the numerical solutions approach to the numerical solution of fuzzy differential equation as β approaches 1. Numerical solutions at $t = 5$ at different values of β are listed in Table 3.

5. Conclusions

In this paper, we have studied a fuzzy fractional differential equation and presented its solution using Zadeh's extension principle. The classical fractional Euler method has also been extended in the fuzzy setting in order to approximate the solutions of linear and nonlinear fuzzy fractional differential equations. Final results showed that the solution of fuzzy fractional differential equations approaches the solution of fuzzy differential equations as the fractional order approaches the integer order.

Acknowledgments

This research was supported by the Short Term Grant (STG) of Universiti Malaysia Perlis (UniMAP) under the Project code 9001-00319 and partially supported by Research Acculturation Grant Scheme (RAGS) under the Project code 9018-00003.

References

[1] G. Jumarie, "Table of some basic fractional calculus formulae derived from a modified Riemann-Liouville derivative for non-differentiable functions," *Applied Mathematics Letters*, vol. 22, no. 3, pp. 378–385, 2009.

[2] J. J. Trujillo, M. Rivero, and B. Bonilla, "On a Riemann-Liouville generalized Taylor's formula," *Journal of Mathematical Analysis and Applications*, vol. 231, no. 1, pp. 255–265, 1999.

[3] G. A. Anastassiou, "Opial type inequalities involving Riemann-Liouville fractional derivatives of two functions with applications," *Mathematical and Computer Modelling*, vol. 48, no. 3-4, pp. 344–374, 2008.

[4] Y. S. Liang, "Box dimensions of Riemann-Liouville fractional integrals of continuous functions of bounded variation," *Nonlinear Analysis: Theory, Methods & Applications*, vol. 72, no. 11, pp. 4304–4306, 2010.

[5] I. Podlubny, "Riesz potential and Riemann-Liouville fractional integrals and derivatives of Jacobi polynomials," *Applied Mathematics Letters*, vol. 10, no. 1, pp. 103–108, 1997.

[6] R. Almeida and D. F. M. Torres, "Calculus of variations with fractional derivatives and fractional integrals," *Applied Mathematics Letters*, vol. 22, no. 12, pp. 1816–1820, 2009.

[7] Z. Wei, W. Dong, and J. Che, "Periodic boundary value problems for fractional differential equations involving a Riemann-Liouville fractional derivative," *Nonlinear Analysis: Theory, Methods & Applications*, vol. 73, no. 10, pp. 3232–3238, 2010.

[8] Z. Wei, Q. Li, and J. Che, "Initial value problems for fractional differential equations involving Riemann-Liouville sequential fractional derivative," *Journal of Mathematical Analysis and Applications*, vol. 367, no. 1, pp. 260–272, 2010.

[9] D. Qian, C. Li, R. P. Agarwal, and P. J. Y. Wong, "Stability analysis of fractional differential system with Riemann-Liouville derivative," *Mathematical and Computer Modelling*, vol. 52, no. 5-6, pp. 862–874, 2010.

[10] S. G. Samko, A. A. Kilbas, and O. I. Marichev, *Fractional Integrals and Derivatives: Theory and Applications*, Gordon and Breach, Yverdon, Switzerland, 1993.

[11] I. Podlubny, *Fractional Differential Equation*, Academic Press, 1999.

[12] V. Daftardar-Gejji and H. Jafari, "Analysis of a system of nonautonomous fractional differential equations involving Caputo derivatives," *Journal of Mathematical Analysis and Applications*, vol. 328, no. 2, pp. 1026–1033, 2007.

[13] D. A. Murio, "On the stable numerical evaluation of Caputo fractional derivatives," *Computers and Mathematics with Applications*, vol. 51, no. 9-10, pp. 1539–1550, 2006.

[14] Z. M. Odibat, "Computing eigenelements of boundary value problems with fractional derivatives," *Applied Mathematics and Computation*, vol. 215, no. 8, pp. 3017–3028, 2009.

[15] V. S. Ertürk and S. Momani, "Solving systems of fractional differential equations using differential transform method," *Journal of Computational and Applied Mathematics*, vol. 215, no. 1, pp. 142–151, 2008.

[16] V. S. Ertürk, S. Momani, and Z. Odibat, "Application of generalized differential transform method to multi-order fractional differential equations," *Communications in Nonlinear Science and Numerical Simulation*, vol. 13, no. 8, pp. 1642–1654, 2008.

[17] K. S. Miller and B. Ross, *An Introduction to the Fractional Calculus and Differential Equations*, John Wiley & Sons, New York, NY, USA, 1993.

[18] K. Diethelm and N. J. Ford, "Analysis of fractional differential equations," *Journal of Mathematical Analysis and Applications*, vol. 265, no. 2, pp. 229–248, 2002.

[19] V. Lakshmikantham and R. N. Mohapatra, *Theory of Fuzzy Differential Equations and Applications*, Taylor & Francis, London, UK, 2003.

[20] A. A. Kilbas, H. M. Srivastava, and J. J. Trujillo, *Theory and Applications of Fractional Differential Equations*, Elsevier Science B.V, Amsterdam, The Netherlands, 2006.

[21] V. Lakshmikantham and A. S. Vatsala, "Basic theory of fractional differential equations," *Nonlinear Analysis: Theory, Methods & Applications*, vol. 69, no. 8, pp. 2677–2682, 2008.

[22] V. Lakshmikantham and S. Leela, "Nagumo-type uniqueness result for fractional differential equations," *Nonlinear Analysis: Theory, Methods & Applications*, vol. 71, no. 7-8, pp. 2886–2889, 2009.

[23] S. Zhang, "Monotone iterative method for initial value problem involving Riemann-Liouville fractional derivatives," *Nonlinear Analysis: Theory, Methods & Applications*, vol. 71, no. 5-6, pp. 2087–2093, 2009.

[24] H. Jafari, H. Tajadodi, and S. A. Hosseini Matikolai, "Homotopy perturbation pade technique for solving fractional Riccati differential equations," *The International Journal of Nonlinear Sciences and Numerical Simulation*, vol. 11, pp. 271–276, 2011.

[25] H. Jafari, H. Tajadodi, H. Nazari, and C. M. Khalique, "Numerical solution of non-linear Riccati differential equations with fractional order," *The International Journal of Nonlinear Sciences and Numerical Simulation*, vol. 11, pp. 179–182, 2011.

[26] F. B. Adda and J. Cresson, "Fractional differential equations and the Schrödinger equation," *Applied Mathematics and Computation*, vol. 161, no. 1, pp. 323–345, 2005.

[27] Z. E. A. Fellah, C. Depollier, and M. Fellah, "Application of fractional calculus to the sound waves propagation in rigid porous materials: validation via ultrasonic measurements," *Acta Acustica united with Acustica*, vol. 88, no. 1, pp. 34–39, 2002.

[28] M. Rahimy, "Applications of fractional differential equations," *Applied Mathematical Sciences*, vol. 4, no. 49–52, pp. 2453–2461, 2010.

[29] M. Z. Ahmad and B. De Baets, "A predator-prey model with fuzzy initial populations," in *Proceedings of the 13th IFSA World Congress and 6th European Society of Fuzzy Logic and Technology Conference*, pp. 1311–1314, 2009.

[30] L. A. Zadeh, "Fuzzy sets," *Information and Control*, vol. 8, no. 3, pp. 338–353, 1965.

[31] R. P. Agarwal, V. Lakshmikantham, and J. J. Nieto, "On the concept of solution for fractional differential equations with uncertainty," *Nonlinear Analysis: Theory, Methods & Applications*, vol. 72, no. 6, pp. 2859–2862, 2010.

[32] S. Arshad and V. Lupulescu, "On the fractional differential equations with uncertainty," *Nonlinear Analysis: Theory, Methods & Applications*, vol. 74, no. 11, pp. 3685–3693, 2011.

[33] T. Allahviranloo, S. Salahshour, and S. Abbasbandy, "Explicit solutions of fractional differential equations with uncertainty," *Soft Computing*, vol. 16, no. 2, pp. 297–302, 2012.

[34] S. Salahshour, T. Allahviranloo, and S. Abbasbandy, "Solving fuzzy fractional differential equations by fuzzy Laplace transforms," *Communications in Nonlinear Science and Numerical Simulation*, vol. 17, no. 3, pp. 1372–1381, 2012.

[35] T. Allahviranloo and M. B. Ahmadi, "Fuzzy Laplace transforms," *Soft Computing*, vol. 14, no. 3, pp. 235–243, 2010.

[36] O. Kaleva, "Fuzzy differential equations," *Fuzzy Sets and Systems*, vol. 24, no. 3, pp. 301–317, 1987.

[37] S. Seikkala, "On the fuzzy initial value problem," *Fuzzy Sets and Systems*, vol. 24, no. 3, pp. 319–330, 1987.

[38] O. Kaleva, "A note on fuzzy differential equations," *Nonlinear Analysis: Theory, Methods & Applications*, vol. 64, no. 5, pp. 895–900, 2006.

[39] B. Bede, I. J. Rudas, and A. L. Bencsik, "First order linear fuzzy differential equations under generalized differentiability," *Information Sciences*, vol. 177, no. 7, pp. 1648–1662, 2007.

[40] B. Bede and S. G. Gal, "Generalizations of the differentiability of fuzzy-number-valued functions with applications to fuzzy differential equations," *Fuzzy Sets and Systems*, vol. 151, no. 3, pp. 581–599, 2005.

[41] M. Z. Ahmad and M. K. Hasan, "Numerical methods for fuzzy initial value problems under different types of interpretation: a comparison study," in *Informatics Engineering and Information*

Science, vol. 252 of *Communications in Computer and Information Science*, pp. 275–288, Springer, Berlin, Germany, 2011.

[42] J. J. Buckley and T. Feuring, "Fuzzy differential equations," *Fuzzy Sets and Systems*, vol. 110, no. 1, pp. 43–54, 2000.

[43] M. T. Mizukoshi, L. C. Barros, Y. Chalco-Cano, H. Román-Flores, and R. C. Bassanezi, "Fuzzy differential equations and the extension principle," *Information Sciences*, vol. 177, no. 17, pp. 3627–3635, 2007.

[44] Y. Chalco-Cano and H. Román-Flores, "On new solutions of fuzzy differential equations," *Chaos, Solitons & Fractals*, vol. 38, no. 1, pp. 112–119, 2008.

[45] S. C. Palligkinis, G. Papageorgiou, and I. T. Famelis, "Runge-Kutta methods for fuzzy differential equations," *Applied Mathematics and Computation*, vol. 209, no. 1, pp. 97–105, 2009.

[46] Y. Chalco-Cano and H. Román-Flores, "Comparation between some approaches to solve fuzzy differential equations," *Fuzzy Sets and Systems*, vol. 160, no. 11, pp. 1517–1527, 2009.

[47] M. Z. Ahmad and M. K. Hasan, "A new approach to incorporate uncertainty into Euler's method," *Applied Mathematical Sciences*, vol. 4, no. 49–52, pp. 2509–2520, 2010.

[48] M. Z. Ahmad and M. K. Hasan, "A new fuzzy version of Euler's method for solving differential equations with fuzzy initial values," *Sains Malaysiana*, vol. 40, no. 6, pp. 651–657, 2011.

[49] M. Z. Ahmad, M. K. Hasan, and B. De Baets, "Analytical and numerical solutions of fuzzy differential equations," *Information Sciences*, vol. 236, pp. 156–167, 2013.

[50] R. Gorenflo and F. Mainardi, "Fractional calculus: integral and differential equations of fractional order," in *Fractals and Fractional Calculus in Continuum Mechanics*, pp. 223–276, Springer, New York, NY, USA, 1997.

[51] Y. Li, Y. Chen, and I. Podlubny, "Mittag-Leffler stability of fractional order nonlinear dynamic systems," *Automatica*, vol. 45, no. 8, pp. 1965–1969, 2009.

[52] L. X. Wang, *A Course in Fuzzy Systems and Control*, Prentice-Hall, Englewood Cliffs, NJ, USA, 1997.

[53] M. Z. Ahmad, M. K. Hasan, and B. De Baets, "A new method for computing continuous functions with fuzzy variable," *Journal of Applied Sciences*, vol. 11, no. 7, pp. 1143–1149, 2011.

[54] M. Z. Ahmad and M. K. Hasan, "Incorporating optimisation technique into Zadeh's extension principle for computing non-monotone functions with fuzzy variable," *Sains Malaysiana*, vol. 40, no. 6, pp. 643–650, 2011.

[55] H. Román-Flores, L. C. Barros, and R. C. Bassanezi, "A note on Zadeh's extensions," *Fuzzy Sets and Systems*, vol. 117, no. 3, pp. 327–331, 2001.

[56] R. Goetschel Jr. and W. Voxman, "Elementary fuzzy calculus," *Fuzzy Sets and Systems*, vol. 18, no. 1, pp. 31–43, 1986.

[57] M. L. Puri and D. A. Ralescu, "Fuzzy random variables," *Journal of Mathematical Analysis and Applications*, vol. 114, no. 2, pp. 409–422, 1986.

[58] R. P. Agarwal, D. O'Regan, and V. Lakshmikantham, "Viability theory and fuzzy differential equations," *Fuzzy Sets and Systems*, vol. 151, no. 3, pp. 563–580, 2005.

[59] Z. M. Odibat and S. Momani, "An algorithm for the numerical solution of differential equations of fractional order," *Journal of Applied Mathematics & Informatics*, vol. 26, pp. 15–27, 2008.

Exp-Function Method for Solving Fractional Partial Differential Equations

Bin Zheng

School of Science, Shandong University of Technology, Zibo, Shandong 255049, China

Correspondence should be addressed to Bin Zheng; zhengbin2601@126.com

Academic Editors: J. Fernández and J. Liu

We extend the Exp-function method to fractional partial differential equations in the sense of modified Riemann-Liouville derivative based on nonlinear fractional complex transformation. For illustrating the validity of this method, we apply it to the space-time fractional Fokas equation and the nonlinear fractional Sharma-Tasso-Olver (STO) equation. As a result, some new exact solutions for them are successfully established.

1. Introduction

Nonlinear differential equations of integer order (NLDEs) can be used to describe many nonlinear phenomena such as fluid mechanics, plasma physics, optical fibers, biology, solid state physics, chemical kinematics, and chemical physics. In the research of the theory of NLDEs, searching for more explicit exact solutions to NLDEs is one of the most fundamental and significant studies in recent years. With the help of computerized symbolic computation, much work has focused on the various extensions and applications of the known algebraic methods to construct the solutions to NLDEs. There have been a variety of powerful methods. For example, these methods include the generalized Riccati subequation method [1, 2], the Jacobi elliptic function expansion [3, 4], the F-expansion method [5], the Exp-function method [6–9], and the (G'/G)-expansion method [10, 11].

Fractional differential equations are generalizations of classical differential equations of integer order and have recently proved to be valuable tools to the modeling of many physical phenomena and have been the focus of many studies due to their frequent appearance in various applications in physics, biology, engineering, signal processing, systems identification, control theory, finance, and fractional dynamics. In order to obtain exact solutions for fractional differential equations, many powerful and efficient methods have been proposed so far (e.g., see [12–24]). But nobody has researched the application of the Exp-function method to fractional differential equations so far to our best knowledge.

In this paper, we will apply the Exp-function method for solving fractional partial differential equations in the sense of modified Riemann-Liouville derivative by Jumarie [25]. By a certain nonlinear fractional complex transformation $\xi = \xi(t, x_1, x_2, \dots, x_n)$, a given fractional partial differential equation expressed in independent variables $t, x_1, x_2, \dots, x_n$ can be turned into another ordinary differential equation of integer order, whose solutions are supposed to have one of the following forms:

$$u(\xi) = \frac{\sum_{p=-m_1}^{m_2} a_p e^{p\xi}}{\sum_{q=-n_1}^{n_2} b_q e^{q\xi}} \quad \text{or} \quad u(\xi) = \frac{\sum_{p=-m_1}^{m_2} a_p e^{ip\xi}}{\sum_{q=-n_1}^{n_2} b_q e^{iq\xi}}, \tag{1}$$

where m_1, m_2, n_1, and n_2 are unknown positive integers that will be determined by using the balance method. Determine the highest order nonlinear term and the linear term of highest order in the ordinary differential equation and express them in terms of (3). Then, in the resulting terms, balance the highest order Exp-function to determine m_2 and n_2 and balance the lowest order Exp-function to determine m_1 and n_1.

The Jumarie's modified Riemann-Liouville derivative of order α is defined by the following expression [25]:

$$D_t^\alpha f(t) = \begin{cases} \dfrac{1}{\Gamma(1-\alpha)} \dfrac{d}{dt} \displaystyle\int_0^t (t-\xi)^{-\alpha} (f(\xi) - f(0))\, d\xi, & 0 < \alpha < 1, \\ \left(f^{(n)}(t)\right)^{(\alpha-n)}, & n \le \alpha < n+1,\ n \ge 1. \end{cases} \quad (2)$$

We list some important properties for the modified Riemann-Liouville derivative as follows [25, (3.10)–(3.13)] (see also in [22–24, 26]):

$$D_t^\alpha t^r = \frac{\Gamma(1+r)}{\Gamma(1+r-\alpha)} t^{r-\alpha}, \quad (3)$$

$$D_t^\alpha (f(t) g(t)) = g(t) D_t^\alpha f(t) + f(t) D_t^\alpha g(t), \quad (4)$$

$$D_t^\alpha f[g(t)] = f_g'[g(t)] D_t^\alpha g(t) = D_g^\alpha f[g(t)] \left(g'(t)\right)^\alpha. \quad (5)$$

In the following, we will apply the Exp-function method to find exact solutions for the space-time fractional Fokas equation and the nonlinear fractional Sharma-Tasso-Olver (STO) equation.

2. Applications of the Exp-Function Method to Fractional Partial Differential Equations

2.1. Space-Time Fractional Fokas Equation. We consider the space-time fractional Fokas equation [22]

$$4\frac{\partial^{2\alpha} q}{\partial t^\alpha \partial x_1^\alpha} - \frac{\partial^{4\alpha} q}{\partial x_1^{3\alpha} \partial x_2^\alpha} + \frac{\partial^{4\alpha} q}{\partial x_2^{3\alpha} \partial x_1^\alpha} + 12 \frac{\partial^\alpha q}{\partial x_1^\alpha} \frac{\partial^\alpha q}{\partial x_2^\alpha} + 12q \frac{\partial^{2\alpha} q}{\partial x_1^\alpha \partial x_2^\alpha} - 6 \frac{\partial^{2\alpha} q}{\partial y_1^\alpha \partial y_2^\alpha} = 0, \quad 0 < \alpha \le 1. \quad (6)$$

In [22], the author solved (6) by a proposed fractional sub-equation method based on the fractional Riccati equation and established some exact solutions for it. Now we will apply the Exp-function method to this equation. Suppose $q(x, y, t) = U(\xi)$, where $\xi = (k_1 x_1^\alpha / \Gamma(1+\alpha)) + (k_2 x_2^\alpha / \Gamma(1+\alpha)) + (l_1 y_1^\alpha / \Gamma(1+\alpha)) + (l_2 y_2^\alpha / \Gamma(1+\alpha)) + (ct^\alpha / \Gamma(1+\alpha)) + \xi_0$, $k_1, k_2, l_1, l_2, c, \xi_0$ are all constants with $k_1, k_2, l_1, l_2, c \neq 0$. Then by the use of (5), (6) can be turned into

$$4ck_1 U'' - k_1^3 k_2 U^{(4)} + k_2^3 k_1 U^{(4)} + 12 k_1 k_2 U'^2 + 12 k_1 k_2 U U'' - 6 l_1 l_2 U'' = 0. \quad (7)$$

First we suppose that the solution of (7) can be expressed by

$$U(\xi) = \frac{\sum_{p=-m_1}^{m_2} a_p e^{p\xi}}{\sum_{q=-n_1}^{n_2} b_q e^{q\xi}}. \quad (8)$$

To determine m_2, n_2, we need to balance the linear term of the highest order with the highest order nonlinear term in (7). By simple calculation, we have

$$U^{(4)} = \frac{c_1 e^{(15n_2+m_2)x} + \cdots + c_2 e^{-(15n_1+m_1)x}}{c_3 e^{16n_2 x} + \cdots + c_4 e^{-16n_1 x}},$$
$$UU'' = \frac{c_5 e^{(3n_2+2m_2)x} + \cdots + c_6 e^{-(3n_1+2m_1)x}}{c_7 e^{5n_2 x} + \cdots + c_8 e^{-5n_1 x}}, \quad (9)$$

where $c_i, i = 1, 2, 3, 4$ are coefficients determined by $a_{m_2}, a_{-m_1}, b_{n_2}, b_{-n_1}$. Balancing the highest order Exp-function in $U^{(4)}$ and UU'', we have $15n_2 + m_2 = 14n_2 + 2m_2$, which implies $m_2 = n_2$. Similarly, balancing the lowest order Exp-function in $U^{(4)}$ and UU'', we have $15n_1 + m_1 = 14n_1 + 2m_1$, which implies $m_1 = n_1$. For simplicity, we will proceed in two selected cases.

Case 1. Let $m_1 = n_1 = 1, m_2 = n_2 = 1$. Then

$$U(\xi) = \frac{a_{-1} e^{-\xi} + a_0 + a_1 e^\xi}{b_{-1} e^{-\xi} + b_0 + b_1 e^\xi}. \quad (10)$$

Substituting (10) into (7), eliminating the denominator, and collecting all the terms with the same power of $e^{p\xi}$ together, equating each coefficient to zero, yield a set of algebraic equations. Solving these equations with the aid of the mathematical software Maple yields the following family of values of $a_i, b_i, i = -1, 0, 1$.

Family 1. Consider

$$a_1 = -\frac{b_1 \left(k_1 k_2^3 - k_2 k_1^3 + 4ck_1 - 6 l_1 l_2\right)}{12 k_1 k_2},$$
$$a_0 = -\frac{b_0 \left(-5 k_1 k_2^3 + 5 k_2 k_1^3 + 4ck_1 - 6 l_1 l_2\right)}{12 k_1 k_2},$$
$$a_{-1} = -\frac{b_0^2 \left(k_1 k_2^3 - k_2 k_1^3 + 4ck_1 - 6 l_1 l_2\right)}{48 b_1 k_1 k_2}, \quad b_{-1} = \frac{b_0^2}{4b_1}, \quad (11)$$

where b_1, b_0 are arbitrary constants.

Substituting the result above into (10), we can obtain the following exact solutions to (7):

$$U_1(\xi) = \Big(-\left(k_1 k_2^3 - k_2 k_1^3 + 4ck_1 - 6 l_1 l_2\right)\left(b_0^2 e^{-\xi} + 4 b_1^2 e^\xi\right) - 4 b_0 b_1 \left(-5 k_1 k_2^3 + 5 k_2 k_1^3 + 4ck_1 - 6 l_1 l_2\right)\Big) \times \left(12 k_1 k_2 \left(b_0^2 e^{-\xi} + 4 b_1^2 e^\xi\right) + 48 b_0 b_1 k_1 k_2\right)^{-1}, \quad (12)$$

where $\xi = (k_1 x_1^\alpha / \Gamma(1+\alpha)) + (k_2 x_2^\alpha / \Gamma(1+\alpha)) + (l_1 y_1^\alpha / \Gamma(1+\alpha)) + (l_2 y_2^\alpha / \Gamma(1+\alpha)) + (ct^\alpha / \Gamma(1+\alpha)) + \xi_0$. If we especially take

$b_0 = 2b_1$, then we obtain the following hyperbolic function solitary wave solution:

$$U_2(\xi) = -\frac{\left(k_1k_2^3 - k_2k_1^3 + 4ck_1 - 6l_1l_2\right)}{12k_1k_2} + \frac{\left(k_2^2 - k_1^2\right)}{4}\operatorname{sech}^2\frac{\xi}{2}. \tag{13}$$

Case 2. Let $m_1 = n_1 = m$, $m_2 = n_2 = m$, and for simplicity, we take

$$U(\xi) = \frac{a_{-m}e^{-m\xi} + a_0 + a_m e^{m\xi}}{b_{-m}e^{-m\xi} + b_0 + b_m e^{m\xi}}. \tag{14}$$

Substituting (14) into (7), eliminating the denominator, and collecting all the terms with the same power of $e^{pn\xi}$ together, equating each coefficient to zero, yield a set of algebraic equations. Solving these equations yields the following result.

Family 2. Consider

$$\begin{aligned} a_m &= -\frac{b_m\left(m^2k_1k_2^3 - m^2k_2k_1^3 + 4ck_1 - 6l_1l_2\right)}{12k_1k_2}, \\ a_0 &= -\frac{b_0\left(-5m^2k_1k_2^3 + 5m^2k_2k_1^3 + 4ck_1 - 6l_1l_2\right)}{12k_1k_2}, \\ a_{-m} &= -\frac{b_0^2\left(m^2k_1k_2^3 - m^2k_2k_1^3 + 4ck_1 - 6l_1l_2\right)}{48b_mk_1k_2}, \\ b_{-m} &= \frac{b_0^2}{4b_m}, \end{aligned} \tag{15}$$

where b_m, b_0 are arbitrary constants.

Substituting the result above into (14), we can obtain the following exact solutions to (7):

$$\begin{aligned} U_3(\xi) &= \left(-\left(m^2k_1k_2^3 - m^2k_2k_1^3 + 4ck_1 - 6l_1l_2\right)\left(b_0^2e^{-\xi} + 4b_m^2e^{\xi}\right)\right. \\ &\quad \left. -4b_0b_m\left(-5m^2k_1k_2^3 + 5m^2k_2k_1^3 + 4ck_1 - 6l_1l_2\right)\right) \\ &\quad \times\left(12k_1k_2(b_0^2e^{-\xi} + 4b_m^2e^{\xi}) + 48b_0b_mk_1k_2\right)^{-1}, \end{aligned} \tag{16}$$

where $\xi = (k_1x_1^\alpha/\Gamma(1+\alpha)) + (k_2x_2^\alpha/\Gamma(1+\alpha)) + (l_1y_1^\alpha/\Gamma(1+\alpha)) + (l_2y_2^\alpha/\Gamma(1+\alpha)) + (ct^\alpha/\Gamma(1+\alpha)) + \xi_0$. If we especially take $b_0 = 2b_m$, then we obtain the following solitary wave solution:

$$\begin{aligned} U_4(\xi) &= -\frac{\left(m^2k_1k_2^3 - m^2k_2k_1^3 + 4ck_1 - 6l_1l_2\right)}{12k_1k_2} \\ &\quad + \frac{m^2\left(k_2^2 - k_1^2\right)}{4}\operatorname{sech}^2\left(\frac{m}{2}\xi\right). \end{aligned} \tag{17}$$

Now we suppose that the solution of (7) can be expressed by

$$U(\xi) = \frac{\sum_{p=-m_1}^{m_2} a_p e^{ip\xi}}{\sum_{q=-n_1}^{n_2} b_q e^{iq\xi}}. \tag{18}$$

Similar to the balancing process for (8), we have $m_1 = n_1$ and $m_2 = n_2$. Similar to the above, we only consider the following two cases.

Case 1. Let $m_1 = n_1 = 1$, $m_2 = n_2 = 1$. Then

$$U(\xi) = \frac{a_{-1}e^{-i\xi} + a_0 + a_1e^{i\xi}}{b_{-1}e^{-i\xi} + b_0 + b_1e^{i\xi}}. \tag{19}$$

Substituting (19) into (7), eliminating the denominator, and collecting all the terms with the same power of $e^{ip\xi}$ together, equating each coefficient to zero, yield a set of algebraic equations. Solving these equations yields the following.

Family 3. Consider

$$\begin{aligned} a_1 &= -\frac{b_1\left(-k_1k_2^3 + k_2k_1^3 + 4ck_1 - 6l_1l_2\right)}{12k_1k_2}, \\ a_0 &= -\frac{b_0\left(5k_1k_2^3 - 5k_2k_1^3 + 4ck_1 - 6l_1l_2\right)}{12k_1k_2}, \\ a_{-1} &= -\frac{b_0^2\left(-k_1k_2^3 + k_2k_1^3 + 4ck_1 - 6l_1l_2\right)}{48b_1k_1k_2}, \quad b_{-1} = \frac{b_0^2}{4b_1}, \end{aligned} \tag{20}$$

where b_1, b_0 are arbitrary constants.

Substituting the result above into (19), we can obtain the following exact solutions to (7):

$$\begin{aligned} U_5(\xi) &= \left(-\left(-k_1k_2^3 + k_2k_1^3 + 4ck_1 - 6l_1l_2\right)\left(b_0^2e^{-i\xi} + 4b_1^2e^{i\xi}\right)\right. \\ &\quad \left. -4b_0b_1\left(5k_1k_2^3 - 5k_2k_1^3 + 4ck_1 - 6l_1l_2\right)\right) \\ &\quad \times\left(12k_1k_2\left(b_0^2e^{-i\xi} + 4b_1^2e^{i\xi}\right) + 48b_0b_1k_1k_2\right)^{-1}, \end{aligned} \tag{21}$$

where $\xi = (k_1x_1^\alpha/\Gamma(1+\alpha)) + (k_2x_2^\alpha/\Gamma(1+\alpha)) + (l_1y_1^\alpha/\Gamma(1+\alpha)) + (l_2y_2^\alpha/\Gamma(1+\alpha)) + (ct^\alpha/\Gamma(1+\alpha)) + \xi_0$. If we especially take $b_0 = 2b_1$, then we obtain the following trigonometric function solution:

$$\begin{aligned} U_6(\xi) &= -\frac{\left(-k_1k_2^3 + k_2k_1^3 + 4ck_1 - 6l_1l_2\right)}{12k_1k_2} \\ &\quad - \frac{\left(k_2^2 - k_1^2\right)}{4}\sec^2\frac{\xi}{2}. \end{aligned} \tag{22}$$

Case 2. Let

$$U(\xi)=\frac{a_{-m}e^{-im\xi}+a_0+a_m e^{im\xi}}{b_{-m}e^{-im\xi}+b_0+b_m e^{im\xi}}. \quad (23)$$

Substituting (23) into (7), eliminating the denominator, and collecting all the terms with the same power of $e^{ipn\xi}$ together, equating each coefficient to zero, yield a set of algebraic equations. Solving these equations, we obtain another family of values of $a_i, b_i, i=-m,0,m$.

Family 4. Consider

$$a_m=-\frac{b_m\left(-m^2k_1k_2^3+m^2k_2k_1^3+4ck_1-6l_1l_2\right)}{12k_1k_2},$$

$$a_0=-\frac{b_0\left(5m^2k_1k_2^3-5m^2k_2k_1^3+4ck_1-6l_1l_2\right)}{12k_1k_2},$$

$$a_{-m}=-\frac{b_0^2\left(-m^2k_1k_2^3+m^2k_2k_1^3+4ck_1-6l_1l_2\right)}{48b_mk_1k_2}, \quad (24)$$

$$b_{-m}=\frac{b_0^2}{4b_m}.$$

Substituting the result above into (23), we can obtain the following exact solutions to (7):

$$U_7(\xi)=\left(-\left(-m^2k_1k_2^3+m^2k_2k_1^3+4ck_1-6l_1l_2\right)\left(b_0^2e^{-\xi}+4b_m^2e^{\xi}\right)-4b_0b_m\left(5m^2k_1k_2^3-5m^2k_2k_1^3+4ck_1-6l_1l_2\right)\right)\times\left(12k_1k_2\left(b_0^2e^{-\xi}+4b_m^2e^{\xi}\right)+48b_0b_mk_1k_2\right)^{-1}, \quad (25)$$

where $\xi=(k_1x_1^{\alpha}/\Gamma(1+\alpha))+(k_2x_2^{\alpha}/\Gamma(1+\alpha))+(l_1y_1^{\alpha}/\Gamma(1+\alpha))+(l_2y_2^{\alpha}/\Gamma(1+\alpha))+(ct^{\alpha}/\Gamma(1+\alpha))+\xi_0$. If we especially take $b_0=2b_m$, then we obtain the following solitary wave solution:

$$U_8(\xi)=-\frac{\left(-m^2k_1k_2^3+m^2k_2k_1^3+4ck_1-6l_1l_2\right)}{12k_1k_2}-\frac{m^2\left(k_2^2-k_1^2\right)}{4}\sec^2\left(\frac{m}{2}\xi\right). \quad (26)$$

Remark 1. Compared with the results in [22], the established solutions by (12), (16), (21), and (25) are new exact solutions to the space-time fractional Fokas equation, while the solutions by (13), (17), (22), and (26) are new solitary wave solutions.

2.2. Nonlinear Fractional Sharma-Tasso-Olver (STO) Equation. We consider the nonlinear fractional Sharma-Tasso-Olver (STO) equation with time-fractional derivative [26, 27]:

$$D_t^{\alpha}u+3au_x^2+3au^2u_x+3auu_{xx}+au_{xxx}=0, \quad 0<\alpha\le 1. \quad (27)$$

In [26], the author solved (27) by the first integral method and obtained some hyperbolic function and trigonometric function solutions, while in [27], the variational iteration method, the Adomian decomposition method, and the homotopy perturbation method are used for obtaining a rational approximation solution for (27). Now we will apply the Exp-function method to (27). To begin with, we suppose $u(x,t)=U(\xi)$, where $\xi=kx+(ct^{\alpha}/\Gamma(1+\alpha))+\xi_0$, k,c,ξ_0 are all constants with $k,c\neq 0$. Then by use of (5), (27) can be turned into

$$cU'+3ak^2U'^2+3akU^2U'+3aUU''+aU'''=0. \quad (28)$$

First we suppose that the solution of (28) can be expressed by

$$U(\xi)=\frac{\sum_{p=-m_1}^{m_2}a_pe^{p\xi}}{\sum_{q=-n_1}^{n_2}b_qe^{q\xi}}. \quad (29)$$

Similar to the previous subsection, by balancing the highest order Exp-function in U''' and U^2U', we have $m_2=n_2$, while we obtain $m_1=n_1$ by balancing the lowest order Exp-function in U''' and U^2U'. Similar to the solving process for the space-time fractional Fokas equation, we will also proceed the computation under two selected cases.

Case 1. Let $m_1=n_1=1, m_2=n_2=1$. Then

$$U(\xi)=\frac{a_{-1}e^{-\xi}+a_0+a_1e^{\xi}}{b_{-1}e^{-\xi}+b_0+b_1e^{\xi}}. \quad (30)$$

Substituting (30) into (28), eliminating the denominator, and collecting all the terms with the same power of $e^{p\xi}$ together, equating each coefficient to zero, yield a set of algebraic equations. Solving these equations yields two families of values for $a_i, b_i, i=-1,0,1$.

Family 1. Consider

$$a_1=\frac{b_1\left(3ak^2\mp\sqrt{-3a^2k^4-3akc}\right)}{3ak}, \quad a_0=0,\ b_0=0,$$

$$a_{-1}=-\frac{b_{-1}\left(2ak^3\pm 2k\sqrt{-3a^2k^4-3akc}-c\right)}{3ak^2\pm\sqrt{-3a^2k^4-3akc}}, \quad (31)$$

where b_1, b_{-1} are arbitrary constants.

Family 2. Consider

$$a_1 = \frac{b_1\left(3ak^2 \mp \sqrt{-3a^2k^4 - 3akc}\right)}{3ak},$$
$$a_0 = \frac{b_0\left(3ak^2 \mp \sqrt{-3a^2k^4 - 3akc}\right)}{3ak}, \quad (32)$$
$$a_{-1} = \frac{b_{-1}\left(9ak^3 \pm k\sqrt{-3a^2k^4 - 3akc} + 3c\right)}{6ak^2 \pm 3\sqrt{-3a^2k^4 - 3akc}},$$

where b_1, b_0, b_{-1} are arbitrary constants.

Substituting the results above into (30), we can obtain the following exact solutions to (28):

$$U_1(\xi) = \left(-\frac{b_{-1}\left(2ak^3 \pm 2k\sqrt{-3a^2k^4 - 3akc} - c\right)}{3ak^2 \pm \sqrt{-3a^2k^4 - 3akc}}e^{-\xi} + \frac{b_1\left(3ak^2 \mp \sqrt{-3a^2k^4 - 3akc}\right)}{3ak}e^{\xi}\right) \times \left(b_{-1}e^{-\xi} + b_1e^{\xi}\right)^{-1}, \quad (33)$$

$$U_2(\xi) = \left(\frac{b_{-1}\left(9ak^3 \pm k\sqrt{-3a^2k^4 - 3akc} + 3c\right)}{6ak^2 \pm 3\sqrt{-3a^2k^4 - 3akc}}e^{-\xi} + \frac{b_0\left(3ak^2 \mp \sqrt{-3a^2k^4 - 3akc}\right)}{3ak} + \frac{b_1\left(3ak^2 \mp \sqrt{-3a^2k^4 - 3akc}\right)}{3ak}e^{\xi}\right) \times \left(b_{-1}e^{-\xi} + b_0 + b_1e^{\xi}\right)^{-1}, \quad (34)$$

where $\xi = kx + (ct^{\alpha}/\Gamma(1+\alpha)) + \xi_0$. If we especially take $c = -ak^3$, $b_1 = b_{-1}$ or $c = -ak^3$, $b_1 = -b_{-1}$ in (33), then we obtain the hyperbolic function solitary wave solutions:

$$U_3(\xi) = k\tanh\xi,$$
$$U_4(\xi) = k\coth\xi. \quad (35)$$

Case 2. Let

$$U(\xi) = \frac{a_{-m}e^{-m\xi} + a_0 + a_me^{m\xi}}{b_{-m}e^{-m\xi} + b_0 + b_me^{m\xi}}. \quad (36)$$

Substituting (36) into (28), eliminating the denominator, and collecting all the terms with the same power of $e^{pn\xi}$ together, equating each coefficient to zero, yield a set of algebraic equations. Solving these equations yields the following results.

Family 3. Consider

$$a_m = \frac{b_m\left(3amk^2 \mp \sqrt{-3a^2m^2k^4 - 3akc}\right)}{3ak}, \qquad a_0 = 0, \; b_0 = 0,$$
$$a_{-m} = -\frac{b_{-m}\left(2am^2k^3 \pm 2mk\sqrt{-3a^2m^2k^4 - 3akc} - c\right)}{3amk^2 \pm \sqrt{-3a^2m^2k^4 - 3akc}}, \quad (37)$$

where b_m, b_{-m} are arbitrary constants.

Family 4. Consider

$$a_m = \frac{b_m\left(3amk^2 \mp \sqrt{-3a^2k^4 - 3akc}\right)}{3ak},$$
$$a_0 = \frac{b_0\left(3amk^2 \mp \sqrt{-3a^2m^2k^4 - 3akc}\right)}{3ak}, \quad (38)$$
$$a_{-m} = \frac{b_{-m}\left(9am^2k^3 \pm mk\sqrt{-3a^2m^2k^4 - 3akc} + 3c\right)}{6amk^2 \pm 3\sqrt{-3a^2m^2k^4 - 3akc}},$$

where b_m, b_0, b_{-m} are arbitrary constants.

Substituting the results above into (36), we can obtain the following exact solutions to (28):

$$U_5(\xi) = \left(-\frac{b_{-m}\left(2am^2k^3 \pm 2mk\sqrt{-3a^2m^2k^4 - 3akc} - c\right)}{3amk^2 \pm \sqrt{-3a^2m^2k^4 - 3akc}}e^{-m\xi} + \frac{b_m\left(3amk^2 \mp \sqrt{-3a^2m^2k^4 - 3akc}\right)}{3ak}e^{m\xi}\right) \times \left(b_{-m}e^{-m\xi} + b_me^{m\xi}\right)^{-1}, \quad (39)$$

$$U_6(\xi) = \left(\frac{b_{-m}\left(9am^2k^3 \pm mk\sqrt{-3a^2m^2k^4 - 3akc} + 3c\right)}{6amk^2 \pm 3\sqrt{-3a^2m^2k^4 - 3akc}}e^{-\xi} + \frac{b_m\left(3amk^2 \mp \sqrt{-3a^2k^4 - 3akc}\right)}{3ak}e^{m\xi} + \frac{b_0\left(3amk^2 \mp \sqrt{-3a^2m^2k^4 - 3akc}\right)}{3ak}\right) \times \left(b_{-m}e^{-m\xi} + b_0 + b_me^{m\xi}\right)^{-1}, \quad (40)$$

where $\xi = kx + (ct^{\alpha}/\Gamma(1+\alpha)) + \xi_0$. If we especially take $c = -am^2k^3$, $b_m = b_{-m}$ or $c = -am^2k^3$, $b_m = -b_{-m}$ in (39), then we obtain the hyperbolic function solitary wave solutions:

$$U_7(\xi) = mk\tanh\xi,$$
$$U_8(\xi) = mk\coth\xi. \quad (41)$$

Now we suppose that the solution of (28) can be expressed by

$$U(\xi) = \frac{\sum_{p=-m_1}^{m_2} a_p e^{ip\xi}}{\sum_{q=-n_1}^{n_2} b_q e^{iq\xi}}. \quad (42)$$

By the balancing process for (28), we have $m_1 = n_1$ and $m_2 = n_2$. Similar to the above, we will also select two cases for computation.

Case 1. Let $m_1 = n_1 = 1, m_2 = n_2 = 1$. Then

$$U(\xi) = \frac{a_{-1}e^{-i\xi} + a_0 + a_1 e^{i\xi}}{b_{-1}e^{-i\xi} + b_0 + b_1 e^{i\xi}}. \quad (43)$$

Substituting (43) into (28), eliminating the denominator, and collecting all the terms with the same power of $e^{ip\xi}$ together, equating each coefficient to zero, yield a set of algebraic equations. Solving these equations, we obtain the following two families of values.

Family 5. Consider

$$a_1 = \frac{b_1\left(3iak^2 \mp \sqrt{3a^2k^4 - 3akc}\right)}{3ak}, \quad a_0 = 0,$$
$$a_{-1} = -\frac{b_{-1}\left(-2ak^3 \pm 2ik\sqrt{3a^2k^4 - 3akc} - c\right)}{3iak^2 \pm \sqrt{3a^2k^4 - 3akc}}, \quad (44)$$

where b_1, b_{-1} are arbitrary constants.

Family 6. Consider

$$a_1 = \frac{b_1\left(3iak^2 \mp \sqrt{3a^2k^4 - 3akc}\right)}{3ak},$$
$$a_0 = \frac{b_0\left(3iak^2 \mp \sqrt{3a^2k^4 - 3akc}\right)}{3ak}, \quad (45)$$
$$a_{-1} = \frac{b_{-1}\left(-9ak^3 \pm ik\sqrt{3a^2k^4 - 3akc} + 3c\right)}{6iak^2 \pm 3\sqrt{3a^2k^4 - 3akc}},$$

where b_1, b_0, b_{-1} are arbitrary constants.

Substituting the results above into (43), we can obtain the following exact solutions to (28):

$$U_9(\xi) = \left(-\frac{b_{-1}\left(-2ak^3 \pm 2ik\sqrt{3a^2k^4 - 3akc} - c\right)}{3iak^2 \pm \sqrt{3a^2k^4 - 3akc}}e^{-i\xi} + \frac{b_1\left(3iak^2 \mp \sqrt{3a^2k^4 - 3akc}\right)}{3ak}e^{i\xi}\right) \times \left(b_{-1}e^{-i\xi} + b_1 e^{i\xi}\right)^{-1}, \quad (46)$$

$$U_{10}(\xi) = \left(\frac{b_{-1}\left(-9ak^3 \pm ik\sqrt{3a^2k^4 - 3akc} + 3c\right)}{6iak^2 \pm 3\sqrt{3a^2k^4 - 3akc}}e^{-i\xi} + \frac{b_0\left(3iak^2 \mp \sqrt{3a^2k^4 - 3akc}\right)}{3ak} + \frac{b_1\left(3iak^2 \mp \sqrt{3a^2k^4 - 3akc}\right)}{3ak}e^{i\xi}\right) \times \left(b_{-1}e^{-i\xi} + b_0 + b_1 e^{i\xi}\right)^{-1} \quad (47)$$

If we especially take $c = -ak^3$, $b_1 = b_{-1}$ or $c = -ak^3$, $b_1 = -b_{-1}$ in (46), then we obtain the trigonometric function solutions:

$$U_{11}(\xi) = -k\tan\xi,$$
$$U_{12}(\xi) = -k\cot\xi. \quad (48)$$

Case 2. Let

$$U(\xi) = \frac{a_{-m}e^{-im\xi} + a_0 + a_m e^{im\xi}}{b_{-m}e^{-im\xi} + b_0 + b_m e^{im\xi}}. \quad (49)$$

Substituting (49) into (28), eliminating the denominator, and collecting all the terms with the same power of $e^{ipm\xi}$ together, equating each coefficient to zero, yield a set of algebraic equations. Solving these equations, we obtain the following two families of values.

Family 7. Consider

$$a_m = \frac{b_m\left(3iamk^2 \mp \sqrt{3a^2m^2k^4 - 3akc}\right)}{3ak}, \quad a_0 = 0,$$
$$a_{-m} = -\frac{b_{-m}\left(-2am^2k^3 \pm 2imk\sqrt{3a^2m^2k^4 - 3akc} - c\right)}{3iamk^2 \pm \sqrt{3a^2m^2k^4 - 3akc}}, \quad (50)$$

where b_m, b_{-m} are arbitrary constants.

Family 8. Consider

$$a_m = \frac{b_m\left(3iamk^2 \mp \sqrt{3a^2k^4 - 3akc}\right)}{3ak},$$
$$a_0 = \frac{b_0\left(3iamk^2 \mp \sqrt{3a^2m^2k^4 - 3akc}\right)}{3ak},$$
$$a_{-m} = \frac{b_{-m}\left(-9am^2k^3 \pm mk\sqrt{3a^2m^2k^4 - 3akc} + 3c\right)}{6iamk^2 \pm 3\sqrt{3a^2m^2k^4 - 3akc}}, \quad (51)$$

where b_m, b_0, b_{-m} are arbitrary constants.

Substituting the results above into (49), we can obtain the following exact solutions to (28):

$$U_{13}(\xi) = \left(-\frac{b_{-m}\left(-2am^2k^3 \pm 2imk\sqrt{3a^2m^2k^4 - 3akc} - c\right)}{3iamk^2 \pm \sqrt{3a^2m^2k^4 - 3akc}} e^{-im\xi} + \frac{b_m\left(3iamk^2 \mp \sqrt{3a^2m^2k^4 - 3akc}\right)}{3ak} e^{im\xi} \right) \times \left(b_{-m}e^{-im\xi} + b_m e^{im\xi}\right)^{-1}, \quad (52)$$

$$U_{14}(\xi) = \left(\frac{b_{-m}\left(-9am^2k^3 \pm mk\sqrt{3a^2m^2k^4 - 3akc} + 3c\right)}{6iamk^2 \pm 3\sqrt{3a^2m^2k^4 - 3akc}} e^{-im\xi} + \frac{b_m\left(3iamk^2 \mp \sqrt{3a^2k^4 - 3akc}\right)}{3ak} e^{im\xi} + \frac{b_0\left(3iamk^2 \mp \sqrt{3a^2m^2k^4 - 3akc}\right)}{3ak} \right) \times \left(b_{-m}e^{-im\xi} + b_0 + b_m e^{im\xi}\right)^{-1}. \quad (53)$$

If we especially take $c = -am^2k^3$, $b_m = b_{-m}$ or $c = -am^2k^3$, $b_m = -b_{-m}$ in (53), then we obtain the trigonometric function solutions:

$$U_{11}(\xi) = -mk\tan\xi, \qquad U_{12}(\xi) = -mk\cot\xi. \quad (54)$$

Remark 2. If we let $m = 2$, then the solutions by (41), (54) reduce to the solutions established by Lu [26, equations (55), (56), (53), (54)]. So in this way, our solutions (41), (54) are of more general forms. Moreover, we note that the solutions denoted by (33), (34), (39), (40), (46), (47), (52), and (53) are essentially different from the results in [26, 27] and are new exact solutions to the nonlinear fractional Sharma-Tasso-Olver equation so far in the literature, while the solutions (35), (48) are new solitary wave solutions to it.

3. Conclusions

We have extended the Exp-function method to solve fractional partial differential equations successfully. As applications, some generalized and new exact solutions for the space-time fractional Fokas equation and the nonlinear fractional Sharma-Tasso-Olver (STO) equation have been successfully found. As one can see, this method is based on the homogeneous balancing principle. So it can also be applied to other fractional partial differential equations where the homogeneous balancing principle is satisfied.

References

[1] L. Song and H. Zhang, "New exact solutions for the Konopelchenko-Dubrovsky equation using an extended Riccati equation rational expansion method and symbolic computation," *Applied Mathematics and Computation*, vol. 187, no. 2, pp. 1373–1388, 2007.

[2] W. Li and H. Zhang, "Generalized multiple Riccati equations rational expansion method with symbolic computation to construct exact complexiton solutions of nonlinear partial differential equations," *Applied Mathematics and Computation*, vol. 197, no. 1, pp. 288–296, 2008.

[3] S. Liu, Z. Fu, S. Liu, and Q. Zhao, "Jacobi elliptic function expansion method and periodic wave solutions of nonlinear wave equations," *Physics Letters A*, vol. 289, no. 1-2, pp. 69–74, 2001.

[4] Z. Yan, "Abundant families of Jacobi elliptic function solutions of the (2 + 1)-dimensional integrable Davey-Stewartson-type equation via a new method," *Chaos, Solitons and Fractals*, vol. 18, no. 2, pp. 299–309, 2003.

[5] M. Wang and X. Li, "Applications of F-expansion to periodic wave solutions for a new Hamiltonian amplitude equation," *Chaos, Solitons and Fractals*, vol. 24, no. 5, pp. 1257–1268, 2005.

[6] J. H. He and X. H. Wu, "Exp-function method for nonlinear wave equations," *Chaos, Solitons and Fractals*, vol. 30, no. 3, pp. 700–708, 2006.

[7] J. H. He and L. N. Zhang, "Generalized solitary solution and compacton-like solution of the Jaulent-Miodek equations using the Exp-function method," *Physics Letters A*, vol. 372, no. 7, pp. 1044–1047, 2008.

[8] X. H. Wu and J. H. He, "Solitary solutions, periodic solutions and compacton-like solutions using the Exp-function method," *Computers and Mathematics with Applications*, vol. 54, no. 7-8, pp. 966–986, 2007.

[9] J. H. He and M. A. Abdou, "New periodic solutions for nonlinear evolution equations using Exp-function method," *Chaos, Solitons and Fractals*, vol. 34, no. 5, pp. 1421–1429, 2007.

[10] M. Wang, X. Li, and J. Zhang, "The (G'/G)-expansion method and travelling wave solutions of nonlinear evolution equations in mathematical physics," *Physics Letters A*, vol. 372, no. 4, pp. 417–423, 2008.

[11] M. Wang, J. Zhang, and X. Li, "Application of the fenced(G'/G)-expansion to travelling wave solutions of the Broer-Kaup and the approximate long water wave equations," *Applied Mathematics and Computation*, vol. 206, no. 1, pp. 321–326, 2008.

[12] A. M. A. El-Sayed and M. Gaber, "The Adomian decomposition method for solving partial differential equations of fractal order in finite domains," *Physics Letters A*, vol. 359, no. 3, pp. 175–182, 2006.

[13] A. M. A. El-Sayed, S. H. Behiry, and W. E. Raslan, "Adomian's decomposition method for solving an intermediate fractional advection-dispersion equation," *Computers and Mathematics with Applications*, vol. 59, no. 5, pp. 1759–1765, 2010.

[14] J. He, "A new approach to nonlinear partial differential equations," *Communications in Nonlinear Science and Numerical Simulation*, vol. 2, no. 4, pp. 230–235, 1997.

[15] G. C. Wu and E. W. M. Lee, "Fractional variational iteration method and its application," *Physics Letters A*, vol. 374, no. 25, pp. 2506–2509, 2010.

[16] S. Guo and L. Mei, "The fractional variational iteration method using He's polynomials," *Physics Letters A*, vol. 375, no. 3, pp. 309–313, 2011.

[17] J. H. He, "Homotopy perturbation technique," *Computer Methods in Applied Mechanics and Engineering*, vol. 178, no. 3-4, pp. 257–262, 1999.

[18] J. H. He, "Coupling method of a homotopy technique and a perturbation technique for non-linear problems," *International Journal of Non-Linear Mechanics*, vol. 35, no. 1, pp. 37–43, 2000.

[19] Z. Odibat and S. Momani, "A generalized differential transform method for linear partial differential equations of fractional order," *Applied Mathematics Letters*, vol. 21, no. 2, pp. 194–199, 2008.

[20] M. Cui, "Compact finite difference method for the fractional diffusion equation," *Journal of Computational Physics*, vol. 228, no. 20, pp. 7792–7804, 2009.

[21] Q. Huang, G. Huang, and H. Zhan, "A finite element solution for the fractional advection-dispersion equation," *Advances in Water Resources*, vol. 31, no. 12, pp. 1578–1589, 2008.

[22] S. Zhang and H. Q. Zhang, "Fractional sub-equation method and its applications to nonlinear fractional PDEs," *Physics Letters A*, vol. 375, no. 7, pp. 1069–1073, 2011.

[23] S. M. Guo, L. Q. Mei, Y. Li, and Y. F. Sun, "The improved fractional sub-equation method and its applications to the spaceC-time fractional differential equations in fluid mechanics," *Physics Letters A*, vol. 376, pp. 407–411, 2012.

[24] B. Lu, "Bäcklund transformation of fractional Riccati equation and its applications to nonlinear fractional partial differential equations," *Physics Letters A*, vol. 376, pp. 2045–2048, 2012.

[25] G. Jumarie, "Modified Riemann-Liouville derivative and fractional Taylor series of nondifferentiable functions further results," *Computers and Mathematics with Applications*, vol. 51, no. 9-10, pp. 1367–1376, 2006.

[26] B. Lu, "The first integral method for some time fractional differential equations," *Journal of Mathematical Analysis and Applications*, vol. 395, pp. 684–693, 2012.

[27] L. N. Song, Q. Wang, and H. Q. Zhang, "Rational approximation solution of the fractional SharmaCTassoCOlever equation," *Journal of Computational and Applied Mathematics*, vol. 224, pp. 210–218, 2009.

11

On Multivalued Contractions in Cone Metric Spaces without Normality

Muhammad Arshad[1] and Jamshaid Ahmad[2]

[1] *Department of Mathematics, International Islamic University, H-10, Islamabad 44000, Pakistan*
[2] *Department of Mathematics, COMSATS Institute of Information Technology, Chak Shahzad, Islamabad 44000, Pakistan*

Correspondence should be addressed to Jamshaid Ahmad; jamshaid_jasim@yahoo.com

Academic Editors: A. Agouzal, A. Ibeas, F. Khani, and F. Kittaneh

Wardowski (2011) in this paper for a normal cone metric space (X, d) and for the family $\mathscr{A}$ of subsets of X established a new cone metric $H : \mathscr{A} \times \mathscr{A} \to E$ and obtained fixed point of set-valued contraction of Nadler type. Further, it is noticed in the work of Janković et al., 2011 that the fixed-point problem in the setting of cone metric spaces is appropriate only in the case when the underlying cone is nonnormal. In the present paper we improve Wardowski's result by proving the same without the assumption of normality on cones.

1. Introduction and Preliminaries

Huang and Zhang [1] generalized the notion of metric space by replacing the set of real numbers by ordered Banach space and defined cone metric space and extended Banach type fixed-point theorems for contractive type mappings. Subsequently, some other authors (e.g., see [2–15] and references therein) studied properties of cone metric spaces and fixed points results of mappings satisfying contractive type condition in cone metric spaces. Recently, Choa et al. [9], Kadelburg and Radenović [16], Klim and Wardowski [17], Latif and Shaddad [18], Radenović and Kadelburg [19], Rezapour and Haghi [20], and Wardowski [14, 21] obtained fixed points of set-valued mappings in normal cone metric spaces. On the other hand, it is shown in [11] that most of the fixed points results of mappings satisfying contractive type condition in cone metric spaces with a normal cone can be reduced to the corresponding results from metric space theory. The fixed-point problem in the setting of cone metric spaces is appropriate only in the case when the underlying cone is nonnormal, because the results concerning fixed points and other results in the case of cone metric spaces with nonnormal solid cones cannot be proved by reducing to metric spaces. In this paper, we prove the result of Wardowski [14] without the assumption of normality of cones. We need the following definitions and results, consistent with [1, 11, 14].

Let E be a Banach Space and P a subset of E. Then, P is called a cone whenever

(i) P is closed, nonempty, and $P \neq \{\theta\}$,

(ii) $ax + by \in P$ for all $x, y \in P$ and nonnegative real numbers a, b,

(iii) $P \cap (-P) = \{\theta\}$.

Each cone P induces a partial ordering $\preccurlyeq$ on E by $x \preccurlyeq y$ if and only if $y - x \in P$. So $x \prec y$ will stand for $x \preccurlyeq y$ and $x \neq y$, while $x \ll y$ will stand for $y - x \in \operatorname{int} P$, where $\operatorname{int} P$ denotes the interior of P. The cone P is called normal if there is a number $K > 0$ such that, for all $x, y \in E$,

$$\theta \preccurlyeq x \preccurlyeq y \Longrightarrow \|x\| \leq K \|y\|. \quad (1)$$

The least positive number K satisfying (1) is called the *normal constant* of P.

Definition 1. Let X be a nonempty set. Suppose the mapping $d : X \times X \to E$ satisfies

(d_1) $\theta \preccurlyeq d(x, y)$ for all $x, y \in X$ and $d(x, y) = \theta$ if and only if $x = y$,

(d_2) $d(x,y)=d(y,x)$ for all $x,y\in X$,

(d_3) $d(x,y)\preccurlyeq d(x,z)+d(z,y)$ for all $x,y,z\in X$.

Then, d is called a cone metric on X, and (X,d) is called a cone metric space.

Let (X,d) be a cone metric space, $x\in X$ and $\{x_n\}_{n\geq 1}$ a sequence in X. Then, $\{x_n\}_{n\geq 1}$ converges to x whenever for every $c\in E$ with $\theta\ll c$ there is a natural number N such that $d(x_n,x)\ll c$ for all $n\geq N$. We denote this by $\lim_{n\to\infty}x_n=x$ or $x_n\to x$. $\{x_n\}_{n\geq 1}$ is a Cauchy sequence whenever for every $c\in E$ with $\theta\ll c$ there is a natural number N such that $d(x_n,x_m)\ll c$ for all $n,m\geq N$. (X,d) is called a complete cone metric space if every Cauchy sequence in X is convergent.

A set $A\subset(X,d)$ is called closed if, for any sequence $\{x_n\}\subset A$ convergent to x, we have $x\in A$. Denote by $N(X)$ the collection of all nonempty subsets of X and by $C(X)$ a collection of all nonempty closed subsets of X. Denote by $\operatorname{Fix}T$ a set of all fixed points of a mapping T. In the present paper, we assume that E is a real Banach space, P is a cone in E with nonempty interior (such cones are called solid), and $\preccurlyeq$ is a partial ordering with respect to P. In accordance with [14, Definition 3.1 and Lemma 3.1], we minutely modify the idea of H-cone metric to make it more comparable with a standard metric.

Definition 2. Let (X,d) be a cone metric space and $\mathscr{A}$ be a collection of nonempty subsets of X. A map $H:\mathscr{A}\times\mathscr{A}\to E$ is called an H-cone metric on $\mathscr{A}$ induced by d if the following conditions hold:

(H_1) $\theta\preccurlyeq H(A,B)$ for all $A,B\in\mathscr{A}$ and $H(A,B)=\theta$ if and only if $A=B$,

(H_2) $H(A,B)=H(B,A)$ for all $A,B\in\mathscr{A}$,

(H_3) $H(A,B)\preccurlyeq H(A,C)+H(C,B)$ for all $A,B,C\in\mathscr{A}$,

(H_4) If $A,B\in\mathscr{A}$, $\theta\prec\varepsilon\in E$ with $H(A,B)\prec\varepsilon$, then for each $a\in A$ there exists $b\in B$ such that $d(a,b)\prec\varepsilon$.

Examples can be seen in [14, examples 3.1 and 3.2].

2. Main Result

Theorem 3. *Let (X,d) be a complete cone metric space. Let $\mathscr{A}$ be a nonempty collection of nonempty closed subsets of X, and let $H:\mathscr{A}\times\mathscr{A}\to E$ be an H-cone metric induced by d. If for a map $T:X\to\mathscr{A}$ there exists $\lambda\in(0,1)$ such that for all $x,y\in X$*

$$H(Tx,Ty)\preccurlyeq\lambda d(x,y),\tag{2}$$

then $\operatorname{Fix}T\neq\emptyset$.

Proof. Let x_0 be an arbitrary but fixed element of X and $x_1\in Tx_0$. If $x_0=x_1$, then $x_0\in\operatorname{Fix}T$, and if $x_0\neq x_1$, using the fact that

$$H(Tx_0,Tx_1)\preccurlyeq\lambda d(x_0,x_1)\prec\sqrt{\lambda}d(x_0,x_1),\tag{3}$$

we may choose $x_2\in X$ such that $x_2\in Tx_1$ and

$$d(x_1,x_2)\prec\sqrt{\lambda}d(x_0,x_1).\tag{4}$$

Similarly, in case $x_1\neq x_2$, we may choose $x_3\in X$ such that $x_3\in Tx_2$ and

$$d(x_2,x_3)\prec\sqrt{\lambda}d(x_1,x_2)\prec\left(\sqrt{\lambda}\right)^2d(x_0,x_1).\tag{5}$$

We can continue this process to find a sequence $\{x_n\}$ of points of X such that

$$\begin{aligned}&x_{n+1}\in Tx_n,\quad n=0,1,2,\ldots,\\ &d(x_n,x_{n+1})\prec\sqrt{\lambda}d(x_{n-1},x_n)\\ &\qquad\prec\left(\sqrt{\lambda}\right)^2d(x_{n-2},x_{n-1})\prec\cdots\prec\left(\sqrt{\lambda}\right)^nd(x_0,x_1).\end{aligned}\tag{6}$$

Now for any $m>n$,

$$\begin{aligned}d(x_m,x_n)&\preccurlyeq d(x_n,x_{n+1})+d(x_{n+1},x_{n+2})+\cdots+d(x_{m-1},x_m)\\ &\preccurlyeq\left[\lambda^{n/2}+\lambda^{(n+1)/2}+\cdots+\lambda^{(m-1)/2}\right]d(x_0,x_1)\\ &\preccurlyeq\left[\frac{\lambda^{n/2}}{1-\lambda^{1/2}}\right]d(x_0,x_1).\end{aligned}\tag{7}$$

Let $\theta\ll c$ be given. Choose a symmetric neighborhood V of θ such that $c+V\subseteq\operatorname{int}P$. Also, choose a natural number N_1 such that $[\lambda^{n/2}/(1-\lambda^{1/2})]d(x_0,x_1)\in V$, for all $n\geq N_1$. Then, $(\lambda^{n/2}/(1-\lambda^{1/2}))d(x_1,x_0)\ll c$, for all $n\geq N_1$. Thus,

$$d(x_m,x_n)\preccurlyeq\left[\frac{\lambda^{n/2}}{1-\lambda^{1/2}}\right]d(x_0,x_1)\ll c,\tag{8}$$

for all $m>n$. Therefore, $\{x_n\}_{n\geq 1}$ is a Cauchy sequence. Since X is complete, there exists $u\in X$ such that $x_n\to u$. Since

$$H(Tx_n,Tu)\preccurlyeq\lambda d(x_n,u)\prec\sqrt{\lambda}d(x_n,u),\tag{9}$$

for each n, $x_{n+1}\in Tx_n$, we have $y_n\in Tu$ such that $d(x_{n+1},y_n)\prec\sqrt{\lambda}d(x_n,u)$. Now, choose a natural number N_2 such that

$$d(x_n,u)\ll\frac{c}{2}\quad\forall n\geq N_2.\tag{10}$$

Then for all $n\geq N_2$,

$$\begin{aligned}d(u,y_n)&\preccurlyeq d(u,x_{n+1})+d(x_{n+1},y_n)\\ &\preccurlyeq d(u,x_{n+1})+\sqrt{\lambda}d(x_n,u)\\ &\preccurlyeq d(u,x_{n+1})+d(x_n,u)\ll\frac{c}{2}+\frac{c}{2}=c.\end{aligned}\tag{11}$$

It follows that $y_n\to u$, and it implies that $u\in Tu$. □

Example 4. Suppose $X = [0,1]$, $E = C_R^2[0,1]$ with the norm $\|x\| = \|x\|_\infty + \|x'\|_\infty$, $P = \{x \in E : x \geq 0\}$, $x(t) = t$, and $y(t) = t^{2K}$. Then, $0 \leq x \leq y$, $\|x\| = 2$, and $\|y\| = 1 + 2K$. For all $K \geq 1$, since $K\|x\| < \|y\|$. Therefore, P is non-normal. Define $d : X \times X \to E$ as follows:

$$(d(x,y))(t) = |x - y| e^t. \tag{12}$$

Let $\mathscr{A}$ be a family of subsets of X of the form $\mathscr{A} = \{[0,x] : x \in X\} \cup \{\{x\} : x \in X\}$, and define $H : \mathscr{A} \times \mathscr{A} \to E$ as follows:

$$H(A,B) = \begin{cases} |x-y| e^t & \text{for } A = [0,x],\ B = [0,y], \\ |x-y| e^t & \text{for } A = \{x\},\ B = \{y\}, \\ \operatorname{Max}\{y, |x-y|\} e^t & \text{for } A = [0,x],\ B = \{y\}, \\ \operatorname{Max}\{x, |x-y|\} e^t & \text{for } A = \{x\},\ B = [0,y]. \end{cases} \tag{13}$$

It is easy to observe that H satisfies (H_1)–(H_4) of Definition 2. Define $T : X \to \mathscr{A}$ as

$$Tx = \begin{cases} \{0\}, & \text{for } x \in \left[0, \frac{1}{2}\right], \\ \left[0, \frac{1}{2}\left(x - \frac{1}{2}\right)^2\right], & \text{for } x \in \left(\frac{1}{2}, 1\right]. \end{cases} \tag{14}$$

Note that T satisfies the conditions of Theorem 3 with $\lambda = 1/2$ and $0 \in \operatorname{Fix} T$.

References

[1] L.-G. Huang and X. Zhang, "Cone metric spaces and fixed point theorems of contractive mappings," *Journal of Mathematical Analysis and Applications*, vol. 332, no. 2, pp. 1468–1476, 2007.

[2] T. Abdeljawad, P. Murthy, and K. Tas, "A Gregus type common fixed point theorem of set-valued mappings in cone metric spaces," *Journal of Computational Analysis and Applications*, vol. 13, no. 4, pp. 622–628, 2011.

[3] J. Ahmad, M. Arshad, and C. Vetro, "On a theorem of khan in a generalized metric space," *International Journal of Analysis*, vol. 2013, Article ID 852727, 6 pages, 2013.

[4] M. Arshad, J. Ahmad, and E. Karapınar, "Some common fixed point results in rectangular metric spaces," *Journal of Analysis*, vol. 2013, Article ID 307234, 7 pages, 2013.

[5] A. G. Ahmad, Z. M. Fadail, M. Abbas, Z. Kadelburg, and S. Radenović, "Some fixed and periodic points in abstract metric spaces," *Abstract and Applied Analysis*, vol. 2012, Article ID 908423, 15 pages, 2012.

[6] M. Arshad, A. Azam, and P. Vetro, "Some common fixed point results in cone metric spaces," *Fixed Point Theory and Applications*, vol. 2009, Article ID 493965, 2009.

[7] S. H. Cho, J. S. Bae, and K. S. Na, "Fixed point theorems for multivalued contractive mappings and multivalued Caristi type mappings in cone metric spaces," *Fixed Point Theory and Applications*, vol. 2012, article 133, 10 pages, 2012.

[8] S. H. Cho and J. S. Bae, "Fixed point theorems for multivalued maps in cone metric spaces," *Fixed Point Theory and Applications*, vol. 2011, article 87, 2011.

[9] Y. J. Choa, S. Hirunworakit, and N. Petrot, "Set-valued fixed-point theorems for generalized contractive mappings without the Hausdorff metric," *Applied Mathematics Letters*, vol. 24, no. 11, pp. 1959–1967, 2011.

[10] L. Gajić and V. Rakočević, "Qusai-contractions on a nonnormal cone metric space," *Fund Analyzer Application*, vol. 46, pp. 75–79, 2012.

[11] S. Janković, Z. Kadelburg, and S. Radenović, "On cone metric spaces: a survey survey," *Nonlinear Analysis. Theory, Methods & Applications*, vol. 74, pp. 2591–2601, 2011.

[12] G. Jungck, S. Radenović, S. Radojević, and V. Rakočević, "Common fixed point theorems for weakly compatible pairs on cone metric spaces," *Fixed Point Theory and Applications*, vol. 2009, Article ID 643840, 2009.

[13] S. R. Kumar, "Common fixed point theorems for sub compatible and sub sequentially continuous maps in 2 metric spaces," *International MatheMatical Forum*, vol. 7, no. 24, pp. 1187–1200, 2012.

[14] D. Wardowski, "On set-valued contractions of Nadler type in cone metric spaces," *Applied Mathematics Letters*, vol. 24, no. 3, pp. 275–278, 2011.

[15] Z. Kadelburg, S. Radenović, and V. Rakočević, "A note on the equivalence of some metric and cone metric fixed point results," *Applied Mathematics Letters*, vol. 24, pp. 370–374, 2011.

[16] Z. Kadelburg and S. Radenović, "Some results on set-valued contractions in abstract metric spaces," *Computers and Mathematics with Applications*, vol. 62, no. 1, pp. 342–350, 2011.

[17] D. Klim and D. Wardowski, "Dynamic processes and fixed points of set-valued nonlinear contractions in cone metric spaces," *Nonlinear Analysis, Theory, Methods and Applications*, vol. 71, no. 11, pp. 5170–5175, 2009.

[18] A. Latif and F. Y. Shaddad, "Fixed point results for multivalued maps in cone metric spaces," *Fixed Point Theory and Applications*, vol. 2010, article 87, Article ID 941371, 2010.

[19] S. Radenović and Z. Kadelburg, "Some results on fixed points of multifunctions on abstract metric spaces," *Mathematical and Computer Modelling*, vol. 53, no. 5-6, pp. 746–754, 2011.

[20] Sh. Rezapour and R. H. Haghi, "Fixed point of multifunctions on cone metric spaces," *Numerical Functional Analysis and Optimization*, vol. 30, no. 7-8, pp. 1–8, 2009.

[21] D. Wardowski, "Endpoints and fixed points of set-valued contractions in cone metric spaces," *Nonlinear Analysis, Theory, Methods and Applications*, vol. 71, no. 1-2, pp. 512–516, 2009.

Existence of Multiple Solutions for a p-Kirchhoff Problem with Nonlinear Boundary Conditions

Zonghu Xiu[1,2] and Caisheng Chen[2]

[1] *Science and Information College, Qingdao Agricultural University, Qingdao 266109, China*
[2] *College of Science, Hohai University, Nanjing 210098, China*

Correspondence should be addressed to Zonghu Xiu; qingda@163.com

Academic Editors: H. Du and S. Momani

The paper considers the existence of multiple solutions of the singular nonlocal elliptic problem $-M(\int_\Omega |x|^{-ap}|\nabla u|^p)$ $\operatorname{div}(|x|^{-ap}|\nabla u|^{p-2}\nabla u) = \lambda h(x)|u|^{r-2}u$, $x \in \Omega$, $M(\int_\Omega |x|^{-ap}|\nabla u|^p)|x|^{-ap}|\nabla u|^{p-2}(\partial u/\partial \nu) = g(x)|u|^{q-2}u$, on $\partial\Omega$, where $1 < (N+1)/2 < p < N$, $a < (N-p)/p$. By the variational method on the Nehari manifold, we prove that the problem has at least two positive solutions when some conditions are satisfied.

1. Introduction and Main Result

In this paper, we consider the existence of multiple solutions for the singular elliptic problem:

$$\begin{aligned} &-M\left(\int_\Omega |x|^{-ap}|\nabla u|^p dx\right)\operatorname{div}\left(|x|^{-ap}|\nabla u|^{p-2}\nabla u\right) \\ &\quad = \lambda h(x)|u|^{r-2}u, \quad x \in \Omega, \\ &M\left(\int_\Omega |x|^{-ap}|\nabla u|^p\right)|x|^{-ap}|\nabla u|^{p-2}\frac{\partial u}{\partial \nu} \\ &\quad = g(x)|u|^{q-2}u, \quad \text{on } \partial\Omega, \end{aligned} \tag{1}$$

where $1 < (N+1)/2 < p < N$, $a < (N-p)/p$, $\lambda > 0$, Ω is an exterior domain of $\mathbb{R}^N$: that is, and $\Omega = \mathbb{R}^N \setminus D$, where D is a bounded domain in $\mathbb{R}^N$ with the smooth boundary $\partial D(= \partial\Omega)$, and $0 \in \Omega$. $g(x)$ and $h(x)$ are continuous functions, $M(s) = \alpha s + \beta$ with the parameters $\alpha, \beta > 0$.

Problem like (1) is usually called nonlocal problem because of the presence of the integral over the entire domain, and this implies that (1) is no longer a pointwise identity. In fact, such kind of problem can be traced back to the work of Kirchhoff. In [1], Kirchhoff investigated the model of the form

$$\rho\frac{\partial^2 u}{\partial t^2} - \left(\frac{P_0}{h} + \frac{E}{2L}\int_0^L \left|\frac{\partial u}{\partial x}\right|^2 dx\right)\frac{\partial^2 u}{\partial x^2} = 0, \tag{2}$$

where ρ, P_0, h, E, and L are all positive constants. This equation extends the classical d'Alembert's wave equation by considering the effects of changes in the length of the strings during the vibrations. Problem (1) is related to the stationary analogue of problem (2). After Kirchhoff's work, various models of Kirchhoff-type have been studied by many authors: we refer the readers to [2–9]. In [4], by the variational methods, Bensedik and Bouchekif considered the problem

$$\begin{aligned} -M\left(\int_\Omega |\nabla u|^2 dx\right)\Delta u &= f(x,u), \quad x \in \Omega, \\ u &= 0, \quad \text{on } \partial\Omega, \end{aligned} \tag{3}$$

where Ω is a bounded domain in $\mathbb{R}^N$. One of the assumptions made on $f(x,t)$ in (3) is that $(\mathbf{f}_1)$ $f(x,t)$ is continuous function on $\overline{\Omega} \times \mathbb{R}$ such that

$$\begin{aligned} f(x,t) &\geq 0, \quad \forall t \geq 0,\ x \in \overline{\Omega}, \\ f(x,t) &= 0, \quad \forall t \leq 0,\ x \in \overline{\Omega}. \end{aligned} \tag{4}$$

The authors proved that problem (3) has a positive solution or has no solution when some other assumptions are fulfilled. In our paper, however, the weight functions $h(x)$ and $g(x)$ are permitted to change sign. Thus, the methods in [4] cannot be directly applied on (1).

In recent years, some other authors considered the Kirchhoff-type equations with p-Laplacian [10–13]. In fact, motivated by [4, 5] and our previous work [14], we consider the existence of multiple solutions for problem (1) on the Nehari manifold by variational methods. We prove that problem (1) has at least two positive solutions. Since $\Omega \subset \mathbb{R}^N$ is an unbounded domain and the problem is singular, the loss of compactness of the Sobolev embedding renders variational technique more delicate.

In order to state our result, we introduce a weighted Sobolev space $E = W_a^{1,p}(\Omega)$, which is the completion of the space $C_0^\infty(\Omega)$ with the norm of

$$\|u\|_E = \left(\int_\Omega |x|^{-ap}|\nabla u|^p dx\right)^{1/p}. \quad (5)$$

For $\rho \geq 1$ and $f = f(x) > 0$ in Ω, we define the space $L^\rho(\Omega, f)$ as being the set of Lebesgue measurable functions $u : \Omega \to \mathbb{R}^1$, which satisfies

$$\|u\|_{L^\rho(\Omega,f)} = \left(\int_\Omega f(x)|u|^\rho dx\right)^{1/\rho} < \infty. \quad (6)$$

The following weighted Sobolev-Hardy inequality is due to Caffarelli et al. [15], which is called the Caffarelli-Kohn-Nirenberg inequality. There is a constant $S_1 > 0$ such that

$$\left(\int_{\mathbb{R}^N} |x|^{-bp^*}|u|^{p^*} dx\right)^{1/p^*} \leq S_1\left(\int_{\mathbb{R}^N} |x|^{-ap}|\nabla u|^p dx\right)^{1/p}, \quad \forall u \in C_0^\infty\left(\mathbb{R}^N\right), \quad (7)$$

where $-\infty < a < (N-p)/p$, $a \leq b < a+1$, $d = a+1-b$, and $p^* = pN/(N-pd)$. Throughout this paper, we make the following assumptions:

(A_1) $h(x)|x|^{br} \in L^\eta(\Omega) \cap L^\infty(\Omega)$ with $\eta = p^*/(p^*-r)$, $h^\pm = \max\{\pm h(x), 0\} \not\equiv 0$,

(A_2) $g(x) \in C(\partial\Omega) \cap L^\infty(\partial\Omega)$,

(A_3) $1 < r < p < 2p \leq q < p_* = p(N-1)/(N-p)$.

Now, we give the definition of weak solution for problem (1).

Definition 1. A function $u \in E$ is said to be a weak solution of problem (1) if for any $\varphi \in C_0^\infty(\Omega)$

$$\left(\alpha\|u\|_E^p + \beta\right)\int_\Omega |x|^{-ap}|\nabla u|^{p-2}\nabla u \cdot \nabla\varphi dx - \int_{\partial\Omega} g(x)|u|^{q-2}u\varphi d\sigma = \lambda\int_\Omega h(x)|u|^{r-2}u\varphi dx. \quad (8)$$

The assumptions (A_1)-(A_3) mean that all the integrals in (8) are well defined and convergent.

In view of (A_3), it follows from the compact trace embedding $W_a^{1,p}(\Omega) \hookrightarrow L^q(\partial\Omega)$ [16] that

$$\int_{\partial\Omega} g(x)|u|^q d\sigma \leq S_2\|g\|_\infty\|u\|_E^q, \quad (9)$$

for some constant $S_2 > 0$, and $\|g\|_\infty \triangleq \|g\|_{L^\infty(\partial\Omega)}$.

Our main result is in the following.

Theorem 2. *Assume (A_1)-(A_3); there exists $\lambda^* > 0$ such that problem (1) has at least two positive solutions for all $\lambda > 0$ satisfying $0 < \lambda < \lambda^*$.*

This paper is organized as follows. In Section 2, we give some properties of the Nehari manifold and set up the variational framework for problem (1). In Section 3, we consider the multiplicity results and prove Theorem 2.

2. Preliminary Results

It is clear that problem (1) has a variational structure. Let $J_\lambda(u) : E \to \mathbb{R}^1$ be the corresponding Euler functional of problem (1), which is defined by

$$J_\lambda(u) = \frac{1}{p}\widehat{M}\left(\|u\|_E^p\right) - \frac{1}{q}\int_{\partial\Omega} g(x)|u|^q d\sigma - \frac{1}{r}\int_\Omega h(x)|u|^r dx, \quad (10)$$

where $\widehat{M}(t) = \int_0^t M(s)ds$. Then, we see that the functional $J_\lambda(u) \in C^1(E, \mathbb{R}^1)$, and for $\forall\varphi \in E$, there holds

$$\left\langle J_\lambda'(u), \varphi\right\rangle = M\left(\|u\|_E^p\right)\int_\Omega |x|^{-ap}|\nabla u|^{p-2}\nabla u \cdot \nabla\varphi dx - \int_{\partial\Omega} g|u|^{q-2}u\varphi d\sigma - \lambda\int_\Omega h|u|^{r-2}u\varphi dx. \quad (11)$$

Particularly, we have

$$\left\langle J_\lambda'(u), u\right\rangle = M\left(\|u\|_E^p\right)\int_\Omega |x|^{-ap}|\nabla u|^p dx - \int_{\partial\Omega} g(x)|u|^q d\sigma - \lambda\int_\Omega h(x)|u|^r dx. \quad (12)$$

It is well known that the weak solution of problem (1) is the critical point of $J_\lambda(u)$. Thus, to prove the existence of weak solutions for problem (1), it is sufficient to show that $J_\lambda(u)$ admits a sequence of critical points. Since $J_\lambda(u)$ is not bounded below on E, it is useful to consider the functional $J_\lambda(u)$ on the Nehari manifold [17, 18]:

$$N_\lambda = \left\{u \in E \setminus \{0\} \mid \left\langle J_\lambda'(u), u\right\rangle = 0\right\}, \quad (13)$$

where $\langle , \rangle$ denotes the usual duality. Then, it follows from (12) that $u \in N_\lambda$ if and only if

$$M\left(\|u\|_E^p\right)\|u\|_E^p - \int_{\partial\Omega} g(x)|u|^q d\sigma - \lambda\int_\Omega h(x)|u|^r dx = 0. \quad (14)$$

Then, we get from (10)–(14) that

$$J_\lambda(u) = \left(\frac{1}{2p} - \frac{1}{r}\right)\alpha\|u\|_E^{2p} - \left(\frac{1}{r} - \frac{1}{p}\right)\beta\|u\|_E^p + \left(\frac{1}{r} - \frac{1}{q}\right)\int_{\partial\Omega} g(x)|u|^q d\sigma \quad (15)$$

$$= \left(\frac{1}{2p} - \frac{1}{q}\right)\alpha\|u\|_E^{2p} + \left(\frac{1}{p} - \frac{1}{q}\right)\beta\|u\|_E^p - \left(\frac{1}{r} - \frac{1}{q}\right)\lambda\int_\Omega h(x)|u|^r dx. \quad (16)$$

We define

$$\Phi_\lambda(u) = \left\langle J'_\lambda(u), u\right\rangle. \quad (17)$$

Then, (14) implies that

$$\left\langle \Phi'_\lambda(u), u\right\rangle = pM'\left(\|u\|_E^p\right)\|u\|_E^{2p} + pM\left(\|u\|_E^p\right)\|u\|_E^p - q\int_{\partial\Omega} g(x)|u|^q d\sigma - r\lambda\int_\Omega h(x)|u|^r dx \quad (18)$$

$$= (2p-q)\alpha\|u\|_E^{2p} + (p-q)\beta\|u\|_E^p + (q-r)\lambda\int_\Omega h(x)|u|^r dx \quad (19)$$

$$= (2p-r)\alpha\|u\|_E^{2p} + (p-r)\beta\|u\|_E^p - (q-r)\int_{\partial\Omega} g(x)|u|^q d\sigma. \quad (20)$$

It is natural to split N_λ into three parts:

$$N_\lambda^+ = \left\{u \in N_\lambda \mid \left\langle \Phi'_\lambda(u), u\right\rangle > 0\right\},$$
$$N_\lambda^- = \left\{u \in N_\lambda \mid \left\langle \Phi'_\lambda(u), u\right\rangle < 0\right\}, \quad (21)$$
$$N_\lambda^0 = \left\{u \in N_\lambda \mid \left\langle \Phi'_\lambda(u), u\right\rangle = 0\right\}.$$

Now, we give some important properties of N_λ^+, N_λ^-, and N_λ^0.

Lemma 3. *Let $1 < r < p < q$ and $q \geq 2p$. Then, $J_\lambda(u)$ is coercive and bounded below on N_λ.*

Proof. For $1 < r < p < q$, we obtain from (A_1), the Hölder and Caffarelli-Kohn-Nirenberg inequalities that

$$\int_\Omega h(x)|u|^r dx \leq \left(\int_\Omega \left(|h(x)||x|^{br}\right)^\eta dx\right)^{1/\eta} \times \left(\int_\Omega |x|^{-bp^*}|u|^{p^*}dx\right)^{r/p^*} \leq S_1^r h_\eta\left(\int_\Omega |x|^{-ap}|\nabla u|^p dx\right)^{r/p} \leq S_1^r h_\eta \|u\|_E^r, \quad (22)$$

where $h_\eta = (\int_\Omega(|g(x)||x|^{br})^\eta dx)^{1/\eta}$, $\eta = p^*/(p^*-r)$. Thus, we get from (16) that

$$J_\lambda(u) = \left(\frac{1}{2p} - \frac{1}{q}\right)\alpha\|u\|_E^{2p} + \left(\frac{1}{p} - \frac{1}{q}\right)\beta\|u\|_E^p - \left(\frac{1}{r} - \frac{1}{q}\right)\int_\Omega h(x)|u|^r dx \geq \|u\|_E^p\left[\left(\frac{1}{2p} - \frac{1}{q}\right)\alpha\|u\|_E^p + \left(\frac{1}{p} - \frac{1}{q}\right)\beta\right] - \left(\frac{1}{r} - \frac{1}{q}\right)\lambda S_1^r h_\eta\|u\|_E^r. \quad (23)$$

Then, one can obtain by the Young inequality and $q \geq 2p$ that $J_\lambda(u)$ is coercive and bounded below on N_λ.

Let

$$\lambda_0 = \begin{cases} \dfrac{2\sqrt{(q-2p)(q-p)}}{(q-r)S_2^r h_\eta} \times \left[\dfrac{2\sqrt{\alpha\beta(2p-r)(p-r)}}{(q-r)\|g\|_\infty}\right]^{(3p-2r)/(2q-3p)}, & q>2p, \\ \dfrac{p\beta}{(2p-r)S_1^r h_\eta} \times \left[\dfrac{2\sqrt{\alpha\beta(2p-r)(p-r)}}{(2p-r)\|g\|_\infty}\right]^{2(p-r)/p}, & q=2p. \end{cases} \quad (24)$$

Then, we have the following result. □

Lemma 4. *Let $r < p < q$, $q \geq 2p$. Then, $N_\lambda^0 = \emptyset$ for $0 < \lambda < \lambda_0$.*

Proof. Suppose that there exists $u \in N_\lambda^0$. If $q > 2p$, then it follows from (19) and (21) that

$$2\sqrt{\beta(q-2p)(q-p)}\|u\|_E^{3p/2} \leq (q-2p)\alpha\|u\|_E^{2p} + \beta(q-p)\|u\|_E^p = (q-r)\lambda\int_\Omega h(x)|u|^r dx \leq (q-r)\lambda S_1^r h_\eta\|u\|_E^r, \quad (25)$$

which implies that

$$\|u\|_E \leq \left[\frac{(q-r)\lambda S_1^r h_\eta}{2\sqrt{(q-2p)(q-p)}}\right]^{2/(3p-2r)}. \quad (26)$$

On the other hand, we can similarly get from (20) and (21) that

$$\begin{aligned}2\sqrt{\alpha\beta(2p-r)(p-r)\|u\|_E^{3p/2}} &\le (2p-r)\alpha\|u\|_E^{2p} + (p-r)\|u\|_E^p \\ &= (q-r)\int_{\partial\Omega} g(x)|u|^q d\sigma \\ &\le (q-r)\|g\|_\infty \|u\|_E^q,\end{aligned} \tag{27}$$

which yields that

$$\|u\|_E \ge \left[\frac{2\sqrt{\alpha\beta(2p-r)(p-r)}}{(q-r)\|g\|_\infty}\right]^{2/(2q-3p)}. \tag{28}$$

Thus, inequalities (26) and (28) show that $\lambda \ge \lambda_0$, which contradicts the hypothesis of λ.

For $q = 2p$, we can similarly obtain from (19) and (21) that

$$\|u\|_E \le \left[\frac{(q-r)\lambda S_1^r h_\eta}{p\beta}\right]^{1/(p-r)}. \tag{29}$$

Thus, (28) and (29) imply that $\lambda > \lambda_0$, which is also a contradiction. Therefore, we complete the proof. □

Lemma 5. *If u_0 is a local minimizer of $J_\lambda(u)$ on N_λ and $u_0 \notin N_\lambda^0$, then u_0 is a critical point of $J_\lambda(u)$.*

Proof. Let

$$\begin{aligned}F(u) = M\left(\|u\|_E^p\right)\|u\|_E^p - \int_{\partial\Omega} g(x)|u|^q d\sigma \\ - \lambda\int_\Omega h(x)|u|^r dx, \quad \forall u \in E.\end{aligned} \tag{30}$$

Consider the following minimizing problem:

$$\min_{u\in N_\lambda} J_\lambda(u), \quad \text{subject to } F(u) = 0. \tag{31}$$

By the Lagrange multiplier principle, there exists $\gamma \in \mathbb{R}^1$ such that

$$J_\lambda'(u_0) = \gamma F'(u_0), \tag{32}$$

$$\left\langle J_\lambda'(u_0), u_0\right\rangle = \gamma\left\langle F'(u_0), u_0\right\rangle = \gamma\left\langle \Phi_\lambda'(u_0), u_0\right\rangle. \tag{33}$$

Since $u_0 \in N_\lambda$ and $u_0 \notin N_\lambda^0$, it follows from (33) that $\gamma = 0$; furthermore, $J_\lambda'(u_0) = 0$. □

Now, we write $N_\lambda = N_\lambda^+ \cup N_\lambda^-$ for $0 < \lambda < \lambda_0$ and define

$$\delta_\lambda^+ = \inf_{u\in N_\lambda^+} J_\lambda(u), \qquad \delta_\lambda^- = \inf_{u\in N_\lambda^-} J_\lambda(u). \tag{34}$$

Denote

$$\lambda_1 = \frac{r(q-p)\beta}{p(q-r)S_1^r h_\eta}\left[\frac{\beta(p-r)}{S_2(q-r)\|g\|_\infty}\right]^{(p-r)/(q-p)}. \tag{35}$$

Then, the following results on δ_λ^+ and δ_λ^- are established.

Lemma 6. *Let $r < p < q$, $q \ge 2p$, and $0 < \lambda < \lambda_1$; then*

(i) $\delta_\lambda^+ < 0$,

(ii) *there exists constant $k_0 > 0$ such that $\delta_\lambda^- > k_0$.*

Proof. (i) For $\forall u \in N_\lambda^+$, we have from (19) and (21) that

$$\begin{aligned}\frac{(q-2p)\alpha}{2pq}\|u\|_E^{2p} + \frac{(q-p)\beta}{pq}\|u\|_E^p \\ \le \frac{(q-r)\lambda}{pq}\int_{\partial\Omega} g(x)|u|^q d\sigma,\end{aligned} \tag{36}$$

$$\lambda\int_{\partial\Omega} g(x)|u|^q d\sigma \ge \frac{(q-p)\beta}{q-r}\|u\|_E^p.$$

Thus, (36) and (16) give that

$$\begin{aligned}J_\lambda(u) &< \left(\frac{q-r}{pq} - \frac{q-r}{qr}\right)\lambda\int_\Omega h(x)|u|^r dx \\ &\le -\frac{\beta(p-r)(q-p)}{qpr}\|u\|_E^p < 0,\end{aligned} \tag{37}$$

which implies that

$$\delta_\lambda^+ = \inf_{u\in N_\lambda^+} J_\lambda(u) < 0. \tag{38}$$

(ii) Let $u \in N_\lambda^+$; one can deduce from (20) and (21) that

$$\|u\|_E \ge \left[\frac{(p-r)\beta}{S_2(q-r)\|g\|_\infty}\right]^{1/(q-p)}. \tag{39}$$

On the other hand, we obtain from (16), (26), and (39) that

$$\begin{aligned}J_\lambda(u) &\ge \left(\frac{1}{2p} - \frac{1}{q}\right)\alpha\|u\|_E^{2p} + \left(\frac{1}{p} - \frac{1}{q}\right)\beta\|u\|_E^p \\ &\quad - \frac{(q-r)\lambda}{rq}S_1^r h_\eta\|u\|_E^r \\ &\ge \|u\|_E^r\left[\frac{(q-p)\beta}{pq}\|u\|_E^{p-r} - \frac{(q-r)\lambda S_1^r h_\eta}{qr}\right] \\ &\ge \|u\|_E^r\left[\frac{(q-p)\beta}{pq}\left(\frac{(p-r)\beta}{S_2(q-r)\|g\|_\infty}\right)^{(p-r)/(q-p)} \right. \\ &\quad \left. - \frac{(q-r)\lambda S_1^r h_\eta}{rq}\right].\end{aligned} \tag{40}$$

Therefore, if $0 < \lambda < \lambda_1$, there exists $k_0 > 0$ such that $\delta_\lambda^- > k_0$. □

For each $u \in E$ with $\int_{\partial\Omega} g(x)|u|^q d\sigma > 0$, we define

$$m_\alpha(t) = \alpha t^{2p-r}\|u\|_E^{2p} + \beta t^{p-r}\|u\|_E^p - t^{q-r}\int_{\partial\Omega} g|u|^q d\sigma. \tag{41}$$

Then, $m_\alpha(0) = 0$, $m_\alpha(t) \to -\infty$ as $t \to +\infty$, and $m_\alpha(t)$ gets its unique maximum at the critical point $t_{\alpha,\max}$. Particularly, $m_0(t)$ gets its unique maximum at

$$t_{0,\max} = \left[\frac{\beta(p-r)\|u\|_E^p}{(q-r)\int_{\partial\Omega} g|u|^q d\sigma}\right]^{1/(q-p)}, \tag{42}$$

and we have the following results.

Lemma 7. *Let $r < p < q < p_*$, $q \geq 2p$, and $0 < \lambda < (p/r)\lambda_1$. For each $u \in E$ with $\int_{\partial\Omega} g(x)|u|^q d\sigma > 0$, one has the following:*

(i) *if $\int_\Omega h(x)|u|^r dx \leq 0$, then there exists $t^- > t_{\alpha,\max}$ such that $t^- u \in N_\lambda^-$ and*

$$J_\lambda(t^- u) = \sup_{t\geq 0} J_\lambda(tu); \tag{43}$$

(ii) *if $\int_\Omega h(x)|u|^r dx > 0$, then there exists $0 < t^+ < t_{\alpha,\max} < t^-$ such that $t^+ u \in N_\lambda^+$, $t^- u \in N_\lambda^-$, and*

$$\begin{aligned} J_\lambda(t^+ u) &= \inf_{0<t\leq t_{\alpha,\max}} J_\lambda(tu), \\ J_\lambda(t^- u) &= \sup_{t\geq t_{\alpha,\max}} J_\lambda(tu). \end{aligned} \tag{44}$$

Proof. (i) For $\forall u \in E$, we define

$$\begin{aligned} \phi_0(t) &= \langle J_\lambda'(tu), tu\rangle \\ &= t^{2p}\alpha\|u\|_E^{2p} + t^p\beta\|u\|_E^p - t^q \\ &\quad \times \int_{\partial\Omega} g(x)|u|^q d\sigma - t^r\lambda\int_\Omega h(x)|u|^r dx, \end{aligned} \tag{45}$$

$$\begin{aligned} \phi_1(t) &= \langle \Phi_\lambda'(tu), tu\rangle \\ &= 2p\alpha t^{2p}\|u\|_E^{2p} + t^p\beta\|u\|_E^p \\ &\quad - t^q\int_{\partial\Omega} g|u|^q d\sigma - t^r\lambda\int_\Omega h(x)|u|^r dx, \end{aligned} \tag{46}$$

$$\begin{aligned} \phi_2(t) &= J_\lambda(tu) \\ &= \frac{\alpha}{2p}t^{2p}\|u\|_E^{2p} + \frac{\beta}{p}t^p\|u\|_E^p \\ &\quad - \frac{1}{q}t^q\int_{\partial\Omega} g|u|^q d\sigma - \frac{\lambda}{r}t^r\int_\Omega h(x)|u|^r dx. \end{aligned} \tag{47}$$

Since $\int_\Omega h(x)|u|^r dx \leq 0$, there exists unique $t^- > t_{\alpha,\max}$ such that

$$m_\alpha'(t^-) < 0, \qquad m_\alpha(t^-) = \int_\Omega h(x)|u|^r dx. \tag{48}$$

Thus, $\phi_0(t^-) = t^r(m_\alpha(t^-) - \int_\Omega h(x)|u|^r dx) = 0$, which implies that $t^- u \in N_\lambda$. It follows from (20) that

$$\begin{aligned} \phi_1(t^-) &= (2p-r)\alpha(t^-)^{2p}\|u\|_E^{2p} \\ &\quad + (p-r)\beta(t^-)^p\|u\|_E^p \\ &\quad - (q-r)(t^-)^q\int_{\partial\Omega} g|u|^q d\sigma \\ &= (t^-)^{r+1}m_\alpha'(t^-) < 0; \end{aligned} \tag{49}$$

that is, $t^- u \in N_\lambda^-$. By (47), we obtain that

$$\begin{aligned} \phi_2'(t) &= \alpha t^{2p-1}\|u\|_E^{2p} + \beta t^{p-1}\|u\|_E^p \\ &\quad - t^{q-1}\int_{\partial\Omega} g|u|^q d\sigma - \lambda t^{r-1}\int_\Omega h(x)|u|^r dx \\ &= t^{r-1}\left[m_\alpha(t) - \lambda\int_\Omega h(x)|u|^r dx\right], \end{aligned} \tag{50}$$

which shows that $\phi_2(t)$ increases for $t \in [0, t^-]$ and decreases for $t \in [t^-, +\infty)$. Therefore,

$$J_\lambda(t^- u) = \phi_2(t^-) = \sup_{t\geq 0} J_\lambda(tu). \tag{51}$$

(ii) We firstly want to prove that $0 < \int_\Omega h(x)|u|^r dx \leq m_\alpha(t_{\alpha,\max})$. In fact,

$$\begin{aligned} m_0(t_{0,\max}) &= \beta\left[\frac{\beta(p-r)\|u\|_E^p}{(q-r)\int_{\partial\Omega} g|u|^q d\sigma}\right]^{(p-r)/(q-p)} \|u\|_E^p \\ &\quad - \left[\frac{\beta(p-r)\|u\|_E^p}{(q-r)\int_{\partial\Omega} g|u|^q d\sigma}\right]^{(q-r)/(q-p)} \\ &\quad \times \int_{\partial\Omega} g|u|^q d\sigma \\ &= \beta^{(q-r)/(q-p)}\|u\|_E^{p(q-r)/(q-p)}\left(\frac{p-r}{q-r}\right)^{(p-r)/(q-p)} \\ &\quad \times \left(\frac{q-p}{q-r}\right)\left[\frac{1}{\|g\|_\infty S_2\|u\|_E^q}\right]^{(p-r)/(q-p)} \\ &\geq \beta^{(q-r)/(q-p)}\|u\|_E^r\left(\frac{q-p}{q-r}\right) \\ &\quad \times \left[\frac{p-r}{(q-r)\|g\|_\infty}\right]^{(p-r)/(q-p)}. \end{aligned} \tag{52}$$

Then, if $0 < \lambda < (p/r)\lambda_1$, we have that

$$m_\alpha(0) = 0 < \lambda \int_\Omega h(x)|u|^r dx \le \beta^{(q-r)/(q-p)} \|u\|_E^r \left(\frac{q-p}{q-r}\right) \left[\frac{p-r}{(q-r)\|g\|_\infty}\right]^{(p-r)/(q-p)} \le m_0(t_{0,\max}) < m_\alpha(t_{\alpha,\max}). \tag{53}$$

Since $\int_\Omega h(x)|u|^r dx > 0$, there exists $0 < t^+ < t_{\alpha,\max} < t^-$ such that

$$m_\alpha(t^+) = m_\alpha(t^-) = \lambda \int_\Omega h(x)|u|^r dx, \tag{54}$$

and $m'_\alpha(t^+) > 0$, $m'_\alpha(t^-) < 0$. Similar to the proof of (i), we get that $t^+u \in N_\lambda^+$, $t^-u \in N_\lambda^-$. We can deduce from (50) that $\phi_2(t)$ decreases for $t \in [0, t^+]$ and increases for $t \in [t^+, t_{\alpha,\max})$. Therefore,

$$J_\lambda(t^+u) = \phi_2(t^+) = \inf_{0\le t\le t_{\alpha,\max}} J_\lambda(tu). \tag{55}$$

Similarly,

$$J_\lambda(t^-u) = \phi_2(t^-) = \sup_{t\ge t_{\alpha,\max}} J_\lambda(tu). \tag{56}$$

Then, we complete the proof. □

For each $u \in E$ with $\int_\Omega h(x)|u|^r dx > 0$, we define

$$\overline{m}_\alpha(t) = \alpha t^{2p-q}\|u\|_E^{2p} + \beta t^{p-q}\|u\|_E^p - \lambda t^{r-q}\int_\Omega h(x)|u|^r dx. \tag{57}$$

Then, $\overline{m}_\alpha(t) \to -\infty$ as $t \to 0^+$ and $\overline{m}_\alpha(t) \to 0$ as $t \to \infty$. Furthermore, $\overline{m}_\alpha(t)$ gets its unique maximum at some certain point $\bar{t}_{\alpha,\max}$. Particularly, $\overline{m}_0(t)$ gets its unique maximum at the critical point

$$\bar{t}_0 = \bar{t}_{0,\max}(u) = \left[\frac{\lambda(q-r)\int_\Omega h(x)|u|^r dx}{\beta(q-p)\|u\|_E^p}\right]^{1/(p-r)}, \tag{58}$$

then we have the following results.

Lemma 8. *Let $1 < r < p < q$, $q \ge 2p$, and $0 < \lambda < p\lambda_1/r$. For each $u \in E$ with $\int_\Omega h(x)|u|^r dx > 0$ one has the following:*

(i) *if $\int_{\partial\Omega} g|u|^q d\sigma < 0$, then there exists unique $0 < t^+ < \bar{t}_{\alpha,\max}$ such that $t^+u \in N_\lambda^+$ and*

$$J_\lambda(t^+u) = \inf_{t\ge 0} J_\lambda(tu); \tag{59}$$

(ii) *if $\int_{\partial\Omega} g|u|^q d\sigma \ge 0$, then there exists $0 < t^+ < \bar{t}_{\alpha,\max} < t^-$ such that $t^+u \in N_\lambda^+$, $t^-u \in N_\lambda^-$, and*

$$J_\lambda(t^+u) = \inf_{0<t\le \bar{t}_{\alpha,\max}} J_\lambda(tu), \qquad J_\lambda(t^-u) = \sup_{t\ge \bar{t}_{\alpha,\max}} J_\lambda(tu). \tag{60}$$

Proof. (i) Since $\int_{\partial\Omega} g|u|^q d\sigma < 0$, there exists unique $0 < t^+ < \bar{t}_{\alpha,\max}$ such that

$$\overline{m}'_\alpha(t^+) > 0, \qquad \overline{m}_\alpha(t^+) = \int_{\partial\Omega} g|u|^q d\sigma. \tag{61}$$

Note that (45) can be rewritten as

$$\phi_0(t) = t^q\left[\overline{m}_\alpha(t) - \int_{\partial\Omega} g|u|^q d\sigma\right]. \tag{62}$$

Thus, (61) shows that $t^+u \in N_\lambda$. By virtue of (14) and (46), we get that

$$\begin{aligned}\phi_1(t^+) &= \langle \Phi'_\lambda(t^+u), t^+u\rangle \\ &= (t^+)^{q+1}\Big[(2p-q)\alpha(t^+)^{2p-q-1} + (p-q)\beta(t^+)^{p-q-1}\|u\|_E^p \\ &\quad -\lambda(r-q)(t^+)^{r-q-1}\int_\Omega h(x)|u|^r dx\Big] \\ &= (t^+)^{q+1}\overline{m}'_\alpha(t^+) > 0,\end{aligned} \tag{63}$$

which implies that $t^+u \in N_\lambda^+$. We have from (47) that

$$\phi'_2(t) = t^{q-1}\left[\overline{m}_\alpha(t) - \int_{\partial\Omega} g|u|^q d\sigma\right]. \tag{64}$$

Then, $\phi_2(t)$ decreases for $t \in [0, t^+]$ and increases for $t \in [t^+, +\infty)$; that is,

$$J_\lambda(t^+u) = \inf_{t\ge 0} J_\lambda(tu). \tag{65}$$

(ii) We need also to prove that $0 < \int_{\partial\Omega} g|u|^q d\sigma < \overline{m}_\alpha(t)$. In fact, we have that

$$\begin{aligned}\overline{m}_\alpha(t_{\alpha,\max}) &\ge \overline{m}_0(\bar{t}_0) \\ &= \beta\left[\frac{\lambda(q-r)\int_\Omega h|u|^r dx}{\beta(q-p)}\|u\|_E^p\right]^{(p-q)/(p-r)}\|u\|_E^{2p} \\ &\quad - \lambda\left[\frac{\lambda(q-r)\int_\Omega h|u|^r dx}{\beta(q-p)}\|u\|_E^p\right]^{(r-q)/(p-r)} \\ &= \lambda^{(p-q)/(p-r)}\beta^{(q-r)/(p-r)}\|u\|_E^{p(q-r)/(p-r)} \\ &\quad \times \left(\frac{q-p}{q-r}\right)^{(q-p)/(p-r)} \\ &\quad \times \left(\frac{p-r}{q-r}\right)\left(\int_\Omega h|u|^r dx\right)^{(p-q)/(p-r)} \\ &\ge \lambda^{(p-q)/(p-r)}\beta^{(q-r)/(p-r)}\left(\frac{q-r}{p-r}\right) \\ &\quad \times \left(\frac{q-p}{(q-r)S_1^r h_\eta}\right)^{(q-p)/(p-r)}\|u\|_E^q \\ &\ge \int_{\partial\Omega} g|u|^q d\sigma \ge 0\end{aligned} \tag{66}$$

for $0 < \lambda \le (p/r)\lambda_1$.

The rest of the proof is similar to that of (ii) in Lemma 7, and here we omit the proof. □

Lemma 9. *Assume* (A_1)*–*(A_3)*. If* $u_n \rightharpoonup u$ *in* E*, then there exists a subsequence of* $\{u_n\}$*, still denoted by* $\{u_n\}$*, such that*

$$\int_\Omega h(x)|u_n|^r dx \longrightarrow \int_\Omega h(x)|u|^r dx, \qquad \int_{\partial\Omega} g(x)|u_n|^q d\sigma \longrightarrow \int_{\partial\Omega} g(x)|u|^q d\sigma. \quad (67)$$

Proof. By the assumption (A_2), we have $h(x)|x|^{br} \in L^\eta(\Omega) \cap L^\infty(\Omega)$, and then for any $\varepsilon > 0$, there exists $R_\varepsilon > 0$ large enough such that

$$\int_{B^c_{R_\varepsilon}} \left(|h(x)|\,|x|^{br}\right)^\eta dx \le \varepsilon^\eta, \quad (68)$$

where $B_R = \{x \in \mathbb{R}^N \mid |x| \le R\}$, $B^c_R = \{x \in \mathbb{R}^N \mid |x| > R\}$ for $R > 0$.

The compact embedding theorem $W^{1,p}_a(\Omega) \hookrightarrow L^r(\Omega, |x|^{-\alpha})$ (Theorem 2.1 in [19]) implies that

$$u_n \longrightarrow u \quad \text{a.e. in } \Omega. \quad (69)$$

Inequality (7) shows that $\{u_n\}$ is bounded in $L^{p^*}(\mathbb{R}^N, |x|^{br})$, which implies that $u_n \rightharpoonup u$ in $L^{p^*}(\mathbb{R}^N, |x|^{br})$. Thus, we can obtain that

$$\int_{B_{R_\varepsilon}\setminus D} |x|^{-bp^*}|u_n - u|^{p^*} dx < \varepsilon^{p^*}, \qquad \int_\Omega |x|^{-bp^*}|u_n|^{p^*} dx \le M^{p^*}, \qquad \int_\Omega |x|^{-bp^*}|u|^{p^*} dx \le M^{p^*} \quad (70)$$

for some $M > 0$. On the other hand, we get from the Höder inequality and (68)–(70) that

$$\int_{B^c_{R_\varepsilon}} h|u_n - u|^r dx \le \left(\int_{B^c_{R_\varepsilon}} \left(|h|\,|x|^{br}\right)^\eta dx\right)^{1/\eta} \times \left(\int_{B^c_{R_\varepsilon}} |x|^{-bp^*}|u_n - u|^{p^*} dx\right)^{r/p^*} \le 2^r M^r \varepsilon, \quad (71)$$

$$\int_{B_{R_\varepsilon}\setminus D} h|u_n - u|^r dx \le \left(\int_{B_{R_\varepsilon}\setminus D} \left(|h|\,|x|^{br}\right)^\eta dx\right)^{1/\eta} \times \left(\int_{B_{R_\varepsilon}\setminus D} |x|^{-bp^*}|u_n - u|^{p^*} dx\right)^{r/p^*} \le c_0\varepsilon^r$$

for some constant $c_0 > 0$ and large n.

Thus, (71) implies that

$$\int_\Omega h(x)|u_n - u|^r dx \longrightarrow 0 \quad \text{as } n \longrightarrow \infty; \quad (72)$$

that is,

$$\int_\Omega h(x)|u_n|^r dx \longrightarrow \int_\Omega h(x)|u|^r dx \quad \text{as } n \longrightarrow \infty. \quad (73)$$

Since $\partial\Omega$ is compact and $g \in C(\partial\Omega) \cap L^\infty(\partial\Omega)$, we obtain by the trace embedding theorem in [20] that

$$\lambda\int_{\partial\Omega} g(x)|u_n|^q d\sigma \longrightarrow \lambda\int_{\partial\Omega} g(x)|u|^q d\sigma \quad \text{as } n \longrightarrow \infty. \quad (74)$$

This concludes the proof. □

3. Existence of Solutions

In this part, we will give the proof of the existence of nonnegative and nontrivial solutions. Before this, we need to prove the following two important lemmas.

Lemma 10. *Assume* (A_3) *and* $0 < \lambda < (p/r)\lambda_1$*. Then, the functional* $J_\lambda(u)$ *has a minimizer* $u_0^+ \in N_\lambda^+$ *and*

(i) $J_\lambda(u_0^+) = \delta_\lambda^+$,

(ii) u_0^+ *is a positive weak solution of problem* (1).

Proof. Since $J_\lambda(u)$ is bounded in N_λ^+, there exists a minimizing sequence of $\{u_k\} \in N_\lambda^+$ such that

$$\lim_{k\to\infty} J_\lambda(u_k) = \inf_{u\in N_\lambda^+} J_\lambda(u). \quad (75)$$

We can get that $\{u_k\}$ is bounded in E and $u_k \rightharpoonup u_0^+$ weakly in E since $J_\lambda(u)$ is coercive. It follows from Lemma 6 that

$$J_\lambda(u_k) \longrightarrow \delta_\lambda^+ < 0 \quad \text{as } k \longrightarrow +\infty. \quad (76)$$

Furthermore, equality (16) and Lemma 9 imply that $\int_\Omega h(x)|u_0^+|^r dx > 0$. In the following, we prove that $u_k \to u_0^+$ in E. Suppose otherwise that

$$\|u_0^+\|_E < \lim_{k\to+\infty}\inf \|u_k\|_E. \quad (77)$$

Let

$$m_{\alpha,u_0^+}(t) = \alpha t^{2p-q}\|u_0^+\|_E^{2p} + \beta t^{p-q}\|u_0^+\|_E^p - \lambda t^{r-q}\int_\Omega h(x)|u_0^+|^r dx,$$

$$\varphi_0(t) = \alpha t^{2p-q}\|u_0^+\|_E^{2p} + \beta t^{p-q}\|u_0^+\|_E^p - \lambda t^{r-q}\int_\Omega h(x)|u_0^+|^r dx - \int_{\partial\Omega} g(x)|u_0^+|^q d\sigma. \quad (78)$$

It is obvious that $m_{\alpha,u_0^+}(t)$ has the unique critical point $t_{\alpha,\max}(u_0^+)$ and $m_{\alpha,u_0^+}(t)$ increases for $t \in [0, t_{\alpha,\max}(u_0^+)]$ and decreases for $t \in [t_{\alpha,\max}(u_0^+), \infty)$. Particularly, $m_{0,u_0^+}(t)$ has the unique critical point at $t_0(u_0^+) = \bar{t}_{0,\max}(u_0^+)$. Since $\int_\Omega h(x)|u_0^+|^r dx > 0$, it follows from Lemma 8 that there exists $t_0^+ < t_{\alpha,\max}(u_0^+)$ such that $t_0^+ u_0^+ \in N_\lambda^+$ and

$$J_\lambda(t_0^+ u_0^+) = \inf_{0 \le t \le t_{\alpha,\max}(u_0^+)} J_\lambda(t u_0^+). \quad (79)$$

By virtue of $t_0^+ u_0^+ \in N_\lambda^+$, we obtain that

$$\varphi_0(t_0^+) = (t_0^+)^{-q} \langle J_\lambda(t_0^+ u_0^+), t_0^+ u_0^+ \rangle = 0. \quad (80)$$

Similarly, we set

$$\varphi_k(t) = \alpha t^{2p-q} \|u_k\|_E^{2p} + \beta t^{p-q} \|u_k\|_E^p - \lambda t^{r-q} \int_\Omega h(x)|u_k|^r dx - \int_{\partial\Omega} g(x)|u_k|^q d\sigma. \quad (81)$$

In view of (77), we get that $\varphi_k(t_0^+) > \varphi_0(t_0^+) = 0$ for large k. Since $\varphi_k(1) = 0$ and $\varphi_k'(t) \ge m'_{\alpha,u_0^+}(t) > 0$ for $t \in (0, t_{\alpha,\max}(u_0^+)]$, we obtain that $1 < t_0^+ < t_{\alpha,\max}(u_0^+)$ and $\varphi_k(t) < 0$ for $t \in [0,1]$. Then, by (77), (79), and Lemma 9, we get that

$$J_\lambda(t_0^+ u_0^+) \le J_\lambda(u_0^+) < \lim_{k\to+\infty} J_\lambda(u_k) = \delta_\lambda^+, \quad (82)$$

which is a contradiction. Hence, $u_k \to u_0^+$ strongly in E, and

$$J_\lambda(u_k) \longrightarrow J_\lambda(u_0^+) = \delta_\lambda^+ \quad \text{as } k \longrightarrow +\infty. \quad (83)$$

Thus, u_0^+ is a minimum of $J_\lambda(u)$ on N_λ^+. Since $J_\lambda(u_0^+) = J_\lambda(|u_0^+|)$ and $|u_0^+| \in N_\lambda^+$, we may assume by Lemma 5 that u_0^+ is a positive solution of problem (1). □

Lemma 11. *Assume (A_3) and $0 < \lambda < (p/r)\lambda_1$. Then, the functional $J_\lambda(u)$ has a minimizer $u_0^- \in N_\lambda^-$ and*

(i) $J_\lambda(u_0^-) = \delta_\lambda^-$,

(ii) *u_0^- is a positive weak solution of problem* (1).

Proof. Since $J_\lambda(u)$ is bounded on N_λ^-, there exists a minimizing sequence $\{u_k\} \in N_\lambda^-$ such that

$$\lim_{k\to\infty} J_\lambda(u_k) = \inf_{u\in N_\lambda^-} J_\lambda(u). \quad (84)$$

Similar to the proof of Lemma 10, we may assume that $u_k \rightharpoonup u_0^-$ weakly in E. For $u_k \in N_\lambda^-$, we deduce by Lemma 6 and (15) that $\int_{\partial\Omega} g(x)|u_k|^q d\sigma > 0$; furthermore, $\int_{\partial\Omega} g(x)|u_0^-|^q d\sigma > 0$. We also want to prove that $u_k \to u_0^-$ strongly in E. In fact, if not, we have

$$\|u_0^+\|_E < \lim_{k\to+\infty} \inf \|u_k\|_E. \quad (85)$$

By virtue of Lemma 7, there exists $t_0^- > 0$ such that

$$J_\lambda(t_0^- u) = \sup_{t\ge 0} J_\lambda(t u_k), \quad (86)$$

and a simple transformation shows that

$$J_\lambda(u_k) \ge J_\lambda(t u_k), \quad \forall u_k \in N_\lambda^-. \quad (87)$$

Therefore, Lemma 9 together with (84)–(87) gives that

$$J_\lambda(t_0^- u_0^-) < \lim_{k\to+\infty} J_\lambda(t_0^- u_k) \le \lim_{k\to+\infty} J_\lambda(u_k) = \delta_\lambda^-, \quad (88)$$

which is a contradiction, and we complete the proof. □

Proof of Theorem 2. We set $\lambda^* = \min\{\lambda_1, \lambda_0\}$. When $0 < \lambda < \lambda^*$, by Lemmas 10 and 11, we obtain that problem (1) has two nontrivial nonnegative solutions $u_0^+ \in N_\lambda^+$, $u_0^- \in N_\lambda^-$. Lemma 4 and the assumptions of Theorem 2 imply that $N_\lambda^+ \cap N_\lambda^- = \emptyset$; then u_0^+ and u_0^- are distinct. Furthermore, since $J_\lambda(u_0^\pm) = J_\lambda(|u_0^\pm|)$ and $|u_0^\pm| \in N_\lambda^\pm$, we can assume that the solutions u_0^+ and u_0^- are positive. This completes the proof. □

4. Conclusions

The object of this paper is to prove the existence of multiple solutions for the nonlinear Kirchhoff-type problem (1). By the variational methods, we discuss the problem on the Nehari manifold and give the sufficient conditions for the existence of solutions. We overcome the difficulty due to the loss of compactness on the unbounded domain. In the future work, we are interested to consider similar problems, but the term on the right will be replaced by abstract functions.

Acknowledgment

The authors would like to express their sincere gratitude to the anonymous reviewers for the valuable comments and suggestions.

References

[1] G. Kirchhoff, *Mechanik*, Teubner, Leipzig, Germany, 1883.

[2] G. Anello, "A uniqueness result for a nonlocal equation of Kirchhoff type and some related open problem," *Journal of Mathematical Analysis and Applications*, vol. 373, no. 1, pp. 248–251, 2011.

[3] C. O. Alves, F. J. S. A. Corrêa, and G. M. Figueiredo, "On a class of nonlocal elliptic problems with critical growth," *Differential Equations & Applications*, vol. 2, pp. 409–417, 2010.

[4] A. Bensedik and M. Bouchekif, "On an elliptic equation of Kirchhoff-type with a potential asymptotically linear at infinity," *Mathematical and Computer Modelling*, vol. 49, no. 5-6, pp. 1089–1096, 2009.

[5] C.-Y. Chen, Y.-C. Kuo, and T.-F. Wu, "The Nehari manifold for a Kirchhoff type problem involving sign-changing weight functions," *Journal of Differential Equations*, vol. 250, no. 4, pp. 1876–1908, 2011.

[6] Y. H. Li, F. Y. Li, and J. P. Shi, "Existence of a posit solutions to Kirchhoff type problems without compactness conditons," *Journal of Differential Equations*, vol. 253, pp. 2285–2294, 2012.

[7] K. Perera and Z. Zhang, "Nontrivial solutions of Kirchhoff-type problems via the Yang index," *Journal of Differential Equations*, vol. 221, no. 1, pp. 246–255, 2006.

[8] F. Wang and Y. An, "Existence of nontrivial solution for a nonlocal elliptic equation with nonlinear boundary condition," *Boundary Value Problems*, vol. 2009, Article ID 540360, 8 pages, 2009.

[9] Z. Zhang and K. Perera, "Sign changing solutions of Kirchhoff type problems via invariant sets of descent flow," *Journal of Mathematical Analysis and Applications*, vol. 317, no. 2, pp. 456–463, 2006.

[10] F. Cammaroto and L. Vilasi, "Multiple solutions for a Kirchhoff-type problem involving the $p(x)$-Laplacian operator," *Nonlinear Analysis: Theory, Methods and Applications*, vol. 74, no. 5, pp. 1841–1852, 2011.

[11] F. J. S. A. Corrêa and G. M. Figueiredo, "On a p-Kirchhoff equation via Krasnoselskii's genus," *Applied Mathematics Letters*, vol. 22, no. 6, pp. 819–822, 2009.

[12] G. Dai and J. Wei, "Infinitely many non-negative solutions for a $p(x)$-Kirchhoff-type problem with Dirichlet boundary condition," *Nonlinear Analysis: Theory, Methods and Applications*, vol. 73, no. 10, pp. 3420–3430, 2010.

[13] C. Ji, "Infinitely many radial solutions for the $p(x)$-Kirchhoff-type equation with oscillatory nonlinearities in RN," *Journal of Mathematical Analysis and Applications*, vol. 388, no. 2, pp. 727–738, 2012.

[14] Z. H. Xiu, "Existence of multiple solutions for a singular elliptic problem with critical sobolev expoent," *Abstract and Applied Analysis*, vol. 2012, Article ID 806397, 15 pages, 2012.

[15] Caffarelli, R. Kohn, and L. Nirenberg, "First order interpolation inequalities with weights," *Compositio Mathematica*, vol. 53, pp. 437–477, 1984.

[16] J. I. Díaz, "Nonlinear partial differential equations and free boundaries," *Elliptic Equations*, vol. 323, pp. 44–95, 1985.

[17] K. J. Brown, "The Nehari manifold for a semilinear elliptic equation involving a sublinear term," *Calculus of Variations and Partial Differential Equations*, vol. 22, no. 4, pp. 483–494, 2005.

[18] Z. Nehari, "On a class of nonlinear second-order differential equations," *Transactions of the American Mathematical Society*, vol. 95, pp. 101–123, 1960.

[19] B. Xuan, "The solvability of quasilinear Brezis-Nirenberg-type problems with singular weights," *Nonlinear Analysis: Theory, Methods and Applications*, vol. 62, no. 4, pp. 703–725, 2005.

[20] R. D. S. Rodrigues, "On elliptic problems involving critical Hardy-Sobolev exponents and sign-changing function," *Nonlinear Analysis: Theory, Methods and Applications*, vol. 73, no. 4, pp. 857–880, 2010.

13

Positive Solutions of Advanced Differential Systems

Josef Diblík[1,2] and Mária Kúdelčíková[3]

[1] *Department of Mathematics and Descriptive Geometry, Faculty of Civil Engineering, Brno University of Technology, Veveří 331/95, 602 00 Brno, Czech Republic*
[2] *Department of Mathematics, Faculty of Electrical Engineering and Communication, Brno University of Technology, Technická 3058/10, 616 00 Brno, Czech Republic*
[3] *Department of Mathematics, University of Žilina, Univerzitná 8215/1, 010 26 Žilina, Slovakia*

Correspondence should be addressed to Josef Diblík; diblik@feec.vutbr.cz

Academic Editors: J. Banaś, M. M. Cavalcanti, N. Kallur, and M. Mansour

We study asymptotic behavior of solutions of general advanced differential systems $\dot{y}(t) = F(t, y_t)$, where $F : \Omega \to \mathbb{R}^n$ is a continuous quasi-bounded functional which satisfies a local Lipschitz condition with respect to the second argument and Ω is a subset in $\mathbb{R} \times C_r^n$, $C_r^n := C([0,r], \mathbb{R}^n)$, $y_t \in C_r^n$, and $y_t(\theta) = y(t+\theta)$, $\theta \in [0,r]$. A monotone iterative method is proposed to prove the existence of a solution defined for $t \to \infty$ with the graph coordinates lying between graph coordinates of two (lower and upper) auxiliary vector functions. This result is applied to scalar advanced linear differential equations. Criteria of existence of positive solutions are given and their asymptotic behavior is discussed.

1. Introduction

Let $C([a,b], \mathbb{R}^n)$, where $a, b \in \mathbb{R}$, $a < b$, be the Banach space of the continuous mappings from the interval $[a,b]$ to $\mathbb{R}^n$ equipped with the supremum norm

$$\|\psi\|_C = \sup_{\theta \in [a,b]} \|\psi(\theta)\|, \quad \psi \in C([a,b], \mathbb{R}^n), \tag{1}$$

where $\|\cdot\|$ is the maximum norm in $\mathbb{R}^n$. In the case of $a = 0$ and $b = r > 0$, we will denote this space as C_r^n; that is,

$$C_r^n := C([0,r], \mathbb{R}^n). \tag{2}$$

If $\sigma \in \mathbb{R}$, $A \geq 0$, and $y \in C([\sigma, \sigma + r + A], \mathbb{R}^n)$, then, for each $t \in [\sigma, \sigma + A]$, we define $y_t \in C_r^n$ by $y_t(\theta) = y(t + \theta)$, $\theta \in [0,r]$.

Let us consider a system of functional differential equations of advanced type:

$$\dot{y}(t) = F(t, y_t), \tag{3}$$

where $F : \Omega \to \mathbb{R}^n$ is a continuous quasi-bounded functional which satisfies a local Lipschitz condition with respect to the second argument and Ω is a subset in $\mathbb{R} \times C_r^n$. We recall that the functional F is quasi-bounded if F is bounded on every set of the form $[t_1, t_2] \times C_{rL}^n \subset \Omega$, where $t_1 < t_2$, $C_{rL}^n := C([0,r], L)$, and L is a closed bounded subset of $\mathbb{R}^n$ (cf. [1, page 305]).

A function y is said to be a *solution of* (3) *on* $[\sigma, \sigma+A)$ with $A > 0$ if $y \in C([\sigma, \sigma + r + A], \mathbb{R})$, $(t, y_t) \in \Omega$ for $t \in [\sigma, \sigma + A)$, and $y(t)$ satisfies (3) for $t \in [\sigma, \sigma + A)$.

The paper is organized as follows. In Section 2 necessary iterative technique is formulated. In Section 3 it is applied to general nonlinear advanced differential system. By monotone iterative method, we will prove a general criterion on existence of bounded solutions of the nonlinear system (3) or, more exactly, we give necessary conditions for the existence of a solution defined for $t \to \infty$ with the graph coordinates lying between graph coordinates of two (lower and upper) auxiliary vector functions. If the lower function can be taken with positive coordinates, then the statement of theorem concerns positive solutions.

Section 4 is devoted to scalar linear cases. Nonlinear result proved in Section 3 is applied to the scalar advanced linear differential equation

$$\dot{y}(t) = (c + d(t))\, y(t + r), \tag{4}$$

where $c, r \in \mathbb{R}^+ = (0, \infty)$ are constants and $d(t)$ is a locally Lipschitz continuous function satisfying $|d(t)| < \varepsilon < c$ for $t \in [t_0, \infty)$. We will assume $(c + \varepsilon) < 1/(re)$ as well. The case when there exist positive solutions is studied. With the aid of two auxiliary equations, constructed using lower and upper estimates of the right-hand side of (4), and under the supposition of the existence of two real (positive) different roots of corresponding transcendental equations, the existence of a positive solution of (4) is proved. Simultaneously its asymptotic behavior is derived. Next, a linear advanced equation, more general than (4),

$$\dot{y}(t) = \gamma(t)\, y(t + r), \tag{5}$$

where $\gamma : [t_0, \infty) \to \mathbb{R}^+$ is bounded, locally Lipschitz continuous function and r is a positive constant, is considered and a criterion of existence of positive solutions is given.

The fact of existence of positive solutions of an advanced differential equation can be documented, for example, by the equation

$$\dot{y}(t) = \frac{1}{\tau e} y(t + \tau) \tag{6}$$

which admits a pair of positive and asymptotically different for $t \to \infty$ solutions:

$$y_1(t) = t \exp\left(\frac{t}{\tau}\right), \qquad y_2(t) = \exp\left(\frac{t}{\tau}\right). \tag{7}$$

In the literature one can find some results on existence of positive solutions of advanced equations. For example, in accordance with [2, page 31] and [3, page 21], the first-order advanced type differential equation

$$\dot{y}(t) = p(t)\, y(t + \tau(t)), \tag{8}$$

where $p, \tau \in C([t_0, \infty), \mathbb{R}^+)$, has a positive solution in the case when $\tau(t) \equiv \tau > 0$ if and only if there exists a continuous function $\lambda(t)$ such that

$$\int_t^{t+\tau} p(u)\, du \le \Lambda^{-1}(\Lambda(t) + \ln \lambda(t)) - t, \tag{9}$$

where $\Lambda(t) = \int_{t_0}^{t} \lambda(s) ds$ and Λ^{-1} is corresponding inverse function.

As a consequence of our result for nonlinear advanced equations we get a sufficient and necessary criterion of positivity different from the above-mentioned criterion (9).

Concluding remarks and open problems are formulated in Section 5.

2. Preliminaries

In this section we introduce some definitions and theorems which will be used later.

Definition 1 (see [4, page 276]). Let $\mathbb{L}$ be a Banach space and let $\mathbb{K} \subseteq \mathbb{L}$. Then $\mathbb{K}$ is called an order cone if and only if

(i) $\mathbb{K}$ is closed, nonempty, and $\mathbb{K} \neq \{0\}$;

(ii) $\alpha, \beta \in R$, $\alpha, \beta \ge 0$, $x, y \in K \Rightarrow \alpha x + \beta y \in \mathbb{K}$;

(iii) $x \in \mathbb{K}$ and $-x \in \mathbb{K} \Rightarrow x = 0$.

On this basis, we define for $x, y \in \mathbb{K}$

$$\begin{aligned} x &\le y \quad \text{iff } y - x \in \mathbb{K}, \\ x &< y \quad \text{iff } x \le y \text{ and } x \neq y, \\ x &\ll y \quad \text{iff } y - x \in \operatorname{int}(\mathbb{K}), \\ x &\nleq y \quad \text{iff } x \le y \text{ is false}, \\ [x, y] &:= \{z \in \mathbb{L} : x \le z \le y\} \ (\text{order interval}). \end{aligned} \tag{10}$$

Definition 2 (see [4, page 277]). Let $\mathbb{L}$, $\mathbb{Y}$ be Banach spaces and let $\mathbb{K} \subseteq \mathbb{L}$ be an order cone.

(1) The order cone $\mathbb{K}$ is called *normal* if and only if there is a number $c > 0$ such that, for all $x, y \in \mathbb{L}$,

$$\text{If } 0 \le x \le y \quad \text{then } \|x\| \le c\|y\|. \tag{11}$$

(2) The operator $T : D(T) \subseteq \mathbb{L} \to \mathbb{Y}$ is called *monotone increasing* if and only if it is true for all $x, y \in D(T)$ that

$$x < y \text{ implies } Tx \le Ty, \tag{12}$$

where $D(T)$ denotes domain of definition of the operator T. The operator is called *strictly* or *strongly* monotone increasing if and only if the symbol "$\le$" is replaced by "$<$" or "$\ll$," respectively. If we replace "$\le$" by "$\ge$," then T is monotone decreasing. Similarly, we have operators which are strictly or strongly monotone decreasing.

2.1. Supersolutions, Subsolutions, and Iterative Methods. We consider the operator equation

$$x = Tx \tag{13}$$

together with two iterative methods

$$u_{n+1} = Tu_n, \qquad v_{n+1} = Tv_n. \tag{14}$$

Definition 3 (see [4, page 282]). The point x is called a supersolution, a strict supersolution, or a strong supersolution of (13) if and only if $x \ge Tx$, $x > Tx$, or $x \gg Tx$, respectively. The prefix "super" is replaced by "sub" when the respective inequalities are reversed.

Definition 4 (see [4, page 54]). Let $\mathbb{L}$, $\mathbb{M}$ be Banach spaces, and $T : D(T) \subseteq \mathbb{L} \to \mathbb{M}$ an operator. T is called compact if and only if

(i) T is continuous;

(ii) T maps bounded sets into relatively compact sets.

Let us recall that a set $\mathbb{A}$ is *bounded* if and only if there is a number $R > 0$ such that $\|x\| \le R$ for all $x \in \mathbb{A}$ and is *relatively compact* (resp., *compact*) if and only if every sequence in $\mathbb{A}$ contains a convergent subsequence (resp., the limit of which also belongs to $\mathbb{A}$).

The following theorem is a part of Theorem 7.A in [4, page 283].

Theorem 5. *Suppose that $T : [u_0, v_0] \subseteq \mathbb{L} \to \mathbb{L}$ is a compact monotone increasing operator on a real Banach space $\mathbb{L}$ with normal order cone $\mathbb{K}$. If u_0 is a subsolution of (13) and if v_0 is a supersolution of (13) with $u_0 \le v_0$, then the iterative sequence $\{v_n\}$ in (14) converges to a fixed point of T, namely, to the greatest fixed point v of T in $[u_0, v_0]$, and $\{u_n\}$ converges to the smallest fixed point u of T in $[u_0, v_0]$. Furthermore, one has the error estimates*

$$u_n \le u \le v \le v_n, \quad \forall n = 0, 1, \ldots. \tag{15}$$

3. Nonlinear Case

In this part the advanced system of differential equations (3) is considered. Using the method of monotone sequences we prove that under certain additional assumptions there exists a solution of (3) satisfying half-infinity interval inequalities formulated in Theorem 6 (inequalities (26)).

By $\mathbb{R}^n_{\ge 0}$ ($\mathbb{R}^n_{>0}$) we denote the set of all component-wise nonnegative (positive) vectors v in $\mathbb{R}^n$; that is, $v = (v^1, \ldots, v^n)$ with $v^i \ge 0$ ($v^i > 0$) for $i = 1, \ldots, n$. For $u, v \in \mathbb{R}^n$, we say that $u \le v$ if $v - u \in \mathbb{R}^n_{\ge 0}$, $u \ll v$ if $v - u \in \mathbb{R}^n_{>0}$, $u < v$ if $u \le v$, and $u \ne v$.

For a fixed $t^* \in \mathbb{R}$ we put $\Omega := (t^*, \infty) \times C^n_r$. In the following, let $t_0 \in \mathbb{R}$ and $t_0 > t^*$.

Now for a given $k \in \mathbb{R}^n_{>0}$, we consider two systems of the integrofunctional inequalities

$$\begin{aligned} \lambda_1(t) &\le (I(k,\lambda_1)(t))^{-1} F(t, I(k,\lambda_1)_t), \\ \lambda_2(t) &\ge (I(k,\lambda_2)(t))^{-1} F(t, I(k,\lambda_2)_t) \end{aligned} \tag{16}$$

on $[t_0, \infty)$, $t_0 \in \mathbb{R}^n$ for $\lambda_1, \lambda_2 \in C([t_0,\infty), \mathbb{R}^n)$, $\lambda_1(t) < \lambda_2(t)$, where

$$I : \mathbb{R}^n_{>0} \times C([t_0,\infty), \mathbb{R}^n) \longrightarrow C([t_0,\infty), \mathbb{R}^n) \tag{17}$$

is defined by

$$I^i(k,\lambda)(t) := k^i \exp\left(\int_{t_0}^t \lambda^i(s)\, ds\right), \tag{18}$$

where $i = 1, \ldots, n$, $t \in [t_0, \infty)$, $\lambda \in C([t_0,\infty), \mathbb{R}^n)$.

The next theorem provides conditions under which it is possible to construct (with the aid of auxiliary given functions λ_1, λ_2 satisfying (16)) monotone sequences of functions converging to a solution of the operator equation

$$\lambda(t) = (T\lambda)(t) \tag{19}$$

on $[t_0, \infty)$ with

$$(T\lambda)(t) := (I(k,\lambda)(t))^{-1} F(t, I(k,\lambda)_t). \tag{20}$$

It is easy to verify that the system (3) is related to operator equation (19) through the substitution

$$y(t) = I(k,\lambda)(t). \tag{21}$$

A function λ is said to be a solution of operator equation (19) on $[\sigma, \sigma+A)$ with $A > 0$ if $\lambda \in C([\sigma, \sigma+r+A), \mathbb{R}^n)$, $(t, \lambda_t) \in \Omega$ for $t \in [\sigma, \sigma + A)$, and $\lambda(t)$ satisfies (19) for $t \in [\sigma, \sigma + A)$.

Theorem 6. *Let one assumes the following.*

(i) *For any $M \ge 0$, $\vartheta \ge t_0$ there are K_1, K_2 such that*

$$\|(T\lambda)(t)\| \le K_1, \tag{22}$$

$$\left\|(T\lambda)(t) - (T\lambda)(t')\right\| \le K_2 \left|t - t'\right| \tag{23}$$

for any $t, t' \in [t_0, \vartheta]$ and any function $\lambda \in C([t_0, \vartheta + r], \mathbb{R}^n)$ with $\|\lambda\|_C \le M$, where the norm is defined by (1) with $a = t_0$ and $b = \vartheta + r$.

(ii) *There is a $k \in \mathbb{R}^n_{>0}$ and functions $\lambda_j \in C([t_0,\infty), \mathbb{R}^n)$, $j = 1, 2$ satisfying $\lambda_1(t) \le \lambda_2(t)$ on $[t_0, \infty)$ and inequalities (16) on $[t_0, \infty)$, that is, the inequalities*

$$\begin{aligned} \lambda_1(t) &\le (T\lambda_1)(t), \\ \lambda_2(t) &\ge (T\lambda_2)(t). \end{aligned} \tag{24}$$

(iii) *For any $\Lambda_j \in C([t_0, s + r), \mathbb{R}^n)$, $j = 1, 2$ with $\Lambda_1(t) \le \Lambda_2(t)$ for $t \in [t_0, s + r)$, $s \ge t_0$, one has*

$$(T\Lambda_1)(t) \le (T\Lambda_2)(t), \quad t \in [t_0, s). \tag{25}$$

Then there exists a solution $y = y(t)$ of the system (3) on $[t_0, \infty)$ satisfying

$$I(k,\lambda_1)(t) \le y(t) \le I(k,\lambda_2)(t) \tag{26}$$

for $t \in [t_0, \infty)$.

Proof. We prove that (19) has a continuous solution $\lambda : [t_0, \infty) \to \mathbb{R}^n$ satisfying $\lambda_1(t) \le \lambda(t) \le \lambda_2(t)$ for $t \ge t_0$. For every fixed $\vartheta \in (t_0 + r, \infty)$, we introduce the Banach space

$$\mathbb{L}_\vartheta := C([t_0, \vartheta], \mathbb{R}^n) \tag{27}$$

of the continuous functions taking $[t_0, \vartheta]$ into $\mathbb{R}^n$ equipped with the maximum norm and the normal cone

$$\mathbb{K}_\vartheta := C([t_0, \vartheta], \mathbb{R}^n_{\ge 0}) \tag{28}$$

of the continuous functions taking $[t_0, \vartheta]$ into $\mathbb{R}^n_{\ge 0}$. By the cone $\mathbb{K}_\vartheta$, a partial ordering $\le$ in $\mathbb{L}_\vartheta$ is given. For $\lambda, \mu \in \mathbb{L}_\vartheta$ we say that $\lambda \le \mu$ if and only if $\mu - \lambda \in \mathbb{K}_\vartheta$. We introduce the operator $T_\vartheta : \mathbb{L}_\vartheta \to \mathbb{L}_\vartheta$ defined by

$$(T_\vartheta \lambda)(t) = \begin{cases} (T\lambda)(t) & \text{for } t \in [t_0, \vartheta - r], \\ (T\lambda)(\vartheta - r) & \text{for } t \in [\vartheta - r, \vartheta]. \end{cases} \tag{29}$$

Let $\lambda, \mu \in \mathbb{L}_\vartheta$ with $\lambda \le \mu$. Condition (iii) implies that $T_\vartheta \lambda \le T_\vartheta \mu$ if $\lambda \le \mu$. Let

$$\mu_{\vartheta,j}(t) := \begin{cases} \lambda_j(t) & \text{for } t \in [t_0, \vartheta - r], \\ \lambda_j(\vartheta - r) & \text{for } t \in [\vartheta - r, \vartheta], \ j = 1, 2. \end{cases} \tag{30}$$

Condition (ii) implies that

$$\mu_{\vartheta,1}(t) \le (T_\vartheta\mu_{\vartheta,1})(t) \le (T_\vartheta^2\mu_{\vartheta,1})(t) \le \cdots \le (T_\vartheta^2\mu_{\vartheta,2})(t) \le (T_\vartheta\mu_{\vartheta,2})(t) \le \mu_{\vartheta,2}(t), \quad t \in [t_0, \vartheta]. \tag{31}$$

In order to show the convergence of the sequences

$$(T_\vartheta^m \mu_{\vartheta,1})_{m=0}^{\infty}, \qquad (T_\vartheta^m \mu_{\vartheta,2})_{m=0}^{\infty} \tag{32}$$

(we set $T_\vartheta^0\mu_{\vartheta,i} = \mu_{\vartheta,i}$ and $T_\vartheta^1\mu_{\vartheta,i} = T\mu_{\vartheta,i}$, $i = 1,2$) to the fixed points of operator T_ϑ, we need to prove that T_ϑ is continuous and compact. The first property is obvious (due to continuity of F and I). Let us prove compactness. To this end, let $\mathscr{L}$ be a bounded subset of $\mathbb{L}_\vartheta$. We have to show that $T_\vartheta\mathscr{L}$ is a relatively compact subset of $\mathbb{L}_\vartheta$. Due to the theorem of Arzelà and Ascoli, it suffices to show that $T_\vartheta\mathscr{L}$ is bounded and equicontinuous. That $T_\vartheta\mathscr{L}$ is bounded follows from condition (22), and the equicontinuity of $T_\vartheta\mathscr{L}$ is assumed by condition (23).

Now we are in a position to apply Theorem 5 (about the monotone iterative method) in order to show the existence of the fixed points $\overline{\mu}_{\vartheta,1}$ and $\overline{\mu}_{\vartheta,2}$ of T_ϑ with

$$\overline{\mu}_{\vartheta,1} = \lim_{m\to\infty} T_\vartheta^m \mu_{\vartheta,1}, \qquad \overline{\mu}_{\vartheta,2} = \lim_{m\to\infty} T_\vartheta^m \mu_{\vartheta,2},$$
$$\mu_{\vartheta,1}(t) \le \overline{\mu}_{\vartheta,1}(t) \le \overline{\mu}_{\vartheta,2}(t) \le \mu_{\vartheta,2}(t), \quad t \in [t_0, \vartheta]. \tag{33}$$

The subsolution and supersolution necessary for application of Theorem 5 are equal to $u_0 := \mu_{\vartheta,1}$ and $v_0 := \mu_{\vartheta,2}$, respectively. Since it is easy to see that

$$T_\vartheta\lambda\big|_{[t_0,\vartheta-r]} = T_\Theta\lambda\big|_{[t_0,\vartheta-r]} \tag{34}$$

for $\Theta \ge \vartheta$ and $\lambda \in C([t_0,\Theta],\mathbb{R}^n)$, we have

$$\overline{\mu}_{\vartheta,j}\big|_{[t_0,\vartheta-r]} = \overline{\mu}_{\Theta,j}\big|_{[t_0,\vartheta-r]} \tag{35}$$

for $\Theta \ge \vartheta$, $j = 1,2$. Thus the functions $\overline{\lambda}_1, \overline{\lambda}_2 \in C([t_0,\infty),\mathbb{R}^n)$ defined, for example, by

$$\overline{\lambda}_j(t) := \begin{cases} \overline{\mu}_{\vartheta,j}(t) & \text{for } t \in [t_0, \vartheta - r], \\ \overline{\mu}_{t+r,j}(t) & \text{for } t \in (\vartheta - r, \infty), \end{cases} \tag{36}$$

where $j = 1,2$, $t \in [t_0,\infty)$, satisfy

$$\lambda_1(t) \le \overline{\lambda}_1(t) \le \overline{\lambda}_2(t) \le \lambda_2(t) \tag{37}$$

on $[t_0,\infty)$ and

$$\overline{\lambda}_j(t) \equiv \left(I\left(k,\overline{\lambda}_j\right)(t)\right)^{-1} f\left(t, I\left(k,\overline{\lambda}_j\right)_t\right) \tag{38}$$

for $j = 1,2$, $t \in [t_0,\infty)$. Now choosing, for example, $\lambda := \overline{\lambda}_1$, and $y = I(k,\lambda)$ (see substitution (21)), the proof is completed since I is strictly increasing with respect to the second argument. □

Remark 7. Tracing the proof of Theorem 6 one can see that assumptions on functional F (continuity, quasi-boundedness, and local Lipschitz condition) are not explicitly mentioned. Nevertheless these assumptions can be viewed as almost necessary for validity of inequalities (22) and (23).

4. Linear Cases

In this section we apply the nonlinear result given in Section 3 to linear advanced equations. We will prove existence of positive solutions, and we derive asymptotic behavior of positive solutions.

4.1. Equation $\dot y(t) = (c + d(t))y(t + r)$. Let us consider linear advanced equation (4):

$$\dot y(t) = (c + d(t))\, y(t + r), \tag{39}$$

where c and r are positive constants, function $d : [t_0,\infty) \to \mathbb{R}$ is locally Lipschitz continuous, $|d(t)| < \varepsilon < c$, and $(c+\varepsilon) < 1/(re)$.

To apply Theorem 6 we consider two auxiliary scalar linear advanced equations:

$$\dot x(t) = (c - \varepsilon)\, x(t + r),$$
$$\dot z(t) = (c + \varepsilon)\, z(t + r). \tag{40}$$

Looking for solutions in the form $x(t) = \exp(\lambda t)$ or $z(t) = \exp(\lambda t)$ we get corresponding transcendental equations:

$$\lambda = (c - \varepsilon)\, e^{\lambda r}, \tag{41}$$

$$\lambda = (c + \varepsilon)\, e^{\lambda r}. \tag{42}$$

The following lemma about the real roots of transcendental equations is a simple consequence of Lemma 1 in [5, page 13] (assumption $(c+\varepsilon) < 1/(re)$ is here substantial).

Lemma 8. *The following is true:*

(a) *equation* (41) *has just two different positive roots* λ_x^1, λ_x^2, $\lambda_x^1 < \lambda_x^2$, *and* (42) *has just two different positive roots* λ_z^1, λ_z^2, $\lambda_z^1 < \lambda_z^2$,

(b) $\lambda_x^1 r < 1$ *and* $\lambda_z^1 r < 1$,

(c) $\lambda_x^2 r > 1$ *and* $\lambda_z^2 r > 1$,

(d) $\lambda_x^1 < \lambda_z^1$ *and* $\lambda_x^2 > \lambda_z^2$.

As a tool for detecting a positive solution of (4), a linear corollary of Theorem 6 is used. We will reformulate the theorem with respect to the case in question; that is, we put

$$n = 1, \qquad F(t, y_t) := (c + d(t))\, y(t + r) \tag{43}$$

(we omit index 1 due to scalar case).

Using substitution (21) we obtain operator $(T\lambda)(t)$ defined by formula (20) in the form

$$(T\lambda)(t) := (c + d(t)) \exp\left(\int_t^{t+r} \lambda(s)\, ds\right). \tag{44}$$

Theorem 9. *Let there be continuous functions* $\lambda_j^* : [t_0,\infty) \to \mathbb{R}$, $j = 1,2$, $\lambda_1^*(t) \le \lambda_2^*(t)$ *for* $t \in [t_0,\infty)$, *satisfying the inequalities*

$$\lambda_1^*(t) \le (c + d(t)) \exp\left(\int_t^{t+r} \lambda_1^*(s)\, ds\right), \tag{45}$$

$$\lambda_2^*(t) \ge (c + d(t)) \exp\left(\int_t^{t+r} \lambda_2^*(s)\, ds\right) \tag{46}$$

on $[t_0, \infty)$. *Then there exists a solution* $y = y(t)$ *of* (4) *on* $[t_0, \infty)$ *such that*

$$\exp\left(\int_{t_0}^{t} \lambda_1^*(s)\,ds\right) \le y(t) \le \exp\left(\int_{t_0}^{t} \lambda_2^*(s)\,ds\right). \quad (47)$$

Proof. Firstly we verify conditions (22) and (23) of Theorem 6. Due to properties of functions d, λ, inequalities

$$|d(t)| < c, \quad |d(t) - d(t')| \le L_d |t - t'| \quad (48)$$

hold for every $t, t' \in [t_0, \infty)$, where L_d is a suitable constant. The boundedness of operator T (condition (22)) is on $[t_0, \infty)$ obvious:

$$|(T\lambda)(t)| = \left|(c + d(t)) \exp\left(\int_{t}^{t+r} \lambda(s)\,ds\right)\right| < K_1 := 2ce^{rM}. \quad (49)$$

Now, condition (22) holds. Let us verify condition (23). At first we estimate the difference

$$\begin{aligned} &\left|\int_{t}^{t+r} \lambda(s)\,ds - \int_{t'}^{t'+r} \lambda(s)\,ds\right| \\ &= \left|\int_{t}^{t+r} \lambda(s)\,ds + \int_{t'+r}^{t'} \lambda(s)\,ds\right| \\ &= \left|\int_{t'+r}^{t+r} \lambda(s)\,ds - \int_{t'}^{t} \lambda(s)\,ds\right| \\ &\le M|t - t'| + M|t - t'| = 2M|t - t'|. \end{aligned} \quad (50)$$

Now we have

$$\begin{aligned} &|(T\lambda)(t) - (T\lambda)(t')| \\ &= \left|(c + d(t)) \exp\left(\int_{t}^{t+r} \lambda(s)\,ds\right) - (a + b(t')) \exp\left(\int_{t'}^{t'+r} \lambda(s)\,ds\right)\right| \\ &\le |d(t) - d(t')| \exp\left(\int_{t}^{t+r} \lambda(s)\,ds\right) \\ &\quad + d(t') \left|\exp\left(\int_{t}^{t+r} \lambda(s)\,ds\right) - \exp\left(\int_{t'}^{t'+r} \lambda(s)\,ds\right)\right| \\ &\quad + c \left|\exp\left(\int_{t}^{t+r} \lambda(s)\,ds\right) - \exp\left(\int_{t'}^{t'+r} \lambda(s)\,ds\right)\right| \\ &< \left[L_d e^{rM} + 2ce^{\xi} 2M\right] |t - t'| \end{aligned} \quad (51)$$

with the value ξ lying between

$$\int_{t}^{t+r} \lambda(s)\,ds, \quad \int_{t'}^{t'+r} \lambda(s)\,ds. \quad (52)$$

Since obviously

$$e^{\xi} \le e^{rM}, \quad (53)$$

we get finally

$$|(T\lambda)(t) - (T\lambda)(t')| \le K_2 |t - t'| \quad (54)$$

with

$$K_2 := (L_d + 4cM) e^{rM}. \quad (55)$$

Inequality (23) is verified. Validity of condition (ii) follows from inequalities (45), and (46), and condition (iii) holds in view of the form of the operator T (see (44)). □

Theorem 10. *Equation* (4) *has a positive solution* $y = y(t)$ *on* $[t_0, \infty)$ *satisfying inequalities*

$$e^{\lambda_x^1 t} \le y(t) \le e^{\lambda_z^1 t}, \quad (56)$$

where λ_x^1, λ_z^1 *are defined in Lemma 8.*

Proof. Without loss of generality we may assume that t_0 is large enough for the asymptotic relations and inequalities to be valid. Now we employ Theorem 9. Let

$$\lambda_1^*(t) := \lambda_x^1, \quad \lambda_2^*(t) := \lambda_z^1(t), \quad t \in [t_0, \infty). \quad (57)$$

First we show that $\lambda_1^*(t)$, $\lambda_2^*(t)$ satisfy inequalities (45), (46). We have to check whether

$$\lambda_x^1 \le (c + d(t)) \exp\left(\int_{t}^{t+r} \lambda_x^1 ds\right). \quad (58)$$

Let us simplify and estimate the right-hand side:

$$\mathscr{R} = (c + d(t)) e^{\lambda_x^1 r} > (c - \varepsilon) e^{\lambda_x^1 r} = \lambda_x^1. \quad (59)$$

That is, inequality (45) is fulfilled. Similarly

$$\lambda_z^1 = (c + \varepsilon) e^{\lambda_z^1 r} > (c + d(t)) \exp\left(\int_{t}^{t+r} \lambda_z^1 ds\right). \quad (60)$$

Thus inequality (46) is valid too. Since all the assumptions of Theorem 9 are fulfilled, for solution $y(t)$ inequalities (56) hold. □

4.2. Equation $\dot{y}(t) = \gamma(t)y(t + r)$. Let us consider linear advanced equation (5):

$$\dot{y}(t) = \gamma(t)\, y(t + r), \quad (61)$$

where $\gamma : [t_0, \infty) \to \mathbb{R}^+$ is bounded, locally Lipschitz continuous and r is a positive constant. This equation can be viewed in a sense as a generalization of (4) (with $c = 0$ and $d(t) = \gamma(t)$). In the following criterion of existence of positive solutions of (5) it is important that only one of inequalities similar to inequalities (45) and (46) must be valid, namely, the inequality similar to (46).

Theorem 11. *For existence of a positive solution $y = y(t)$ of (5) on $[t_0, \infty)$, the existence of a continuous function $\lambda : [t_0, \infty) \to \mathbb{R}$ is necessary and sufficient, satisfying inequality*

$$\lambda(t) \ge \gamma(t) \exp\left(\int_t^{t+r} \lambda(s)\, ds\right) \tag{62}$$

on $[t_0, \infty)$. The mentioned positive solution $y = y(t)$ satisfies inequalities

$$1 < y(t) \le \exp\left(\int_{t_0}^{t} \lambda(s)\, ds\right). \tag{63}$$

Proof

Sufficiency. The proof uses Theorem 6 and can be performed similarly as the proof of Theorem 9. Therefore we omit the full version of the proof and underline only main points. Inequalities relevant to inequalities (45), (46) are

$$\begin{aligned} \lambda_1^*(t) &\le \gamma(t) \exp\left(\int_t^{t+r} \lambda_1^*(s)\, ds\right), \\ \lambda_2^*(t) &\ge \gamma(t) \exp\left(\int_t^{t+r} \lambda_2^*(s)\, ds\right) \end{aligned} \tag{64}$$

on $[t_0, \infty)$. We set $\lambda_1^*(t) \equiv 0$ and $\lambda_2^*(t) \equiv \lambda(t)$. Then inequalities (64) turn into

$$0 \le \gamma(t) \exp\left(\int_t^{t+r} 0\, ds\right), \tag{65}$$

$$\lambda(t) \ge \gamma(t) \exp\left(\int_t^{t+r} \lambda(s)\, ds\right). \tag{66}$$

The first inequality (65) obviously holds, and the second inequality (66) coincides with inequality (62). Then inequality (63) is a consequence of (47). Since the solution is increasing, inequality on the left-hand side of (63) must be sharp. The solution satisfying (63) is obviously positive.

Necessity. If a positive solution $y = y(t)$ of (5) on $[t_0, \infty)$ exists, we set

$$\lambda_1^*(t) \equiv \lambda_2^*(t) = \frac{\dot{y}(t)}{y(t)}. \tag{67}$$

Then both inequalities (64) hold simultaneously since for $i = 1, 2$

$$\begin{aligned} \lambda_i^*(t) &= \frac{\dot{y}(t)}{y(t)} = \frac{\gamma(t)\, y(t+r)}{y(t)} = \gamma(t) \exp\left(\int_t^{t+r} \frac{\dot{y}(s)}{y(s)} ds\right) \\ &= \gamma(t) \exp\left(\int_t^{t+r} \lambda_i^*(s)\, ds\right). \end{aligned} \tag{68}$$

Taking $\lambda(t)$ in a suitable way we can get sufficient conditions of existence of positive solutions of (5) on $[t_0, \infty)$. The following corollary illustrates such possibility. □

Corollary 12. *For existence of a positive solution $y = y(t)$ of (5) on $[t_0, \infty)$ inequality*

$$\int_t^{t+r} \gamma(s)\, ds \le \frac{1}{e} \tag{69}$$

is on $[t_0, \infty)$ sufficient.

Proof. The corollary is a consequence of Theorem 11, where $\lambda(t) = e\gamma(t)$. □

5. Concluding Remarks

Note that criterion (62) of existence of positive solution of advanced equation (5) given by Theorem 11, as a particular case of result for nonlinear system, seems to be independent of the mentioned criterion (9). In recent book [6] there are considered some linear classes of advanced differential equations in Chapter 5. Sufficient conditions of existence of positive solutions are derived. Inequality [6, formula (5.2.2)] contains, in particular, inequality (62) and inequality [6, formula (5.2.3)] contains, in particular, inequality (69).

Results obtained in the paper are sharp in a meaning. Indeed, in accordance with [2, page 31], [3, page 22] all solutions of (5) oscillate if

$$\liminf_{t \to \infty} \int_t^{t+r} \gamma(t) > \frac{1}{e} \tag{70}$$

in spite of the fact that inequality (69), guaranteeing the existence of a positive solution, admits the possibility

$$\liminf_{t \to \infty} \int_t^{t+r} \gamma(t) = \frac{1}{e}. \tag{71}$$

This occurs, for example, for $\gamma(t) = 1/er$. It means that inequality (69) in terms of integral average of function γ is the best possible.

In connection with the statement of Theorem 10 a problem about the role played by all real roots $\lambda_x^1, \lambda_x^2, \lambda_z^1$, and λ_z^2 of transcendental equations (41) and (42) arises. Roots λ_x^1, λ_z^1, $\lambda_x^1 < \lambda_z^1$ with properties given in Lemma 8 were used in Theorem 10 to detect a positive solution of (4) satisfying inequalities (56). The role played by the roots $\lambda_x^2, \lambda_z^2, \lambda_x^2 > \lambda_z^2$ with properties given in Lemma 8 in the discussion on existence of positive solutions of (4) is not clarified yet. Obviously λ_x^2 and λ_z^2 cannot be used in Theorem 9 to replace λ_j^*, $j = 1, 2$, in inequalities (45) and (46) (i.e., it is not possible to set $\lambda_1^* = \lambda_z^2, \lambda_2^* = \lambda_x^2$). Nevertheless, as it was demonstrated by advanced equation (6), it has two classes of asymptotically different positive solutions given by inequalities (7). This is the reason why we formulate the following claim.

Claim 1. Equation (4) has a positive solution $y = y_2(t)$ on $[t_0, \infty)$ satisfying inequalities

$$e^{\lambda_z^2 t} \le y_2(t) \le e^{\lambda_x^2 t}, \tag{72}$$

where λ_x^2, λ_z^2 are defined in Lemma 8.

Acknowledgment

The first author was supported by Grant no. P201/11/0768 of Czech Grant Agency (Prague).

References

[1] R. D. Driver, *Ordinary and Delay Differential Equations*, Springer, New York, NY, USA, 1977.

[2] L. H. Erbe, Q. Kong, and B. G. Zhang, *Oscillation Theory for Functional Differential Equations*, Marcel Dekker, New York, NY, USA, 1994.

[3] R. P. Agarwal, M. Bohner, and W.-T. Li, *Nonoscillation and Oscillation: Theory for Functional Differential Equations*, Marcel Dekker, New York, NY, USA, 2004.

[4] E. Zeidler, *Nonlinear Functional Analysis and Its Applications, Part I, Fixed-Point Theorems*, Springer, New York, NY, USA, 1985.

[5] J. Diblík and M. Kúdelčíková, "Nonoscillating solutions of the equation $\dot{x}(t) = -(a + b/t)x(t - \tau)$," *Studies of the University of Žilina*, vol. 15, pp. 11–24, 2002.

[6] R. P. Agarwal, L. Berezansky, E. Braverman, and A. Domoshnitsky, *Nonoscillation Theory of Functional Differential Equations with Applications*, Springer, New York, NY, USA, 2012.

On Satnoianu-Wu's Inequality

Bo-Yan Xi

College of Mathematics, Inner Mongolia University for Nationalities, Tongliao City, Inner Mongolia 028043, China

Correspondence should be addressed to Bo-Yan Xi; baoyintu78@qq.com

Academic Editors: F. J. Garcia-Pacheco, Y. Sawano, and S. Tikhonov

By applying techniques in the theory of convex functions and Schur-geometrically convex functions, the author investigates a conjecture of Satnoianu on an algebraic inequality and generalizes some known results in recent years.

1. Introduction

For $n \in \mathbb{N} = \{1, 2, 3, \ldots\}$ and $x_i \in \mathbb{R}_+ = (0, \infty)$ with $i = 1, 2, \ldots, n$, let

$$A_n(x) = \frac{1}{n}\sum_{i=1}^{n} x_i, \qquad G_n(x) = \left(\prod_{i=1}^{n} x_i\right)^{1/n}, \qquad H_n(x) = \frac{n}{\sum_{i=1}^{n}(1/x_i)}. \tag{1}$$

These quantities are, respectively, called the arithmetic, geometric, and harmonic means of a positive sequence $x = (x_1, x_2, \ldots, x_n)$. For more information on the theory of means, see the monograph [1] or the papers [2–5] and plenty of references therein.

For convenience, in what follows, we use the notation $x^q = (x_1^q, x_2^q, \ldots, x_n^q)$ for

$$0 < m \le x_1^q, x_2^q, \ldots, x_n^q \le M, \tag{2}$$

and $q \in \mathbb{R} = (-\infty, \infty)$.

In [6], Satnoianu posed the following conjecture.

Conjecture 1. *For $n \ge 2$, $x_i > 0$, and $\lambda \ge n^{n-1} - 1$, it is valid that*

$$\sum_{i=1}^{n}\left(\frac{x_i^{n-1}}{x_i^{n-1} + \lambda\prod_{k=1, k\neq i}^{n} x_k}\right)^{1/(n-1)} \ge \frac{n}{(1+\lambda)^{1/(n-1)}}. \tag{3}$$

This conjecture has been solved and researched in [7–10]. Among them, Wu obtained in [10] the following result.

Theorem 2 (see [10, Theorem 1]). *Let n be a positive integer and let α, β, and x_i for $i = 1, 2, \ldots, n$ be positive numbers. If $p \ne 0$ and*

$$\beta \ge \left(n^{\max\{p,1\}} - 1\right)\alpha, \tag{4}$$

then

$$\sum_{i=1}^{n}\left[\frac{x_i^q}{\alpha x_i^q + \beta G(x^q)}\right]^{1/p} \ge \frac{n}{(\alpha+\beta)^{1/p}}. \tag{5}$$

It is clear that inequality (5) generalizes (3). Therefore, we would like to call inequality (5) Satnoianu-Wu's inequality.

Now we naturally pose the following questions:

(1) Can one improve the condition (4)?

(2) Does the reversed inequality of (5) exist?

(3) Are there other types of inequalities of Satnoianu-Wu type?

The goal of this paper is to answer these questions.

2. Definitions and Lemmas

We need the following definitions and lemmas.

Definition 3 (see [11, page 8]). Let $x = (x_1, x_2, \ldots, x_n) \in \mathbb{R}^n$ and $y = (y_1, y_2, \ldots, y_n) \in \mathbb{R}^n$. We say that x is majorized by y (in symbols $x \prec y$) if

$$\sum_{i=1}^{k} x_{[i]} \le \sum_{i=1}^{k} y_{[i]}, \qquad \sum_{i=1}^{n} x_i = \sum_{i=1}^{n} y_i, \tag{6}$$

for $k = 1, 2, \ldots, n-1$, where $x_{[1]} \geq \cdots \geq x_{[n]}$ and $y_{[1]} \geq \cdots \geq y_{[n]}$ are rearrangements of x and y in a descending order.

Definition 4 (see [12]). Let $\Omega \subseteq \mathbb{R}_+^n$.

(1) The set Ω is said to be geometrically convex if $(x_1^\lambda y_1^{1-\lambda}, \ldots, x_n^\lambda y_n^{1-\lambda}) \in \Omega$ for every $x, y \in \Omega$ and $\lambda \in [0,1]$.

(2) A function $\varphi : \Omega \to \mathbb{R}_+$ is said to be Schur-geometrically convex on Ω if $\ln x = (\ln x_1, \ldots, \ln x_n) \prec \ln y = (\ln y_1, \ldots, \ln y_n)$ implies $\varphi(x) \leq \varphi(y)$ for every $x, y \in \Omega$.

(3) A function $\varphi : \Omega \to \mathbb{R}_+$ is said to be Schur-geometrically concave on Ω if $\ln x = (\ln x_1, \ldots, \ln x_n) \prec \ln y = (\ln y_1, \ldots, \ln y_n)$ implies $\varphi(x) \geq \varphi(y)$ for every $x, y \in \Omega$.

Lemma 5 (see [12]). *Let $\Omega \subseteq \mathbb{R}_+^n$ be a symmetric and geometrically convex set with inner points and $\varphi : \Omega \to \mathbb{R}_+$ a symmetric and differentiable function in Ω°. Then φ is a Schur-geometrically convex (or Schur-geometrically concave, resp.) function on Ω if and only if*

$$(\ln x_1 - \ln x_2)\left[x_1 \frac{\partial \varphi(x)}{\partial x_1} - x_2 \frac{\partial \varphi(x)}{\partial x_2}\right] \geq 0 \quad (\text{or} \leq 0, \text{resp.}), \quad x \in \Omega^\circ. \tag{7}$$

Lemma 6 (see [1, page 4, Bernoulli's inequality]). *The inequality*

$$(1+t)^r \geq 1 + rt \tag{8}$$

holds for $r \geq 1$ and $t \geq -1$ or for $r \leq 0$ and $t > -1$. If $0 < r < 1$ and $t \geq -1$, inequality (8) *is reversed.*

3. Main Results

Now we start off to demonstrate our main results.

Theorem 7. *Let m, M be defined as in* (2) *and $n \geq 2$, let $\alpha, \beta, x_i \in \mathbb{R}_+$ for $i = 1, 2, \ldots, n$, and let $p, q \in \mathbb{R}$ with $p \neq 0$. If $(1-p)m\beta \geq 2pM\alpha$, then*

$$\sum_{i=1}^n \left[\frac{x_i^q}{\alpha x_i^q + \beta A_n(x^q)}\right]^{1/p} \geq \frac{n}{(\alpha+\beta)^{1/p}}. \tag{9}$$

If $(1-p)M\beta \leq 2pm\alpha$, inequality (9) *is reversed.*

Proof. For $M' > m' > 0$ and $r \in \mathbb{R}$ with $r \neq 0$, since

$$\left[\left(\frac{t}{1+t}\right)^r\right]'' = \frac{r(r-1-2t)t^{r-1}}{(1+t)^{r+2}}, \quad t \in [m', M'], \tag{10}$$

then

(1) for $r < 0$, or for $r > 1$ and $r - 1 - 2M' > 0$, the function $(t/(1+t))^r$ is convex on $[m', M']$;

(2) for $0 < r \leq 1$, or for $r > 1$ and $r - 1 - 2m' < 0$, the function $(t/(1+t))^r$ is concave on $[m', M']$.

By Jensen's inequality, when $r < 0$, or when $r > 1$ and $r - 1 - 2M' > 0$, we have

$$\sum_{i=1}^n \left(\frac{u_i}{1+u_i}\right)^r \geq n\left[\frac{A_n(u)}{1+A_n(u)}\right]^r, \tag{11}$$

where $u = (u_1, u_2, \ldots, u_n) \in [m', M']^n$. When $0 < r \leq 1$, or when $r > 1$ and $r - 1 - 2m' < 0$, inequality (11) is reversed.

Let $u_i = \alpha x_i^q / \beta A_n(x^q)$ for $i = 1, 2, \ldots, n$, $r = 1/p$, $m' = \alpha m/\beta M$, and $M' = \alpha M/\beta m$. Then it is easy to show that $A_n(u) = \alpha/\beta$. Making use of (11), we obtain inequality (9). The proof of Theorem 7 is complete. □

Corollary 8. *Under the conditions of Theorem 7 and when $M = (n-1)m$, if $(1-p)\beta > 2p(n-1)\alpha$, then*

$$\sum_{i=1}^n \left[\frac{x_i^q}{\alpha x_i^q + \beta A_n(x^q)}\right]^{1/p} \geq \frac{n}{(\alpha+\beta)^{1/p}}; \tag{12}$$

if $(1-p)(n-1)\beta < 2p\alpha$, inequality (12) *is reversed.*

Corollary 9. *Let $n \geq 2$, $\alpha, \beta, x_i \in \mathbb{R}_+$ for $i = 1, 2, \ldots, n$, and $p, q \in \mathbb{R}$. If $p < 0$, then*

$$\sum_{i=1}^n \left[\frac{x_i^q}{\alpha x_i^q + \beta A_n(x^q)}\right]^{1/p} \geq \frac{n}{(\alpha+\beta)^{1/p}}; \tag{13}$$

if $p \geq 1$, inequality (13) *is reversed.*

Proof. This follows from Theorem 7 by considering that if $p < 0$, we have $(1-p)m\beta > 2pM\alpha$ and that if $p \geq 1$, we have $(1-p)M\beta < 2pm\alpha$. □

Theorem 10. *Let $n \geq 2$, $\alpha, \beta, x_i \in \mathbb{R}_+$ for $i = 1, 2, \ldots, n$, $p, q \in \mathbb{R}$ with $p \neq 0$, and m, M defined by* (2)*. If $m\beta \geq pM\alpha$, then*

$$\sum_{i=1}^n \left[\frac{x_i^q}{\alpha x_i^q + \beta G_n(x^q)}\right]^{1/p} \geq \frac{n}{(\alpha+\beta)^{1/p}}; \tag{14}$$

if $M\beta \leq mp\alpha$, inequality (14) *is reversed.*

Proof. For $M' > m' > 0$, let

$$f(u) = \sum_{i=1}^n (1+u_i)^r, \quad u \in [m', M']^n. \tag{15}$$

Then

$$\frac{\partial f(u)}{\partial u_i} = r(1+u_i)^{r-1}, \qquad \frac{\partial}{\partial u_i}\left[u_i \frac{\partial f(u)}{\partial u_i}\right] = r(1+u_i)^{r-2}(1+ru_i), \tag{16}$$

for $1 \leq i \leq n$. When $r > 0$, or when $r < 0$ and $1 + rm' < 0$, we have

$$(\ln u_1 - \ln u_2)\left[u_1 \frac{\partial f(u)}{\partial u_1} - u_2 \frac{\partial f(u)}{\partial u_2}\right] \geq 0, \quad u \in [m', M']^n; \tag{17}$$

when $r < 0$ and $1 + rM' > 0$, inequality (17) is reversed.

Using Lemma 5, we have the following conclusions:

(1) if $r > 0$, or if $r < 0$ and $1 + rm' < 0$, the function $f(u)$ is Schur-geometrically convex on $[m', M']^n$;

(2) if $r < 0$ and $1 + rM' > 0$, the function $f(u)$ is Schur-geometrically concave on $[m', M']^n$.

By the fact that

$$(\ln G_n(u), \ln G_n(u), \ldots, \ln G_n(u)) \prec (\ln u_1, \ln u_2, \ldots, \ln u_n), \quad u \in \mathbb{R}_+^n, \tag{18}$$

and by Definition 4, if $r > 0$, or if $r < 0$ and $1 + rm' < 0$, we have

$$\sum_{i=1}^{n} (1 + u_i)^r \geq n(1 + G_n(u))^r, \quad u \in [m', M']^n; \tag{19}$$

if $r < 0$ and $1 + rM' > 0$, inequality (19) is reversed.

Letting $u_i = \beta G_n(x^q)/\alpha x_i^q$ for $i = 1, 2, \ldots, n$ leads to $G_n(u) = \beta/\alpha$. Putting $r = -1/p$, $m' = \beta m/\alpha M$ and $M' = \beta M/\alpha m$ in inequality (19) results in inequality (14). The proof of Theorem 10 is complete. □

Corollary 11. *Under the conditions of Theorem 10 and when $M = (n-1)m$, if $\beta \geq p(n-1)\alpha$, then*

$$\sum_{i=1}^{n} \left[\frac{x_i^q}{\alpha x_i^q + \beta A_n(x^q)} \right]^{1/p} \geq \frac{n}{(\alpha + \beta)^{1/p}}; \tag{20}$$

if $(n-1)\beta \leq p\alpha$, inequality (20) is reversed.

Remark 12. When $n \geq 2$ and $p > 1$, from Lemma 6 and (4), it follows that

$$\beta \geq \left(n^{\max\{p,1\}} - 1\right)\alpha > p(n-1)\alpha. \tag{21}$$

Theorem 13. *Let $n \geq 2$, $\alpha, \beta, x_i \in \mathbb{R}_+$ for $i = 1, 2, \ldots, n$, and $p, q \in \mathbb{R}$ with $p \neq 0$. If $p \geq -1$, then*

$$\sum_{i=1}^{n} \left[\frac{x_i^q}{\alpha x_i^q + \beta H_n(x^q)} \right]^{1/p} \geq \frac{n}{(\alpha + \beta)^{1/p}}; \tag{22}$$

if $p \leq -1$, inequality (22) is reversed.

Proof. Since $(1 + t)^r$ is a convex (or concave, resp.) function on $\mathbb{R}_+$ for $r \geq 1$ or $r < 0$ (or for $0 < r \leq 1$, resp.), by Jensen's inequality, if $r \geq 1$ or $r < 0$, we have

$$\sum_{i=1}^{n} (1 + u_i)^r \geq n(1 + A_n(u))^r, \quad u \in \mathbb{R}_+^n; \tag{23}$$

if $0 < r \leq 1$, inequality (23) is reversed.

Letting $u_i = \beta H_n(x^q)/\alpha x_i^q$ and $r = -1/p$ shows $A_n(u) = \beta/\alpha$. Further from (23), we obtain inequality (22). The proof of Theorem 13 is complete. □

Remark 14. It is clear that inequalities (9) and (22) both generalize inequality (3).

Corollary 15. *Let $n \geq 2$, $\alpha, \beta, x_i \in \mathbb{R}_+$ for $i = 1, 2, \ldots, n$, $p, q \in \mathbb{R}$ with $p \neq 0$, and m, M defined as in (2).*

(1) *When $-1 \leq p < 0$, one has*

$$\begin{aligned} \sum_{i=1}^{n} \left[\frac{x_i^q}{\alpha x_i^q + \beta A_n(x^q)} \right]^{1/p} &\geq \sum_{i=1}^{n} \left[\frac{x_i^q}{\alpha x_i^q + \beta G_n(x^q)} \right]^{1/p} \\ &\geq \sum_{i=1}^{n} \left[\frac{x_i^q}{\alpha x_i^q + \beta H_n(x^q)} \right]^{1/p} \\ &\geq \frac{n}{(\alpha + \beta)^{1/p}}. \end{aligned} \tag{24}$$

(2) *When $0 < p < 1$ and $(1-p)m\beta > 2pM\alpha$, one has*

$$\begin{aligned} \sum_{i=1}^{n} \left[\frac{x_i^q}{\alpha x_i^q + \beta H_n(x^q)} \right]^{1/p} &\geq \sum_{i=1}^{n} \left[\frac{x_i^q}{\alpha x_i^q + \beta G_n(x^q)} \right]^{1/p} \\ &\geq \sum_{i=1}^{n} \left[\frac{x_i^q}{\alpha x_i^q + \beta A_n(x^q)} \right]^{1/p} \\ &\geq \frac{n}{(\alpha + \beta)^{1/p}}. \end{aligned} \tag{25}$$

(3) *When $p > 0$ and $m\beta > pM\alpha$, one has*

$$\begin{aligned} \sum_{i=1}^{n} \left[\frac{x_i^q}{\alpha x_i^q + \beta H_n(x^q)} \right]^{1/p} &\geq \sum_{i=1}^{n} \left[\frac{x_i^q}{\alpha x_i^q + \beta G_n(x^q)} \right]^{1/p} \\ &\geq \frac{n}{(\alpha + \beta)^{1/p}}. \end{aligned} \tag{26}$$

(4) *When $p > 0$ and $M\beta < mp\alpha$, one has*

$$\begin{aligned} \sum_{i=1}^{n} \left[\frac{x_i^q}{\alpha x_i^q + \beta A_n(x^q)} \right]^{1/p} &\leq \sum_{i=1}^{n} \left[\frac{x_i^q}{\alpha x_i^q + \beta G_n(x^q)} \right]^{1/p} \\ &\leq \frac{n}{(\alpha + \beta)^{1/p}}. \end{aligned} \tag{27}$$

Proof. This follows from utilizing the well-known harmonic-geometric-arithmetic mean inequality

$$H_n(x) \leq G_n(x) \leq A_n(x) \tag{28}$$

and Theorems 7 to 13. □

Corollary 16. *Under the conditions of Corollary 15 and when* $M = (n-1)m$,

(1) *if* $0 < p < 1$ *and* $(1-p)\beta > 2p(n-1)\alpha$, *one has*

$$\sum_{i=1}^{n}\left[\frac{x_i^q}{\alpha x_i^q + \beta H_n(x^q)}\right]^{1/p} \ge \sum_{i=1}^{n}\left[\frac{x_i^q}{\alpha x_i^q + \beta G_n(x^q)}\right]^{1/p} \ge \sum_{i=1}^{n}\left[\frac{x_i^q}{\alpha x_i^q + \beta A_n(x^q)}\right]^{1/p} \ge \frac{n}{(\alpha+\beta)^{1/p}}; \quad (29)$$

(2) *if* $p > 0$ *and* $\beta > p(n-1)\alpha$, *one has*

$$\sum_{i=1}^{n}\left[\frac{x_i^q}{\alpha x_i^q + \beta H_n(x^q)}\right]^{1/p} \ge \sum_{i=1}^{n}\left[\frac{x_i^q}{\alpha x_i^q + \beta G_n(x^q)}\right]^{1/p} \ge \frac{n}{(\alpha+\beta)^{1/p}}; \quad (30)$$

(3) *if* $p > 0$ *and* $(n-1)\beta < p\alpha$, *one has*

$$\sum_{i=1}^{n}\left[\frac{x_i^q}{\alpha x_i^q + \beta A_n(x^q)}\right]^{1/p} \le \sum_{i=1}^{n}\left[\frac{x_i^q}{\alpha x_i^q + \beta G_n(x^q)}\right]^{1/p} \le \frac{n}{(\alpha+\beta)^{1/p}}. \quad (31)$$

Acknowledgments

The author thanks four anonymous referees for their careful corrections to and valuable comments on the original version of this paper. This work was partially supported by the Foundation of the Research Program of Science and Technology at Universities of Inner Mongolia Autonomous Region under Grant no. NJZY13159, China.

References

[1] P. S. Bullen, *Handbook of Means and Their Inequalities*, vol. 560 of *Mathematics and Its Applications*, Kluwer Academic Publishers, Dordrecht, The Netherlands, 2003.

[2] B.-N. Guo and F. Qi, "A simple proof of logarithmic convexity of extended mean values," *Numerical Algorithms*, vol. 52, no. 1, pp. 89–92, 2009.

[3] B.-N. Guo and F. Qi, "The function $(b^x-a^x)/x$: logarithmic convexity and applications to extended mean values," *Filomat*, vol. 25, no. 4, pp. 63–73, 2011.

[4] F. Qi, "A note on Schur-convexity of extended mean values," *Rocky Mountain Journal of Mathematics*, vol. 35, no. 5, pp. 1787–1793, 2005.

[5] J.-L. Zhao, Q.-M. Luo, B.-N. Guo, and F. Qi, "Logarithmic convexity of Gini means," *Journal of Mathematical Inequalities*, vol. 6, no. 4, pp. 509–516, 2012.

[6] R. A. Satnoianu, "The proof of the conjectured inequality from the Mathematical Olympiad, Washington DC 2001," *Gazeta Matematica*, vol. 106, pp. 390–393, 2001.

[7] W. Janous, "On a conjecture of Razvan Satnoianu," *Gazeta Matematica Seria A*, vol. 20, no. 99, pp. 97–103, 2002.

[8] Y. Miao, S.-F. Xu, and Y.-X. Chen, "An answer to the conjecture of Satnoianu," *Applied Mathematics E-Notes*, vol. 9, pp. 262–265, 2009.

[9] R. A. Satnoianu, "Improved GA-convexity inequalities," *Journal of Inequalities in Pure and Applied Mathematics*, vol. 3, no. 5, article 82, 2002.

[10] S.-H. Wu, "Note on a conjecture of R. A. Satnoianu," *Mathematical Inequalities and Applications*, vol. 12, no. 1, pp. 147–151, 2009.

[11] A. W. Marshall, I. Olkin, and B. C. Arnold, *Inequalities: Theory of Majorization and Its Applications*, Springer, New York, NY, USA, 2nd edition, 2011.

[12] Y.-M. Chu, X.-M. Zhang, and G.-D. Wang, "The schur geometrical convexity of the extended mean values," *Journal of Convex Analysis*, vol. 15, no. 4, pp. 707–718, 2008.

15

On the Connection Coefficients of the Chebyshev-Boubaker Polynomials

Paul Barry

School of Science, Waterford Institute of Technology, Waterford, Ireland

Correspondence should be addressed to Paul Barry; pbarry@wit.ie

Academic Editors: A. Bekir, S. Hristova, and T.-z. Huang

The Chebyshev-Boubaker polynomials are the orthogonal polynomials whose coefficient arrays are defined by ordinary Riordan arrays. Examples include the Chebyshev polynomials of the second kind and the Boubaker polynomials. We study the connection coefficients of this class of orthogonal polynomials, indicating how Riordan array techniques can lead to closed-form expressions for these connection coefficients as well as recurrence relations that define them.

1. Introduction

In this paper, we will show that the usual recurrence relations that define elements of Riordan arrays can be used to give a set of recurrence relations for the connection coefficients of families of orthogonal polynomials defined by Riordan arrays. It will also be seen that we can give closed-form expressions for the connection coefficients themselves, at least so far, as it is easy to give closed-form expressions for the Riordan arrays concerned. Thus, in the special case of orthogonal polynomials that are defined by Riordan arrays, we can achieve two of the main goals involved in the study of connection coefficients for orthogonal polynomials [1–7].

If $P_n(x)$ and $Q_n(x)$ are two families of orthogonal polynomials, the connection coefficients $a_{n,k}$ of P_n in terms of Q_n, defined by

$$P_n = \sum_{k=0}^{n} a_{n,k} Q_k, \tag{1}$$

are often of interest. In terms of matrices, we have

$$\mathbf{P} = \mathbf{A}\mathbf{Q}, \tag{2}$$

where $\mathbf{P}$ is the coefficient array of the polynomials $P_n(x)$, with elements $p_{n,k}$, and $\mathbf{Q}$ is the coefficient array of the polynomials $Q_n(x)$, with elements $q_{n,k}$. Thus, we have

$$P_n(x) = \sum_{k=0}^{n} p_{n,k} x^k, \qquad Q_n(x) = \sum_{k=0}^{n} q_{n,k} x^k. \tag{3}$$

The matrices $\mathbf{P}$ and $\mathbf{Q}$ are lower-triangular matrices. Clearly, the matrix $\mathbf{A}$ of connection coefficients is then given by

$$\mathbf{A} = \mathbf{P}\mathbf{Q}^{-1}. \tag{4}$$

We immediately note that

$$\mathbf{A}^{-1} = \mathbf{Q}\mathbf{P}^{-1}, \tag{5}$$

so that the elements of $\mathbf{A}^{-1}$ are the connection coefficients of Q_n in terms of P_n.

The matrix $\mathbf{A}$ is lower triangular since both $\mathbf{P}$ and $\mathbf{Q}$ are.

We will define a family of orthogonal polynomials $P_n(x)$ to be of *Chebyshev-Boubaker* type, if the coefficient array $\mathbf{P} = (p_{n,k})$ is a Riordan array of the type

$$\left(\frac{1+ux+vx^2}{1+ax+bx^2}, \frac{\lambda x}{1+ax+bx^2}\right). \tag{6}$$

This means that

$$p_{n,k} = [x^n] \frac{1+ux+vx^2}{1+ax+bx^2}\left(\frac{\lambda x}{1+ax+bx^2}\right)^k. \tag{7}$$

In the case that $\lambda = 1$, the polynomials are monic. An array of the type

$$\left(\frac{1+ux+vx^2}{1+ax+bx^2}, \frac{\lambda x}{1+ax+bx^2}\right) \tag{8}$$

will be called a *Chebyshev-Boubaker array* (with parameters (a, b, u, v, λ)). This is a generalization of the notion of a Chebyshev-Boubaker array considered in [8]. The Boubaker polynomials [8] correspond to the parameters $(0, 1, 0, 3, 1)$.

Example 1. The Chebyshev polynomials of the second kind $U_n(x)$ are defined by

$$U_n(x) = \sum_{k=0}^{\lfloor n/2 \rfloor} (-1)^k \binom{n-k}{k} (2x)^{n-2k} \quad (9)$$

and have coefficient array $\mathbf{U}$ given by A053117

$$\mathbf{U} = \left(\frac{1}{1+x^2}, \frac{2x}{1+x^2} \right). \quad (10)$$

Example 2. We let $S_n(x) = U_n(x/2)$ be the monic Chebyshev polynomials of the second kind. The coefficient array $\mathbf{S}$ of these polynomials is given by A049310

$$\mathbf{S} = \left(\frac{1}{1+x^2}, \frac{x}{1+x^2} \right). \quad (11)$$

This is the case $u = v = a = 0$, $b = 1$, and $\lambda = 2$.

The moments m_n of this family of orthogonal polynomials are the aerated Catalan numbers that begin with

$$1, 0, 1, 0, 2, 0, 5, 0, 14, 0, 42, \ldots. \quad (12)$$

Their moment representation is given by

$$m_n = \frac{1}{2\pi} \int_{-2}^{2} x^n \sqrt{4-x^2}\, dx. \quad (13)$$

Very often, orthogonal polynomials are specified in terms of the three-term recurrence that they satisfy. In the monic case, this will take the form

$$P_n(x) = (x - \alpha_n) P_{n-1}(x) - \beta_n P_{n-2}(x), \quad (14)$$

with suitable initial conditions. For instance, we have

$$S_n(x) = (x-2) S_{n-1}(x) - S_{n-2}(x), \qquad S_0(x) = 1, \quad S_1(x) = x. \quad (15)$$

The recurrence coefficients α_n and β_n for the family P_n of orthogonal polynomials are the entries in the production matrix (or Stieltjes matrix [9]) of the matrix $\mathbf{P}^{-1}$. This production matrix is equal to $\mathbf{P}\overline{\mathbf{P}^{-1}}$, where the notation $\overline{\mathbf{M}}$ denotes the matrix $\mathbf{M}$ with its first row removed. For instance, the production matrix of $\mathbf{U}^{-1} = (1/(1+x^2), 2x/(1+x^2))^{-1}$ is given by

$$\begin{pmatrix} 0 & \frac{1}{2} & 0 & 0 & 0 & 0 & \cdots \\ \frac{1}{2} & 0 & \frac{1}{2} & 0 & 0 & 0 & \cdots \\ 0 & \frac{1}{2} & 0 & \frac{1}{2} & 0 & 0 & \cdots \\ 0 & 0 & \frac{1}{2} & 0 & \frac{1}{2} & 0 & \cdots \\ 0 & 0 & 0 & \frac{1}{2} & 0 & \frac{1}{2} & \cdots \\ \vdots & \vdots & \vdots & \vdots & \vdots & \vdots & \ddots \end{pmatrix}, \quad (16)$$

corresponding to the three-term recurrence

$$\frac{1}{2} U_n(x) = (x-0) U_{n-1}(x) - \frac{1}{2} U_{n-2}(x), \quad (17)$$

or

$$U_n(x) = 2x U_{n-1}(x) - U_{n-2}(x). \quad (18)$$

Similarly, the production matrix of $\mathbf{S}^{-1} = (1/(1+x^2), x/(1+x^2))^{-1}$ is given by

$$\begin{pmatrix} 0 & 1 & 0 & 0 & 0 & 0 & \cdots \\ 1 & 0 & 1 & 0 & 0 & 0 & \cdots \\ 0 & 1 & 0 & 1 & 0 & 0 & \cdots \\ 0 & 0 & 1 & 0 & 1 & 0 & \cdots \\ 0 & 0 & 0 & 1 & 0 & 1 & \cdots \\ \vdots & \vdots & \vdots & \vdots & \vdots & \vdots & \ddots \end{pmatrix}, \quad (19)$$

corresponding to the three-term recurrence

$$S_n(x) = x S_{n-1}(x) - S_{n-2}(x). \quad (20)$$

More generally, the production matrix of $\mathbf{S}_b^{-1} = (1/(1+bx^2), x/(1+bx^2))^{-1}$ is given by

$$\begin{pmatrix} 0 & 1 & 0 & 0 & 0 & 0 & \cdots \\ b & 0 & 1 & 0 & 0 & 0 & \cdots \\ 0 & b & 0 & 1 & 0 & 0 & \cdots \\ 0 & 0 & b & 0 & 1 & 0 & \cdots \\ 0 & 0 & 0 & b & 0 & 1 & \cdots \\ \vdots & \vdots & \vdots & \vdots & \vdots & \vdots & \ddots \end{pmatrix}, \quad (21)$$

corresponding to the three-term recurrence

$$S_n^{(b)}(x) = x S_{n-1}^{(b)}(x) - b S_{n-2}^{(b)}(x), \qquad S_0^{(b)} = 1, \quad S_1^{(b)}(x) = x, \quad (22)$$

where

$$S_n^{(b)}(x) = \sum_{k=0}^{\lfloor n/2 \rfloor} \binom{n-k}{k} (-b)^k x^{n-2k}. \quad (23)$$

Example 3. The Boubaker polynomials [8, 10–15] have coefficient array

$$\left(\frac{1+3x^2}{1+x^2}, \frac{x}{1+x^2} \right). \quad (24)$$

They can be expressed as

$$P_n(x) = \sum_{k=0}^{\lfloor n/2 \rfloor} \binom{n-k}{k} \frac{n-4k}{n-k} (-1)^k x^{n-2k}. \quad (25)$$

We have

$$\left(\frac{1+3x^2}{1+x^2}, \frac{x}{1+x^2} \right)^{-1} = \left(\frac{1+x^2 c(x^2)^2}{1+3x^2 c(x^2)^2}, xc(x^2) \right) = \left(\frac{1}{1+2x^2 c(x^2)}, xc(x^2) \right), \quad (26)$$

where

$$c(x) = \frac{1-\sqrt{1-4x}}{2x} \tag{27}$$

is the generating function of the Catalan numbers

$$C_n = \frac{1}{n+1}\binom{2n}{n}. \tag{28}$$

Hence, the moments m_n of this family of orthogonal polynomials have g.f. $1/(1+2x^2c(x^2))$. The sequence m_n begins with

$$1, 0, -2, 0, 2, 0, -4, 0, 2, 0, -12, 0, -12, 0, -72, \ldots \tag{29}$$

and has the moment representation

$$m_n = -\frac{1}{\pi}\int_{-2}^{2} x^n \frac{\sqrt{4-x^2}}{4+3x^2}\,dx + \frac{2}{3}\left(-\frac{2}{\sqrt{3}}i\right)^n + \frac{2}{3}\left(\frac{2}{\sqrt{3}}i\right)^n. \tag{30}$$

We have

$$\begin{aligned} m_{2n} &= \sum_{k=0}^{n} \frac{k+0^{n+k}}{n+0^{nk}}\binom{2n-k-1}{n-k}(-2)^k \\ &= \sum_{k=0}^{n} \frac{2k+1}{n+k+1}\binom{2n}{n-k}(-3)^k. \end{aligned} \tag{31}$$

The sequence m_{2n} is A126984. We note that the Hankel transforms of m_n and m_{2n} are both equal to $(-2)^n$. The production matrix of $((1+3x^2)/(1+x^2), x/(1+x^2))^{-1}$ is given by

$$\begin{pmatrix} 0 & 1 & 0 & 0 & 0 & 0 & \cdots \\ -2 & 0 & 1 & 0 & 0 & 0 & \cdots \\ 0 & 1 & 0 & 1 & 0 & 0 & \cdots \\ 0 & 0 & 1 & 0 & 1 & 0 & \cdots \\ 0 & 0 & 0 & 1 & 0 & 1 & \cdots \\ \vdots & \vdots & \vdots & \vdots & \vdots & \vdots & \ddots \end{pmatrix}. \tag{32}$$

In this paper, we will give a closed-form expression for the connection coefficients of the Chebyshev-Boubaker polynomials $P_n(x)$ with coefficient array

$$\mathbf{P} = \left(\frac{1+ux+vx^2}{1+ax+bx^2}, \frac{x}{1+ax+bx^2}\right) \tag{33}$$

in terms of the Chebyshev-Boubaker polynomials $Q_n(x)$ with coefficient array

$$\mathbf{Q} = \left(\frac{1+sx+tx^2}{1+cx+dx^2}, \frac{x}{1+cx+dx^2}\right). \tag{34}$$

We note that the polynomials $P_n(x)$ satisfy the recurrence

$$P_n(x) = (x-a)P_{n-1}(x) - bP_{n-2}(x), \quad n \geq 3, \tag{35}$$

with

$$\begin{aligned} &P_0(x) = 1, \qquad P_1(x) = x + u - a, \\ &P_2(x) = x^2 + (u-2a)x - ua + a^2 + v - b, \end{aligned} \tag{36}$$

with a similar expression for $Q_n(x)$.

In the next section, we recall notational elements that we will use in the sequel, along with a brief introduction to Riordan arrays.

2. Preliminaries on Integer Sequences and Riordan Arrays

In this section, we define terms used later to discuss integer sequences, Riordan arrays, production matrices, orthogonal polynomials, and Hankel transforms. Readers familiar with these areas and the links between them may skip this section.

For an integer sequence a_n, that is, an element of $\mathbb{Z}^{\mathbb{N}}$, the power series $f(x) = \sum_{n=0}^{\infty} a_n x^n$ is called the *ordinary generating function* or g.f. of the sequence. a_n is thus the coefficient of x^n in this series. We denote this by $a_n = [x^n]f(x)$. For instance, $F_n = [x^n](x/(1-x-x^2))$ is the nth Fibonacci number A000045, while $C_n = [x^n]((1-\sqrt{1-4x})/2x)$ is the nth Catalan number A000108. We use the notation $0^n = [x^n]1$ for the sequence $1, 0, 0, 0, \ldots$, A000007. Thus, $0^n = [n=0] = \delta_{n,0} = \binom{0}{n}$. Here, we have used the Iverson bracket notation [16], defined by $[\mathcal{P}] = 1$ if the proposition $\mathcal{P}$ is true and $[\mathcal{P}] = 0$ if $\mathcal{P}$ is false.

For a power series $f(x) = \sum_{n=0}^{\infty} a_n x^n$ with $f(0) = 0$, we define the reversion or compositional inverse of f to be the power series $\overline{f}(x)$ such that $f(\overline{f}(x)) = x$.

For a lower triangular matrix $(a_{n,k})_{n,k\geq 0}$, the row sums give the sequence with general term $\sum_{k=0}^{n} a_{n,k}$. For a sequence a_n, the sequence of determinants $|a_{i+j}|_{0\leq i,j\leq n}$ is called the Hankel transform of a_n [17–27].

The *Riordan group* [28, 29] is a set of infinite lower-triangular integer matrices, where each matrix is defined by a pair of generating functions $g(x) = 1 + g_1x + g_2x^2 + \cdots$ and $f(x) = f_1x + f_2x^2 + \cdots$ where $f_1 \neq 0$ [29]. We often require in addition that $f_1 = 1$. The associated matrix is the matrix whose ith column is generated by $g(x)f(x)^i$ (the first column being indexed by 0). The matrix corresponding to the pair g, f is denoted by (g, f) or $\mathcal{R}(g, f)$. The group law is then given by

$$(g, f)\cdot(h, l) = (g, f)(h, l) = (g(h\circ f), l\circ f). \tag{37}$$

The identity for this law is $I = (1, x)$, and the inverse of (g, f) is

$$(g, f)^{-1} = \left(\frac{1}{g\circ\overline{f}}, \overline{f}\right), \tag{38}$$

where $\overline{f}$ is the compositional inverse of f.

If $\mathbf{M}$ is the matrix (g, f) and $\mathbf{a} = (a_0, a_1, \ldots)^T$ is an integer sequence (expressed as an infinite column vector) with ordinary generating function $\mathcal{A}(x)$, then the sequence $\mathbf{Ma}$ has ordinary generating function $g(x)\mathcal{A}(f(x))$. The (infinite) matrix (g, f) can thus be considered to act on the ring of integer sequences $\mathbb{Z}^{\mathbb{N}}$ by multiplication, where a sequence is regarded as a (infinite) column vector. We can extend this action to the ring of power series $\mathbb{Z}[[x]]$ by

$$(g, f) : \mathcal{A}(x) \longmapsto (g, f)\cdot\mathcal{A}(x) = g(x)\mathcal{A}(f(x)). \tag{39}$$

Example 4. The so-called *binomial matrix* B is the element $(1/(1-x), x/(1-x))$ of the Riordan group. It has general element $\binom{n}{k}$ and hence as an array coincides with Pascal's triangle. More generally, B^m is the element $(1/(1-mx), x/(1-mx))$ of the Riordan group, with general term $\binom{n}{k}m^{n-k}$. It is easy to show that the inverse B^{-m} of B^m is given by $(1/(1+mx), x/(1+mx))$.

The row sums of the matrix (g, f) have generating function

$$(g, f) \cdot \frac{1}{1-x} = \frac{g(x)}{1-f(x)}. \tag{40}$$

Example 5. The inverse of the Riordan array $(1/(1+x), x/(1+x)^2)$ is the Riordan array $(c(x), xc(x)^2)$. This follows since the solution of the equation

$$\frac{z}{(1+z)^2} = x \tag{41}$$

is given by

$$z = xc(x)^2. \tag{42}$$

We then have

$$1 + xc(x)^2 = c(x). \tag{43}$$

Many interesting examples of sequences and Riordan arrays can be found in Neil Sloane's On-Line Encyclopedia of Integer Sequences (OEIS), [30, 31]. Sequences are frequently referred to by their OEIS number. For instance, the binomial matrix B ("Pascal's triangle") is A007318.

For an invertible matrix M, its production matrix (also called its Stieltjes matrix) [32, 33] is the matrix

$$\mathcal{S}_M = M^{-1}\overline{M}, \tag{44}$$

where $\overline{M}$ is the matrix M with its first row removed. A Riordan array M is the inverse of the coefficient array of a family of orthogonal polynomials [34–36] if and only if $\mathcal{S}_M$ is tridiagonal [8, 17]. Necessarily, the Jacobi coefficients (i.e., the coefficients of the three-term recurrence [34] that defines the polynomials) of these orthogonal polynomials are then constant.

Example 6. The production matrix of $(c(x), xc(x)^2)$ is given by

$$\begin{pmatrix} 1 & 1 & 0 & 0 & 0 & 0 & \cdots \\ 1 & 2 & 1 & 0 & 0 & 0 & \cdots \\ 0 & 1 & 2 & 1 & 0 & 0 & \cdots \\ 0 & 0 & 1 & 2 & 1 & 0 & \cdots \\ 0 & 0 & 0 & 1 & 2 & 1 & \cdots \\ 0 & 0 & 0 & 0 & 1 & 2 & \cdots \\ \vdots & \vdots & \vdots & \vdots & \vdots & \vdots & \ddots \end{pmatrix}. \tag{45}$$

An important feature of Riordan arrays is that they have a number of sequence characterizations [37, 38]. The simplest of these is as follows.

Proposition 7 (see [38], Theorems 2.1, and 2.2). *Let $D = [d_{n,k}]$ be an infinite triangular matrix. Then, D is a Riordan array if and only if there exist two sequences $A = [\alpha_0, \alpha_1, \alpha_2, \ldots]$ and $Z = [z_0, z_1, z_2, \ldots]$ with $\alpha_0 \neq 0$, $z_0 \neq 0$ such that*

(i) $d_{n+1,k+1} = \sum_{j=0}^{\infty} \alpha_j d_{n,k+j}$, $(k, n = 0, 1, \ldots)$;

(ii) $d_{n+1,0} = \sum_{j=0}^{\infty} z_j d_{n,j}$, $(n = 0, 1, \ldots)$.

The coefficients $\alpha_0, \alpha_1, \alpha_2, \ldots$ and $z_0, z_1, z_2, \ldots$ are called the A-sequence and the Z-sequence of the Riordan array $M = (g(x), f(x))$, respectively. Letting $A(x)$ be the generating function of the A-sequence and $Z(x)$ the generating function of the Z-sequence, we have

$$A(x) = \frac{x}{\bar{f}(x)}, \qquad Z(x) = \frac{1}{\bar{f}(x)}\left(1 - \frac{d_{0,0}}{g(\bar{f}(x))}\right). \tag{46}$$

The first column of $\mathcal{S}_M$ is then generated by $Z(x)$, while the kth column is generated by $x^{k-1}A(x)$ (taking the first column to be indexed by 0).

Proposition 8. *The inverse of the Chebyshev-Boubaker array*

$$\mathbf{P} = \left(\frac{1+ux+vx^2}{1+ax+bx^2}, \frac{1}{1+ax+bx^2}\right)^{-1} \tag{47}$$

has production matrix which begins with

$$\begin{pmatrix} a-u & 1 & 0 & 0 & 0 & 0 & \cdots \\ b-v & a & 1 & 0 & 0 & 0 & \cdots \\ 0 & b & a & 1 & 0 & 0 & \cdots \\ 0 & 0 & b & a & 1 & 0 & \cdots \\ 0 & 0 & 0 & b & a & 1 & \cdots \\ 0 & 0 & 0 & 0 & b & a & \cdots \\ \vdots & \vdots & \vdots & \vdots & \vdots & \vdots & \ddots \end{pmatrix}. \tag{48}$$

Proof. We let $\mathbf{M} = (g(x), f(x)) = \mathbf{P}^{-1}$. Then,

$$A(x) = \frac{x}{\bar{f}(x)} = \frac{x}{x/(1+ax+bx^2)} = 1 + ax + bx^2. \tag{49}$$

We also have

$$Z(x) = \frac{1+ax+bx^2}{x}\left(1 - \frac{1+ux+vx^2}{1+ax+bx^2}\right) = (a-u) + (b-v)x, \tag{50}$$

which proves the assertion. □

3. Chebyshev-Boubaker Arrays

We can characterize the Chebyshev-Boubaker polynomials $P_n(x)$ with coefficient array $\mathbf{P} = ((1+ux+vx^2)/(1+ax+bx^2), \lambda x/(1+ax+bx^2))$ as follows.

Proposition 9 (see [8]). *The Chebyshev-Boubaker coefficient array $((1+ux+vx^2)/(1+ax+bx^2), \lambda x/(1+ax+bx^2))$ is the coefficient array of the generalized Chebyshev polynomials $P_n(x)$ of the second kind given by*

$$P_n(x) = \left(\sqrt{b}\right)^n U_n\left(\frac{\lambda x - a}{2\sqrt{b}}\right) + u\left(\sqrt{b}\right)^{n-1} U_{n-1}\left(\frac{\lambda x - a}{2\sqrt{b}}\right) + v\left(\sqrt{b}\right)^{n-2} U_{n-2}\left(\frac{\lambda x - a}{2\sqrt{b}}\right). \tag{51}$$

The following variant is also true. The array $((1 + ux + vx^2)/(1 + \lambda ax + bx^2), \lambda x/(1 + \lambda ax + bx^2))$ is the coefficient array of the generalized Chebyshev polynomials of the second kind $P_n^{(\lambda)}(x)$ given by

$$P_n^{(\lambda)}(x) = \left(\sqrt{b}\right)^n U_n\left(\frac{\lambda(x-a)}{2\sqrt{b}}\right) + u\left(\sqrt{b}\right)^{n-1} U_{n-1} \times \left(\frac{\lambda(x-a)}{2\sqrt{b}}\right) + v\left(\sqrt{b}\right)^{n-2} U_{n-2}\left(\frac{\lambda(x-a)}{2\sqrt{b}}\right). \quad (52)$$

The following factorization of the Chebyshev-Boubaker array will be useful when it comes to giving a closed-form expression for the connection coefficients.

Proposition 10. *(The canonical factorization of Chebyshev-Boubaker arrays). Given the Chebyshev-Boubaker array*

$$\mathbf{P} = \left(\frac{1 + ux + vx^2}{1 + a\lambda x + bx^2}, \frac{\lambda x}{1 + a\lambda x + bx^2}\right), \quad (53)$$

one has

$$\mathbf{P} = \left(1 + ux + vx^2, x\right) \cdot \left(\frac{1}{1 + bx^2}, \frac{\lambda x}{1 + bx^2}\right) \cdot \left(\frac{1}{1 + ax}, \frac{x}{1 + ax}\right). \quad (54)$$

Proof. The proof is a straight-forward application of the multiplication rule for Riordan arrays. For instance, we have

$$\begin{aligned}\left(\frac{1}{1 + bx^2}, \frac{\lambda x}{1 + bx^2}\right) \cdot \left(\frac{1}{1 + ax}, \frac{x}{1 + ax}\right) &= \left(\frac{1}{1 + bx^2} \frac{1}{1 + a\lambda x/(1 + bx^2)}, \frac{\lambda x/(1 + bx^2)}{1 + a\lambda x/(1 + bx^2)}\right) \\ &= \left(\frac{1}{1 + a\lambda x + bx^2}, \frac{\lambda x}{1 + a\lambda x + bx^2}\right).\end{aligned} \quad (55)$$

□

For the monic case $\lambda = 1$, we have

$$\left(\frac{1 + ux + vx^2}{1 + ax + bx^2}, \frac{x}{1 + ax + bx^2}\right) = \left(1 + ux + vx^2, x\right) \cdot \left(\frac{1}{1 + bx^2}, \frac{x}{1 + bx^2}\right) \cdot \left(\frac{1}{1 + ax}, \frac{x}{1 + ax}\right). \quad (56)$$

Corollary 11. *If*

$$\mathbf{P} = \left(\frac{1 + ux + vx^2}{1 + a\lambda x + bx^2}, \frac{\lambda x}{1 + a\lambda x + bx^2}\right), \quad (57)$$

then

$$\mathbf{P}^{-1} = \left(\frac{1}{1 - ax}, \frac{x}{1 - ax}\right) \cdot \left(c\left(bx^2\right), \frac{1}{\lambda} xc\left(bx^2\right)\right) \cdot \left(\frac{1}{1 + ax + bx^2}, x\right). \quad (58)$$

In the monic case $\lambda = 1$, we then have

$$\left(\frac{1 + ux + vx^2}{1 + ax + bx^2}, \frac{x}{1 + ax + bx^2}\right)^{-1} = \left(\frac{1}{1 - ax}, \frac{x}{1 - ax}\right) \cdot \left(c\left(bx^2\right), xc\left(bx^2\right)\right) \cdot \left(\frac{1}{1 + ax + bx^2}, x\right). \quad (59)$$

We can characterize the Jacobi coefficients of $P_n(x)$ in terms of those of $Q_n(x)$ as follows. We let $\mathcal{S}_P = \overline{\mathbf{P}\mathbf{P}^{-1}}$ be the production matrix of $\mathbf{P}^{-1}$, and we let $\mathcal{S}_Q = \overline{\mathbf{Q}\mathbf{Q}^{-1}}$ be the production matrix of $\mathbf{Q}^{-1}$. We then have the following proposition.

Proposition 12. *One has,*

$$\mathcal{S}_P = \mathbf{A}\mathcal{S}_Q\mathbf{A}^{-1}. \quad (60)$$

Proof. We have $\mathcal{S}_P = \overline{\mathbf{P}\mathbf{P}^{-1}} = \mathbf{P}\mathbb{S}\mathbf{P}^{-1}$, where $\mathbb{S}$ is the shift matrix with ones on the super-diagonal and zeros elsewhere, with a similar expression for $\mathcal{S}_Q$. We then have

$$\begin{aligned}\mathcal{S}_P &= \mathbf{P}\mathbb{S}\mathbf{P}^{-1} \\ &= \mathbf{A}\mathbf{Q}\mathbb{S}(\mathbf{A}\mathbf{Q})^{-1} \\ &= \mathbf{A}\mathbf{Q}\mathbb{S}\mathbf{Q}^{-1}\mathbf{A}^{-1} \\ &= \mathbf{A}\mathcal{S}_Q\mathbf{A}^{-1}.\end{aligned} \quad (61)$$

□

4. The Chebyshev-Boubaker Connection Coefficients

We let

$$\begin{aligned}\mathbf{P} &= \left(\frac{1 + ux + vx^2}{1 + ax + bx^2}, \frac{x}{1 + ax + bx^2}\right), \\ \mathbf{Q} &= \left(\frac{1 + sx + tx^2}{1 + cx + dx^2}, \frac{x}{1 + cx + dx^2}\right).\end{aligned} \quad (62)$$

For the matrix of connection coefficients $\mathbf{A}$, we recall that we have $\mathbf{A} = \mathbf{P}\mathbf{Q}^{-1}$. Now, $\mathbf{P}$ and $\mathbf{Q}^{-1}$ have the following factorizations:

$$\begin{aligned}\mathbf{P} &= \left(1 + ux + vx^2, x\right) \cdot \left(\frac{1}{1 + bx^2}, \frac{x}{1 + bx^2}\right) \cdot \left(\frac{1}{1 + ax}, \frac{x}{1 + ax}\right), \\ \mathbf{Q}^{-1} &= \left(\frac{1}{1 - cx}, \frac{x}{1 - cx}\right) \cdot \left(c\left(dx^2\right), xc\left(dx^2\right)\right) \cdot \left(\frac{1}{1 + sx + tx^2}, x\right).\end{aligned} \quad (63)$$

We now proceed to list the elements of the factors involved. The matrix $(1/(1+ax), x/(1+ax))$ is the generalized inverse binomial matrix with general element

$$\left(\frac{1}{1+ax}, \frac{x}{1+ax}\right) = \left(\binom{n}{k}(-a)^{n-k}\right)_{0\le n,k\le\infty}. \quad (64)$$

The matrix $(1/(1+bx^2), x/(1+bx^2))$ gives the coefficients of the Chebyshev polynomials $U_n(\sqrt{b}x)$. We have

$$\left(\frac{1}{1+bx^2}, \frac{x}{1+bx^2}\right) = \left(\binom{\frac{n+k}{2}}{k}(-b)^{(n-k)/2}\frac{1+(-1)^{n-k}}{2}\right)_{0\le n,k\le\infty}. \quad (65)$$

The matrix $(1+ux+vx^2, x)$ is the banded matrix with the elements $(1, u, v)$ occupying the three diagonals from the main diagonal down (with zeros elsewhere). The elements of the matrix product $(1+ux+vx^2, x)\cdot(1/(1+bx^2), x/(1+bx^2))$ are then given by

$$\binom{\frac{n+k}{2}}{k}(-b)^{(n-k)/2}\frac{1+(-1)^{n-k}}{2} + u\binom{\frac{n+k-1}{2}}{k}(-b)^{(n-k-1)/2}\frac{1-(-1)^{n-k}}{2} + v\binom{\frac{n+k-2}{2}}{\frac{n-k-2}{2}}(-b)^{(n-k-2)/2}\frac{1+(-1)^{n-k}}{2}. \quad (66)$$

Let us denote these elements by $T_{n,k}$. Then, the matrix $\mathbf{P}$ has elements given by

$$p_{n,k} = \sum_{j=0}^{n} T_{n,j}\binom{j-k}{k}(-a)^{j-k}. \quad (67)$$

A similar analysis of the factors in the expression for $\mathbf{Q}^{-1}$ shows that the general element of $\mathbf{Q}^{-1}$ is given by

$$q^*_{n,k} = \sum_{i=0}^{n}\sum_{j=0}^{n}\binom{n}{j}c^{n-j}\frac{i+1}{j+1}\binom{j+1}{\frac{j-i}{2}} \times d^{(j-i)/2}\frac{1+(-1)^{j-i}}{2}\sum_{r=0}^{\lfloor (i-k)/2\rfloor}\binom{i-k-r}{r} \times (-s)^{i-k-r}(-t)^r, \quad (68)$$

since we have

$$\left(\frac{1}{1-cx}, \frac{x}{1-cx}\right) = \left(\binom{n}{k}c^{n-k}\right)_{0\le n,k\le\infty},$$

$$\left(c\left(dx^2\right), xc\left(dx^2\right)\right) = \left(\frac{k+1}{n+1}\binom{n+1}{\frac{n-k}{2}}d^{(n-k)/2}\frac{1+(-1)^{n-k}}{2}\right)_{0\le n,k\le\infty},$$

$$\left(\frac{1}{1+sx+tx^2}, x\right) = \left(\sum_{j=0}^{\lfloor (n-k)/2\rfloor}\binom{n-k-j}{j}(-s)^{n-k-j}(-t)^j\right)_{0\le n,k\le\infty}. \quad (69)$$

We gather our results in the following proposition.

Proposition 13. *Let*

$$\mathbf{P} = \left(\frac{1+ux+vx^2}{1+ax+bx^2}, \frac{x}{1+ax+bx^2}\right), \quad \mathbf{Q} = \left(\frac{1+sx+tx^2}{1+cx+dx^2}, \frac{x}{1+cx+dx^2}\right) \quad (70)$$

be the coefficient arrays of two families of Chebyshev-Boubaker polynomials. Then, the connection coefficients $a_{n,k}$ for P_n in terms of Q_n are given by

$$a_{n,k} = \sum_{j=0}^{n} p_{n,j}q^*_{j,k}, \quad (71)$$

where

$$p_{n,k} = \sum_{j=0}^{n} T_{n,j}\binom{j-k}{k}(-a)^{j-k}, \quad (72)$$

with $T_{n,k}$ given by the expression

$$\binom{\frac{n+k}{2}}{k}(-b)^{(n-k)/2}\frac{1+(-1)^{n-k}}{2} + u\binom{\frac{n+k-1}{2}}{k}(-b)^{(n-k-1)/2}\frac{1-(-1)^{n-k}}{2} + v\binom{\frac{n+k-2}{2}}{\frac{n-k-2}{2}}(-b)^{(n-k-2)/2}\frac{1+(-1)^{n-k}}{2},$$

$$q^*_{n,k} = \sum_{i=0}^{n}\sum_{j=0}^{n}\binom{n}{j}c^{n-j}\frac{i+1}{j+1}\binom{j+1}{\frac{j-i}{2}} \times d^{(j-i)/2}\frac{1+(-1)^{j-i}}{2}\sum_{r=0}^{\lfloor (i-k)/2\rfloor}\binom{i-k-r}{r} \times (-s)^{i-k-r}(-t)^r. \quad (73)$$

The connection coefficient elements in the Chebyshev-Boubaker class of polynomials are given by Riordan arrays. The elements of these arrays can be expressed in terms of recurrences defined by the associated production matrices. Thus, we are naturally interested in the structure of $\mathcal{S}_{\mathbf{A}}$, the production matrix of $\mathbf{A}$.

Proposition 14. *Let* $\mathbf{A}$ *be the matrix of connection coefficients that express the polynomials* $P_n(x)$ *with coefficient array* $\mathbf{P}$ *in terms of the polynomials* $Q_n(x)$ *with coefficient array* $\mathbf{Q}$. *Then, one has*

$$\mathcal{S}_{\mathbf{A}} = \mathbf{Q}\mathcal{S}_{\mathbf{P}}\mathbf{Q}^{-1}. \tag{74}$$

Proof. We have

$$\begin{aligned}\mathcal{S}_{\mathbf{A}} &= \mathbf{A}^{-1}\mathbb{S}\mathbf{A}\\ &= \left(\mathbf{P}\mathbf{Q}^{-1}\right)^{-1}\mathbb{S}\left(\mathbf{P}\mathbf{Q}^{-1}\right)\\ &= \mathbf{Q}\mathbf{P}^{-1}\mathbb{S}\mathbf{P}\mathbf{Q}^{-1}\\ &= \mathbf{Q}\mathcal{S}_{\mathbf{P}}\mathbf{Q}^{-1}.\end{aligned} \tag{75}$$

□

Proposition 7 can now be interpreted in this context as follows, giving a set of recurrence relations for the connection coefficients.

Proposition 15. *If the A and Z sequences of* $\mathcal{S}_{\mathbf{A}}$ *are denoted, respectively, by* α_n *and* z_n, *then the connection coefficients* $a_{n,k}$ *satisfy the following recurrences*

$$\begin{aligned}a_{n+1,k+1} &= \sum_{j=0}^{n-k}\alpha_j a_{n,k+j},\\ a_{n+1,0} &= \sum_{j=0}^{n} z_j a_{n,j}.\end{aligned} \tag{76}$$

5. Examples

In this section, we give examples of the foregoing, using simple examples that are close to the Chebyshev polynomials of the second kind. This enables us to illustrate the theory without an unnecessarily high overhead.

Example 16. We take the example of the families of orthogonal polynomials that have the aerated Catalan and the ordinary Catalan numbers as moments, respectively. Thus, let $\mathbf{P} = \mathbf{S}$, the coefficient array of the monic Chebyshev polynomials of the second kind. We have

$$\mathbf{S} = \left(\frac{1}{1+x^2}, \frac{x}{1+x^2}\right). \tag{77}$$

We will have

$$\mathbf{Q} = \left(\frac{1}{1+x}, \frac{x}{(1+x)^2}\right). \tag{78}$$

This is the array of the Morgan-Voyce polynomials $B_n(x) = \sum_{k=0}^{n}\binom{n+k}{2k}(-1)^{n-k}x^k$. Then,

$$\begin{aligned}\mathbf{A} = \mathbf{S}\mathbf{Q}^{-1} &= \left(\frac{1}{1+x^2}, \frac{x}{1+x^2}\right)\cdot\left(\frac{1}{1+x}, \frac{x}{(1+x)^2}\right)^{-1}\\ &= \left(\frac{1}{1+x^2}, \frac{x}{1+x^2}\right)\cdot\left(c(x), xc(x)^2\right).\end{aligned} \tag{79}$$

This implies that

$$\mathbf{A} = \left(\frac{1}{1+x^2}c\left(\frac{x}{1+x^2}\right), \frac{x}{1+x^2}c\left(\frac{x}{1+x^2}\right)^2\right). \tag{80}$$

The matrix $\mathbf{A}$ begins with

$$\begin{pmatrix} 1 & 0 & 0 & 0 & 0 & 0 & \cdots\\ 1 & 1 & 0 & 0 & 0 & 0 & \cdots\\ 1 & 3 & 1 & 0 & 0 & 0 & \cdots\\ 3 & 7 & 5 & 1 & 0 & 0 & \cdots\\ 9 & 19 & 17 & 7 & 1 & 0 & \cdots\\ 25 & 57 & 55 & 31 & 9 & 1 & \cdots\\ \vdots & \vdots & \vdots & \vdots & \vdots & \vdots & \ddots \end{pmatrix}. \tag{81}$$

The first column terms $a_n = a_{n,0}$ have generating function

$$\frac{1}{1+x^2}c\left(\frac{x}{1+x^2}\right) = \frac{1+x^2-\sqrt{1-4x+2x^2-4x^3+x^4}}{2x\left(1+x^2\right)}. \tag{82}$$

This form of g.f. suggests that this sequence may have an interesting Hankel transform. In fact, the sequence $a_{n,0}$ is A101499, a sequence which gives the number of peakless Motzkin paths of length n in which the $(1,0)$-steps at levels greater than level 0 come in two colors (Emeric Deutsch). The Hankel transform of $a_{n,0}$ starts with

$$\begin{aligned}&1, 0, -4, -16, -64, 0, 4096, 65536, 1048576, 0,\\ &-1073741824, \ldots.\end{aligned} \tag{83}$$

This is A162547, which is a Somos-4 variant [27, 39] in the sense that we have

$$a_n = \frac{\left(4a_{n-1}a_{n-3} - 4a_{n-2}^2\right)}{a_{n-4}}, \quad (n \neq 4k+1). \tag{84}$$

We have

$$a_{n,k} = \sum_{j=0}^{n-k}\sum_{i=0}^{\lfloor j/2\rfloor}(-1)^i\binom{j-i}{i}C_{j-2i}\binom{n-j-1}{n-k-j}2^{n-k-j}. \tag{85}$$

We note that due to the combinatorial interpretation of $a_{n,0}$ and the positivity of the Catalan numbers, we can conclude that all the connection coefficients $a_{n,k}$ are positive in this case.

We now find expressions for the elements α_n and z_n of the production matrix $\mathcal{S}_{\mathbf{A}}$ of $\mathbf{A}$.

We use the values

$$\begin{aligned} g(x) &= \frac{1}{1+x^2}c\left(\frac{x}{1+x^2}\right) \\ &= \frac{1+x^2-\sqrt{1-4x+2x^2-4x^3+x^4}}{2x(1+x^2)}, \\ f(x) &= \frac{x}{1+x^2}c\left(\frac{x}{1+x^2}\right)^2 \\ &= \frac{(1-x)^2-\sqrt{1-4x+2x^2-4x^3+x^4}}{2x}, \end{aligned} \tag{86}$$

along with (46) to find that

$$\begin{aligned} Z(x) &= \frac{\sqrt{1-4x+2x^2-4x^3+x^4}+x^2-1}{2x}, \\ A(x) &= \frac{(1+x)^2+\sqrt{1-4x+2x^2-4x^3+x^4}}{2}. \end{aligned} \tag{87}$$

This implies that

$$\begin{aligned} z_n &= \binom{1}{n} - \sum_{k=0}^{\lfloor (n-1)/2 \rfloor} \frac{1}{k+1}\binom{n-k-2}{k} \\ &\quad \times \binom{n-k-1}{k}(-2)^{n-1-2k}, \\ \alpha_n &= \binom{2}{n} - \sum_{k=0}^{\lfloor (n-2)/2 \rfloor} \frac{1}{k+1}\binom{n-k-3}{k} \\ &\quad \times \binom{n-k-2}{k}(-2)^{n-2-2k}. \end{aligned} \tag{88}$$

With these values for z_n and α_n, we thus obtain the following recurrences for the connection coefficients:

$$\begin{aligned} a_{n+1,k+1} &= \sum_{j=0}^{n-k} \alpha_j a_{n,k+j}, \quad (k, n = 0, 1, \ldots), \\ a_{n+1,0} &= \sum_{j=0}^{n} z_j a_{n,j}, \quad (n = 0, 1, \ldots). \end{aligned} \tag{89}$$

We note that, in this case, the sequence elements α_n and z_n are essentially diagonal sums of generalized Narayana triangles [40].

Example 17. In this example, we let P_n be the family of orthogonal polynomials with the Catalan numbers C_n as moments, and we let Q_n be the family of orthogonal polynomials with the central binomial coefficients $\binom{2n}{n}$ A000984 as moments. We find that

$$\begin{aligned} \mathbf{P} &= \left(\frac{1}{1+x^2}, \frac{x}{1+x^2}\right), \\ \mathbf{Q} &= \left(\frac{1-x}{1+x^2}, \frac{x}{1+x^2}\right). \end{aligned} \tag{90}$$

We obtain

$$\mathbf{A} = \mathbf{P}\mathbf{Q}^{-1} = \left(\frac{1}{1-x}, x\right), \tag{91}$$

which is the partial sum matrix (the lower-triangular matrix all of whose nonzero elements are 1). This corresponds to

$$P_n(x) = \sum_{k=0}^{n} Q_k(x). \tag{92}$$

In fact, we have

$$\begin{aligned} Q_n(x) &= \binom{n+k}{n-k}(-1)^{n-k} - \binom{n+k-1}{n-k-1}(-1)^{n-k-1}, \\ P_n(x) &= \binom{n+k}{n-k}(-1)^{n-k} = \binom{n+k}{2k}(-1)^{n-k} = \sum_{k=0}^{n} Q_k(x). \end{aligned} \tag{93}$$

Example 18. Let $P_n(x) = S_n(x)$, the monic Chebyshev polynomials of the second kind with the aerated Catalan numbers as moments, and let $Q_n(x)$ be the family of orthogonal polynomials with $\binom{2n}{n}$ as moments. We get

$$\begin{aligned} \mathbf{A} &= \left(\frac{1}{1+x^2}, \frac{x}{1+x^2}\right) \cdot \left(\frac{1-x}{1+x}, \frac{x}{(1+x)^2}\right)^{-1} \\ &= \left(\frac{1}{1+x^2}, \frac{x}{1+x^2}\right) \cdot \left(\frac{1}{\sqrt{1-4x}}, xc(x)^2\right) \\ &= \left(\frac{1}{1+x^2}\frac{1}{\sqrt{1-4(x/(1+x^2))}}, \frac{x}{1+x^2}c\left(\frac{x}{1+x^2}\right)^2\right) \\ &= \left(\frac{1}{\sqrt{1-4x+2x^2-4x^3+x^4}}, \right. \\ &\quad \left. \frac{(1-x)^2-\sqrt{1-4x+2x^2-4x^3+x^4}}{2x}\right). \end{aligned} \tag{94}$$

The sequence $a_n = a_{n,0}$ is then A101500, with

$$a_n = \sum_{k=0}^{\lfloor n/2 \rfloor} \binom{n-k}{k}\binom{2(n-2k)}{n-2k}(-1)^k. \tag{95}$$

In this example, we have

$$\mathbf{Q}^{-1} = \left(\frac{1-x}{1+x}, \frac{x}{(1+x)^2}\right)^{-1} = \left(\binom{2n}{n-k}\right), \tag{96}$$

which is A094527. Hence, we have

$$a_{n,k} = \sum_{j=0}^{n} \binom{\frac{n+j}{2}}{j}(-1)^{(n-j)/2}\frac{1+(-1)^{n-j}}{2}\binom{2j}{j-k}. \tag{97}$$

Note that this implies that

$$a_n = \sum_{j=0}^{n} \binom{\frac{n+j}{2}}{j}(-1)^{(n-j)/2}\frac{1+(-1)^{n-j}}{2}\binom{2j}{j}. \tag{98}$$

Example 19. In this final example, we let $\mathbf{P}$ be the coefficient array of the Boubaker polynomials. Thus,

$$\mathbf{P} = \left(\frac{1+3x^2}{1+x^2}, \frac{x}{1+x^2} \right). \tag{99}$$

We recall that the Boubaker polynomials can be expressed as

$$P_n(x) = \sum_{k=0}^{\lfloor n/2 \rfloor} \binom{n-k}{k} \frac{n-4k}{n-k} (-1)^k x^{n-2k}. \tag{100}$$

We let $\mathbf{Q}$ be the coefficient array of the Morgan-Voyce polynomials

$$Q_n(x) = \sum_{k=0}^{n} \binom{n+k}{2k} (-1)^{n-k} x^k. \tag{101}$$

Then,

$$\mathbf{Q} = \left(\frac{1}{1+x}, \frac{x}{(1+x)^2} \right). \tag{102}$$

The connection coefficients of the Boubaker polynomials in terms of the Morgan-Voyce polynomials are the elements of the array

$$\mathbf{A} = \left(\frac{1+3x^2}{1+x^2}, \frac{x}{1+x^2} \right) \cdot \left(\frac{1}{1+x}, \frac{x}{(1+x)^2} \right)^{-1}. \tag{103}$$

Thus,

$$\mathbf{A} = \left(\frac{\left(1+3x^2\right)\left(1+x^2-\sqrt{1-4x+2x^2-4x^3+x^4}\right)}{2x\left(1+x^2\right)}, \frac{(1-x)^2-\sqrt{1-4x+2x^2-4x^3+x^4}}{2x} \right). \tag{104}$$

Letting

$$T_{n,k} = \sum_{j=0}^{n-k} \sum_{i=0}^{\lfloor j/2 \rfloor} (-1)^i \binom{j-i}{i} C_{j-2i} \binom{n-j-1}{n-k-j} 2^{n-k-j}, \tag{105}$$

(see (85)), we obtain

$$a_{n,k} = T_{n,k} + 3T_{n-2,k}. \tag{106}$$

Again, we see that all the coefficients $a_{n,k}$ are positive.

6. Conclusions

We have shown that in the case of orthogonal polynomials defined by Riordan arrays, we can give both closed form expressions and recurrence relations for the connection coefficients, using the machinery from the theory of Riordan arrays.

References

[1] E. Godoy, A. Ronveaux, A. Zarzo, and I. Area, "Minimal recurrence relations for connection coefficients between classical orthogonal polynomials: continuous case," *Journal of Computational and Applied Mathematics*, vol. 84, no. 2, pp. 257–275, 1997.

[2] S. Lewanowicz, "The hypergeometric functions approach to the connection problem for the classical orthogonal polynomials," Tech. Rep., Institute of Computer Science; University of Wroclaw, 2003.

[3] P. Maroni and Z. da Rocha, "Connection coefficients for orthogonal polynomials and the canonical sequence," *Centro de Matema'tica da Universidade do Porto*, vol. 29, pp. 1–18, 2007.

[4] A. Ronveaux, I. Area, E. Godoy, and A. Zarzo, "Lectures on recursive approach to connection and linearization coefficients between polynomials: introduction," in *Proceedings of the Workshop on Special Functions and Differential Equations*, K. Srinivasa Rao et al., Ed., pp. 83–101, Allied Publishers, Madras, India, January 1997.

[5] A. Ronveaux, A. Zarzo, and E. Godoy, "Recurrence relations for connection coefficients between two families of orthogonal polynomials," *Journal of Computational and Applied Mathematics*, vol. 62, no. 1, pp. 67–73, 1995.

[6] R. Szwarc, "Connection coefficients of orthogonal polynomials," *Canadian Mathematical Bulletin*, vol. 35, pp. 548–556, 1992.

[7] J. Wimp, "Connection coefficients, orthogonal polynomials and the WZ-algorithms," *Numerical Algorithms*, vol. 21, no. 1–4, pp. 377–386, 1999.

[8] P. Barry and A. Hennessy, "Meixner-type results for Riordan arrays and associated integer sequences," *Journal of Integer Sequences*, vol. 13, no. 9, Article ID 10.9.4, 2010.

[9] H. S. Wall, *Analytic Theory of Continued Fractions*, AMS Chelsea, 1967.

[10] M. Agida and A. S. Kumar, "A Boubaker polynomials expansion scheme solution to random Love's equation in the case of a rational Kernel," *Electronic Journal of Theoretical Physics*, vol. 7, no. 24, pp. 319–326, 2010.

[11] K. Boubaker, A. Chaouachi, M. Amlouk, and H. Bouzouita, "Enhancement of pyrolysis spray disposal performance using thermal time-response to precursor uniform deposition," *EPJ Applied Physics*, vol. 37, no. 1, pp. 105–109, 2007.

[12] O. M. Dada, O. B. Awojoyogbe, M. Agida, and K. Boubaker, "Variable separation and Boubaker polynomial expansion scheme for solving the neutron transport equation," *Physics International*, vol. 2, pp. 25–30, 2011.

[13] H. Labadiah and K. Boubaker, "A Sturm-Liouville shaped characteristic differential equation as a guide to establish a quasi-polynomial expression to the Boubaker polynomials," *Journal of Differential Equations and Control Processes*, pp. 117–133, 2007.

[14] H. Labadiah, M. Dada, B. Awojoyogbe, B. Mahmoud, and A. Bannour, "Establishment of an ordinary generating function and a Christoffel-Darboux type first-order differential equation for the heat equation related Boubaker-Turki polynomials," *Journal of Differential Equations and Control Processes*, pp. 51–66, 2008.

[15] A. Luzón and M. A. Morón, "Recurrence relations for polynomial sequences via Riordan matrices," *Linear Algebra and Its Applications*, vol. 433, no. 7, pp. 1422–1446, 2010.

[16] I. Graham, D. E. Knuth, and O. Patashnik, *Concrete Mathematics*, Addison-Wesley, New York, NY, USA, 1994.

[17] P. Barry, "Riordan arrays, orthogonal polynomials as moments, and Hankel transforms," *Journal of Integer Sequences*, vol. 14, no. 2, Article ID 11.2.2, 2011.

[18] P. Barry, P. Rajković, and M. Petković, "An application of Sobolev orthogonal polynomials to the computation of a special Hankel determinant," in *Approximation and Computation: In Honor of Gradimir V. Milovanovic, Optimization and Its Applications*, W. Gautschi, G. Rassias, and M. Themistocles, Eds., vol. 42, pp. 53–60, 2011.

[19] R. A. Brualdi and S. Kirkland, "Aztec diamonds and digraphs, and Hankel determinants of Schröder numbers," *Journal of Combinatorial Theory B*, vol. 94, no. 2, pp. 334–351, 2005.

[20] W. Chammam, F. Marcellán, and R. Sfaxi, "Orthogonal polynomials, Catalan numbers, and a general Hankel determinant evaluation," *Linear Algebra and Its Applications*, vol. 436, no. 7, pp. 2105–2116, 2012.

[21] J. Cigler, "Some nice Hankel determinants," http://homepage.univie.ac.at/johann.cigler/preprints/hankel-conjectures.pdf.

[22] C. Krattenthaler, "Advanced determinant calculus: a complement," *Linear Algebra and Its Applications*, vol. 411, no. 1–3, pp. 68–166, 2005.

[23] J. W. Layman, "The Hankel transform and some of its properties," *Journal of Integer Sequences*, vol. 4, no. 1, Article ID 01.1.5, 2001.

[24] Ch. Radoux, "Calcul effectif de certains déterminants de Hankel," *Bulletin de la Société Mathematique de Belgique. Série B*, vol. 31, pp. 49–55, 1979.

[25] R. A. Sulanke and G. Xin, "Hankel determinants for some common lattice paths," *Advances in Applied Mathematics*, vol. 40, no. 2, pp. 149–167, 2008.

[26] R. Vein and P. Dale, *Determinants and Their Applications in Mathematical Phyiscs*, Springer, 1998.

[27] G. Xin, "Proof of the Somos-4 Hankel determinants conjecture," *Advances in Applied Mathematics*, vol. 42, pp. 152–156, 2009.

[28] L. W. Shapiro, S. Getu, W.-J. Woan, and L. C. Woodson, "The Riordan group," *Discrete Applied Mathematics*, vol. 34, no. 1–3, pp. 229–239, 1991.

[29] R. Sprugnoli, "Riordan arrays and combinatorial sums," *Discrete Mathematics*, vol. 132, no. 1–3, pp. 267–290, 1994.

[30] N. J. A. Sloane, "The On-Line Encyclopedia of Integer Sequences," 2012, http://oeis.org/.

[31] N. J. A. Sloane, "The on-line encyclopedia of integer sequences," *Notices of the AMS—American Mathematical Society*, vol. 50, pp. 912–915, 2003.

[32] E. Deutsch, L. Ferrari, and S. Rinaldi, "Production matrices," *Advances in Applied Mathematics*, vol. 34, no. 1, pp. 101–122, 2005.

[33] E. Deutsch, L. Ferrari, and S. Rinaldi, "Production matrices and Riordan arrays," *Annals of Combinatorics*, vol. 13, no. 1, pp. 65–85, 2009.

[34] T. S. Chihara, *An Introduction to Orthogonal Polynomials*, Gordon and Breach Science, 1978.

[35] W. Gautschi, *Orthogonal Polynomials: Computation and Approximation*, Clarendon Press, 2003.

[36] G. Szegö, *Orthogonal Polynomials*, American Mathematical Society, Providence, RI, USA, 4th edition, 1975.

[37] G.-S. Cheon, H. Kim, and L. W. Shapiro, "Riordan group involutions," *Linear Algebra and Its Applications*, vol. 428, no. 4, pp. 941–952, 2008.

[38] T.-X. He and R. Sprugnoli, "Sequence characterization of Riordan arrays," *Discrete Mathematics*, vol. 309, no. 12, pp. 3962–3974, 2009.

[39] D. Gale, "The strange and surprising saga of the Somos sequence," *Mathematical Intelligencer*, vol. 13, pp. 40–42, 1991.

[40] P. Barry, "On a generalization of the Narayana triangle," *Journal of Integer Sequences*, vol. 14, no. 4, Article ID 11.4.5, 2011.

Nonoscillatory Central Schemes for Hyperbolic Systems of Conservation Laws in Three-Space Dimensions

Andrew N. Guarendi and Abhilash J. Chandy

Department of Mechanical Engineering, The University of Akron, Akron, OH 44325-3903, USA

Correspondence should be addressed to Abhilash J. Chandy; achandy@uakron.edu

Academic Editors: Z. Guo and S. A. Mohiuddine

We extend a family of high-resolution, semidiscrete central schemes for hyperbolic systems of conservation laws to three-space dimensions. Details of the schemes, their implementation, and properties are presented together with results from several prototypical applications of hyperbolic conservation laws including a nonlinear scalar equation, the Euler equations of gas dynamics, and the ideal magnetohydrodynamic equations. Parallel scaling analysis and grid-independent results including contours and isosurfaces of density and velocity and magnetic field vectors are shown in this study, confirming the ability of these types of solvers to approximate the solutions of hyperbolic equations efficiently and accurately.

1. Introduction

Over the past couple of decades, much work has gone into the construction, analysis, and implementation of modern numerical algorithms for the approximate solution of systems of nonlinear hyperbolic conservation laws of the form

$$(u)_t + f(u)_x + g(u)_y + h(u)_z = 0. \quad (1)$$

Numerical solutions of these equations are of tremendous practical importance as they govern a variety of physical phenomena in natural and engineering applications. In particular, a number of high-resolution schemes have been developed and tested for this purpose [1–3]. The first-order Lax-Friedrichs scheme [4] is actually the forerunner for a large class of central high-resolution schemes that have seen much development in recent years. This paper presents the formulation and testing of such 3D central high-resolution schemes.

These central schemes enjoy a major advantage of simplicity over methods like upwind schemes, in that no approximate Riemann solvers are involved in their construction. To provide a brief background, in 1990, Nessayahu and Tadmor introduced a fully discrete second-order nonoscillatory central scheme (NT scheme), [1], which was further extended to higher orders of accuracy [5–7], as well as to multidimensional systems [8–10]. The main ingredient in the construction of the NT method was a second-order nonoscillatory, monotonic upstream scheme for conservation laws-(MUSCL-) type [11], piecewise linear interpolant (instead of the piecewise constant one employed in the Lax-Friedrichs scheme) in combination with a higher-order solver for the time evolution [1]. However, applying the fully discrete NT scheme to the second-order convection-diffusion equations did not provide the desired resolution of discontinuities (see [12–14]) due to the accumulation of excessive numerical dissipation [12, 15]. This led Kurganov and Tadmor [12] to the development of a set of second-order semi-discrete central schemes, which had smaller dissipation than the original NT scheme, and, unlike the fully discrete central schemes, they could be efficiently used with time steps as small as required by the CFL stability restriction due to the diffusion term.

The schemes presented in Kurganov and Tadmor's study saw further developments in the form of third-order extensions [15] and central weighted essentially nonoscillatory (CWENO) reconstruction [8] (also see [16, 17] for essentially nonoscillatory (ENO) implementations). Such weighted essentially nonoscillatory (WENO) reconstructions were introduced first in an upwind framework [18] after which they were extended to a central framework [5, 8, 9, 19]. More

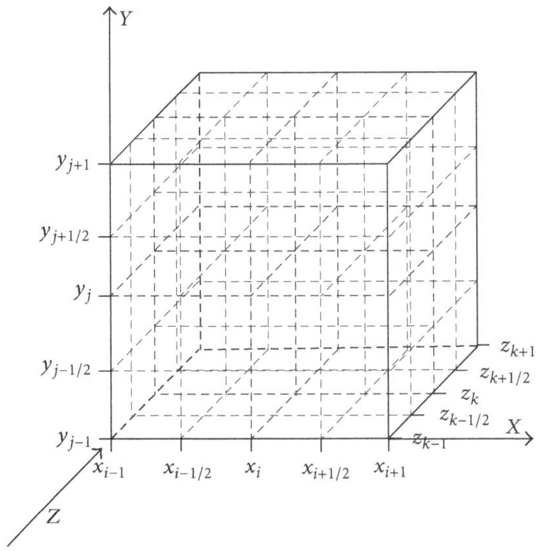

FIGURE 1: Modified central differencing in three-space dimensions.

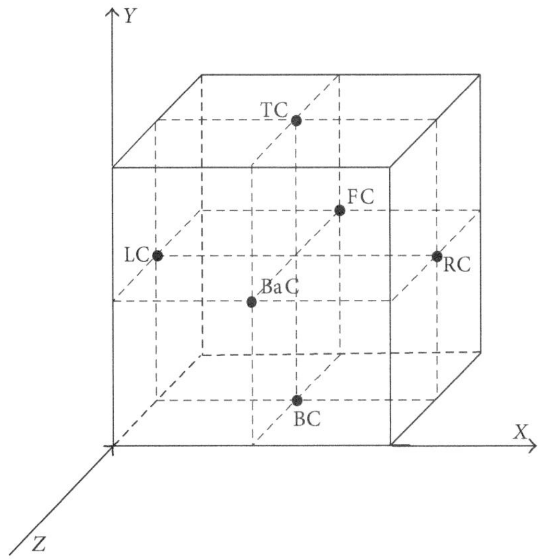

FIGURE 2: Reconstructions in the x-, y-, and z-directions.

recently Balbás and Tadmor [20, 21] presented extensions of the semidiscrete central schemes of [12] with arbitrary order, r, specifically third- and fourth-order reconstructions with the possibility of additional reconstructions in the diagonal directions. In this paper, we present and test the third-order accurate semidiscrete central schemes of Balbás and Tadmor [21] for 3D systems of equations.

2. Formulation

Starting with a general hyperbolic conservation law in three-space dimensions, (1), let the sliding averages of u over the cells $C_{i,j,k} := [x_{i-1/2}, x_{i+1/2}] \times [y_{j-1/2}, y_{j+1/2}] \times [z_{k-1/2}, z_{k+1/2}]$ (see Figure 1) at time level, n, be

$$\overline{u}^n_{i,j,k} := \frac{1}{\Delta x \Delta y \Delta z} \int_{C_{i,j,k}} u\left(x, y, z, t^n\right)\, dx\, dy\, dz, \quad (2)$$

where $\Delta x, \Delta y$, and Δz are the cell widths in the x-, y-, and z-directions, respectively.

Following [21], the first step passing from a fullydiscrete to a semi-discrete formulation consists of reducing the size of the staggered cells covering the Riemann fans. So cells with variable width of order $O(\Delta t)$ are used, by incorporating the maximum local speeds of propagation across the cell interfaces [12, 22], which are given by

$$\begin{aligned} a^n_{i+1/2,j,k} &:= \max\left[\rho\left(\frac{\partial f}{\partial u}\left(u^{\mathrm{LC}}_{i+1,j,k}\right)\right), \rho\left(\frac{\partial f}{\partial u}\left(u^{\mathrm{RC}}_{i,j,k}\right)\right)\right], \\ b^n_{i,j+1/2,k} &:= \max\left[\rho\left(\frac{\partial g}{\partial u}\left(u^{\mathrm{BC}}_{i,j+1,k}\right)\right), \rho\left(\frac{\partial g}{\partial u}\left(u^{\mathrm{TC}}_{i,j,k}\right)\right)\right], \\ c^n_{i,j,k+1/2} &:= \max\left[\rho\left(\frac{\partial h}{\partial u}\left(u^{\mathrm{BaC}}_{i,j,k+1}\right)\right), \rho\left(\frac{\partial h}{\partial u}\left(u^{\mathrm{FC}}_{i,j,k}\right)\right)\right], \end{aligned} \quad (3)$$

where the superscripts LC, RC, BC, TC, BaC and FC stand for the left, right, bottom, top, back, and front centers, respectively (Figure 2). The previous equations are exact for genuinely nonlinear and linearly degenerate fields [21].

The cell interface values in the x-, y-, and z-directions, shown in the previous equations, are defined as

$$\begin{aligned} u^{\mathrm{RC}}_{i,j,k} &:= p^n_{i,j,k}\left(x_{i+1/2}, y_j, z_k\right), \\ u^{\mathrm{LC}}_{i,j,k} &:= p^n_{i,j,k}\left(x_{i-1/2}, y_j, z_k\right), \\ u^{\mathrm{TC}}_{i,j,k} &:= p^n_{i,j,k}\left(x_i, y_{j+1/2}, z_k\right), \\ u^{\mathrm{BC}}_{i,j,k} &:= p^n_{i,j,k}\left(x_i, y_{j-1/2}, z_k\right), \\ u^{\mathrm{FC}}_{i,j,k} &:= p^n_{i,j,k}\left(x_i, y_j, z_{k+1/2}\right), \\ u^{\mathrm{BaC}}_{i,j,k} &:= p^n_{i,j,k}\left(x_i, y_j, z_{k-1/2}\right). \end{aligned} \quad (4)$$

They are calculated via a nonoscillatory piecewise polynomial reconstruction given by

$$R\left(x, y, z; \overline{u}^n\right) = \sum_{i,j,k} p^n_{i,j,k}\left(x, y, z\right) \mathbf{1}_{C_{i,j,k}}; \quad (5)$$

the polynomials $p^n_{i,j,k}(x, y, z)$ are determined so that $R(x, y, z; \overline{u}^n)$ satisfies properties such as conservation of cell averages, accuracy, and nonoscillatory behavior (see [21]). For more details about the derivation of the polynomials and their properties, see [21]. This information allows the separation of regions of smoothness and regions of nonsmoothness, where the discontinuities propagate in one or more directions. In the nonsmooth regions, staggered evolution over the respective control volumes is used to obtain cell averages, while, for the smooth regions, the polynomial $p^n_{i,j,k}(x, y, z)$ and the corresponding fluxes are integrated over the respective control volume to obtain cell averages in these regions. This way, a new polynomial, say $R(x, y, z; \overline{w}^n)$, is formed, which is reconstructed from these smooth and non-smooth portions

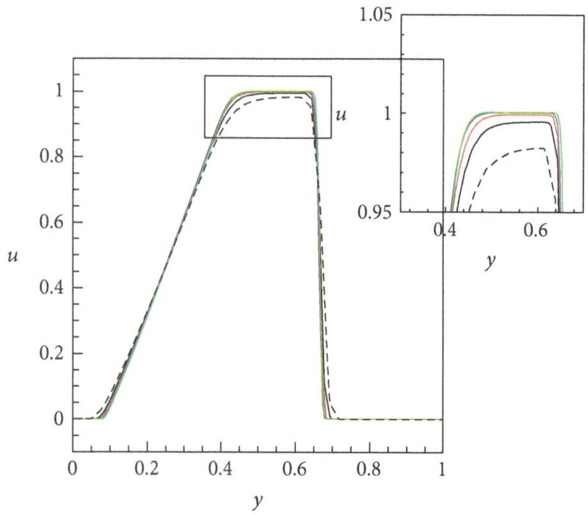

FIGURE 3: Instantaneous solutions, at $X = 0.5$ and $Y = 0.5$, of the nonlinear scalar equation (19) using the CWENO polynomial reconstruction at time $t = 0.2$ for various grid resolutions. Lines: dashed black : 40^3, solid black : 60^3, solid red : 80^3, solid blue : 100^3, and solid green : 120^3.

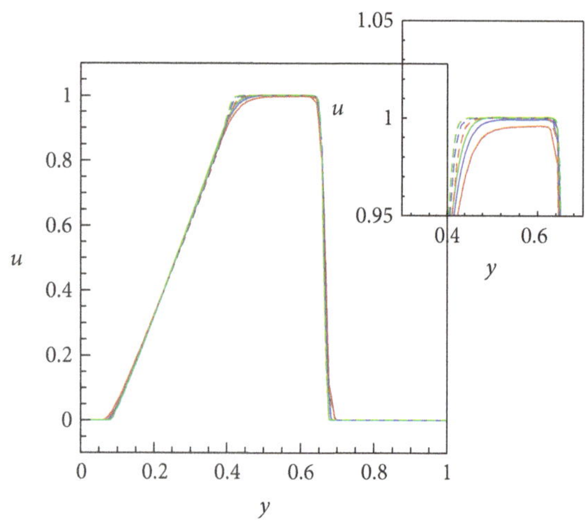

FIGURE 4: Instantaneous solutions, at $X = 0.5$ and $Y = 0.5$, of the nonlinear scalar equation (19) at time $t = 0.2$ for 3rd- and 2nd-order CWENO reconstructions and at various grid resolutions. Solid lines: 3rd order, dashed lines: 2nd order, red: 60^3, blue: 80^3, and green: 100^3.

of the solution and it is reprojected back onto the original grid cells.

The resulting semi-discrete scheme in the limit as $\Delta t \to 0$ is as follows:

$$\begin{aligned}\frac{d}{dt}\overline{u}_{i,j,k}(t) = &-\frac{H^x_{i+1/2,j,k}(t) - H^x_{i-1/2,j,k}(t)}{\Delta x}\\ &-\frac{H^y_{i,j+1/2,k}(t) - H^y_{i,j-1/2,k}(t)}{\Delta y}\\ &-\frac{H^z_{i,j,k+1/2}(t) - H^z_{i,j,k-1/2}(t)}{\Delta z}\end{aligned} \tag{6}$$

with numerical fluxes

$$\begin{aligned}H^x_{i+1/2,j,k}(t) &= \frac{1}{2}\left[f\left(u^{\mathrm{LC}}_{i+1,j,k}\right) + f\left(u^{\mathrm{RC}}_{i,j,k}\right)\right]\\ &\quad - \frac{a_{i+1/2,j,k}(t)}{2}\left[u^{\mathrm{LC}}_{i+1,j,k} - u^{\mathrm{RC}}_{i,j,k}\right],\\ H^y_{i,j+1/2,k}(t) &= \frac{1}{2}\left[g\left(u^{\mathrm{TC}}_{i,j+1,k}\right) + g\left(u^{\mathrm{BC}}_{i,j,k}\right)\right]\\ &\quad - \frac{b_{i,j+1/2,k}(t)}{2}\left[u^{\mathrm{TC}}_{i,j+1,k} - u^{\mathrm{BC}}_{i,j,k}\right],\\ H^z_{i,j,k+1/2}(t) &= \frac{1}{2}\left[h\left(u^{\mathrm{FC}}_{i,j,k+1}\right) + h\left(u^{\mathrm{BaC}}_{i,j,k}\right)\right]\\ &\quad - \frac{c_{i,j,k+1/2}(t)}{2}\left[u^{\mathrm{FC}}_{i,j,k+1} - u^{\mathrm{BaC}}_{i,j,k}\right]\end{aligned} \tag{7}$$

for the third-order CWENO reconstruction without any diagonal smoothing (diagonal smoothing described in Section 3). There is also an option of carrying out diagonal smoothing, the equations of which are not provided here.

3. Implementation of Semidiscrete Central Schemes: Cell Interface Reconstructions

This section provides a third-order nonoscillatory reconstruction in three-space dimensions, that was implemented for computing the solutions of hyperbolic conservation laws (see (19), (21), and (22)–(25)). The reconstruction of the point values of u presented in this section is the third-order CWENO polynomial reconstruction of Kurganov and Levy [15]. The properties of the piecewise quadratic polynomial, that is used here, were presented in [15, 21]. In each cell $C_{i,j,k}$ (defined before), the polynomials $[p^n_{i,j,k}(x,y,z)]$ in (4) are written as a convex combination of three polynomials $P_{-1}(x,y,z)$, $P_0(x,y,z)$, and $P_1(x,y,z)$. In the x-direction they are as follows:

$$\begin{aligned}P^n_{i,j,k}(x,y,z) &= w_{-1}P_{-1}(x,y,z) + w_0P_0(x,y,z)\\ &\quad + w_1P_1(x,y,z), \quad \sum_{m\in -1,0,1} w_m = 1,\end{aligned} \tag{8}$$

where the linear polynomials

$$\begin{aligned}P_{-1}(x,y,z) &= \overline{u}^n_{i,j,k} + \frac{\overline{u}^n_{i,j,k} - \overline{u}^n_{i-1,j,k}}{\Delta x}(x - x_i),\\ P_1(x,y,z) &= \overline{u}^n_{i,j,k} + \frac{\overline{u}^n_{i+1,j,k} - \overline{u}^n_{i,j,k}}{\Delta x}(x - x_i)\end{aligned} \tag{9}$$

conserve the pair of cell averages $\overline{u}^n_{i-1,j,k}$, $\overline{u}^n_{i,j,k}$ and $\overline{u}^n_{i,j,k}$, $\overline{u}^n_{i+1,j,k}$, respectively, and the parabola centered around x_i,

$$\begin{aligned}P_0(x,y,z) &= \overline{u}^n_{i,j,k} - \frac{1}{12}\left(\left(\overline{u}^n_{i-1,j,k} - 2\overline{u}^n_{i,j,k} + \overline{u}^n_{i+1,j,k}\right)\right.\\ &\quad \left. + \left(\overline{u}^n_{i,j,k-1} - 2\overline{u}^n_{i,j,k} + \overline{u}^n_{i,j,k+1}\right)\right)\\ &\quad + \frac{\overline{u}^n_{i+1,j,k} - \overline{u}^n_{i-1,j,k}}{2\Delta x}(x - x_i)\\ &\quad + \frac{\overline{u}^n_{i+1,j,k} - \overline{u}^n_{i,j,k} + \overline{u}^n_{i-1,j,k}}{\Delta x^2}(x - x_i)^2,\end{aligned} \tag{10}$$

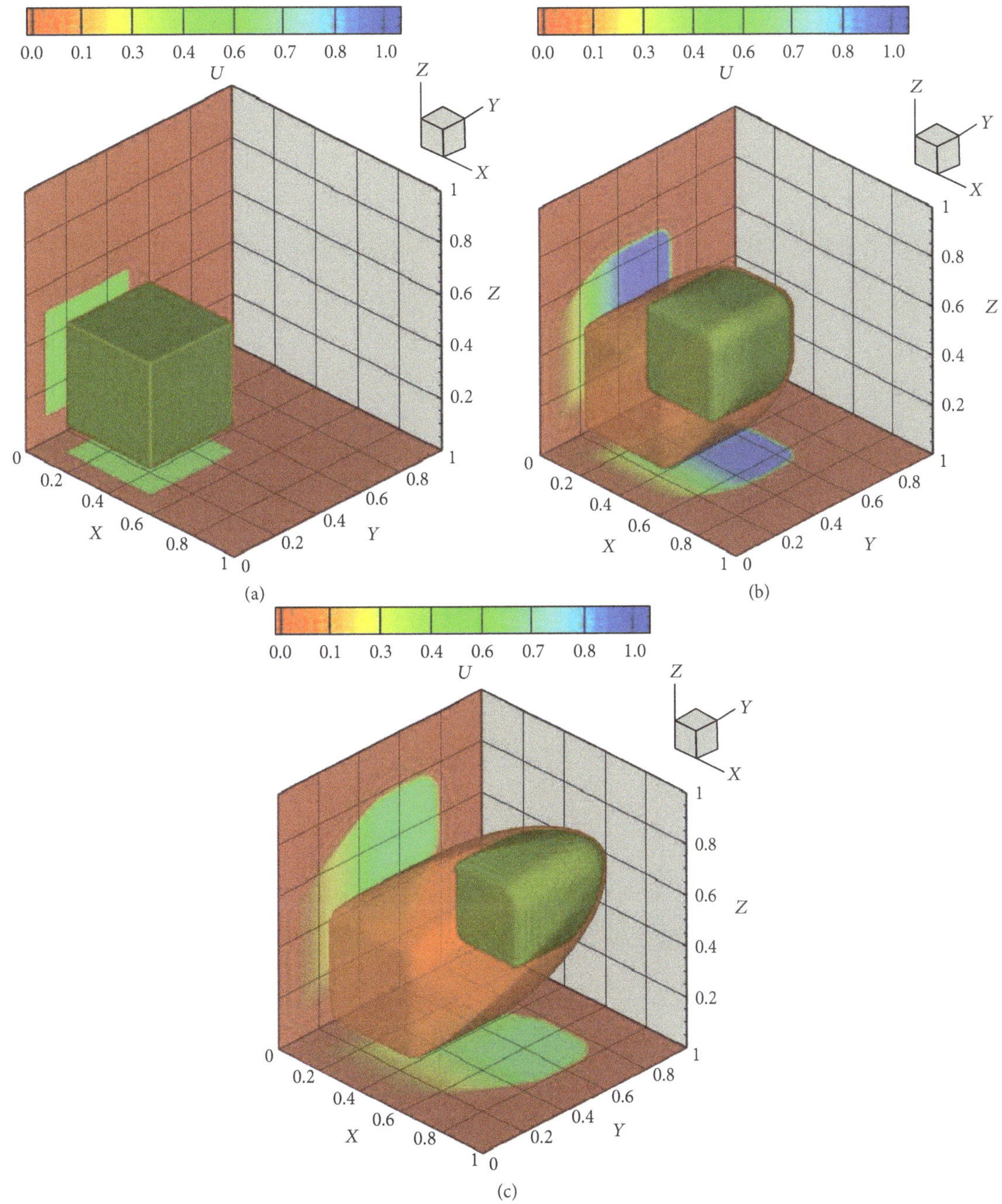

FIGURE 5: Instantaneous solutions of the nonlinear scalar equation (19) using the CWENO polynomial reconstruction diagonal smoothing at times (a) $t = 0$, (b) $t = 0.4$, and (c) $t = 0.8$. Shown are isosurfaces at two values 0.5 and 0.05, along with slices along $X = 0.3$ and $Z = 0.3$, which are projected to the end of the domain. Initial condition—$u = 0.5$, for $0.1 \leq x \leq 0.5$, $0.1 \leq y \leq 0.5$, and $0.1 \leq z \leq 0.5$.

is determined so as to satisfy

$$\begin{aligned} c_{-1}P_{-1}(x,y,z) &+ c_1 P_1(x,y,z) \\ &+ (1 - c_{-1} - c_1) P_0(x,y,z) \\ &= u^n_{i,j,k} + u_{x|i,j,k}(x - x_i) + \frac{1}{2} u_{xx|i,j,k}(x - x_i)^2, \end{aligned} \tag{11}$$

where

$$\begin{aligned} u^n_{i,j,k} := \overline{u}^n_{i,j,k} - \frac{1}{24}\Big(&\left(\overline{u}^n_{i-1,j,k} - 2\overline{u}^n_{i,j,k} + \overline{u}^n_{i+1,j,k}\right) \\ &+ \left(\overline{u}^n_{i,j-1,k} - 2\overline{u}^n_{i,j,k} + \overline{u}^n_{i,j+1,k}\right) \\ &+ \left(\overline{u}^n_{i,j,k-1} - 2\overline{u}^n_{i,j,k} + \overline{u}^n_{i,j,k+1}\right)\Big), \end{aligned}$$

$$\begin{aligned} u_{x|i,j,k} &:= \frac{\overline{u}^n_{i+1,j,k} - \overline{u}^n_{i-1,j,k}}{2\Delta x}, \\ u_{xx|i,j,k} &:= \frac{\overline{u}^n_{i+1,j,k} - \overline{u}^n_{i,j,k} + \overline{u}^n_{i-1,j,k}}{\Delta x^2}. \end{aligned} \tag{12}$$

Note that for the central parabola equation (10) additional corrections in the y- and z-directions guarantee third-order accuracy when the point value $u(x_i, y_j, z_k, t^n)$ is recovered from the neighboring cell averages.

The conservation of the cell averages $\overline{u}^n_{i-1,j,k}, \overline{u}^n_{i,j,k}, \overline{u}^n_{i+1,j,k}$ and the accuracy property (property $\mathbf{P_2}$ in [8, 13]) are

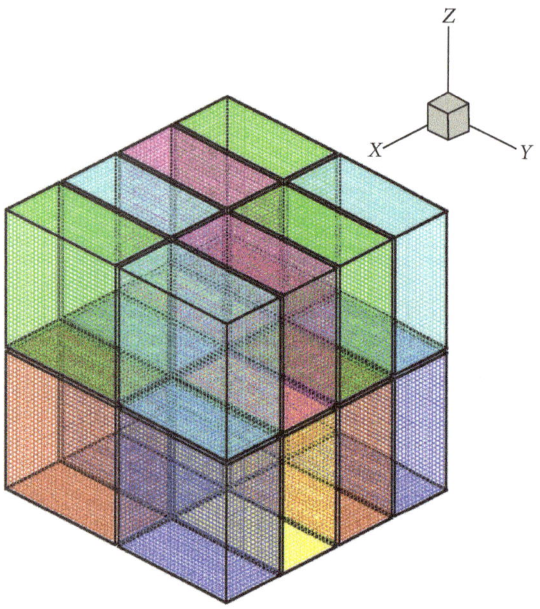

Figure 6: Mesh and domain decomposition for 64 × 64 × 64 grid with a 4 × 2 × 2 processor configuration.

guaranteed [8, 21] by any symmetric choice of weights c_m (e.g., $c_{-1} = c_1 = 1/4$, $c_0 = 1/2$). The nonoscillatory behavior (property $\mathbf{P}_3$ in [8, 21]) is attained with nonlinear weights

$$w_m = \frac{\alpha_m}{\sum_l \alpha_l}, \quad \text{with } \alpha_m = \frac{c_m}{(\epsilon + IS_m)^2}, \; m, l \in \{-1, 0, 1\}. \tag{13}$$

$\epsilon \ll 1$ prevents the denominator from being zero ($\epsilon = 10^{-6}$), and the smoothness indicators provide a local measure of the derivatives $P_m(x, y, z)$, switching automatically to the second-order reconstructions P_{-1} and P_1 in the presence of steep gradients and avoiding the onset of spurious oscillations [9, 15, 18], and in this case they read

$$\begin{aligned} IS_{-1} &= \left(\overline{u}^n_{i,j,k} - \overline{u}^n_{i-1,j,k}\right)^2, \\ IS_1 &= \left(\overline{u}^n_{i+1,j,k} - \overline{u}^n_{i,j,k}\right)^2, \\ IS_0 &= \frac{13}{3}\left(\overline{u}^n_{i+1,j,k} - 2\overline{u}^n_{i,j,k} + \overline{u}^n_{i-1,j,k}\right)^2, \\ &\quad + \frac{1}{4}\left(\overline{u}^n_{i+1,j,k} - \overline{u}^n_{i-1,j,k}\right)^2. \end{aligned} \tag{14}$$

In the case of systems of equations, the smoothness indicators are given by the *norm-scaled* average of the componentwise indicators, $IS_m^{(nq)}$, given by

$$IS_m = \frac{1}{EQ}\sum_{nq=1}^{EQ} \frac{1}{\left\|u^{(nq)}\right\|_2 + \epsilon} IS_m^{(nq)}, \quad m \in \{-1, 0, 1\}, \tag{15}$$

where $u^{(nq)}$ stands for the (nq)th component of u and

$$\left\|u^{(nq)}\right\|_2 = \sum_{i,j,k} \left|u^{(nq)}_{i,j,k}\right|^2 \Delta x \Delta y \Delta z \tag{16}$$

represents its l_2 norm over the discretized solution domain.

The interface values can now be calculated from (8)–(10) as follows:

$$\begin{aligned} u^{\mathrm{RC}}_{i,j,k} &:= p^n_{i+1/2,j,k}(x, y, z) \\ &= w_{-1}\left[\overline{u}^n_{i,j,k} + \frac{1}{2}\left(\overline{u}^n_{i,j,k} - \overline{u}^n_{i-1,j,k}\right)\right] \\ &\quad + w_0\left[\overline{u}^n_{i,j,k} - \frac{1}{12}\left(\left(\overline{u}^n_{i-1,j,k} - 2\overline{u}^n_{i,j,k} + \overline{u}^n_{i+1,j,k}\right)\right.\right. \\ &\qquad + \left(\overline{u}^n_{i,j-1,k} - 2\overline{u}^n_{i,j,k} + \overline{u}^n_{i,j+1,k}\right) \\ &\qquad \left. + \left(\overline{u}^n_{i,j,k-1} - 2\overline{u}^n_{i,j,k} + \overline{u}^n_{i,j,k+1}\right)\right) \\ &\qquad + \frac{1}{4}\left(\overline{u}^n_{i+1,j,k} - \overline{u}^n_{i-1,j,k}\right) \\ &\qquad \left. + \frac{1}{4}\left(\overline{u}^n_{i+1,j,k} - 2\overline{u}^n_{i,j,k} + \overline{u}^n_{i-1,j,k}\right)\right] \\ &\quad + w_1\left[\overline{u}^n_{i,j,k} + \frac{1}{2}\left(\overline{u}^n_{i+1,j,k} - \overline{u}^n_{i,j,k}\right)\right], \\ u^{\mathrm{LC}}_{i,j,k} &:= p^n_{i-1/2,j,k}(x, y, z) \\ &= w_{-1}\left[\overline{u}^n_{i,j,k} - \frac{1}{2}\left(\overline{u}^n_{i,j,k} - \overline{u}^n_{i-1,j,k}\right)\right] \\ &\quad + w_0\left[\overline{u}^n_{i,j,k} - \frac{1}{12}\left(\left(\overline{u}^n_{i-1,j,k} - 2\overline{u}^n_{i,j,k} + \overline{u}^n_{i+1,j,k}\right)\right.\right. \\ &\qquad + \left(\overline{u}^n_{i,j-1,k} - 2\overline{u}^n_{i,j,k} + \overline{u}^n_{i,j+1,k}\right) \\ &\qquad \left. + \left(\overline{u}^n_{i,j,k-1} - 2\overline{u}^n_{i,j,k} + \overline{u}^n_{i,j,k+1}\right)\right) \\ &\qquad - \frac{1}{4}\left(\overline{u}^n_{i+1,j,k} - \overline{u}^n_{i-1,j,k}\right) \\ &\qquad \left. + \frac{1}{4}\left(\overline{u}^n_{i+1,j,k} - 2\overline{u}^n_{i,j,k} + \overline{u}^n_{i-1,j,k}\right)\right] \\ &\quad + w_1\left[\overline{u}^n_{i,j,k} - \frac{1}{2}\left(\overline{u}^n_{i+1,j,k} - \overline{u}^n_{i,j,k}\right)\right]. \end{aligned} \tag{17}$$

Similar reconstructions to (8) can be carried out in the y- and z-directions and the interface values $u^{\mathrm{FC}}_{i,j,k}$, $u^{\mathrm{BaC}}_{i,j,k}$, $u^{\mathrm{TC}}_{i,j,k}$, and $u^{\mathrm{BC}}_{i,j,k}$ can be derived in a straightforward manner. Section 4 presents specific test cases related to the governing equations presented before along with a parallel scaling analysis of the implementation of such schemes on multiple platforms.

4. Numerical Test Cases

This section presents results from the solutions of a single scalar equation (19), the Euler equations of gas dynamics (21), and the ideal equations of magnetohydrodynamics (22) and systems of equations and in applications of a scalar convection problem, the Richtmyer-Meshkov instability, and the Orszag-Tang vortex problem, respectively. For all the calculations presented here, we choose a uniform grid in physical space. For temporal discretization, the third-order strong stability-preserving (SSP) Runge-Kutta [23, 24] is used

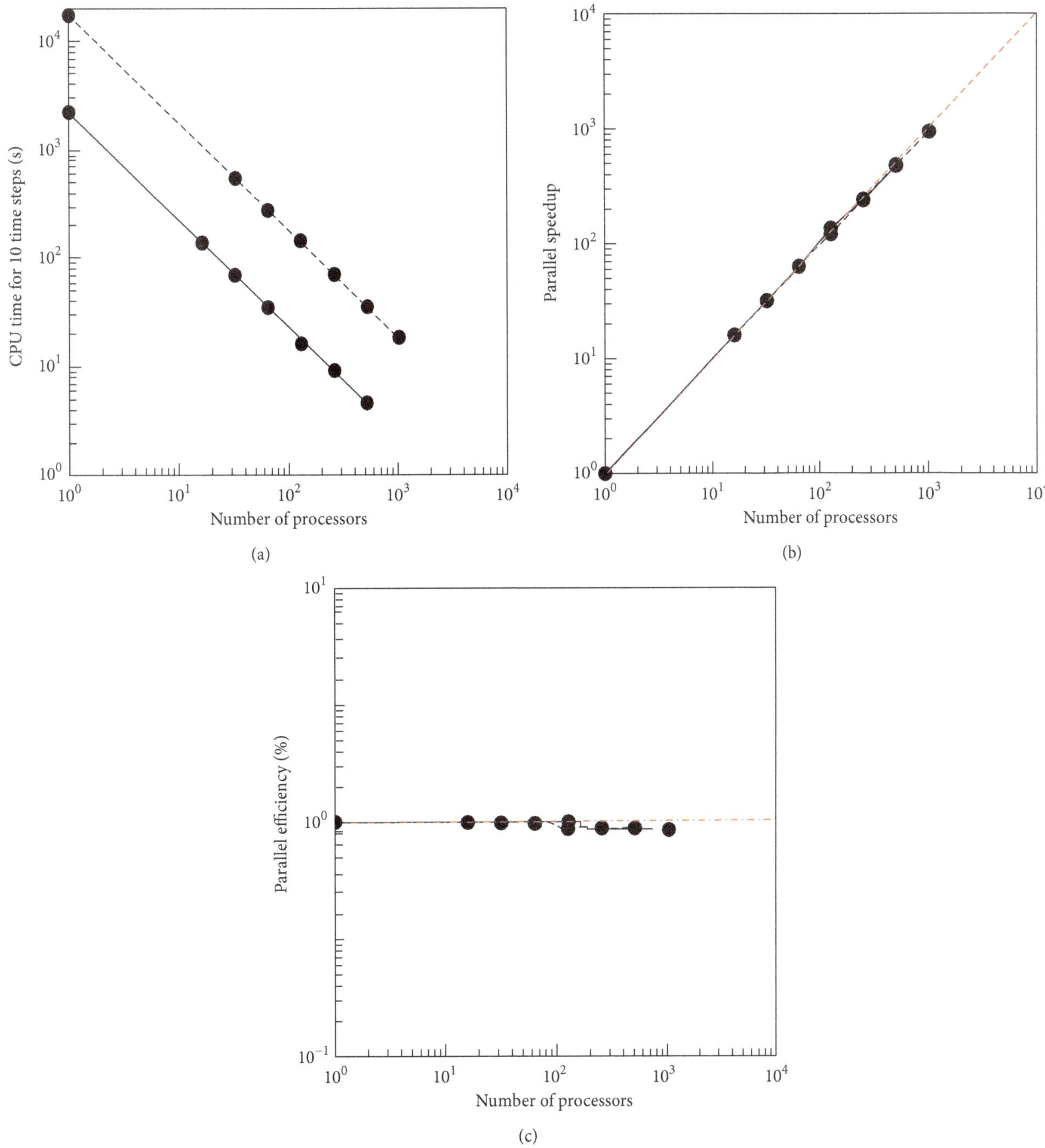

FIGURE 7: Parallel scaling analysis for the solution of the Euler hydrodynamics system (21) on two different problem sizes: (a) CPU time for 10 time steps, (b) speedup, and (c) efficiency; solid line with symbols: 128^3; dashed line with symbols: 256^3. The red dashed line indicates perfect or ideal values.

and the time step is dynamically calculated to satisfy the CFL restriction given by

$$\Delta t = \mathrm{CFL} \times \left(\sqrt{\frac{\left(\max\left|a_{i,j,k}\right|\right)^2}{(\Delta x)^2} + \frac{\left(\max\left|b_{i,j,k}\right|\right)^2}{(\Delta y)^2} + \frac{\left(\max\left|c_{i,j,k}\right|\right)^2}{(\Delta z)^2}} \right)^{-1}, \quad (18)$$

where $a_{i,j,k}$, $b_{i,j,k}$, and $c_{i,j,k}$ are the speeds of propagation (3). The CFL number was chosen to be 0.5 in all the cases presented here.

4.1. Single Scalar Equation: Inviscid Burgers Equation. The presentation of three-dimensional semi-discrete schemes, formulated in Section 3, begins with the solution of a single scalar equation given by

$$(u)_t + \nabla \cdot \left(u^2\right) = 0. \quad (19)$$

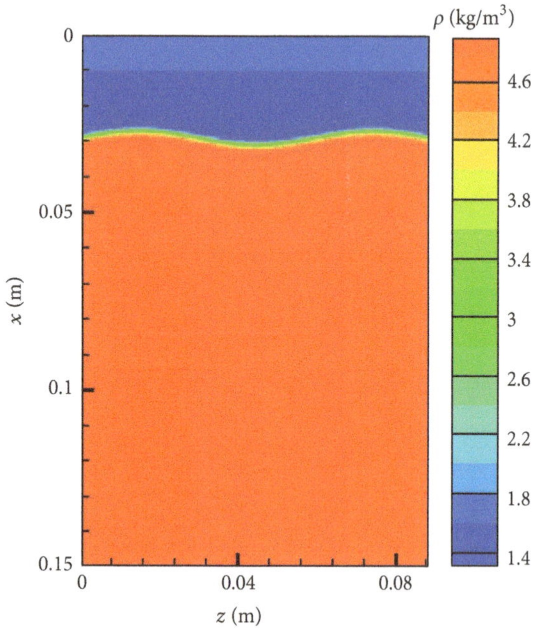

FIGURE 8: Instantaneous contours of initial density across centerline y-direction ($Y = 0$) at $t = 0$ ms.

The equation is solved in a 3D computational domain of size $1 \times 1 \times 1$ with a total of up to 100^3 points, and the initial conditions are such that the variable u is given by

$$u = 0.5; \quad 0.1 \le x \le 0.5,\ 0.1 \le y \le 0.5,\ 0.1 \le z \le 0.5, \tag{20}$$

and zero elsewhere.

Firstly, the solution of the previous equation is presented at various grids to assess the convergence of such a scheme. Also note that no diagonal smoothing is applied in this case. In Figure 3, solutions are obtained at various grid resolutions such as 40^3, 60^3, 80^3, 100^3, and 120^3. The solution with 40^3 resolution is highly inaccurate due to excessive dissipation and can be disregarded in this case. As we go from 60^3 to 120^3, the solution starts to improve, and eventually, it does not change; that is, the difference between the $100^3$and 120^3 solutions is not significant. This also indicates that the solution becomes grid independent.

Next, the schemes presented in this paper are compared to the second-order scheme of Kurganov and Tadmor [12], where a nonoscillatory reconstruction consisting of a minmod limiter was used. Figure 4 shows the 3rd- and 2nd-order solutions at three different grid resolutions in order to assess if there is in fact an improvement in resolution with the order of the scheme. With regard to resolving the discontinuities, the 3rd-order scheme presented here exhibits slightly more dissipation than the 2nd-order method at all resolutions. The solution is expected to dissipate more with the application of diagonal smoothing.

Figure 5 presents the entire 3D solution of (19) using the polynomial reconstructions *without* diagonal smoothing, respectively, for 100^3 grid resolution. Isosurfaces at two different values and the slices in two different directions are shown at various times. The nonzero region or "cube" moves towards the corner $(1, 1, 1)$ and diffuses as well, as time progresses.

4.2. Parallel Performance Analysis for System of Equations. The computational requirements for the solution of hyperbolic problems could become prohibitive in the case of three-dimensional, geometrically complex enclosures. These requirements increase further when realistic fluid flows like viscous or turbulent flows are considered, thereby requiring larger computational effort and memory. Recent developments in high-performance computing promise a substantial increase in computational speed and offer new possibilities for more accurate simulations. Three-dimensional domain decomposition is used to speed the calculations, where the computational domain is decomposed into a number of rectangular blocks with each processor being responsible for a single block. An example of this decomposition can be seen by the gaps in the grid in Figure 6 for the specific case of 16 processors.

Most of the calculations in the interior of each of the subdomains are independent of the domain decomposition and can continue as if they are performed serially. Problems arise near the subdomain boundaries where, for example, finite differences calculated adjacent to the subdomain boundaries may need several points outside the subdomain. To support these circumstances, two rows of "ghost points" are carried along with the interior solutions that contain copies of the interior solution from the neighboring subdomain. These points are exchanged and updated from neighboring processors as needed to ensure that all near-wall calculations are performed with current variable values.

If a uniform grid is used, then the subdomains in each direction will contain equal number of grid points. However, for a nonuniform grid, the the division locations between the subdomains need to be selected to provide good load balancing or an equivalent amount of work for each processor in each time step. Hence, for the purpose of a scaling analysis, Figure 7 illustrates the CPU time, parallel speedup, $S_p = T_{\text{serial}}/T_{\text{parallel}}$, and the parallel efficiency, $E_p = S_p/p$, where T_{serial}, T_{parallel}, and p are the CPU time for serial and parallel runs and the number of processors, respectively. The scaling analysis is presented for the numerical solutions of the Euler hydrodynamic system presented before; (21) diagnostics for two different problem sizes are presented: one is with 128^3, while the other is with 256^3 number of points. The simulations were conducted on the Glenn cluster at Ohio Supercomputing. Due to the high memory requirements of the code, the lowest number of processors on which a 128^3 simulation can be run is 16, while the corresponding number for the 256^3 simulation is 32. In order to present a complete scaling analysis, that is, to calculate speedup and efficiency, it is assumed that these quantities are ideal up to 16 and 32 processors for the 128 and 256^3 simulations, respectively.

Figure 7 shows the simulation time for 10 time steps on a log scale, where the point corresponding to a single processor was in fact extrapolated from the nearest point assuming ideal efficiency (100%). The CPU time decreases linearly with

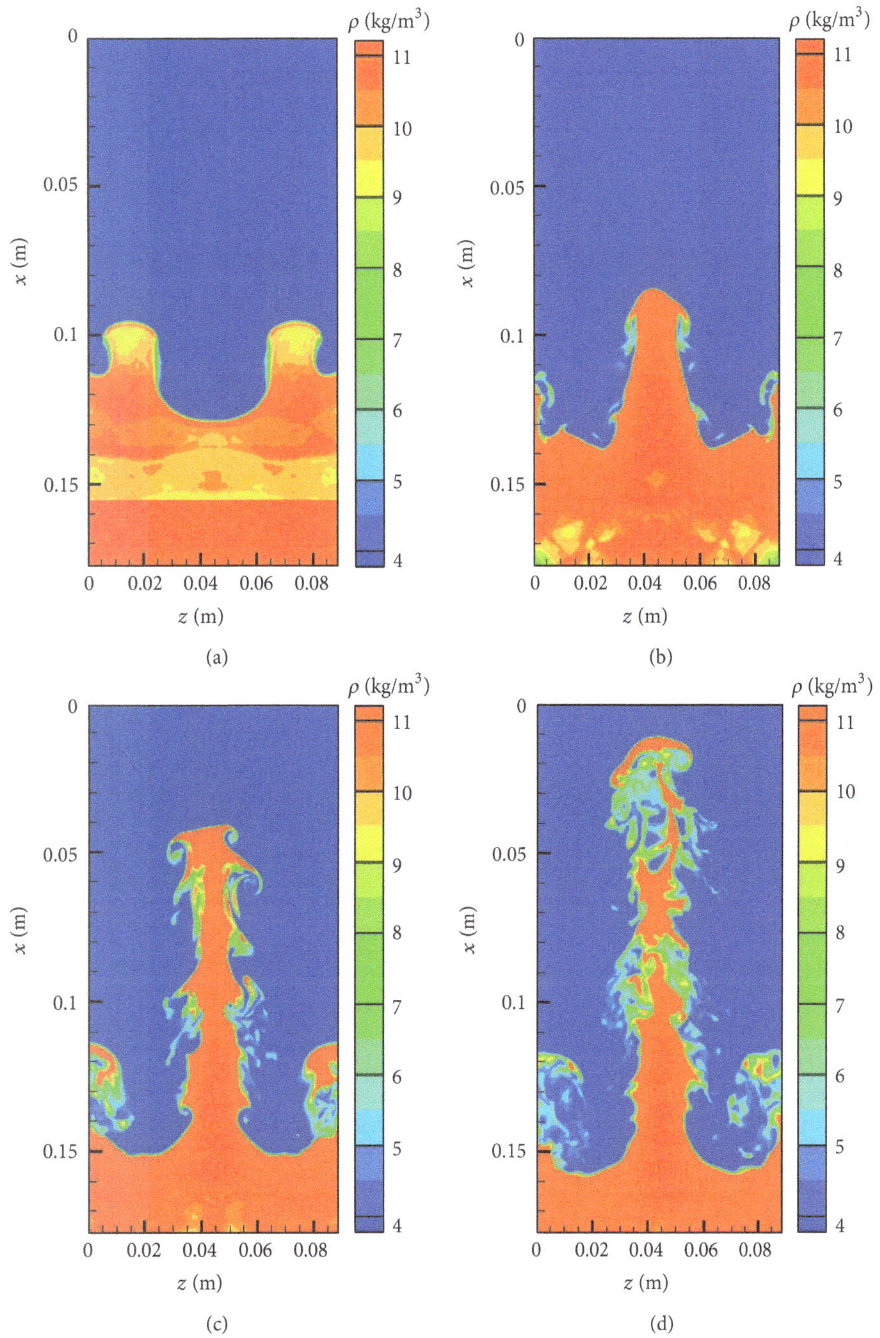

FIGURE 9: Instantaneous contours of density across centerline y-direction ($Y = 0$) at (a) $t = 1$ ms, (b) $t = 2$ ms, (c) $t = 3$ ms, and (d) $t = 4$ ms.

the number of processors which is encouraging. On the same figure speedup and efficiency are close to ideal (red dashed line), with efficiency values ranging between 94% and 100%.

4.3. Euler System of Equations: Richtmyer-Meshkov Instability. Euler equations of gas dynamics are given by

$$\begin{aligned}
(\rho)_t + \nabla \cdot (\rho \boldsymbol{v}) &= 0, \\
(\rho \boldsymbol{v})_t + \nabla \cdot \left[\left(\rho \boldsymbol{v} \boldsymbol{v}^T \right) + p \mathbf{I} \right] &= 0, \\
(E)_t + \nabla \cdot \left[\left(\frac{\gamma}{\gamma - 1} p + \frac{1}{2} \rho v^2 \right) \boldsymbol{v} \right] &= 0.
\end{aligned} \tag{21}$$

Here ρ and E are scalar quantities representing the mass density and total internal energy, respectively. $\boldsymbol{v} = (u, v, w)^T$ is the velocity field with *Euclidean* norm $v^2 := \|\boldsymbol{v}\|^2$. The pressure, p, is coupled to the total internal energy, $E = (1/2)\rho v^2 + p/(\gamma - 1)$.

The evolution of the Richtmyer-Meshkov instability (RMI) [25, 26] is considered in this section. RMI arises when a shock passes through an interface between two fluids of widely ranging densities. A generic feature of these systems, as is the case for fluid turbulence in general, is the existence of fluctuations on multiple length scales. Three-dimensional

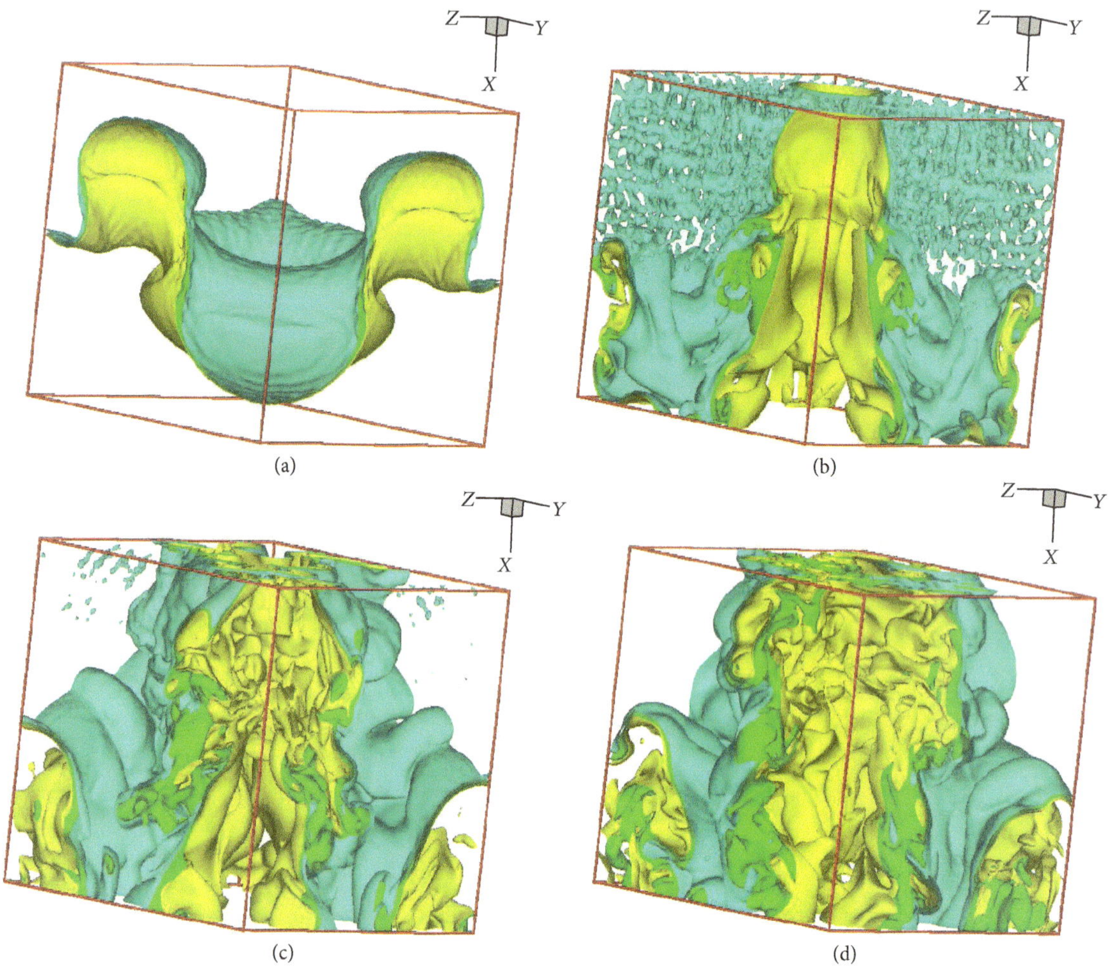

FIGURE 10: Instantaneous isosurfaces of density ($C = 7$ and $C = 9$) on a subdomain of size $4.5 \times 4.5 \times 4.5\,\text{cm}^3$, (a) $t = 1$ ms, (b) $t = 2$ ms, (c) $t = 3$ ms, and (d) $t = 4$ ms.

simulations of the reshocked RMI modeled after the Mach 1.21 experiment of Collins and Jacobs [27] are presented in the present work. The simulations use the 3rd-order CWENO reconstruction method without diagonal smoothing (to avoid excess dissipation resulting from it) using $1024 \times 512 \times 512$ grid points on a domain of $17.8 \times 8.9 \times 8.9\,\text{cm}^3$. For test purposes and in order to have a higher resolution, the domain size here in these simulations is more than 50% smaller in the x-direction as compared to experiments.

The initial conditions were adapted from the Mach 1.21 air/SF_6 experimental shock tube configuration of Collins and Jacobs [27]. The adiabatic exponent $\gamma = 1.24815$ corresponding to an air mixture was used. The ratio of densities is given by $\rho_{SF_6}/\rho_{air} = 4.063$. The initial sinusoidal interface $\eta(y,z) = a_o \sin(2\pi y/\lambda) \sin(2\pi z/\lambda)$ had preshock amplitude $a_o = 0.2\,\text{cm}$ and wavelength $\lambda = 5.933\,\text{cm}$. An initial diffusion layer thickness of $\delta = 0.5\,\text{cm}$ was used [28], where the thickness function is $S(x,y,z) = 1$ if $d \leq 0$, $= \exp(-\alpha|d|^8)$ if $0 < d < 1$ and 0 otherwise. $d = (x_s + \eta(y,z) + \delta - x)/(2\delta)$, and $\alpha = -\ln\beta$ (β is machine zero). Figure 8 shows the initial condition in terms of density for this case. See [28] for a 2D implementation of these initial conditions.

The following boundary conditions were used: (a) inflow at the test section entrance in the streamwise x-direction; (b) reflecting at the end wall of the test section in the streamwise direction; and (c) being periodic in the y- and z-directions corresponding to the cross-section of the test section. The reflecting boundary condition is implemented by reversing the normal component of the velocity vector: $u(x,t) = -u(x,t)$ at $x = 17.8\,\text{cm}$ (maximum in the streamwise direction) and at the ghost points, which is exact and does not generate spurious noise [28].

Figures 9 and 10 show the instantaneous contour slices of density and isosurfaces of density, respectively, at times given by $t = 1$ ms, $t = 2$ ms, $t = 3$ ms, and $t = 4$ ms. As the RMI instability develops, spikes of SF_6 fall into the air. Following this initial growth, the spikes roll-up and additional complex structures begin to appear. The results presented here are qualitatively similar to those of other studies, for example, [28].

4.4. Ideal Magnetohydrodynamic (MHD) Equations: Orszag-Tang Vortex System. The system of equations for ideal magnetohydrodynamics (MHD) is given by

$$(\rho)_t + \nabla \cdot (\rho \boldsymbol{v}) = 0, \tag{22}$$

$$(\rho \boldsymbol{v})_t + \nabla \cdot \left[\left(\rho \boldsymbol{v}\boldsymbol{v}^{\mathrm{T}} \right) + \left(p + \frac{1}{2}B^2 \right) \mathbf{I} - \mathbf{B}\mathbf{B}^T \right] = 0, \tag{23}$$

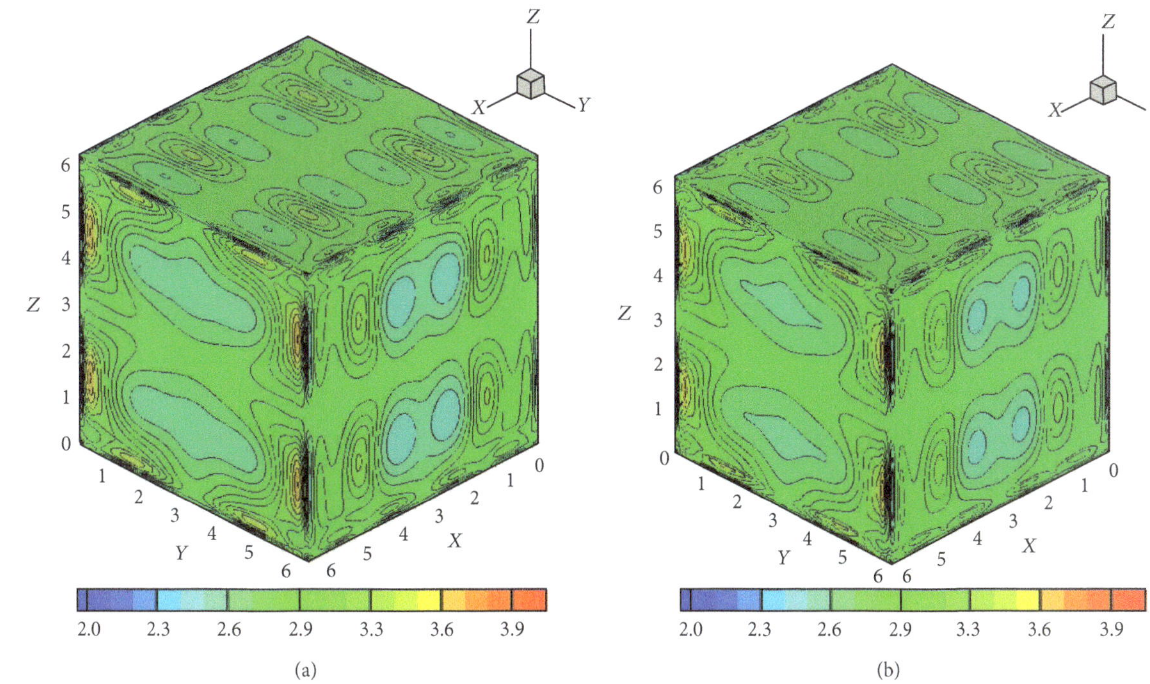

FIGURE 11: Instantaneous contours of density across x-, y-, and z-direction centerlines ($= \pi$) at $t = 0.2$ for two different grid sizes 128^3 (a) and 256^3 (b).

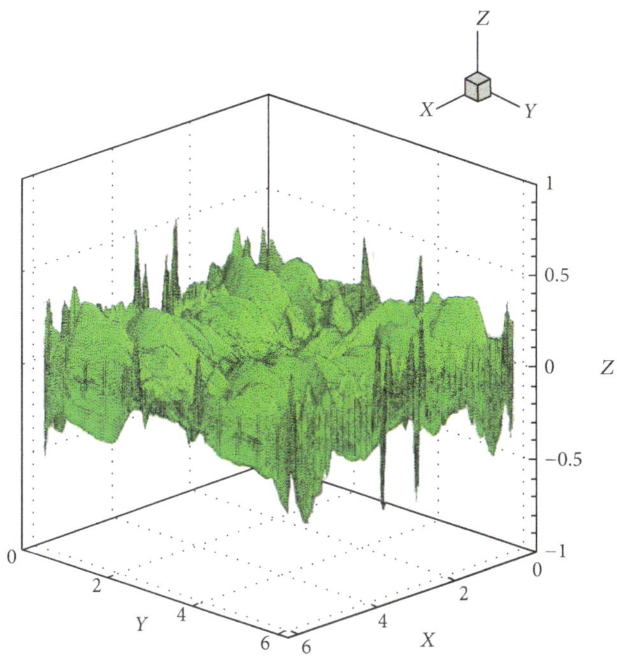

FIGURE 12: Instantaneous surface plot of the divergence of the magnetic field ($\nabla \cdot \mathbf{B}$) on all the Z-surfaces at $t = 0.5$ for the solution of the Orszag-Tang system.

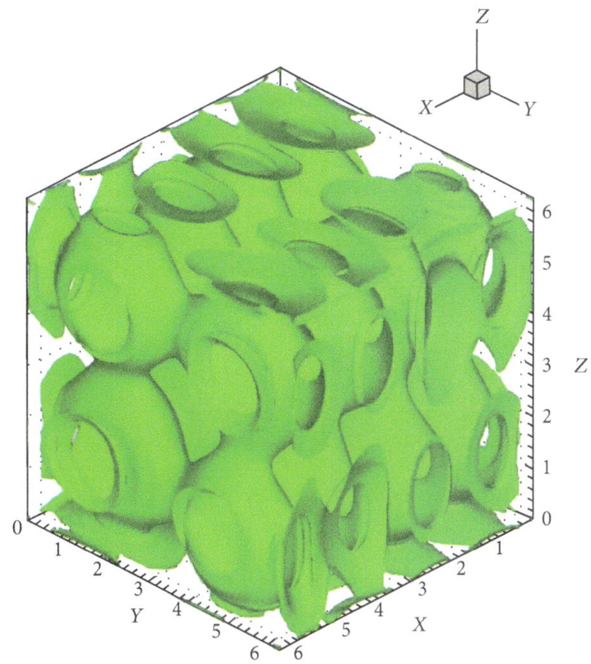

FIGURE 13: Isosurface of the density at a value $\rho_C = 3.0$ at $t = 0.2$.

$$(\mathbf{B})_t - \nabla \times (\boldsymbol{v} \times \mathbf{B}) = 0, \tag{24}$$

$$(E)_t + \nabla \cdot \left[\left(\frac{\gamma}{\gamma - 1} p + \frac{1}{2} \rho v^2 \right) \boldsymbol{v} - (\boldsymbol{v} \times \mathbf{B}) \times \mathbf{B} \right] = 0. \tag{25}$$

Here ρ and E are scalar quantities representing the mass density and total internal energy, respectively, $\boldsymbol{v} = (u, v, \omega)^T$ is the velocity field with *Euclidean* norm $v^2 := \|\boldsymbol{v}\|^2$, and $\mathbf{B} = (B_1, B_2, B_3)^T$ and $B^2 := \|\mathbf{B}\|^2$ represent the magnetic field and its norm, respectively. The pressure, p, is coupled to the total internal energy, $E = (1/2)\rho v^2 + p/(\gamma - 1) + (1/2)B^2$. Furthermore, the system of MHD equations is augmented by the solenoidal constraint; that is, if the condition $\nabla \cdot \mathbf{B} = 0$ is satisfied initially at $t = 0$, then by (24) it remains invariant in time.

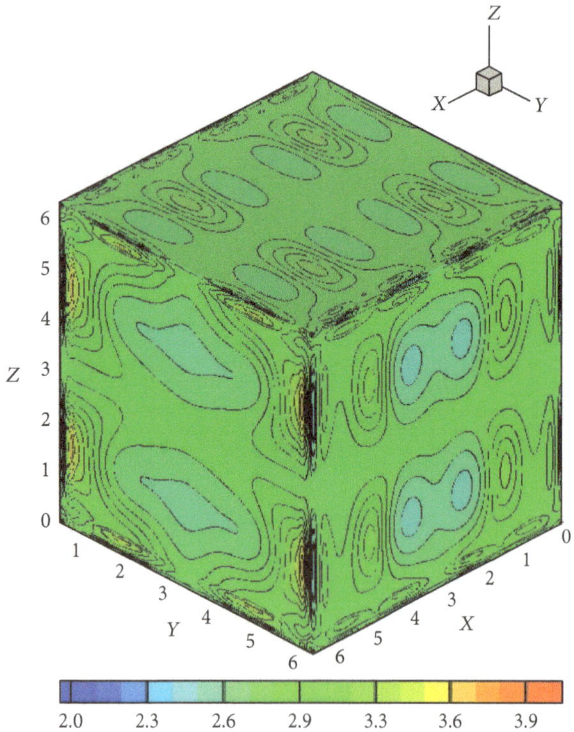

(a)

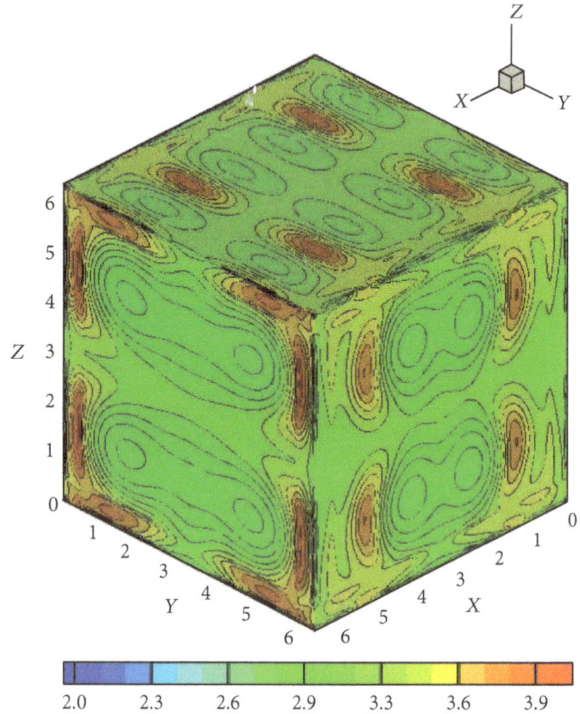

(b)

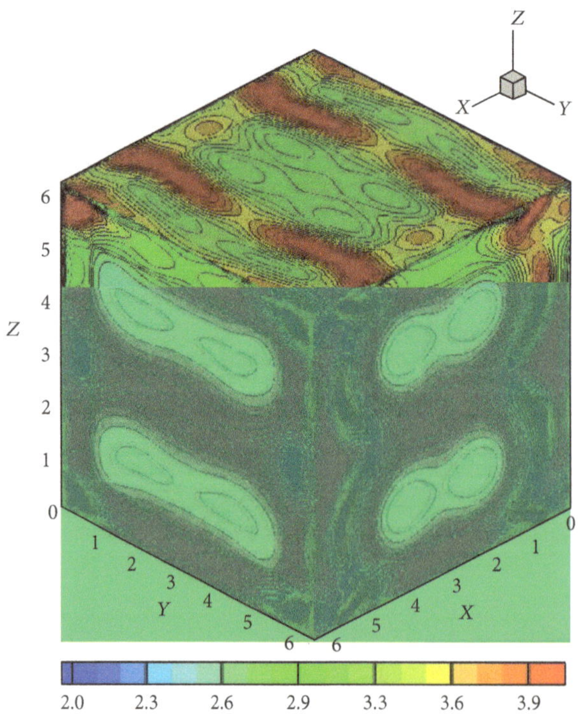

(c)

FIGURE 14: Instantaneous contours of density across x-, y-, and z- direction centerlines ($= \pi$) at (a) $t = 0.2$, (b) $t = 0.4$, and (c) $t = 0.8$ (grid: 256^3).

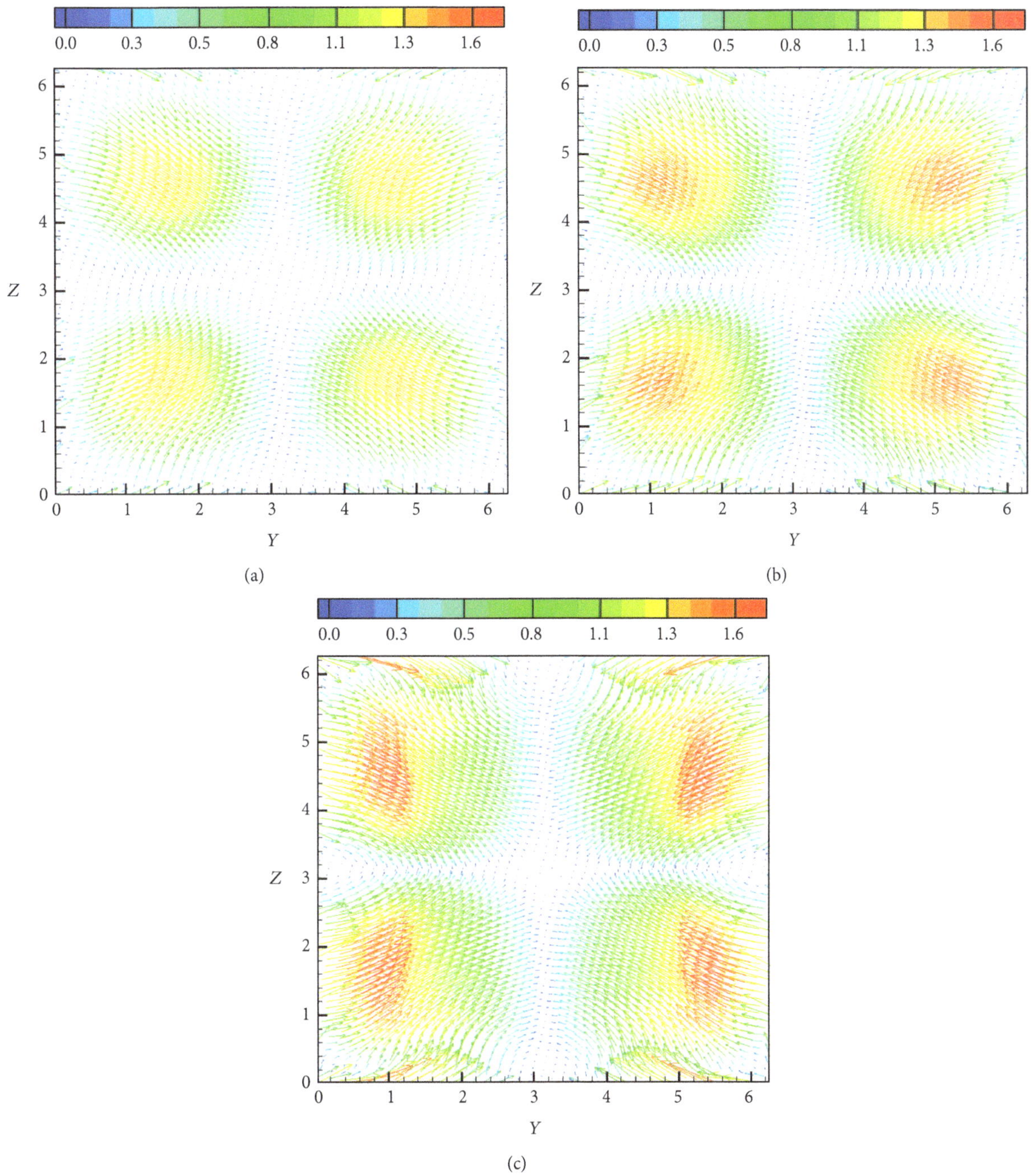

FIGURE 15: Instantaneous slices across the x-direction centerline ($= \pi$) showing vectors of the 2D magnetic field (B_y, B_z) colored by the magnitude of the magnetic field at (a) $t = 0.2$, (b) $t = 0.4$, and (c) $t = 0.8$ (grid: 256^3).

The 3D problem, that is investigated here, is a 3D MHD problem, the Orszag-Tang-type problem [29]. The evolution of the vortex system involves the interaction between several shock waves traveling at various speed regimes [30, 31], which makes the problem especially attractive for numerical experiments. The initial data for this problem are the following:

$$\rho = \gamma^2,$$

$$p = \gamma,$$

$$\mathbf{u} = (-\sin y \sin z, \quad \sin x \ \sin z, \quad 0),$$

$$\mathbf{B} = (-\sin y \sin z, \quad \sin 2x \ \sin z, \quad \sin 2x \ \sin y) \tag{26}$$

with $0 \le x, y, z \le 2\pi$, where $\gamma = 5/3$. Again, grid independence is demonstrated in Figure 11 through density contours on slices across the centerlines planes on coarse (128^3) and fine grids (256^3). The results presented here in this section are those computed on the 256^3 grid using the CWENO reconstruction without diagonal smoothing.

A way of demonstrating the accuracy of a numerical method is to determine whether the solenoidal constraint $\nabla \cdot \mathbf{B} = 0$ is maintained throughout the simulation. Since $\nabla \cdot \mathbf{B} = 0$ initially, theoretically it should remain so throughout the simulation. However, the accumulation of numerical errors can usually lead to nonphysical phenomenon known as magnetic monopoles (when $\nabla \cdot \mathbf{B}$ is not equal to 0). The schemes presented here when first introduced in [21] in 1D and 2D frameworks did not require an explicit enforcement of the solenoidal constraint for producing stable and reasonably accurate solutions, and hence no such treatment is used here either. Figure 12 shows the surface plots of $\nabla \cdot \mathbf{B}$ on all the Z-surfaces at a certain instant in time. The maximum value of the magnitude of $\nabla \cdot \mathbf{B}$ at this instant is 0.41, which is actually representative of the entire simulation. Although these simulations did not show any kind of instabilities, the divergence values are reasonably large for such computations and go to show that in 3D some divergence treatment [32, 33] might be necessary. Figure 13 shows a density isosurface, and Figure 14 shows contours of density on three slices across the $x = y = z = \pi$ planes. Figure 15 shows the 2D magnetic field vector colored by magnetic field magnitude. These results are similar to those of previous studies [32, 33] and demonstrate the ability of such higher-order central schemes to resolve the shocks that the vortex system develops while maintaining the simplicity and ease of implementation typical of this black-box type of finite difference schemes.

5. Conclusions

Extensions of the semi-discrete schemes of Balbás and Tadmor [21] to 3D are presented and tested for the first time in this paper. The numerical test cases chosen include evolution of (a) a single scalar equation or an inviscid Burgers equation, (b) the Richtmeyer-Meshkov instability (RMI) using Euler hydrodynamic equations, and (c) the Orszag-Tang vortex system using ideal magnetohydrodynamic equations. Grid independence was demonstrated for two of the three cases presented here. For the single scalar equation test case, our 3rd-order results exhibited more dissipation than those with corresponding 2nd-order schemes with minmod reconstructions. Parallel scaling analysis showed almost ideal efficiencies and speedups based on the assumption that they hold ideal values up until 16 processors. The results obtained with these schemes for the Richtmeyer-Meshkov instability and the Orszag-Tang vortex system confirm the ability of this type of solver to approximate the discontinuous solutions of Eulerian gas dynamics and ideal magnetohydrodynamics equations.

References

[1] E. Tadmor, "Non-oscillatory central differencing for hyperbolic conservation laws," *Journal of Computational Physics*, vol. 87, no. 2, pp. 408–463, 1990.

[2] R. J. LeVeque, *Numerical Methods for Conservation Laws*, Lectures in Mathematics, Birkhäuser, Basel, Switzerland, 1992.

[3] M. E. Hussaini, B. E. VanLeer, and J. E. VanRosendale, *Upwind and High-Resolution Schemes*, Springer, New York, NY, USA, 1997.

[4] K. O. Friedrichs and P. D. Lax, "Systems of conservation equations with a convex extension," *Proceedings of the National Academy of Sciences of the United States of America*, vol. 68, pp. 1686–1688, 1971.

[5] D. Levy, G. Puppo, and G. Russo, "Central WENO schemes for hyperbolic systems of conservation laws," *Mathematical Modelling and Numerical Analysis*, vol. 33, no. 3, pp. 547–571, 1999.

[6] X.-D. Liu and E. Tadmor, "Third order nonoscillatory central scheme for hyperbolic conservation laws," *Numerische Mathematik*, vol. 79, no. 3, pp. 397–425, 1998.

[7] D. Levy, G. Puppo, and G. Russo, "A fourth-order central weno scheme for multidimensional hyperbolic systems of conservation laws," *SIAM Journal on Scientific Computing*, vol. 24, no. 2, pp. 480–506, 2003.

[8] D. Levy, G. Puppo, and G. Russo, "Compact central WENO schemes for multidimensional conservation laws," *SIAM Journal on Scientific Computing*, vol. 22, no. 2, pp. 656–672, 2001.

[9] D. Levy, G. Puppo, and G. Russo, "Third order central WENO scheme for 2D conservation laws," *Applied Numerical Mathematics*, vol. 33, no. 1, pp. 415–421, 2000.

[10] G.-S. Jiang and E. Tadmor, "Nonoscillatory central schemes for multidimensional hyperbolic conservation laws," *SIAM Journal on Scientific Computing*, vol. 19, no. 6, pp. 1892–1917, 1998.

[11] B. VanLeer, "Towards the ultimate conservative difference scheme. V. A second-order sequel to Godunov's method," *Journal of Computational Physics*, vol. 32, no. 1, pp. 101–136, 1979.

[12] A. Kurganov and E. Tadmor, "New high-resolution central schemes for nonlinear conservation laws and convection-diffusion equations," *Communications on Pure and Applied Mathematics*, vol. 50, pp. 753–771, 1997.

[13] A. Kurganov and P. Rosenau, "Effects of a saturating dissipation in Burgers-type equations," *Communications on Pure and Applied Mathematics*, vol. 50, no. 8, pp. 753–771, 1997.

[14] A. Kurganov, D. Levy, and P. Rosenau, "On Burgers-type equations with nonmonotonic dissipative fluxes," *Communications on Pure and Applied Mathematics*, vol. 51, no. 5, pp. 443–473, 1998.

[15] A. Kurganov and D. Levy, "A third-order semidiscrete central scheme for conservation laws and convection-diffusion equations," *SIAM Journal on Scientific Computing*, vol. 22, no. 4, pp. 1461–1488, 2001.

[16] A. Harten, B. Engquist, S. Osher, and S. R. Chakravarthy, "Uniformly high order accurate essentially non-oscillatory schemes, III," *Journal of Computational Physics*, vol. 71, no. 2, pp. 231–303, 1987.

[17] W. Shu, "Numerical experiments on the accuracy of ENO and modified ENO schemes," *Journal of Scientific Computing*, vol. 5, no. 2, pp. 127–149, 1990.

[18] G.-S. Jiang and C.-W. Shu, "Efficient implementation of weighted ENO schemes," *Journal of Computational Physics*, vol. 126, no. 1, pp. 202–228, 1996.

[19] L. Cai, J.-H. Feng, and W.-X. Xie, "A CWENO-type central-upwind scheme for ideal MHD equations," *Applied Mathematics and Computation*, vol. 168, no. 1, pp. 600–612, 2005.

[20] J. Balbás, E. Tadmor, and C.-C. Wu, "Non-oscillatory central schemes for one- and two-dimensional MHD equations: I," *Journal of Computational Physics*, vol. 201, no. 1, pp. 261–285, 2004.

[21] J. Balbás and E. Tadmor, "Nonoscillatory central schemes for one- and two-dimensional magnetohydrodynamics equations. II; High-order semidiscrete schemes," *SIAM Journal on Scientific Computing*, vol. 28, no. 2, pp. 533–560, 2006.

[22] A. Kurganov and G. Petrova, "A third-order semi-discrete genuinely multidimensional central scheme for hyperbolic conservation laws and related problems," *Numerische Mathematik*, vol. 88, no. 4, pp. 683–729, 2001.

[23] C. W. Shu, "Total-variation-diminishing time discretizations," *SIAM Journal on Scientific Computing*, vol. 9, pp. 1073–1084, 1988.

[24] S. Gottlieb, C.-W. Shu, and E. Tadmor, "Strong stability-preserving high-order time discretization methods," *SIAM Review*, vol. 43, no. 1, pp. 89–112, 2001.

[25] R. Richtmeyer, "Taylor instability in shock acceleration of compressible fluids, Comm," *Pure and Applied Mathematics*, vol. 13, pp. 297–319, 1960.

[26] E. E. Meshkov, "Instability of the interface of two gases accelerated by a shock wave," *Fluid Dynamics*, vol. 4, no. 5, pp. 101–104, 1972.

[27] B. D. Collins and J. W. Jacobs, "PLIF flow visualization and measurements of the Richtmyer-Meshkov instability of an air/SF6 interface," *Journal of Fluid Mechanics*, vol. 464, pp. 113–136, 2002.

[28] M. Latini, O. Schilling, and W. S. Don, "Effects of WENO flux reconstruction order and spatial resolution on reshocked two-dimensional Richtmyer-Meshkov instability," *Journal of Computational Physics*, vol. 221, no. 2, pp. 805–836, 2007.

[29] S. A. Orszag and C.-M. Tang, "Small-scale structure of two-dimensional magnetohydrodynamic turbulence," *Journal of Fluid Mechanics*, vol. 90, no. 1, pp. 129–143, 1979.

[30] R. B. Dahlburg and J. M. Picone, "Evolution of the Orszag-Tang vortex system in a compressible medium. I. Initial average subsonic flow," *Physics of Fluids B*, vol. 1, no. 11, pp. 2153–2165, 1989.

[31] G.-S. Jiang and C.-C. Wu, "A high-order WENO finite difference scheme for the equations of ideal magnetohydrodynamics," *Journal of Computational Physics*, vol. 150, no. 2, pp. 561–594, 1999.

[32] P. Arminjon and R. Touma, "Central finite volume methods with constrained transport divergence treatment for ideal MHD," *Journal of Computational Physics*, vol. 204, no. 2, pp. 737–759, 2005.

[33] R. Touma and P. Arminjon, "Central finite volume schemes with constrained transport divergence treatment for three-dimensional ideal MHD," *Journal of Computational Physics*, vol. 212, no. 2, pp. 617–636, 2006.

A Novel Definition of L-Fuzzy Lattice Based on Fuzzy Set

Jun-Fang Zhang

Department of Basic Science, Beijing University of Agriculture, Beijing 102206, China

Correspondence should be addressed to Jun-Fang Zhang; zhangjunfang@bac.edu.cn

Academic Editors: F. Başar and A. Favini

The concept of L-fuzzy lattice is presented by means of an L-fuzzy partially ordered set. An L-fuzzy partially ordered set A is an L-fuzzy lattice if and only if one of $A_{[a]}$, $A^{[a]}$, and $A^{(a)}$ is a lattice.

1. Introduction

Many concepts of fuzzy algebra were presented, since the concept of fuzzy subgroups was introduced by Rosenfeld. The concept of fuzzy lattice was also introduced by Tepavčević and Trajkovski [1]. But the authors in [1] defined an L-fuzzy lattice based on a fuzzy set of a crisp lattice, so the scope is very limited, and they gave its characterizations by only one kind of its cut sets. To overcome this shortcoming, in this paper we try to present a new definition of an L-fuzzy lattice in an L-fuzzy subset of a general set in terms of an L-poset. Furthermore, its many characterizations are given using the theory of L_β-sets and L_α-sets proposed by Shi Fu-Gui in 1995. In this way, we obtain more generalized conclusion about L-fuzzy lattice in order to make their applications becoming comprehensive.

2. Preliminaries

Throughout this paper L denotes a completely distributive lattice, and $M(L)$ denotes the set of all nonzero $\vee$-irreducible elements in L. $P(L)$ denotes the set of all nonunit prime elements in L. X denotes a nonempty usual sets. L^X is the set of all L-fuzzy sets on X. We will not differ a crisp set from its character function. For empty set $\emptyset \subset L$, we define $\bigwedge\emptyset = 1$ and $\bigvee\emptyset = 0$. According to [2], each element a in L has a greatest maximal family and a greatest minimal family which we, respectively, denote by $\alpha(a)$ and $\beta(a)$. From [2] we know that $\alpha^*(a) = P(L)\bigcap\alpha(a)$ is a maximal family of a, $\alpha(1) = \emptyset$, and $\beta^*(a) = M(L)\bigcap\beta(a)$ is a minimal family of a, $\beta(0) = \emptyset$.

Now we recall some basic concepts and results.

Definition 1 (see [3]). Let $A \in L^X$ and $a \in L$. Define

$$\begin{aligned} A_{[a]} &= \{x \in X \mid A(x) \geq a\}, \\ A_{(a)} &= \{x \in X \mid a \in \beta(A(x))\}, \\ A^{[a]} &= \{x \in X \mid a \notin \alpha(A(x))\}, \\ A^{(a)} &= \{x \in X \mid A(x) \nleq a\}. \end{aligned} \tag{1}$$

From [3] we know that $a \in \beta(b)$ implies $A_{[b]} \subset A_{(a)} \subset A_{[a]}$ and $a \in \alpha(b)$ implies $A^{[a]} \subset A^{(b)} \subset A^{[b]}$. If $L = [0,1]$, then $A_{[a]} = A^{[a]}$ and $A_{(a)} = A^{(a)}$.

Definition 2 (see [3–5]). For each L-fuzzy set A in L^X, we have

(1) $A = \bigvee_{a\in L}(a \wedge A_{[a]}) = \bigvee_{a\in M(L)}(a \wedge A_{[a]}) = \bigvee_{a\in L}(a \wedge A_{(a)}) = \bigvee_{a\in M(L)}(a \wedge A_{(a)})$,

(2) $A = \bigwedge_{a\in L}(a \vee A^{[a]}) = \bigwedge_{a\in P(L)}(a \vee A^{[a]}) = \bigwedge_{a\in L}(a \vee A^{(a)}) = \bigwedge_{a\in P(L)}(a \vee A^{(a)})$.

Definition 3 (see [6]). Let $A \in L^X$ and $B \in L^Y$. Define an L-fuzzy set $A \times B$ on $X \times Y$ by

$$(A \times B)(x, y) = A(x) \wedge B(y), \quad \forall (x, y) \in X \times Y. \tag{2}$$

$A \times B$ is called the product of A and B.

Definition 4 (see [6]). Let $A \in L^X$, $B \in L^Y$, and $R \in L^{X\times Y}$. R is called an L-fuzzy relation from (of) A to B if $R \leq A \times B$.

Definition 5 (see [7]). Let $A \in L^X$, $B \in L^Y$. An L-fuzzy relation f from A to B is called an L-fuzzy mapping from (of)

A into B if $a \in M(L)$, $f_{[a]}$ is a mapping from (of) A into B, and then we write it as $f : A \to B$.

Theorem 6 (see [7]). *Let $A \in L^X, B \in L^Y$, and $f \leq A \times B$, and then the following conditions are equivalent:*

(1) *f is an L-fuzzy mapping from A into B;*

(2) *for each $a \in M(L)$, $f_{(a)}$ is a mapping from $A_{(a)}$ into $B_{(a)}$;*

(3) *for each $a \in P(L)$, $f^{(a)}$ is a mapping from $A^{(a)}$ into $B^{(a)}$.*

Theorem 7 (see [8]). *Let $A \in L^X, B \in L^Y$, and $f \leq A \times B$. If for every $b, c \in L$, $\alpha^*(b \vee c) = \alpha^*(b) \bigcap \alpha^*(c)$, then the following conditions are equivalent:*

(1) *f is an L-fuzzy mapping from A into B;*

(2) *for every $a \in \alpha^*(0)$, $f^{[a]}$ is a mapping from $A^{[a]}$ into $B^{[a]}$.*

Definition 8 (see [9]). Let $A \in L^X, B \in L^Y$, and $f : A \to B$. For $C \leq A$, we define $f(C) = \bigvee_{a \in M(L)}(a \wedge f_{[a]}(C_{[a]}))$. Then $f(C)$ is called the image of C under f.

Theorem 9 (see [9]). *Let $A \in L^X, B \in L^Y$, and $f : A \to B$, and then for $C \leq A$ we have*

(1) *for each $a \in M(L)$, $f(C)_{(a)} \subseteq f_{(a)}(C_{(a)}) \subseteq f_{[a]}(C_{[a]}) \subseteq f(C)_{[a]}$,*

(2) $f(C) = \bigvee_{a \in M(L)}(a \wedge f_{(a)}(C_{(a)}))$,

(3) *for each $a \in P(L)$, $f(C)^{(a)} = f^{(a)}(C^{(a)})$,*

(4) $f(C) = \bigwedge_{a \in P(L)}(a \vee f^{(a)}(C^{(a)}))$.

Definition 10 (see [9]). Let $A \in L^X, B \in L^Y$, and $f : A \to B$. For $D \leq B$, we define $f^{-1}(D) = \bigvee_{a \in M(L)}(a \wedge (f_{[a]})^{-1}(D_{[a]}))$. Then $f^{-1}(D)$ is called the inverse image of D under f.

Theorem 11 (see [9]). *Let $A \in L^X, B \in L^Y$, and $f : A \to B$. Then for $D \leq B$, we have*

(1) *for each $a \in M(L)$, $(f^{-1}(D))_{(a)} \subseteq (f_{(a)})^{-1}(D_{(a)}) \subseteq (f_{[a]})^{-1}(D_{[a]}) \subseteq (f^{-1}(D))_{[a]}$,*

(2) $f^{-1}(D) = \bigvee_{a \in M(L)}(a \wedge (f_{(a)})^{-1}(D_{(a)}))$,

(3) *for each $a \in P(L)$, $(f^{-1}(D))^{(a)} = (f^{(a)})^{-1}(D^{(a)})$,*

(4) $f^{-1}(D) = \bigwedge_{a \in P(L)}(a \vee (f^{(a)})^{-1}(D^{(a)}))$.

Definition 12 (see [10]). Let X be a set, $A \in L^X$, and $A \neq 0$. An L-fuzzy relation R from A to A is called an L-fuzzy partial order on A if R satisfies the following conditions:

(1) for all $x \in A_{(0)}$, $R(x, x) = A(x)$,

(2) $R \circ R \leq R$,

(3) for all $x, y \in A_{(0)}$, $R(x, y) \wedge R(y, x) \neq 0 \Rightarrow x = y$.

When R is an L-fuzzy partial order on A, we call (A, R) an L-fuzzy partial order set or L-poset for short.

Theorem 13 (see [10]). *For an L-fuzzy relation R on A, the following implications (4) $\Rightarrow$ (1) and (1) $\Leftrightarrow$ (2) $\Leftrightarrow$ (3) $\Leftrightarrow$ (5) $\Leftrightarrow$ (6) $\Leftrightarrow$ (7) are true.*

(1) *R is an L-fuzzy partial order on A.*

(2) *For each $a \in L\backslash_{\{0\}}$, if $R_{[a]}$ is not empty set, then it is a partial order on $A_{[a]}$.*

(3) *For each $a \in M(L)$, if $R_{[a]}$ is not empty set, then it is a partial order on $A_{[a]}$.*

(4) *For each $a \in \beta(1)$, if $R_{(a)}$ is not empty set, then it is a partial order on $A_{(a)}$.*

(5) *For each $a \in \alpha(0)$, if $R^{[a]}$ is not empty set, then it is a partial order on $A^{[a]}$.*

(6) *For each $a \in \alpha^*(0)$, if $R^{[a]}$ is not empty set, then it is a partial order on $A^{[a]}$.*

(7) *For each $a \in P(L)$, if $R^{(a)}$ is not empty set, then it is a partial order on $A^{(a)}$.*

Remark 14 (see [10]). In general, (1) $\Rightarrow$ (4) in the previous theorem is not true. This can be seen from the following example.

Example 15 (see [10]). Let $X = \{x, y, z\}$, $L = \{1, a, b, c\} \cup [0, 1/2]$, where $x, y,$ and z are different and $[0, 1/2]$ is an interval. We define the order in L as follow.

For all $e \in [0, 1/2]$, $e \leq c = 1/2$, $c < a$, $c < b$, $a \nleq b$, $b \nleq a$, $a < 1$, and $b < 1$. The order in $[0, 1/2]$ is as usual. Then L is a completely distributive lattice. Take $R \in L^{X \times X}$ such that

$$\begin{gathered} R(x,x) = R(y,y) = R(z,z) = 1, \\ R(y,x) = R(z,y) = R(z,x) = 0, \\ R(x,y) = a, \qquad R(y,z) = b, \\ R(x,z) = \frac{1}{2}. \end{gathered} \tag{3}$$

Obviously we have

$$\begin{gathered} (R \circ R)(x,x) = (R \circ R)(y,y) = (R \circ R)(z,z) = 1, \\ (R \circ R)(z,y) = (R \circ R)(y,x) = (R \circ R)(z,x) = 0, \\ (R \circ R)(x,y) = a, \qquad (R \circ R)(y,z) = b, \\ (R \circ R)(x,z) = \frac{1}{2}. \end{gathered} \tag{4}$$

Thus R is an L-fuzzy partial order on X. But it is easy to check that

$$\begin{gathered} R_{(1/2)} = \{(x,x), (y,y), (z,z), (x,y), (y,z)\}, \\ R_{(1/2)} \circ R_{(1/2)} = \{(x,x), (y,y), (z,z), (x,y), \\ (y,z), (x,z)\} \not\subset R_{(1/2)}. \end{gathered} \tag{5}$$

This shows that $R_{(1/2)}$ is not a partial order on X.

3. L-Fuzzy Lattice

Definition 16. Let $A \in L^X$, $x, y, s \in X$, R be an L-fuzzy partial order on A. s is called an L-fuzzy supremum of x, y if the following conditions are true:

(S1) $A(s) \geq A(x) \wedge A(y)$

(S2) $R(x, s) \geq R(x, x)$; $R(y, s) \geq R(y, y)$,

(S3) $R(s, z) \geq R(x, z) \wedge R(y, z)$.

Definition 17. Let $A \in L^X$, $x, y, t \in X$, R be an L-fuzzy partial order on A. t is called an L-fuzzy infimum of x, y if the following conditions are true:

(T1) $A(t) \geq A(x) \wedge A(y)$,

(T2) $R(t, x) \geq R(x, x)$; $R(t, y) \geq R(y, y)$,

(T3) $R(z, t) \geq R(z, x) \wedge R(z, y)$.

Definition 18. An L-fuzzy partially ordered set (A, R) is called an L-fuzzy lattice on X if for any $x, y \in A_{(0)}$, both L-fuzzy supremum and L-fuzzy infimum of x, y exist.

Theorem 19. *Let (A, R) be an L-fuzzy partially ordered set. Then the following conditions are equivalent.*

(1) *(A, R) is an L-fuzzy lattice on X.*

(2) *For any $a \in L\backslash_{\{0\}}, (A_{[a]}, R_{[a]})$ is a lattice.*

(3) *For any $a \in M(L), (A_{[a]}, R_{[a]})$ is a lattice.*

(4) *For any $a \in \alpha(0), (A^{[a]}, R^{[a]})$ is a lattice.*

(5) *For any $a \in \alpha^*(0), (A^{[a]}, R^{[a]})$ is a lattice.*

(6) *For any $a \in P(L), (A^{(a)}, R^{(a)})$ is a lattice.*

Proof. (1) $\Rightarrow$ (2) $\Rightarrow$ (3) $\Rightarrow$ (6) $\Rightarrow$ (1). Since above-mentioned sets have been posets, we only need to prove that supremum and infimum of x, y exist.

(1) $\Rightarrow$ (2). For any $a \in L\backslash_{\{0\}}$, let $x, y \in A_{(0)}$ and $x, y \in A_{[a]}$, and then $A(x) \geq a, A(y) \geq a$. By Definitions 1 and 2 we know that there exist $s, t \in A_{(0)}$ such that

$$A(S) \geq A(x) \wedge A(y) \geq a, \qquad A(I) \geq A(x) \wedge A(y) \geq a. \tag{6}$$

So $s, t \in A_{[a]}$. By (S2) we know that $R(x, s) \geq R(x, x) = A(x) \geq a$ and $R(y, s) \geq R(y, y) = A(y) \geq a$. Therefore, $(x, s) \in R_{[a]}, (y, s) \in R_{[a]}$. Analogously we can prove that $(t, x) \in R_{[a]}, (t, y) \in R_{[a]}$. For any $(x, z), (y, z) \in R_{[a]}$, we have that $R(s, z) \geq R(x, z) \wedge R(y, z) \geq a$, hence $(s, z) \in R_{[a]}$. Analogously, for any $(z, x), (z, y) \in R_{[a]}$, we can prove that $(z, t) \in R_{[a]}$. Thereby, s, t are, respectively, supremum and infimum of x, y with respect to $R_{[a]}$ in $A_{[a]}$. Then it is proved that $(A_{[a]}, R_{[a]})$ is a lattice.

(2) $\Rightarrow$ (3) is obvious.

(3) $\Rightarrow$ (6). For any $a \in P(L)$, let $x, y \in A^{(a)}$. Since $A^{(a)} = \bigcup_{b \nleq a} A_{[b]}$, there exist $b_1, b_2 \nleq a$ such that $x \in A_{[b_1]}, y \in A_{[b_2]}$. Take $b = b_1 \wedge b_2$, then we have $x, y \in A_{[b]}$ and $b = b_1 \wedge b_2 \nleq a$ (since a is a prime element). By (3) we know that for any $a \in M(L), (A_{[a]}, R_{[a]})$ is a lattice. Then there exist $x \vee_b y, x \wedge_b y \in A_{[b]} \subset A^{(a)}$ such that

$$(x, x \vee_b y) \in R_{[b]} \subset R^{(a)}, \qquad (y, x \vee_b y) \in R_{[b]} \subset R^{(a)},$$
$$\forall (x, z), (y, z) \in R_{[b]} \subset R^{(a)}, \qquad (x \vee_b y, z) \in R_{[b]} \subset R^{(a)}. \tag{7}$$

Thus $x \vee_b y$ is supremum with respect to $R^{(a)}$ of x, y in $A^{(a)}$. Analogously we can prove that infimum exists in $A^{(a)}$ too. Hence $(A^{(a)}, R^{(a)})$ is a lattice.

(6) $\Rightarrow$ (1). Let $x, y \in A_{(0)}$ and $a \in P(L)$, $A(x) \wedge A(y) \nleq a$. Then $A(x) \nleq a$, $A(y) \nleq a$; that is, $x, y \in A^{(a)}$. From (6) there exist $x \vee_a y$ and $x \wedge_a y$ in $A^{(a)}$; that is, $A(x \vee_a y) \nleq a$ and $A(x \wedge_a y) \nleq a$. So we have

$$A(x \vee_a y) \geq A(x) \wedge A(y), \qquad A(x \wedge_a y) \geq A(x) \wedge A(y) \tag{8}$$

and by (6) we obtain that $(x, x \vee_a y) \in R^{(a)}$ and $(y, x \vee_a y) \in R^{(a)}$. This shows that $R(x, x \vee_a y) \nleq a, R(y, x \vee_a y) \nleq a$. Therefore,

$$R(x, x \vee_a y) \geq A(x) = R(x, x),$$
$$R(y, x \vee_a y) \geq A(y) = R(y, y). \tag{9}$$

By (6) we have that for each $(x, z), (y, z) \in R^{(a)}$; that is, $R(x, z) \nleq a, R(y, z) \nleq a$, and we can obtain that $(x \vee_a y, z) \in R^{(a)}$; that is, $R(x \vee_a y, z) \nleq a$. Therefore $R(x \vee_a y, z) \geq R(x, z) \wedge R(y, z)$. As before we can prove that

$$R(x \wedge_a y, x) \geq R(x, x), \qquad R(x \wedge_a y, y) \geq R(y, y),$$
$$R(z, x \wedge_a y) \geq R(z, x) \wedge R(z, y). \tag{10}$$

Let $s = x \vee_a y, t = x \wedge_a y$. Then from Definition 1 we know that s, t are L-supremum and L-infimum of x, y respectively.

Then we prove (1) $\Rightarrow$ (4) $\Rightarrow$ (5) $\Rightarrow$ (1).

(1) $\Rightarrow$ (4). For any $a \in \alpha^*(0)$, let $x, y \in A^{[a]}$; that is, $a \notin \alpha(A(x)), a \notin \alpha(A(y))$. From (1) there exist s, t such that $A(s) \geq A(x) \wedge A(y)$ and $A(t) \geq A(x) \wedge A(y)$. This implies that $a \notin \alpha(A(s))$ and $a \notin \alpha(A(t))$; that is, $s, t \in A^{[a]}$. From (1) we know that $R(x, s) \geq R(x, x) = A(x)$ and $R(y, s) \geq R(y, y) = A(y)$. Then we obtain that

$$a \notin \alpha(R(x, s)), \quad a \notin \alpha(R(x, t)),$$
$$\text{that is } (x, s) \in R^{[a]}, \quad (y, s) \in R^{[a]}. \tag{11}$$

And by (S2) : $R(s, z) \geq R(x, z) \wedge R(y, z)$ we have that for each $(x, z), (y, z) \in R^{[a]}$; that is, $a \notin \alpha(R(x, z)), a \notin \alpha(R(y, z))$, and then $a \notin \alpha(R(s, z))$; that is, $(s, z) \in R^{[a]}$. Analogously we prove that for any $(z, x), (z, y) \in R^{[a]}, (z, t) \in R^{[a]}$. Hence s is supremum of x, y with respect to $R^{[a]}$ in $A^{[a]}$. Similarly we can prove that t is infimum of x, y with respect to $R^{[a]}$ in $A^{[a]}$. So $(A^{[a]}, R^{[a]})$ is a lattice.

(4) $\Rightarrow$ (5) is obvious.

(5) $\Rightarrow$ (1). For any $x, y \in A_{(0)}$, let $a \in \alpha^*(0)$ and $a \notin \alpha(A(x)), a \notin \alpha(A(y))$; that is, $x, y \in A^{[a]}$. From (5) there exist $x \vee_a y, x \wedge_a y \in A^{[a]}$; that is, $a \notin \alpha(A(x \vee_a y)), a \notin \alpha(A(x \wedge_a y))$. This shows that

$$A(x \vee_a y) \geq A(x) \wedge A(y),\\ A(x \wedge_a y) \geq A(x) \wedge A(y). \tag{12}$$

From (5) we have that $(x, x \vee_a y) \in R^{[a]}$ and $(y, x \vee_a y) \in R^{[a]}$; that is, $a \notin \alpha(R(x, x \vee_a y))$ and $a \notin \alpha(R(y, x \vee_a y))$. Therefore,

$$R(x, x \vee_a y) \geq A(x) = R(x, x),\\ R(y, x \vee_a y) \geq A(y) = R(y, y). \tag{13}$$

And by (5), for any $(x, z), (y, z) \in R^{[a]}$; that is, $a \notin \alpha(R(x, z)), a \notin \alpha(R(y, z))$. Then we have $(x \vee_a y, z) \in R^{[a]}$; that is, $a \notin \alpha(R(x \vee_a y, z))$. Hence $R(x \vee_a y, z) \geq R(x, z) \wedge R(y, z)$. Analogously we can conclude that

$$R(x \wedge_a y, x) \geq R(x, x),\\ R(x \wedge_a y, y) \geq R(y, y),\\ R(z, x \wedge_a y) \geq R(z, x) \wedge R(z, y). \tag{14}$$

Let $s = x \vee_a y, t = x \wedge_a y$, and then s, t are L-supremum and L-infimum of x, y, respectively. □

Theorem 20. *Let (A, R) be an L-fuzzy partially ordered set. If for each $a \in \beta^*(1), (A_{(a)}, R_{(a)})$ is a lattice, and then (A, R) is an L-fuzzy lattice on X.*

Proof. For any $x, y \in A_{(0)}$, let $a \in \beta^*(1)$, $a \in \beta(A(x))$ and $a \in \beta(A(y))$; that is, $x, y \in A_{(a)}$. There exist $x \vee_a y$ and $x \wedge_a y \in$; that is, $a \in \beta(A(x \vee_a y))$ and $a \in \beta(A(x \wedge_a y))$. Thereby

$$A(x \vee_a y) \geq A(x) \wedge A(y),\\ A(x \wedge_a y) \geq A(x) \wedge A(y). \tag{15}$$

Since $(A_{(a)}, R_{(a)})$ is a lattice, we have that $(x, x \vee_a y) \in R_{(a)}$ and $(y, x \vee_a y) \in R_{(a)}$; that is, $a \in \beta(R(x, x \vee_a y))$ and $a \in \beta(R(y, x \vee_a y))$. This implies that

$$R(x, x \vee_a y) \geq A(x) = R(x, x),\\ R(y, x \vee_a y) \geq A(y) = R(y, y). \tag{16}$$

Analogously we can conclude that for all $(x, z), (y, z) \in R_{(a)}$; that is, $a \in \beta(R(x, z)), a \in \beta(R(y, z))$, and we have $(x \vee_a y, z) \in R_{(a)}$; that is, $a \in \beta(R(x \vee_a y, z))$. Then we obtain that $R(x \vee_a y, z) \geq R(x, z) \wedge R(y, z)$. Similarly, we can prove that

$$R(x \wedge_a y, x) \geq R(x, x),\\ R(x \wedge_a y, y) \geq R(y, y),\\ R(z, x \wedge_a y) \geq R(z, x) \wedge R(z, y). \tag{17}$$

Let $s = x \vee_a y, t = x \wedge_a y$, so it is proved that s, t are L-supremum and L-infimum of x, y, respectively. □

Remark 21. Inversely the previous theorem is not true when (A, R) is a poset. This can be seen from Remark 14.

Definition 22. Let X be a nonempty set, and let A, B be L-fuzzy lattices of X. B is called an L-fuzzy sublattice of A if $B \leq A$.

Definition 23. Let X, Y be nonempty sets, and let A, B be L-fuzzy lattices of X, Y. An L-fuzzy mapping $f : A \to B$ is called an L-fuzzy lattice homomorphism if for any $a \in P(L)$, $f^{(a)} : A^{(a)} \to B^{(a)}$ is a lattice homomorphism.

From the corresponding theorems in [9] and knowledge in general algebra we can easily obtain the following theorem.

Theorem 24. *Let X, Y be nonempty sets, let A, B be L-fuzzy lattices of X, Y, let $f : A \to B$ be an L-fuzzy lattice homomorphism, and then the following propositions are true.*

(1) *If C is an L-fuzzy sublattice of A, then $f(C)$ is an L-fuzzy sublattice of B.*

(2) *If D is an L-fuzzy sublattice of B, then $f^{-1}(D)$ is an L-fuzzy sublattice of A.*

4. Fuzzy Sublattice

In [1], the author gave an L-valued fuzzy lattice. From the following analysis we can see that L-valued fuzzy lattice is a special case of bifuzzy lattices in fact.

Definition 25 (see [1]). Let L be a complete lattice with the greatest element 1_L and the least element 0_L, and let (X, R) be a lattice, $A \in L^X$. A is called a lattice-valued fuzzy lattice if all the p-cuts of A are sublattices of X.

Remark 26. Here the p-cut of A indicates the case of $A_{[a]}$ in fact.

Theorem 27 (see [1]). *Let L be a complete lattice with the greatest element 1_L and the least element 0_L, let (X, R) be a lattice, $A \in L^X$, and then A is called an L-valued fuzzy lattice if and only if for all $x, y \in X$, the following conditions are true:*

(A1) $A(x \vee y) \geq A(x) \wedge A(y)$,

(A2) $A(x \wedge y) \geq A(x) \wedge A(y)$.

Remark 28. When L is a completely distributive lattice, (X, R) is a lattice, here R is a relation on X. In the Definitions 16–18 of L-fuzzy lattice, let L-fuzzy partial order $R' = R \cap (A \times A) \subseteq A \times A$, $R' = R \cap (A \times A) \subseteq A \times A$, and then the conditions (R1)–(R4) in Definition 25 are true. Thus it can be seen that the L-valued fuzzy lattice is a special case of L-fuzzy lattice. Since A is an fuzzy subset of X, we call this L-valued fuzzy lattice A as fuzzy sublattice of X in the following application.

Now we study the relation between fuzzy sublattice and crisp lattice by means of their four level cut sets, furthermore we give the definition of fuzzy lattice homomorphism and corresponding theorems.

Definition 29. Let L be a completely distributive lattice, let (X, R) be a lattice, and let R be a relation on X, $A \in L^X$, if for all $x, y \in X$, there exist $x \vee y, x \wedge y \in X$ such that

$$A \in L^X, \text{ and } A = \begin{pmatrix} 0 & h & e & f & g & 1 \\ 0 & 1/2 & a & b & 1 & 1 \end{pmatrix}$$

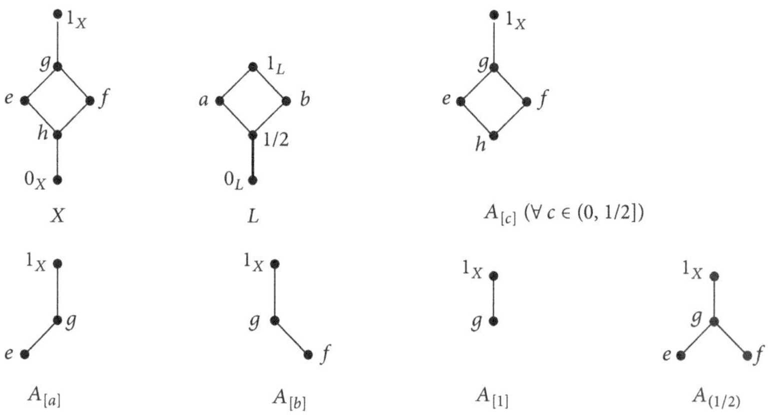

FIGURE 1

(A1) $A(x \vee y) \geq A(x) \wedge A(y)$,

(A2) $A(x \wedge y) \geq A(x) \wedge A(y)$, then we call A as a fuzzy sublattice of X.

Theorem 30. *Let L be a completely distributive lattice, let (X, R) be a lattice, and let R be a relation on X, $A \in L^X$. Then we can obtain that* (1), (2), (3), (4), (5), *and* (8) *are equivalent, and* (6) $\Rightarrow$ (1) *is true.*

(1) *A is a fuzzy sublattice of X.*

(2) *For each $a \in L$, $(A_{[a]}, R)$ is a sublattice of (X, R).*

(3) *For each $a \in M(L)$, $(A_{[a]}, R)$ is a sublattice of (X, R).*

(4) *For each $a \in L$, $(A^{[a]}, R)$ is a sublattice of (X, R).*

(5) *For each $a \in P(L)$, $(A^{[a]}, R)$ is a sublattice of (X, R).*

(6) *For each $a \in L$, $(A_{(a)}, R)$ is a sublattice of (X, R).*

(7) *For each $a \in M(L)$, $(A_{(a)}, R)$ is a sublattice of (X, R).*

(8) *For each $a \in P(L)$, $(A^{(a)}, R)$ is a sublattice of (X, R).*

Proof. (1) $\Rightarrow$ (2). For each $a \in L$, let $x, y \in A_{[a]}$. Then we have $A(x) \geq a$, $A(y) \geq a$. From (1), there exist $x \vee y, x \wedge y \in X$ such that $A(x \vee y) \geq A(x) \wedge A(y) \geq a$ and $A(x \wedge y) \geq A(x) \wedge A(y) \geq a$. Therefore $x \vee y, x \wedge y \in A_{[a]}$. Then it is proved that $(A_{[a]}, R)$ is a sublattice of (X, R).

(2) $\Rightarrow$ (3) is obvious.

(3) $\Rightarrow$ (8). For each $a \in P(L)$, let $x, y \in A^{(a)} = \bigcup_{b \nleq a,\, b \in M(L)} A_{[b]}$. There exist $b_1 \nleq a, b_2 \nleq a$ such that $x \in A_{[b_1]}$, $y \in A_{[b_2]}$. Hence $x, y \in A_{[b_1 \wedge b_2]} = A_{[b]}$. Since a is a prime element, we have that $b = b_1 \wedge b_2 \nleq a$. From (3), we have $x \vee y, x \wedge y \in A_{[b]}$, therefore $x \vee y, x \wedge y \in A^{(a)}$. So we obtain that $(A^{(a)}, R)$ is a sublattice of (X, R).

(8) $\Rightarrow$ (4). For each $a \in L$, let $x, y \in A^{[a]} = \bigcap_{b \in \alpha(a)} A^{(b)}$. Hence for all $b \in \alpha(a)$, and b is a prime element. We know that $x, y \in A^{(b)}$. From (8), $(A^{(b)}, R)$ is a sublattice of (X, R), so we have $x \vee y, x \wedge y \in A^{(b)}$. Thus $x \vee y, x \wedge y \in A^{[a]}$. Therefore, $(A^{[a]}, R)$ is a sublattice of (X, R).

(4) $\Rightarrow$ (5) is obvious.

(5) $\Rightarrow$ (1). For each $x, y \in X$, let $a \in P(L)$ and $x, y \in A^{[a]}$; that is, $a \notin \alpha(A(x)), a \notin \alpha(A(y))$. From (5) there exist $x \vee y, x \wedge y \in A^{[a]}$; that is, $a \notin \alpha(A(x \vee y)), a \notin \alpha(A(x \wedge y))$. Thus $A(x \vee y) \geq A(x) \wedge A(y), A(x \wedge y) \geq A(x) \wedge A(y)$. Therefore, A is a fuzzy lattice of X.

(6) $\Rightarrow$ (7) is obvious.

(7) $\Rightarrow$ (1). For each $x, y \in X$, let $a \in M(L)$ and $x, y \in A_{(a)}$; that is, $a \in \beta(A(x)), a \in \beta(A(y))$. From (7), there exist $x \vee y, x \wedge y \in A_{(a)}$; that is, $a \in \beta(A(x \vee y)), a \in \beta(A(x \wedge y))$. So it is proved that $A(x \vee y) \geq A(x) \wedge A(y)$ and $A(x \wedge y) \geq A(x) \wedge A(y)$. This shows that A is a fuzzy lattice of X. □

Remark 31. Generally, (1) $\Rightarrow$ (6) in the previous theorem is not true. This can be seen from the following example.

Example 32. $X = \{0_X, h, e, f, g, 1_X\}$, where $e \nleq f, f \nleq e$. $L = [0_L, 1/2] \cup \{a, b, 1_X\}$, where $a \nleq b, b \nleq a$. $A \in L^X$ (see Figure 1).

Then $A_{[0_L]} = X$, for all $c \in (0_L, 1/2]$, $A_{[c]} = \{h, e, f, g, 1_X\}$, $A_{[a]} = \{e, g, 1_X\}$, $A_{[b]} = \{f, g, 1_X\}$, and $A_{[1_L]} = \{g, 1_X\}$. This implies for all $a \in L$, $(A_{[a]}, R)$ is a sublattice of X; that is, A is a fuzzy lattice of X. While for $1/2 \in \beta^*(1)$, $A_{(1/2)} = \{e, f, g, 1_X\}$ is not a sublattice of X. This implies that (1) $\nRightarrow$ (6).

Remark 33. The following condition makes (1) $\Rightarrow$ (6) be true.

Theorem 34. *If for all $b, c \in L$, $\beta(b \wedge c) = \beta(b) \bigcap \beta(c)$, then* (1) $\Rightarrow$ (6) *in the previous theorem is true.*

Proof. (1)$\Rightarrow$(6). For each $a \in L, x, y \in A_{(a)}$, and then $a \in \beta(A(x)), a \in \beta(A(y))$. Therefore, $a \in \beta(A(x) \wedge A(y)) = \beta(A(x)) \cap \beta(A(y))$. From (1) we know that $A(x \vee y) \geq A(x) \wedge A(y)$ and $A(x \wedge y) \geq A(x) \wedge A(y)$. So we have that $a \in \beta(A(x \vee y))$ and $a \in \beta(A(x \wedge y))$. This shows that $(A_{(a)}, R)$ is a sublattice of (X, R). □

Definition 35. Let X be a nonempty set, let A, B be fuzzy lattices of X, and if $B \leq A$, we call B as a fuzzy sublattice of A.

Definition 36. Let X, Y be nonempty sets, and let A, B be fuzzy sublattices of X, Y, respectively. L-fuzzy mapping $f : A \to B$ is named a fuzzy lattice homomorphism if it satisfies that for all $a \in P(L)$, $f^{(a)} : A^{(a)} \to B^{(a)}$ is a lattice homomorphism.

The following theorems are easily obtained by the corresponding theorem in [9] and the knowledge in general algebra.

Theorem 37. *Let X, Y be nonempty sets, and let A, B be fuzzy sublattices of X, Y, respectively. $f : A \to B$ is a fuzzy lattice homomorphism, and then the following conditions are true:*

(1) *if C is a fuzzy sublattice of A, then $f(C)$ is a fuzzy sublattice of B;*

(2) *if D is a fuzzy sublattice of B, then $f^{-1}(D)$ is a fuzzy sublattice of A.*

Theorem 38. *Let X, Y be nonempty sets, and let A, B be fuzzy sublattices of X, Y, respectively. $f : A \to B$ is a fuzzy mapping, then the following conditions are true:*

(1) *$f : A \to B$ is a fuzzy lattice homomorphism;*

(2) *for all $a \in M(L)$, $f_{[a]}$ is a lattice homomorphism from $A_{[a]}$ to $B_{[a]}$.*

Proof. (1) $\Rightarrow$ (2). Let $a \in M(L)$ and $x, y \in A_{[a]}$, and then $A(x) \geq a, A(y) \geq a$. For any $b \in P(L), b \ngeq a$, we have $x \in A^{(b)}, y \in A^{(b)}$. From (1) we know that $f : A \to B$ is a fuzzy lattice homomorphism. Thus there exist $u_b \in B^{(b)}, v_b \in B^{(b)}$ such that $(x, u_b) \in f^{(b)}, (y, v_b) \in f^{(b)}$; that is, $u_b = f^{(b)}(x), v_b = f^{(b)}(x)$. From (1) we obtain that

$$f^{(b)}(x \vee y) = f^{(b)}(x) \vee f^{(b)}(y) = u_b \vee v_b, \quad \text{that is } (x \vee y, u_b \vee v_b) \in f^{(b)}. \tag{18}$$

Take $c \in P(L)$ such that $c \ngeq a$, and then $a \nleq b \vee c$. And take $e \in P(L)$ such that $e \geq b \vee c$ and $e \ngeq a$; in this way we have $(x, u_e) \in f^{(e)} \subseteq f^{(b)}, (x, u_e) \in f^{(e)} \subseteq f^{(c)}$. From (1) we know that $u_b = u_c = u_e$. Then take $u = u_c, v = v_c$; we obtain that $(x, u), (y, v) \in \cap\{f^{(b)} \mid b \in P(L), a \nleq b\} = f_{[a]}$, and $(x \vee y, u \vee v) \in \bigcap\{f^{(b)} \mid b \in P(L), a \nleq b\} = f_{[a]}$. Therefore

$$f_{[a]}(x \vee y) = u \vee v = f_{[a]}(x) \vee f_{[a]}(y). \tag{19}$$

Similarly, it is easy to prove that $f_{[a]}(x \wedge y) = f_{[a]}(x) \wedge f_{[a]}(y)$. This implies that $f_{[a]} : A_{[a]} \to B_{[a]}$ is a lattice homomorphism.

(2) $\Rightarrow$ (1). Let $a \in P(L)$ and $x, y \in A^{(a)}$, then $A(x) \nleq a, A(y) \nleq a$. Since a is a prime element we obtain $A(x) \wedge A(y) \nleq a$. Take $b \in M(L)$ such that $b \leq A(x) \wedge A(y)$ and $b \nleq a$, and then we have $x, y \in A_{[b]} \subseteq A^{(a)}$. From (2) we know that $f_{[b]} : A_{[b]} \to B_{[b]}$ is a lattice homomorphism, so there exist $u, v \in B_{[b]} \subseteq B^{(a)}$ such that $(x, u), (y, v) \in f_{[b]} \subseteq f^{(a)}$ and $(x \vee y, u \vee v) \in f_{[b]} \subseteq f^{(a)}$. Hence

$$f^{(a)}(x \vee y) = u \vee v = f^{(a)}(x) \vee f^{(a)}(y). \tag{20}$$

Similarly, $f^{[a]}(x \wedge y) = f^{[a]}(x) \wedge f^{[a]}(y)$; that is, $f^{(a)} : A^{(a)} \to B^{(a)}$ is a lattice homomorphism.

Similar to [11], we can easily prove the following theorems. □

Theorem 39. *Let X, Y be nonempty sets, and let A, B be fuzzy sublattices of X, Y, respectively. $f : A \to B$ is an L-fuzzy mapping, if for all $b, c \in L$, $\beta(b \wedge c) = \beta(b) \bigcap \beta(c)$, and then the following conditions are equivalent:*

(1) *$f : A \to B$ is a fuzzy lattice homomorphism;*

(2) *for all $a \in M(L)$, $f_{(a)}$ is a lattice homomorphism from $A_{(a)}$ to $B_{(a)}$.*

Theorem 40. *Let X, Y be nonempty sets, and let A, B be fuzzy sublattices of X, Y, respectively. $f : A \to B$ is an L-fuzzy mapping, if for all $a \in \alpha^*(0)$, $f^{[a]}$ is a lattice homomorphism from $A^{[a]}$ to $B^{[a]}$, and then $f : A \to B$ is a fuzzy lattice homomorphism.*

Theorem 41. *Let X, Y be nonempty sets, and let A, B be fuzzy sublattices of X, Y, respectively. $f : A \to B$ is an L-fuzzy mapping, if for all $b, c \in L, \alpha(b \vee c) = \alpha(b) \bigcap \alpha(c)$, and then the following conditions are equivalent:*

(1) *$f : A \to B$ is a fuzzy lattice homomorphism;*

(2) *for all $a \in \alpha^*(0)$, $f^{[a]}$ is a lattice homomorphism from $A^{[a]}$ to $B^{[a]}$.*

5. Fuzzy Lattice

In Definitions 16, 17, and 18, let $A = X$, and then the conditions are true obviously. So we can get another special case of bi-fuzzy lattice-fuzzy lattice. Now we give its definition and corresponding theorems.

Definition 42. Let L be a completely distributive lattice, $X \neq \emptyset$, and let R be an L-fuzzy relation on X, if R satisfies for any $x, y, z \in X$, there exist $s, t \in X$ such that

(S1) $R(x, s) \geq R(x, x)$, $R(y, s) \geq R(y, y)$,

(S2) $R(s, z) \geq R(x, z) \wedge R(y, z)$,

(T1) $R(t, x) \geq R(x, x)$, $R(t, y) \geq R(y, y)$,

(T2) $R(z, t) \geq R(z, x) \wedge R(z, y)$.

then we call s, t as supremum and infimum of x, y with respect to R, respectively.

Definition 43. Let L be a completely distributive lattice, $X \neq \emptyset$, and let R be an L-fuzzy relation on X, if for any $x, y \in X$, both supremum and infimum of x, y with respect to R exist, and then we call X as a fuzzy lattice with respect to R.

Same to the corresponding theorems in last section,we have the following theorem.

Theorem 44. *Let L be a completely distributive lattice, $X \neq \emptyset$, and let R be an L-fuzzy relation; then (1), (2), (3), (6), (7), and*

(8) *of the following conditions are equivalent, and* (4) $\Rightarrow$ (5) $\Rightarrow$ (1) *is true.*

(1) *X is a fuzzy lattice with respect to R.*

(2) *For each $a \in L\backslash_{\{0\}}$, $(X, R_{[a]})$ is a lattice.*

(3) *For each $a \in M(L)$, $(X, R_{[a]})$ is a lattice.*

(4) *For each $a \in \beta(1)$, $(X, R_{(a)})$ is a lattice.*

(5) *For each $a \in \beta^*(1)$, $(X, R_{(a)})$ is a lattice.*

(6) *For each $a \in \alpha(0)$, $(X, R^{[a]})$ is a lattice.*

(7) *For each $a \in \alpha^*(0)$, $(X, R^{[a]})$ is a lattice.*

(8) *For each $a \in P(L)$, $(X, R^{(a)})$ is a lattice.*

Acknowledgment

This work was supported by the National Natural Science Foundation of China (no. 61171195).

References

[1] A. Tepavčević and G. Trajkovski, "L-fuzzy lattices: an introduction," *Fuzzy Sets and Systems*, vol. 123, no. 2, pp. 209–216, 2001.

[2] G.-J. Wang, *Theory of L-Fuzzy Topological Spaces*, Shanxi Normal University Press, Xi'an, China, 1988.

[3] F.-G. Shi, "The theory and applications of L_β-nested sets and L_α-nested sets," *Fuzzy Systems and Mathematics*, vol. 9, no. 4, pp. 65–72, 1995 (Chinese).

[4] F.-G. Shi, "L-fuzzy sets and prime element nested sets," *Journal of Mathematical Research and Exposition*, vol. 16, no. 3, pp. 398–402, 1996 (Chinese).

[5] F.-G. Shi, "Theory of molecular nested sets and its applications," *Journal of Yantai Teachers University (Natural Science)*, no. 1, pp. 33–36, 1996 (Chinese).

[6] F.-G. Shi, "L-fuzzy relations and L-fuzzy subgroups," *Journal of Fuzzy Mathematics*, vol. 8, no. 2, pp. 491–499, 2000.

[7] F.-G. Shi, "L-fuzzy mapping of L-fuzzy sets," *Fuzzy Systems and Mathematics*, vol. 14, no. 3, pp. 16–24, 2000.

[8] X.-D. Zhang, "Characterizations of L-fuzzy mappings," *Fuzzy Systems and Mathematics*, vol. 16, no. 2, pp. 48–51, 2002.

[9] X.-D. Zhang, "Representations of image and inverse of L-fuzzy mappings," *Fuzzy Systems and Mathematics*, vol. 16, no. 1, pp. 24–28, 2002.

[10] J. Zhao, "L-fuzzy posets," *Fuzzy Systems and Mathematics*, vol. 15, no. 3, pp. 36–39, 2001.

[11] J. F. Zhang, "L-fuzzy subgroups and L-fuzzy group homomorphisms," *Fuzzy Systems and Mathematics*, vol. 17, no. 3, pp. 61–67, 2003.

Oscillation of a Class of Fractional Differential Equations with Damping Term

Huizeng Qin and Bin Zheng

School of Science, Shandong University of Technology, Zibo, Shandong 255049, China

Correspondence should be addressed to Bin Zheng; zhengbin2601@126.com

Academic Editors: J. Banaś and M. M. Cavalcanti

We investigate the oscillation of a class of fractional differential equations with damping term. Based on a certain variable transformation, the fractional differential equations are converted into another differential equations of integer order with respect to the new variable. Then, using Riccati transformation, inequality, and integration average technique, some new oscillatory criteria for the equations are established. As for applications, oscillation for two certain fractional differential equations with damping term is investigated by the use of the presented results.

1. Introduction

In the investigations of qualitative properties for differential equations, research of oscillation has gained much attention by many authors in the last few decades (e.g., see [1–16]). In these investigations, we notice that very little attention is paid to oscillation of fractional differential equations.

In [17], Jumarie proposed a definition for fractional derivative which is known as the modified Riemann-Liouville derivative in the literature. Since then, many authors have investigated various applications of the modified Riemann-Liouville derivative (e.g., see [18–21]) including various fractional calculus formulae, the fractional variational iteration method, the Bäcklund transformation method, and the fractional subequation method for soling fractional partial differential equations. In this paper, based on the modified Riemann-Liouville derivative, we are concerned with oscillation of a class of fractional differential equations with damping term as follows:

$$D_t^{\alpha}\left[a(t)\, D_t^{\alpha}\left(r(t)\, D_t^{\alpha} x(t)\right)\right] + p(t)\, D_t^{\alpha}\left(r(t)\, D_t^{\alpha} x(t)\right) + q(t)\, x(t) = 0, \quad t \geq t_0 > 0,\ 0 < \alpha < 1, \tag{1}$$

where $D_t^{\alpha}(\cdot)$ denotes the modified Riemann-Liouville derivative with respect to the variable t, the function $a \in C^{\alpha}([t_0,\infty), R_+)$, $r \in C^{2\alpha}([t_0,\infty), R_+)$, $p, q \in C([t_0,\infty), R_+)$ and C^{α} denotes continuous derivative of order α.

The definition and some important properties for the modified Riemann-Liouville derivative of order α are listed as follows (see also in [20–24]):

$$D_t^{\alpha} f(t) = \begin{cases} \dfrac{1}{\Gamma(1-\alpha)} \dfrac{d}{dt} \times \displaystyle\int_0^t (t-\xi)^{-\alpha} \left(f(\xi) - f(0)\right) d\xi, & 0 < \alpha < 1, \\ \left(f^{(n)}(t)\right)^{(\alpha-n)}, & n \leq \alpha < n+1,\ n \geq 1, \end{cases} \tag{2}$$

$$D_t^{\alpha} t^r = \frac{\Gamma(1+r)}{\Gamma(1+r-\alpha)} t^{r-\alpha}, \tag{3}$$

$$D_t^{\alpha}\left(f(t)\, g(t)\right) = g(t)\, D_t^{\alpha} f(t) + f(t)\, D_t^{\alpha} g(t), \tag{4}$$

$$D_t^{\alpha} f[g(t)] = f_g'[g(t)]\, D_t^{\alpha} g(t) = D_g^{\alpha} f[g(t)] \left(g'(t)\right)^{\alpha}. \tag{5}$$

As usual, a solution $x(t)$ of (1) is called oscillatory if it has arbitrarily large zeros; otherwise, it is called nonoscillatory. Equation (1) is called oscillatory if all its solutions are oscillatory.

We organize the next as follows. In Section 2, using Riccati transformation, inequality, and integration average technique, we establish some new oscillatory criteria for (1), while we present some examples for them in Section 3.

2. Oscillatory Criteria for (1)

In the following, we denote $\xi = t^{\alpha}/\Gamma(1+\alpha)$, $\xi_i = t_i^{\alpha}/\Gamma(1+\alpha)$, $i = 0,1,2,3,4,5$, $a(t) = \tilde{a}(\xi)$, $r(t) = \tilde{r}(\xi)$, $p(t) = \tilde{p}(\xi)$, $q(t) = \tilde{q}(\xi)$, $R_+ = (0,\infty)$, $\tilde{\delta}_1(\xi,\xi_i) = \int_{\xi_i}^{\xi}(1/\tilde{a}^{1/\gamma}(s))ds$, $\delta_1(t,t_i) = \tilde{\delta}_1(\xi,\xi_i)$, $\tilde{\delta}_2(\xi,\xi_i) = \int_{\xi_i}^{\xi}(\tilde{\delta}_1(s,\xi_i)/\tilde{r}(s))ds$, $\delta_2(t,t_i) = \tilde{\delta}_2(\xi,\xi_i)$, $A(\xi) = \exp(\int_{\xi_0}^{\xi}(\tilde{p}(s)/\tilde{a}(s))ds)$. Let $h_1, h_2, H \in C([\xi_0,\infty), R)$ satisfy

$$H(\xi,\xi) = 0, \quad H(\xi,s) > 0, \quad \xi > s \ge \xi_0. \tag{6}$$

H has continuous partial derivatives $\partial H(\xi,s)/\partial\xi$ and $\partial H(\xi,s)/\partial s$ on $[\xi_0,\infty)$ such that

$$\begin{aligned}\frac{\partial H(\xi,s)}{\partial \xi} &= -h_1(\xi,s)\sqrt{H(\xi,s)},\\ \frac{\partial H(\xi,s)}{\partial s} &= -h_2(\xi,s)\sqrt{H(\xi,s)}, \quad \xi > s \ge \xi_0.\end{aligned} \tag{7}$$

Lemma 1. *Assume $x(t)$ is an eventually positive solution of (1), and*

$$\int_{\xi_0}^{\infty}\frac{1}{A(s)\tilde{a}(s)}ds = \infty, \tag{8}$$

$$\int_{t_0}^{\infty}\frac{\alpha t^{\alpha-1}}{\Gamma(1+\alpha)r(t)}dt = \infty, \tag{9}$$

$$\int_{\xi_0}^{\infty}\frac{1}{\tilde{r}(\zeta)}\int_{\zeta}^{\infty}\frac{1}{A(\tau)\tilde{a}(\tau)}\int_{\tau}^{\infty}A(s)\tilde{q}(s)\,ds\,d\tau\,d\zeta = \infty. \tag{10}$$

Then, there exists a sufficiently large T such that $D_t^{\alpha}(r(t)D_t^{\alpha}x(t)) > 0$ on $[T,\infty)$ and either $D_t^{\alpha}x(t) > 0$ on $[T,\infty)$ or $\lim_{t\to\infty}x(t) = 0$.

Proof. Let $x(t) = \tilde{x}(\xi)$, where $\xi = t^{\alpha}/\Gamma(1+\alpha)$. Then, by use of (3), we obtain $D_t^{\alpha}\xi(t) = 1$, and furthermore, by use of the first equality in (5), we have

$$D_t^{\alpha}x(t) = D_t^{\alpha}\tilde{x}(\xi) = \tilde{x}'(\xi)D_t^{\alpha}\xi(t) = \tilde{x}'(\xi). \tag{11}$$

Similarly, we have $D_t^{\alpha}a(t) = \tilde{a}'(\xi)$, $D_t^{\alpha}r(t) = \tilde{r}'(\xi)$. So, (1) can be transformed into the following form:

$$\begin{aligned}&\left[\tilde{a}(\xi)\left(\tilde{r}(\xi)\tilde{x}'(\xi)\right)'\right]' + \tilde{p}(\xi)\left(\tilde{r}(\xi)\tilde{x}'(\xi)\right)'\\ &\quad + \tilde{q}(\xi)\tilde{x}(\xi) = 0, \quad \xi \ge \xi_0 > 0.\end{aligned} \tag{12}$$

Since $x(t)$ is an eventually positive solution of (1), then $\tilde{x}(\xi)$ is an eventually positive solution of (12), and there exists $\xi_1 > \xi_0$ such that $\tilde{x}(\xi) > 0$ on $[\xi_1,\infty)$. Furthermore, we have

$$\begin{aligned}&\left[A(\xi)\tilde{a}(\xi)\left(\tilde{r}(\xi)\tilde{x}'(\xi)\right)'\right]'\\ &= A(\xi)\left[\tilde{a}(\xi)\left(\tilde{r}(\xi)\tilde{x}'(\xi)\right)'\right]'\\ &\quad + A'(\xi)\tilde{a}(\xi)\left(\tilde{r}(\xi)\tilde{x}'(\xi)\right)'\\ &= A(\xi)\left[\tilde{a}(\xi)\left(\tilde{r}(\xi)\tilde{x}'(\xi)\right)'\right]'\\ &\quad + A(\xi)\tilde{p}(\xi)\left(\left(\tilde{r}(\xi)\tilde{x}'(\xi)\right)'\right)^{\gamma}\\ &= -A(\xi)\tilde{q}(\xi)\tilde{x}(\xi) < 0, \quad \xi \ge \xi_1.\end{aligned} \tag{13}$$

Then, $A(\xi)\tilde{a}(\xi)(\tilde{r}(\xi)\tilde{x}'(\xi))'$ is strictly decreasing on $[\xi_1,\infty)$, and thus $(\tilde{r}(\xi)\tilde{x}'(\xi))'$ is eventually of one sign. We claim $(\tilde{r}(\xi)\tilde{x}'(\xi))' > 0$ on $[\xi_2,\infty)$, where $\xi_2 > \xi_1$ is sufficiently large. Otherwise, assume that there exists a sufficiently large $\xi_3 > \xi_2$ such that $(\tilde{r}(\xi)\tilde{x}'(\xi))' < 0$ on $[\xi_3,\infty)$. Then, $\tilde{r}(\xi)\tilde{x}'(\xi)$ is strictly decreasing on $[\xi_3,\infty)$, and we have

$$\begin{aligned}\tilde{r}(\xi)\tilde{x}'(\xi) - \tilde{r}(\xi_3)\tilde{x}'(\xi_3) &= \int_{\xi_3}^{\xi}\frac{A(s)\tilde{a}(s)\left(\tilde{r}(s)\tilde{x}'(s)\right)'}{A(s)\tilde{a}(s)}ds\\ &\le A(\xi_3)\tilde{a}(\xi_3)\left(\tilde{r}(\xi_3)\tilde{x}'(\xi_3)\right)'\\ &\quad \times \int_{\xi_3}^{\xi}\frac{1}{A(s)\tilde{a}(s)}ds.\end{aligned} \tag{14}$$

By (8), we have $\lim_{\xi\to\infty}\tilde{r}(\xi)\tilde{x}'(\xi) = -\infty$. So there exists a sufficiently large ξ_4 with $\xi_4 > \xi_3$ such that $\tilde{x}'(\xi) < 0$, $\xi \in [\xi_4,\infty)$. Furthermore,

$$\begin{aligned}\tilde{x}(\xi) - \tilde{x}(\xi_4) &= \int_{\xi_4}^{\xi}\tilde{x}'(s)\,ds = \int_{\xi_4}^{\xi}\frac{\tilde{r}(s)\tilde{x}'(s)}{\tilde{r}(s)}ds\\ &\le \tilde{r}(\xi_4)\tilde{x}'(\xi_4)\int_{\xi_4}^{\xi}\frac{1}{\tilde{r}(s)}ds\\ &= \tilde{r}(\xi_4)\tilde{x}'(\xi_4)\int_{t_4}^{\infty}\frac{\alpha t^{\alpha-1}}{\Gamma(1+\alpha)r(t)}dt.\end{aligned} \tag{15}$$

By (9), we deduce that $\lim_{\xi\to\infty}\tilde{x}(\xi) = -\infty$, which contradicts the fact that $\tilde{x}(\xi)$ is an eventually positive solution of (9). So, $(\tilde{r}(\xi)\tilde{x}'(\xi))' > 0$ on $[\xi_2,\infty)$, and $D_t^{\alpha}(r(t)D_t^{\alpha}x(t)) > 0$ on $[t_2,\infty)$. Thus, $D_t^{\alpha}x(t) = \tilde{x}'(\xi)$ is eventually of one sign. Now we assume $\tilde{x}'(\xi) < 0$, $\xi \in [\xi_5,\infty)$ for some sufficiently large $\xi_5 > \xi_4$. Since $\tilde{x}(\xi) > 0$, furthermore we have $\lim_{\xi\to\infty}\tilde{x}(\xi) = \beta \ge 0$. We claim $\beta = 0$. Otherwise, assume $\beta > 0$. Then $\tilde{x}(\xi) \ge \beta$ on $[\xi_5,\infty)$, and, for $\xi \in [\xi_5,\infty)$, by (12) we have

$$\begin{aligned}\left[A(\xi)\tilde{a}(\xi)\left(\tilde{r}(\xi)\tilde{x}'(\xi)\right)'\right]' &\le -A(\xi)\tilde{q}(\xi)\tilde{x}(\xi)\\ &\le -A(\xi)\tilde{q}(\xi)\beta.\end{aligned} \tag{16}$$

Substituting ξ with s in the previous inequality, an integration with respect to s from ξ to ∞ yields

$$\begin{aligned}-A(\xi)\tilde{a}(\xi)\left(\tilde{r}(\xi)\tilde{x}'(\xi)\right)' &\leq -\lim_{\xi\to\infty}A(\xi)\tilde{a}(\xi)\left(\tilde{r}(\xi)\tilde{x}'(\xi)\right)' \\ &\quad -\beta\int_{\xi}^{\infty}A(s)\tilde{q}(s)\,ds \\ &< -\beta\int_{\xi}^{\infty}A(s)\tilde{q}(s)\,ds,\end{aligned} \tag{17}$$

which means

$$\left(\tilde{r}(\xi)\tilde{x}'(\xi)\right)' > \frac{\beta}{A(\xi)\tilde{a}(\xi)}\int_{\xi}^{\infty}A(s)\tilde{q}(s)\,ds. \tag{18}$$

Substituting ξ with τ in (18), an integration for (18) with respect to τ from ξ to ∞ yields

$$\begin{aligned}-\tilde{r}(\xi)\tilde{x}'(\xi) &> -\lim_{\xi\to\infty}\tilde{r}(\xi)\tilde{x}'(\xi) \\ &\quad +\beta\int_{\xi}^{\infty}\frac{1}{A(\tau)\tilde{a}(\tau)}\int_{\tau}^{\infty}A(s)\tilde{q}(s)\,ds\,d\tau \\ &> \beta\int_{\xi}^{\infty}\frac{1}{A(\tau)\tilde{a}(\tau)}\int_{\tau}^{\infty}A(s)\tilde{q}(s)\,ds\,d\tau;\end{aligned} \tag{19}$$

that is,

$$\tilde{x}'(\xi) < -\beta\frac{1}{\tilde{r}(\xi)}\int_{\xi}^{\infty}\frac{1}{A(\tau)\tilde{a}(\tau)}\int_{\tau}^{\infty}A(s)\tilde{q}(s)\,ds\,d\tau. \tag{20}$$

Substituting ξ with ζ in (20), an integration for (20) with respect to ζ from ξ_5 to ξ yields

$$\begin{aligned}&\tilde{x}(\xi)-\tilde{x}(\xi_5) \\ &\quad < -\beta\int_{\xi_5}^{\xi}\frac{1}{\tilde{r}(\zeta)}\int_{\zeta}^{\infty}\frac{1}{A(\tau)\tilde{a}(\tau)}\int_{\tau}^{\infty}A(s)\tilde{q}(s)\,ds\,d\tau\,d\zeta.\end{aligned} \tag{21}$$

By (10), one can see $\lim_{t\to\infty}\tilde{x}(\xi) = -\infty$, which causes a contradiction. So, the proof is complete. □

Lemma 2. *Assume that x is an eventually positive solution of* (1) *such that*

$$D_t^{\alpha}\left(r(t)D_t^{\alpha}x(t)\right) > 0, \qquad D_t^{\alpha}x(t) > 0 \tag{22}$$

on $[t_1,\infty)_{\mathbb{T}}$, where $t_1 \geq t_0$ is sufficiently large. Then, for $t \geq t_1$, we have

$$D_t^{\alpha}x(t) \geq \frac{A(\xi)\delta_1(t,t_1)a(t)D_t^{\alpha}\left(r(t)D_t^{\alpha}x(t)\right)}{r(t)}, \tag{23}$$

$$x(t) \geq A(\xi)\delta_2(t,t_1)a(t)D_t^{\alpha}\left(r(t)D_t^{\alpha}x(t)\right). \tag{24}$$

Proof. By (13), we obtain that $A(\xi)\tilde{a}(\xi)(\tilde{r}(\xi)\tilde{x}'(\xi))'$ is strictly decreasing on $[\xi_1,\infty)$. So,

$$\begin{aligned}\tilde{r}(\xi)\tilde{x}'(\xi) &\geq \tilde{r}(\xi)\tilde{x}'(\xi)-\tilde{r}(\xi_1)\tilde{x}'(\xi_1) \\ &= \int_{\xi_1}^{\xi}\frac{A(s)\tilde{a}(s)\left(\tilde{r}(s)\tilde{x}'(s)\right)'}{A(s)\tilde{a}(s)}ds \\ &\geq A(\xi)\tilde{a}(\xi)\left(\tilde{r}(\xi)\tilde{x}'(\xi)\right)'\int_{\xi_1}^{\xi}\frac{1}{A(s)\tilde{a}(s)}ds \\ &= A(\xi)\tilde{a}(\xi)\left(\tilde{r}(\xi)\tilde{x}'(\xi)\right)'\tilde{\delta}_1(\xi,\xi_1);\end{aligned} \tag{25}$$

that is

$$r(t)D_t^{\alpha}x(t) \geq A(\xi)\delta_1(t,t_1)a(t)D_t^{\alpha}\left(r(t)D_t^{\alpha}x(t)\right), \tag{26}$$

which admits (23). On the other hand, we have

$$\begin{aligned}\tilde{x}(\xi) &\geq \tilde{x}(\xi)-\tilde{x}(\xi_1) \\ &\geq \int_{\xi_1}^{\xi}\frac{A(s)\tilde{\delta}_1(s,\xi_1)\tilde{a}(s)\left(\tilde{r}(s)\tilde{x}'(s)\right)'}{A(s)\tilde{r}(s)}ds \\ &\geq A(\xi)\tilde{a}(\xi)\left(\tilde{r}(\xi)\tilde{x}'(\xi)\right)'\int_{\xi_1}^{\xi}\frac{\tilde{\delta}_1(s,\xi_1)}{A(s)\tilde{r}(s)}ds \\ &= A(\xi)\tilde{\delta}_2(\xi,\xi_1)\tilde{a}(\xi)\left(\tilde{r}(\xi)\tilde{x}'(\xi)\right)',\end{aligned} \tag{27}$$

which can be rewritten as (24). So the proof is complete. □

Lemma 3 (see [25, Theorem 41]). *Assume that A and B are nonnegative real numbers. Then,*

$$\lambda AB^{\lambda-1}-A^{\lambda} \leq (\lambda-1)B^{\lambda} \tag{28}$$

for all $\lambda > 1$.

Theorem 4. *Assume that* (8)–(10) *hold. If there exists $\phi \in C^{\alpha}([t_0,\infty),R_+)$ such that for any sufficiently large $T \geq \xi_0$, there exist a, b, c with $T \leq a < c < b$ satisfying*

$$\begin{aligned}&\frac{1}{H(b,c)}\int_{c}^{b}H(b,s)A(s)\tilde{\phi}(s)\tilde{q}(s)\,ds \\ &\quad +\frac{1}{H(c,a)}\int_{a}^{c}H(s,a)A(s)\tilde{\phi}(s)\tilde{q}(s)\,ds \\ &\quad > \frac{1}{H(b,c)}\int_{c}^{b}\frac{\tilde{r}(s)\tilde{\phi}(s)}{4\tilde{\delta}_1(s,\xi_2)}Q_2^2(b,s)\,ds \\ &\quad +\frac{1}{H(c,a)}\int_{a}^{c}\frac{\tilde{r}(s)\tilde{\phi}(s)}{4\tilde{\delta}_1(s,\xi_2)}Q_1^2(s,a)\,ds,\end{aligned} \tag{29}$$

where $\tilde{\phi}(\xi) = \phi(t)$, $Q_1(s,\xi) = h_1(s,\xi) - (\tilde{\phi}'(s)/\tilde{\phi}(s))\sqrt{H(s,\xi)}$, $Q_2(\xi,s) = h_2(\xi,s) - (\tilde{\phi}'(s)/\tilde{\phi}(s))\sqrt{H(\xi,s)}$; *then,* (1) *is oscillatory or satisfies* $\lim_{t\to\infty}x(t) = 0$.

Proof. Assume that (1) has a nonoscillatory solution x on $[t_0,\infty)$. Without loss of generality, we may assume that $x(t) > 0$ on $[t_1,\infty)$, where t_1 is sufficiently large. By Lemma 1, we have $D_t^\alpha(r(t)D_t^\alpha x(t)) > 0$, $t \in [t_2,\infty)$, where $t_2 > t_1$ is sufficiently large, and either $D_t^\alpha x(t) > 0$ on $[t_2,\infty)$ or $\lim_{t\to\infty}x(t) = 0$. Now we assume $D_t^\alpha x(t) > 0$ on $[t_2,\infty)$. Define the generalized Riccati function:

$$\omega(t) = \phi(t)\frac{A(\xi)a(t)D_t^\alpha(r(t)D_t^\alpha x(t))}{x(t)}. \tag{30}$$

Then, for $t \in [t_2,\infty)$, we have

$$\begin{aligned}
D_t^\alpha\omega(t) &= D_t^\alpha\phi(t)\frac{A(\xi)a(t)D_t^\alpha(r(t)D_t^\alpha x(t))}{x(t)} \\
&\quad + \phi(t)D_t^\alpha\left\{\frac{A(\xi)a(t)D_t^\alpha(r(t)D_t^\alpha x(t))}{x(t)}\right\} \\
&= \frac{\phi(t)D_t^\alpha(A(\xi)a(t)D_t^\alpha(r(t)D_t^\alpha x(t)))}{x(t)} \\
&\quad - \phi(t)\frac{D_t^\alpha x(t)A(\xi)a(t)D_t^\alpha(r(t)D_t^\alpha x(t))}{x^2(t)} \\
&\quad + \frac{D_t^\alpha\phi(t)}{\phi(t)}\omega(t) \\
&= (\phi(t)[A(\xi)D_t^\alpha(a(t)D_t^\alpha(r(t)D_t^\alpha x(t))) \\
&\qquad + a(t)D_t^\alpha(r(t)D_t^\alpha x(t))D_t^\alpha A(\xi)]) \\
&\quad \times (x(t))^{-1} \\
&\quad - \phi(t)\frac{D_t^\alpha x(t)A(\xi)a(t)D_t^\alpha(r(t)D_t^\alpha x(t))}{x^2(t)} \\
&\quad + \frac{D_t^\alpha\phi(t)}{\phi(t)}\omega(t) \\
&= \left(\phi(t)\left[A(\xi)D_t^\alpha(a(t)D_t^\alpha(r(t)D_t^\alpha x(t)))\right.\right. \\
&\qquad \left.\left. + a(t)D_t^\alpha(r(t)D_t^\alpha x(t))A'(\xi)D_t^\alpha\xi\right]\right) \\
&\quad \times (x(t))^{-1} \\
&\quad - \phi(t)\frac{D_t^\alpha x(t)A(\xi)a(t)D_t^\alpha(r(t)D_t^\alpha x(t))}{x^2(t)} \\
&\quad + \frac{D_t^\alpha\phi(t)}{\phi(t)}\omega(t).
\end{aligned} \tag{31}$$

Using $D_t^\alpha\xi = 1$ and (23), we obtain

$$\begin{aligned}
D_t^\alpha\omega(t) &\le \left(\phi(t)\left[A(\xi)D_t^\alpha(a(t)D_t^\alpha(r(t)D_t^\alpha x(t)))\right.\right. \\
&\qquad \left.\left. + a(t)D_t^\alpha(r(t)D_t^\alpha x(t))A(\xi)\frac{\tilde{p}(\xi)}{\tilde{a}(\xi)}\right]\right) \\
&\quad \times (x(t))^{-1} \\
&\quad - \phi(t)\left((A(\xi)\delta_1(t,t_2)a(t)D_t^\alpha(r(t)D_t^\alpha x(t))\right. \\
&\qquad \times D_t^\alpha x(t)A(\xi)a(t)D_t^\alpha(r(t)D_t^\alpha x(t))) \\
&\qquad \left.\times\left(r(t)x^2(t)\right)^{-1}\right) \\
&\quad + \frac{D_t^\alpha\phi(t)}{\phi(t)}\omega(t) \\
&= (\phi(t)A(\xi)[D_t^\alpha(a(t)D_t^\alpha(r(t)D_t^\alpha x(t))) \\
&\qquad + p(t)D_t^\alpha(r(t)D_t^\alpha x(t))]) \\
&\quad \times (x(t))^{-1} \\
&\quad - \frac{\delta_1(t,t_2)}{\phi(t)r(t)}\omega^2(t) + \frac{D_t^\alpha\phi(t)}{\phi(t)}\omega(t) \\
&= -A(\xi)q(t)\phi(t) - \frac{\delta_1(t,t_2)}{\phi(t)r(t)}\omega^2(t) \\
&\quad + \frac{D_t^\alpha\phi(t)}{\phi(t)}\omega(t).
\end{aligned} \tag{32}$$

Let $\omega(t) = \tilde{\omega}(\xi)$. Then $D_t^\alpha w(t) = \tilde{w}'(\xi)$, and $D_t^\alpha\phi(t) = \tilde{\phi}'(\xi)$. So (32) is transformed into the following form:

$$\begin{aligned}
\tilde{\omega}'(\xi) &\le -A(\xi)\tilde{\phi}(\xi)\tilde{q}(\xi) - \frac{\tilde{\delta}_1(\xi,\xi_2)}{\tilde{r}(\xi)\tilde{\phi}(\xi)}\tilde{\omega}^2(\xi) \\
&\quad + \frac{\tilde{\phi}'(\xi)}{\tilde{\phi}(\xi)}\tilde{\omega}(\xi), \quad \xi \ge \xi_2.
\end{aligned} \tag{33}$$

Choose a, b, c arbitrarily in $[\xi_2,\infty)$ with $b > c > a$. Substituting ξ with s, multiplying both sides of (33) by $H(\xi,s)$, and integrating it with respect to s from c to ξ for $\xi \in [c,b)$, we get that

$$\begin{aligned}
&\int_c^\xi H(\xi,s)A(s)\tilde{\phi}(s)\tilde{q}(s)\,ds \\
&\quad \le -\int_c^\xi H(\xi,s)\tilde{w}'(s)\,ds \\
&\qquad + \int_c^\xi H(\xi,s)\frac{\tilde{\phi}'(s)}{\tilde{\phi}(s)}\tilde{w}(s)\,ds
\end{aligned}$$

$$
\begin{aligned}
&- \int_c^{\xi} H(\xi,s) \frac{\tilde{\delta}_1(s,\xi_2)}{\tilde{r}(s)\tilde{\phi}(s)} \tilde{\omega}^2(s)\, ds \\
&= H(\xi,c)\tilde{w}(c) \\
&- \int_c^{\xi} \left[\left(\frac{\tilde{\delta}_1(s,\xi_2) H(\xi,s)}{\tilde{r}(s)\tilde{\phi}(s)} \right)^{1/2} \tilde{w}(s) + \frac{1}{2} \left(\frac{\tilde{r}(s)\tilde{\phi}(s)}{\tilde{\delta}_1(s,\xi_2)} \right)^{1/2} Q_2(\xi,s) \right]^2 ds \\
&+ \int_c^{\xi} \frac{\tilde{r}(s)\tilde{\phi}(s)}{4\tilde{\delta}_1(s,\xi_2)} Q_2^2(\xi,s)\, ds \\
&\leq H(\xi,c)\tilde{w}(c) + \int_c^{\xi} \frac{\tilde{r}(s)\tilde{\phi}(s)}{4\tilde{\delta}_1(s,\xi_2)} Q_2^2(\xi,s)\, ds.
\end{aligned} \tag{34}
$$

Dividing both sides of the inequality (34) by $H(\xi,c)$ and letting $\xi \to b^-$, we obtain

$$
\begin{aligned}
&\frac{1}{H(b,c)} \int_c^b H(b,s) A(s) \tilde{\phi}(s) \tilde{q}(s)\, ds \\
&\leq \tilde{w}(c) + \frac{1}{H(b,c)} \int_c^b \frac{\tilde{r}(s)\tilde{\phi}(s)}{4\tilde{\delta}_1(s,\xi_2)} Q_2^2(b,s)\, ds.
\end{aligned} \tag{35}
$$

On the other hand, substituting ξ with s, multiplying both sides of (33) by $H(s,\xi)$, and integrating it with respect to s from ξ to c for $\xi \in (a,c]$, we get that

$$
\begin{aligned}
&\int_{\xi}^{c} H(s,\xi) A(s) \tilde{\phi}(s) \tilde{q}(s)\, ds \\
&\leq - \int_{\xi}^{c} H(s,\xi) \tilde{w}'(s)\, ds + \int_{\xi}^{c} H(s,\xi) \frac{\tilde{\phi}'(s)}{\tilde{\phi}(s)} \tilde{w}(s)\, ds \\
&- \int_{\xi}^{c} H(s,\xi) \frac{\tilde{\delta}_1(s,\xi_2)}{\tilde{r}(s)\tilde{\phi}(s)} \tilde{\omega}^2(s)\, ds \\
&= -H(c,\xi)\tilde{w}(c) \\
&- \int_{\xi}^{c} \left[\left(\frac{\tilde{\delta}_1(s,\xi_2) H(s,\xi)}{\tilde{r}(s)\tilde{\phi}(s)} \right)^{1/2} \tilde{w}(s) + \frac{1}{2} \left(\frac{\tilde{r}(s)\tilde{\phi}(s)}{\tilde{\delta}_1(s,\xi_2)} \right)^{1/2} Q_1(\xi,s) \right]^2 ds \\
&+ \int_{\xi}^{c} \frac{\tilde{r}(s)\tilde{\phi}(s)}{4\tilde{\delta}_1(s,\xi_2)} Q_1^2(s,\xi)\, ds \\
&\leq -H(c,\xi)\tilde{w}(c) + \int_{\xi}^{c} \frac{\tilde{r}(s)\tilde{\phi}(s)}{4\tilde{\delta}_1(s,\xi_2)} Q_1^2(s,\xi)\, ds.
\end{aligned} \tag{36}
$$

Dividing both sides of the inequality (36) by $H(c,\xi)$ and letting $\xi \to a^+$, we obtain

$$
\begin{aligned}
&\frac{1}{H(c,a)} \int_a^c H(s,a) A(s) \tilde{\phi}(s) \tilde{q}(s)\, ds \\
&\leq -\tilde{w}(c) + \frac{1}{H(c,a)} \int_a^c \frac{\tilde{r}(s)\tilde{\phi}(s)}{4\tilde{\delta}_1(s,\xi_2)} Q_1^2(s,a)\, ds.
\end{aligned} \tag{37}
$$

A combination of (35) and (37) yields

$$
\begin{aligned}
&\frac{1}{H(b,c)} \int_c^b H(b,s) A(s) \tilde{\phi}(s) \tilde{q}(s)\, ds \\
&+ \frac{1}{H(c,a)} \int_a^c H(s,a) A(s) \tilde{\phi}(s) \tilde{q}(s)\, ds \\
&\leq \frac{1}{H(b,c)} \int_c^b \frac{\tilde{r}(s)\tilde{\phi}(s)}{4\tilde{\delta}_1(s,\xi_2)} Q_2^2(b,s)\, ds \\
&+ \frac{1}{H(c,a)} \int_a^c \frac{\tilde{r}(s)\tilde{\phi}(s)}{4\tilde{\delta}_1(s,\xi_2)} Q_1^2(s,a)\, ds,
\end{aligned} \tag{38}
$$

which contradicts (29). So, the proof is complete. □

Theorem 5. *Under the conditions of Theorem 4, if for any sufficiently large* $l \geq \xi_0$,

$$
\limsup_{\xi \to \infty} \int_l^{\xi} \left[H(s,l) A(s) \tilde{\phi}(s) \tilde{q}(s) - \frac{\tilde{r}(s)\tilde{\phi}(s)}{4\tilde{\delta}_1(s,\xi_2)} Q_1^2(s,l) \right] ds > 0, \tag{39}
$$

$$
\limsup_{\xi \to \infty} \int_l^{\xi} \left[H(\xi,s) A(s) \tilde{\phi}(s) \tilde{q}(s) - \frac{\tilde{r}(s)\tilde{\phi}(s)}{4\tilde{\delta}_1(s,\xi_2)} Q_2^2(\xi,s) \right] ds > 0, \tag{40}
$$

then (1) is oscillatory.

Proof. For any $T \geq \xi_0$, let $a = T$. In (39), we choose $l = a$. Then, there exists $c > a$ such that

$$
\int_a^c \left[H(s,a) A(s) \tilde{\phi}(s) \tilde{q}(s) - \frac{\tilde{r}(s)\tilde{\phi}(s)}{4\tilde{\delta}_1(s,\xi_2)} Q_1^2(s,a) \right] ds > 0. \tag{41}
$$

In (40), we choose $l = c > a$. Then there exists $b > c$ such that

$$
\int_c^b \left[H(b,s) A(s) \tilde{\phi}(s) \tilde{q}(s) - \frac{\tilde{r}(s)\tilde{\phi}(s)}{4\tilde{\delta}_1(s,\xi_2)} Q_2^2(b,s) \right] ds > 0. \tag{42}
$$

Combining (41) and (42), we obtain (29). The conclusion thus comes from Theorem 4, and the proof is complete. □

In Theorems 4 and 5, if we choose $H(\xi, s) = (\xi - s)^{\lambda}$, $\xi \geq s \geq \xi_0$, where $\lambda > 1$ is a constant, then we obtain the following two corollaries.

Corollary 6. *Under the conditions of Theorem 4, if for any sufficiently large $T \geq \xi_0$, there exist a, b, c with $T \leq a < c < b$ satisfying*

$$\begin{aligned}
&\frac{1}{(c-a)^{\lambda}} \int_a^c (s-a)^{\lambda} A(s) \tilde{\phi}(s) \tilde{q}(s)\, ds \\
&\quad + \frac{1}{(b-c)^{\lambda}} \int_c^b (b-s)^{\lambda} A(s) \tilde{\phi}(s) \tilde{q}(s)\, ds \\
&> \frac{1}{(c-a)^{\lambda}} \int_a^c \frac{\tilde{r}(s) \tilde{\phi}(s)}{4\tilde{\delta}_1(s,\xi_2)} (s-a)^{\lambda-2} \left(\lambda + \frac{\tilde{\phi}'(s)}{\tilde{\phi}(s)}(s-a)\right)^2 ds \\
&\quad + \frac{1}{(b-c)^{\lambda}} \int_c^b \frac{\tilde{r}(s) \tilde{\phi}(s)}{4\tilde{\delta}_1(s,\xi_2)} (b-s)^{\lambda-2} \\
&\qquad \times \left(\lambda - \frac{\tilde{\phi}'(s)}{\tilde{\phi}(s)}(b-s)\right)^2 ds,
\end{aligned} \tag{43}$$

then (1) *is oscillatory.*

Corollary 7. *Under the conditions of Theorem 5, if for any sufficiently large $l \geq \xi_0$,*

$$\begin{aligned}
&\limsup_{\xi \to \infty} \int_l^{\xi} \Bigg[(s-l)^{\lambda} A(s) \tilde{\phi}(s) \tilde{q}(s) \\
&\qquad - \frac{\tilde{r}(s) \tilde{\phi}(s)}{4\tilde{\delta}_1(s,\xi_2)} (s-l)^{\lambda-2} \\
&\qquad \times \left(\lambda + \frac{\tilde{\phi}'(s)}{\tilde{\phi}(s)}(s-l)\right)^2 \Bigg] ds > 0,
\end{aligned} \tag{44}$$

$$\begin{aligned}
&\limsup_{\xi \to \infty} \int_l^{\xi} \Bigg[(\xi-s)^{\lambda} A(s) \tilde{\phi}(s) \tilde{q}(s) \\
&\qquad - \frac{\tilde{r}(s) \tilde{\phi}(s)}{4\tilde{\delta}_1(s,\xi_2)} (\xi-s)^{\lambda-2} \\
&\qquad \times \left(\lambda - \frac{\tilde{\phi}'(s)}{\tilde{\phi}(s)}(\xi-s)\right)^2 \Bigg] ds > 0,
\end{aligned}$$

then (1) *is oscillatory.*

Theorem 8. *Assume* (8)–(10) *hold, and*

$$\int_{\xi_0}^{\infty} \left[A(s) \tilde{\phi}(s) \tilde{q}(s) - \frac{\tilde{r}(s) \left[\tilde{\phi}'(s)\right]^2}{4\tilde{\delta}_1(s,\xi_2) \tilde{\phi}(s)} \right] ds = \infty, \tag{45}$$

where $\tilde{\phi}$ is defined as in Theorem 4. Then every solution of (1) *is oscillatory or satisfies $\lim_{t\to\infty} x(t) = 0$.*

Proof. Assume (1) has a nonoscillatory solution x on $[t_0, \infty)$. Without loss of generality, we may assume $x(t) > 0$ on $[t_1, \infty)$, where t_1 is sufficiently large. By Lemma 1, we have $D_t^{\alpha}(r(t) D_t^{\alpha} x(t)) > 0$, $t \in [t_2, \infty)$, where $t_2 > t_1$ is sufficiently large, and either $D_t^{\alpha} x(t) > 0$ on $[t_2, \infty)$ or $\lim_{t\to\infty} x(t) = 0$. Now we assume that $D_t^{\alpha} x(t) > 0$ on $[t_2, \infty)$. Let $\omega(t)$, $\tilde{\omega}(\xi)$ be defined as in Theorem 4. Then we obtain (33), and furthermore,

$$\begin{aligned}
\tilde{\omega}'(\xi) &\leq -A(\xi) \tilde{\phi}(\xi) \tilde{q}(\xi) - \frac{\tilde{\delta}_1(\xi,\xi_2)}{\tilde{r}(\xi) \tilde{\phi}(\xi)} \tilde{\omega}^2(\xi) + \frac{\tilde{\phi}'(\xi)}{\tilde{\phi}(\xi)} \tilde{\omega}(\xi) \\
&\leq -A(\xi) \tilde{\phi}(\xi) \tilde{q}(\xi) + \frac{\tilde{r}(\xi) \left[\tilde{\phi}'(\xi)\right]^2}{4\tilde{\delta}_1(\xi,\xi_2) \tilde{\phi}(\xi)}, \quad \xi \geq \xi_2.
\end{aligned} \tag{46}$$

Substituting ξ with s in (46) and integrating (46) with respect to s from ξ_2 to ξ yield

$$\begin{aligned}
\int_{\xi_2}^{\xi} \left[A(s) \tilde{\phi}(s) \tilde{q}(s) - \frac{\tilde{r}(s) \left[\tilde{\phi}'(s)\right]^2}{4\tilde{\delta}_1(s,\xi_2) \tilde{\phi}(s)} \right] ds &\leq \omega(\xi_2) - \omega(\xi) \\
&\leq \omega(\xi_2) < \infty,
\end{aligned} \tag{47}$$

which contradicts (45). So, the proof is complete. □

Theorem 9. *Assume* (8)–(10) *hold, and there exists a function $G \in C([\xi_0, \infty), \mathbb{R})$ such that $G(\xi, \xi) = 0$, for $\xi \geq \xi_0$, $G(\xi, s) > 0$, for $\xi > s \geq \xi_0$, and G has a nonpositive continuous partial derivative $G_s'(\xi, s)$. If*

$$\begin{aligned}
&\limsup_{\xi \to \infty} \frac{1}{G(\xi,\xi_0)} \\
&\quad \times \Bigg\{ \int_{\xi_0}^{\xi} G(\xi,s) \Bigg\{ K \tilde{\phi}(s) \tilde{q}(s) - \tilde{\phi}(s) \tilde{\varphi}'(s) \\
&\qquad + \frac{\tilde{\phi}(s) \tilde{\delta}_1(s,\xi_2) \tilde{\varphi}^{1+1/\gamma}(s)}{\tilde{r}(s)} \\
&\qquad - \frac{\left[(\gamma+1) \tilde{\varphi}^{1/\gamma}(s) \tilde{\phi}(s) \tilde{\delta}_1(s,\xi_2) + \tilde{r}(s) \tilde{\phi}'(s)\right]^{\gamma+1}}{(\gamma+1)^{\gamma+1} \left[\tilde{\phi}(s) \tilde{\delta}_1(s,\xi_2)\right]^{\gamma} \tilde{r}(s)} \Bigg\} ds \Bigg\} \\
&= \infty,
\end{aligned} \tag{48}$$

where $\tilde{\phi}$ is defined as in Theorem 4, then every solution of (1) *is oscillatory or satisfies $\lim_{t\to\infty} x(t) = 0$.*

Proof. Assume (1) has a nonoscillatory solution x on $[t_0, \infty)$. Without loss of generality, we may assume $x(t) > 0$ on $[t_1, \infty)$, where t_1 is sufficiently large. By Lemma 1, we have $D_t^{\alpha}(r(t) D_t^{\alpha} x(t)) > 0$, $t \in [t_2, \infty)$, where $t_2 > t_1$ is sufficiently large, and either $D_t^{\alpha} x(t) > 0$ on $[t_2, \infty)$ or $\lim_{t\to\infty} x(t) = 0$.

Now we assume $D_t^{\alpha}x(t) > 0$ on $[t_2, \infty)$. Let $\omega(t)$, $\widetilde{\omega}(\xi)$ be defined as in Theorem 4. By (46), we have

$$A(\xi)\widetilde{\phi}(\xi)\widetilde{q}(\xi) - \frac{\widetilde{r}(\xi)\left[\widetilde{\phi}'(\xi)\right]^2}{4\widetilde{\delta}_1(\xi,\xi_2)\widetilde{\phi}(\xi)} \le \widetilde{\omega}'(\xi), \quad \xi \ge \xi_2. \tag{49}$$

Substituting ξ with s in (49), multiplying both sides by $G(\xi, s)$, and then integrating both sides of (49) with respect to s from ξ_2 to ξ yield

$$\begin{aligned}
&\int_{\xi_2}^{\xi} G(\xi,s)\left\{A(s)\widetilde{\phi}(s)\widetilde{q}(s) - \frac{\widetilde{r}(s)\left[\widetilde{\phi}'(s)\right]^2}{4\widetilde{\delta}_1(s,\xi_2)\widetilde{\phi}(s)}\right\}ds \\
&\quad \le -\int_{\xi_2}^{\xi} G(\xi,s)\widetilde{\omega}'(s)\,ds \\
&\quad = G(\xi,\xi_2)\,\omega(\xi_2) + \int_{\xi_2}^{\xi} H_s'(\xi,s)\,\omega(s)\,\Delta s \\
&\quad \le G(\xi,\xi_2)\,\omega(\xi_2) \\
&\quad \le G(\xi,\xi_0)\,\omega(\xi_2).
\end{aligned} \tag{50}$$

Then,

$$\begin{aligned}
&\int_{\xi_0}^{\xi} G(\xi,s)\left\{A(s)\widetilde{\phi}(s)\widetilde{q}(s) - \frac{\widetilde{r}(s)\left[\widetilde{\phi}'(s)\right]^2}{4\widetilde{\delta}_1(s,\xi_2)\widetilde{\phi}(s)}\right\}ds \\
&\quad = \int_{\xi_0}^{\xi_2} G(\xi,s)\left\{A(s)\widetilde{\phi}(s)\widetilde{q}(s) - \frac{\widetilde{r}(s)\left[\widetilde{\phi}'(s)\right]^2}{4\widetilde{\delta}_1(s,\xi_2)\widetilde{\phi}(s)}\right\}ds \\
&\quad + \int_{\xi_2}^{\xi} G(\xi,s)\left\{A(s)\widetilde{\phi}(s)\widetilde{q}(s) - \frac{\widetilde{r}(s)\left[\widetilde{\phi}'(s)\right]^2}{4\widetilde{\delta}_1(s,\xi_2)\widetilde{\phi}(s)}\right\}ds \\
&\le G(\xi,\xi_0)\,\widetilde{\omega}(\xi_2) + G(\xi,\xi_0) \\
&\quad \times \int_{\xi_0}^{\xi_2}\left|A(s)\widetilde{\phi}(s)\widetilde{q}(s) - \frac{\widetilde{r}(s)\left[\widetilde{\phi}'(s)\right]^2}{4\widetilde{\delta}_1(s,\xi_2)\widetilde{\phi}(s)}\right|ds.
\end{aligned} \tag{51}$$

So,

$$\begin{aligned}
&\limsup_{\xi\to\infty}\frac{1}{G(\xi,\xi_0)}\left\{\int_{\xi_0}^{\xi} G(\xi,s)\left\{A(s)\widetilde{\phi}(s)\widetilde{q}(s)\right.\right. \\
&\qquad \left.\left. - \frac{\widetilde{r}(s)\left[\widetilde{\phi}'(s)\right]^2}{4\widetilde{\delta}_1(s,\xi_2)\widetilde{\phi}(s)}\right\}ds\right\} \\
&\quad \le \widetilde{\omega}(\xi_2) + \int_{\xi_0}^{\xi_2}\left|A(s)\widetilde{\phi}(s)\widetilde{q}(s) - \frac{\widetilde{r}(s)\left[\widetilde{\phi}'(s)\right]^2}{4\widetilde{\delta}_1(s,\xi_2)\widetilde{\phi}(s)}\right|ds < \infty,
\end{aligned} \tag{52}$$

which contradicts (48). So the proof is complete. □

3. Applications of the Results

Example 10. Consider the following fractional differential equation:

$$\begin{aligned}
&D_t^{1/3}\left[t^{1/9}D_t^{1/3}D_t^{1/3}x(t)\right] + t^{-1/3}D_t^{1/3}D_t^{1/3}x(t) \\
&\quad + t^{-2/3}x(t) = 0, \quad t \ge 2.
\end{aligned} \tag{53}$$

In (1), if we set $t_0 = 2$, $\alpha = 1/3$, $a(t) = t^{1/9}$, $r(t) \equiv 1$, $(t) = t^{-1/3}$, $q(t) = t^{-2/3}$, then we obtain (53). So $\xi_0 = 2^{1/3}/\Gamma(4/3)$, $\widetilde{a}(\xi) = a(t) = t^{1/9} = [\Gamma(4/3)\xi]^{1/3}$, $\widetilde{r}(\xi) \equiv 1$, $\widetilde{p}(\xi) = [\Gamma(4/3)\xi]^{-1}$, $\widetilde{q}(\xi) = [\Gamma(4/3)\xi]^{-2}$. Furthermore, $A(\xi) = \exp([\Gamma(4/3)]^{-4/3}\int_{\xi_0}^{\xi} s^{-4/3}ds) = \exp(3[\Gamma(4/3)]^{-4/3}[\xi_0^{-1/3} - \xi^{-1/3}])$, which implies $1 \le A(\xi) \le \exp(3[\Gamma(4/3)]^{-4/3}\xi_0^{-1/3})$. On the other hand, $\widetilde{\delta}_1(\xi,\xi_2) = \int_{\xi_2}^{\xi}(1/A(s)\widetilde{a}(s))ds \ge [\Gamma(4/3)]^{-1/3}\exp(-3[\Gamma(4/3)]^{-4/3}\xi_0^{-1/3})\int_{\xi_2}^{\xi} s^{-1/3}ds = (3/2)[\Gamma(4/3)]^{-1/3}\exp(-3[\Gamma(4/3)]^{-(4/3)}\xi_0^{-1/3})(\xi^{2/3} - \xi_2^{2/3})$, which implies $\lim_{\xi\to\infty}\widetilde{\delta}_1(\xi,\xi_2) = \infty$, and then (8) holds. So, there exists a sufficiently large $T > \xi_2$ such that $\widetilde{\delta}_1(\xi,\xi_2) > [\Gamma(4/3)]^2$ on $[T,\infty)$. In (9),

$$\int_{t_0}^{\infty}\frac{\alpha t^{\alpha-1}}{\Gamma(1+\alpha)\,r(t)}dt = \int_{\xi_0}^{\infty}\frac{1}{\widetilde{r}(s)}ds = \int_{\xi_0}^{\infty}ds = \infty. \tag{54}$$

In (10),

$$\begin{aligned}
&\int_{\xi_0}^{\infty}\frac{1}{\widetilde{r}(\zeta)}\int_{\zeta}^{\infty}\frac{1}{A(\tau)\,\widetilde{a}(\tau)}\int_{\tau}^{\infty}A(s)\,\widetilde{q}(s)\,ds\,d\tau\,d\zeta \\
&\quad \ge \left[\Gamma\left(\frac{4}{3}\right)\right]^{-1/3}\exp\left(-3\left[\Gamma\left(\frac{4}{3}\right)\right]^{-4/3}\xi_0^{-1/3}\right) \\
&\quad \times \int_{\xi_0}^{\infty}\int_{\zeta}^{\infty}\tau^{-1/3}\int_{\tau}^{\infty}s^{-1}ds\,d\tau\,d\zeta = \infty.
\end{aligned} \tag{55}$$

In (48), letting $\widetilde{\phi}(\xi) = \xi$, we obtain

$$\begin{aligned}
&\int_{\xi_0}^{\infty}\left[A(s)\widetilde{\phi}(s)\widetilde{q}(s) - \frac{\widetilde{r}(s)\left[\widetilde{\phi}'(s)\right]^2}{4\widetilde{\delta}_1(s,\xi_2)\widetilde{\phi}(s)}\right]ds \\
&\quad = \int_{\xi_0}^{\infty}\left\{A(s)\left[\Gamma\left(\frac{4}{3}\right)\right]^{-2} - \frac{1}{4\widetilde{\delta}_1(s,\xi_2)}\right\}\frac{1}{s}ds \\
&\quad = \int_{\xi_0}^{T}\left\{A(s)\left[\Gamma\left(\frac{4}{3}\right)\right]^{-2} - \frac{1}{4\widetilde{\delta}_1(s,\xi_2)}\right\}\frac{1}{s}ds \\
&\quad + \int_{T}^{\infty}\left\{A(s)\left[\Gamma\left(\frac{4}{3}\right)\right]^{-2} - \frac{1}{4\widetilde{\delta}_1(s,\xi_2)}\right\}\frac{1}{s}ds \\
&\quad \ge \int_{\xi_0}^{T}\left\{A(s)\left[\Gamma\left(\frac{4}{3}\right)\right]^{-2} - \frac{1}{4\widetilde{\delta}_1(s,\xi_2)}\right\}\frac{1}{s}ds \\
&\quad + \int_{T}^{\infty}\frac{3}{4}\left[\Gamma\left(\frac{4}{3}\right)\right]^{-2}\frac{1}{s}ds = \infty.
\end{aligned} \tag{56}$$

Therefore, (53) is oscillatory by Theorem 8.

Example 11. Consider the following fractional differential equation:

$$D_t^{1/2}\left[t^{1/4}D_t^{1/2}D_t^{1/2}x(t)\right]+\frac{\Gamma(3/2)}{\sqrt{t}}D_t^{1/2}D_t^{1/2}x(t)+x(t)=0,\quad t\geq 2. \tag{57}$$

In (1), if we set $t_0 = 2$, $\alpha = 1/2$, $a(t) = t^{1/4}$, $r(t) \equiv 1$, $p(t) = \Gamma(3/2)/\sqrt{t}$, $q(t) \equiv 1$, then we obtain (57). So $\xi_0 = 2^{1/2}/\Gamma(3/2)$, $\widetilde{a}(\xi) = a(t) = t^{1/4} = \sqrt{\Gamma(3/2)\xi}$, $\widetilde{r}(\xi) \equiv 1$, $\widetilde{p}(\xi) = \xi^{-1}$, $\widetilde{q}(\xi) \equiv 1$. Furthermore, $A(\xi) = \exp([\Gamma(3/2)]^{-1/2}\int_{\xi_0}^{\xi} s^{-3/2}ds) = \exp([\Gamma(3/2)]^{-1/2}[2\xi_0^{-1/2} - 2\xi^{-1/2}])$, which implies $1 \leq A(\xi) \leq \exp(2[\Gamma(3/2)]^{-1/2}\xi_0^{-1/2})$. On the other hand, $\widetilde{\delta}_1(\xi,\xi_2) = \int_{\xi_2}^{\xi}(1/A(s)\widetilde{a}(s))ds \geq [\Gamma(3/2)]^{-1/2}\exp(-2[\Gamma(3/2)]^{-1/2}\xi_0^{-1/2})\int_{\xi_2}^{\xi}(1/\sqrt{s})ds = 2[\Gamma(3/2)]^{-1/2}\exp(-2[\Gamma(3/2)]^{-1/2}\xi_0^{-1/2})(\sqrt{\xi} - \sqrt{\xi_2})$, which implies $\lim_{\xi\to\infty}\widetilde{\delta}_1(\xi,\xi_2) = \infty$. So, there exists a sufficiently large $T > \xi_2$ such that $\widetilde{\delta}_1(\xi,\xi_2) > 1$ on $[T,\infty)$.

From the analysis above, one can see the (8) holds. We now test (9) and (10). In (9),

$$\int_{t_0}^{\infty}\frac{\alpha t^{\alpha-1}}{\Gamma(1+\alpha)r(t)}dt=\int_{\xi_0}^{\infty}\frac{1}{\widetilde{r}(s)}ds=\int_{\xi_0}^{\infty}ds=\infty. \tag{58}$$

In (10),

$$\begin{aligned}&\int_{\xi_0}^{\infty}\frac{1}{\widetilde{r}(\zeta)}\int_{\zeta}^{\infty}\frac{1}{A(\tau)\widetilde{a}(\tau)}\int_{\tau}^{\infty}A(s)\widetilde{q}(s)\,ds\,d\tau\,d\zeta\\&\quad\geq\left[\Gamma\left(\frac{3}{2}\right)\right]^{-1/2}\exp\left(-2\left[\Gamma\left(\frac{3}{2}\right)\right]^{-1/2}\xi_0^{-1/2}\right)\\&\quad\times\int_{\xi_0}^{\infty}\int_{\zeta}^{\infty}\frac{1}{\sqrt{\tau}}\int_{\tau}^{\infty}ds\,d\tau\,d\zeta=\infty.\end{aligned} \tag{59}$$

So, (9) and (10) hold. On the other hand, in (44), after putting $\widetilde{\phi}(\xi) \equiv 1$, $\lambda = 2$, for any sufficiently large l, we have

$$\begin{aligned}&\limsup_{\xi\to\infty}\int_l^{\xi}\left[(s-l)^{\lambda}A(s)\widetilde{\phi}(s)\widetilde{q}(s)-\frac{\widetilde{r}(s)\widetilde{\phi}(s)}{4\widetilde{\delta}_1(s,\xi_2)}(s-l)^{\lambda-2}\left(\lambda+\frac{\widetilde{\phi}'(s)}{\widetilde{\phi}(s)}(s-l)\right)^2\right]ds\\&\quad\geq\limsup_{\xi\to\infty}\int_l^{\xi}\left[(s-l)^2s^{-1}-1\right]ds=\infty,\\&\limsup_{\xi\to\infty}\int_l^{\xi}\left[(\xi-s)^{\lambda}A(s)\widetilde{\phi}(s)\widetilde{q}(s)-\frac{\widetilde{r}(s)\widetilde{\phi}(s)}{4\widetilde{\delta}_1(s,\xi_2)}(\xi-s)^{\lambda-2}\left(\lambda-\frac{\widetilde{\phi}'(s)}{\widetilde{\phi}(s)}(\xi-s)\right)^2\right]ds\\&\quad\geq\limsup_{\xi\to\infty}\int_l^{\xi}\left[(\xi-s)^2s^{-1}-1\right]ds=\infty.\end{aligned} \tag{60}$$

So (44) holds, and then by Corollary 7 we deduce that (57) is oscillatory.

References

[1] S. R. Grace, J. R. Graef, and M. A. El-Beltagy, "On the oscillation of third order neutral delay dynamic equations on time scales," *Computers and Mathematics with Applications*, vol. 63, no. 4, pp. 775–782, 2012.

[2] Y. Sun, Z. Han, Y. Sun, and Y. Pan, "Oscillation theorems for certain third order nonlinear delay dynamic equations on time scales," *Electronic Journal of Qualitative Theory of Differential Equations*, vol. 75, pp. 1–14, 2011.

[3] L. Erbe, T. S. Hassan, and A. Peterson, "Oscillation of third order functional dynamic equations with mixed arguments on time scales," *Journal of Applied Mathematics and Computing*, vol. 34, no. 1-2, pp. 353–371, 2010.

[4] T. S. Hassan, "Oscillation of third order nonlinear delay dynamic equations on time scales," *Mathematical and Computer Modelling*, vol. 49, no. 7-8, pp. 1573–1586, 2009.

[5] D. Cakmak and A. Tiryaki, "Oscillation criteria for certain forced second-order nonlinear differential equations," *Applied Mathematics Letters*, vol. 17, no. 3, pp. 275–279, 2004.

[6] Z. Han, T. Li, S. Sun, and C. Zhang, "Oscillation behavior of third-order neutral Emden-Fowler delay dynamic equations on time scales," *Advances in Difference Equations*, vol. 2010, Article ID 586312, 2010.

[7] M. Bohner, "Some oscillation criteria for first order delay dynamic equations," *Far East Journal of Applied Mathematics*, vol. 18, no. 3, pp. 289–304, 2005.

[8] R. P. Agarwal, M. Bohner, and S. H. Saker, "Oscillation of second order delay dynamic equations," *Canadian Applied Mathematics Quarterly*, vol. 13, pp. 1–18, 2005.

[9] R. P. Agarwal and S. R. Grace, "Oscillation of certain third-order difference equations," *Computers and Mathematics with Applications*, vol. 42, no. 3–5, pp. 379–384, 2001.

[10] N. Parhi, "Oscillation and non-oscillation of solutions of second order difference equations involving generalized difference," *Applied Mathematics and Computation*, vol. 218, pp. 458–468, 2011.

[11] S. H. Saker, "Oscillation of nonlinear dynamic equations on time scales," *Applied Mathematics and Computation*, vol. 148, no. 1, pp. 81–91, 2004.

[12] F. Meng and Y. Huang, "Interval oscillation criteria for a forced second-order nonlinear differential equations with damping," *Applied Mathematics and Computation*, vol. 218, no. 5, pp. 1857–1861, 2011.

[13] D. X. Chen, "Oscillation criteria of fractional differential equations," *Advances in Difference Equations*, vol. 2012, article 33, 2012.

[14] D. X. Chen, "Oscillatory behavior of a class of fractional differential equations with damping," *UPB Scientific Bulletin, Series A*, vol. 75, no. 1, pp. 107–118, 2013.

[15] B. Zheng, "Oscillation for a class of nonlinear fractional differential equations with damping term," *Journal of Advanced Mathematical Studies*, vol. 6, no. 1, pp. 107–115, 2013.

[16] A. Aghili and M. R. Masomi, "Integral transform method for solving time fractional systems and fractional heat equation," *Bulletin of Parana's Mathematical Society*, vol. 32, pp. 305–322, 2014.

[17] G. Jumarie, "Modified Riemann-Liouville derivative and fractional Taylor series of nondifferentiable functions further results," *Computers and Mathematics with Applications*, vol. 51, no. 9-10, pp. 1367–1376, 2006.

[18] G. Jumarie, "Table of some basic fractional calculus formulae derived from a modified Riemann-Liouville derivative for non-differentiable functions," *Applied Mathematics Letters*, vol. 22, no. 3, pp. 378–385, 2009.

[19] N. Faraz, Y. Khan, H. Jafari, A. Yildirim, and M. Madani, "Fractional variational iteration method via modified Riemann-Liouville derivative," *Journal of King Saud University*, vol. 23, no. 4, pp. 413–417, 2011.

[20] B. Lu, "Backlund transformation of fractional Riccati equation and its applications to nonlinear fractional partial differential equations," *Physics Letters A*, vol. 376, pp. 2045–2048, 2012.

[21] S. Zhang and H.-Q. Zhang, "Fractional sub-equation method and its applications to nonlinear fractional PDEs," *Physics Letters Section A*, vol. 375, no. 7, pp. 1069–1073, 2011.

[22] S. M. Guo, L. Q. Mei, Y. Li, and Y. F. Sun, "The improved fractional sub-equation method and its applications to the space-time fractional differential equations in fluid mechanics," *Physics Letters A*, vol. 376, pp. 407–411, 2012.

[23] B. Zheng, "(G'/G)-expansion method for solving fractional partial differential equations in the theory of mathematical physics," *Communications in Theoretical Physics*, vol. 58, pp. 623–630, 2012.

[24] B. Lu, "The first integral method for some time fractional differential equations," *Journal of Mathematical Analysis and Applications*, vol. 395, pp. 684–693, 2012.

[25] G. H. Hardy, J. E. Littlewood, and G. P. Pólya, *Inequalities*, Cambridge University Press, Cambridge, UK, 2nd edition, 1988.

Fractional Solutions of Bessel Equation with N-Method

Erdal Bas, Resat Yilmazer, and Etibar Panakhov

Department of Mathematics, Firat University, 23119 Elazig, Turkey

Correspondence should be addressed to Erdal Bas; erdalmat@yahoo.com

Academic Editors: R. M. Guedes, Z. Mukandavire, and R. Wangkeeree

This paper deals with the design fractional solution of Bessel equation. We obtain explicit solutions of the equation with the help of fractional calculus techniques. Using the N-fractional calculus operator N^{ν} method, we derive the fractional solutions of the equation.

1. Introduction, Definitions, and Preliminaries

Fractional calculus has an important place in the field of math. Firstly, L'Hospital and Leibniz were interested in the topic in 1695, [1]. Fractional calculus is an area of applied mathematics that deals with derivatives and integrals of arbitrary orders and their applications in science, engineering, mathematics, economics, and other fields. The seeds of fractional derivatives were planted over 300 years ago. Since then many efficient mathematicians of their times, such as N. H. Abel, M. Caputo, L. Euler, J. Fourier, A. K. Grunwald, J. Hadamard, G. H. Hardy, O. Heaviside, H. J. Holmgren, P. S. Laplace, G. W. Leibniz, A. V. Letnikov, J. Liouville, and B. Riemann, have contributed to this field; all these references can be seen in [1–5]. The mathematics involved appeared very different applications of this field. Fractional calculus has been applied to almost every field of science. They are viscoelasticity, electrical engineering, electrochemistry, biology, biophysics and bioengineering, signal and image processing, mechanics, mechatronics, physics, and control theory. During the last decade, Samko et al. [4], Nishimoto [6–12], and Podlubny [3] have been helpful in introducing the field to engineering, science, economics and finance, and pure and applied field. Furthermore, there were many studies in this field [5, 13–16]. Various scientists have studied that concept. The progress in this field continues [2–4, 6–12, 17–20].

N-Fractional calculus is a very interesting method because this method is applied to singular equation. Note that fractional solutions can be obtained for kinds of singular equation via this method [6–12, 17]. In this paper, our aim is to apply the same way for singular Sturm-Liouville equation with Bessel potential and find fractional solutions of this equation. Furthermore, we give some applications and their graphs of fractional solutions of the equation.

Now, consider the following the Bessel equation:

$$x\frac{d^2z}{dx^2}+\frac{dz}{dx}+\left[\lambda x-\frac{p^2}{x}\right]z=0,\quad 0<x\leq 1,\tag{1}$$

where λ and p are real numbers. By means of the substitution $y=\sqrt{xz}$ (1) reduces to the form

$$\frac{d^2y}{dx^2}+\left(\lambda-\frac{p^2-1/4}{x^2}\right)y=0.\tag{2}$$

Bessel equation for having the analogous singularity is given in [21].

The differintegration operators and their generalizations [6–11, 17, 18] have been used to solve some classes of differential equations and fractional differential equations.

Two of the most commonly encountered tools in the theory and applications of fractional calculus are provided by the Riemann-Liouville operator R_z^{υ} ($\upsilon\in\mathbb{C}$) and the Weyl operator W_z^{υ} ($\upsilon\in\mathbb{C}$), which are defined by [17, 19].

Consider

$$R_z^{\upsilon} f(z) = \begin{cases} \dfrac{1}{\Gamma(\upsilon)} \displaystyle\int_0^z (z-t)^{\upsilon-1} f(t)\, dt & (\mathrm{Re}(\upsilon) > 0), \\ \dfrac{d^n}{dz^n} R_z^{\upsilon+n} f(z) & (-n < \mathrm{Re}(\upsilon) \le 0;\ n \in \mathbb{N}), \end{cases} \tag{3}$$

$$W_z^{\upsilon} f(z) = \begin{cases} \dfrac{1}{\Gamma(\upsilon)} \displaystyle\int_z^{\infty} (t-z)^{\upsilon-1} f(t)\, dt & (\mathrm{Re}(\upsilon) > 0), \\ \dfrac{d^n}{dz^n} W_z^{\upsilon+n} f(z) & (-n < \mathrm{Re}(\upsilon) \le 0;\ n \in \mathbb{N}) \end{cases} \tag{4}$$

provided that the defining integrals in (3) and (4) exist, $\mathbb{N}$ being the set of positive integers.

Definition 1 (cf. [6–10, 12, 20]). Let

$$D = \{D^-, D^+\}, \quad C = \{C^-, C^+\}, \tag{5}$$

where C^- is a curve along the cut joining two points z and $-\infty + i\ \mathrm{Im}(z)$, C^+ is a curve along the cut joining two points z and $\infty + i\ \mathrm{Im}(z)$, D^- is a domain surrounded by C^-, and D^+ is a domain surrounded by C^+. (Here D contains the points over the curve C.)

Moreover, let $f = f(z)$ be a regular function in D ($z \in D$),

$$f_\nu(z) = (f(z))_\nu = \frac{\Gamma(\nu+1)}{2\pi i} \int_C \frac{f(t)\, dt}{(t-z)^{\nu+1}} \quad (\nu \in \mathbb{R} \setminus \mathbb{Z}^-;\ \mathbb{Z}^- = \{\ldots, -3, -2, -1\}),$$
$$f_{-n}(z) = \lim_{\nu \to -n} f_\nu(z) \quad (n \in \mathbb{Z}^+), \tag{6}$$

where $t \ne z$,

$$-\pi \le \arg(t-z) \le \pi \quad \text{for } C^-,$$
$$0 \le \arg(t-z) \le 2\pi \quad \text{for } C^+. \tag{7}$$

Then $f_\nu(z)$ ($\nu > 0$) is said to be the fractional derivative of $f(z)$ of order ν and $f_\nu(z)$ ($\nu < 0$) is said to be the fractional integral of $f(z)$ of order $-\nu$, provided (in each case) that $|f_\nu(z)| < \infty$ ($\nu \in \mathbb{R}$).

Finally, let the fractional calculus operator (Nishimoto's operator) N^ν be defined by (cf. [6–10])

$$N^\nu = \left(\frac{\Gamma(\nu+1)}{2\pi i} \int_C \frac{dt}{(t-z)^{\nu+1}} \right) \quad (\nu \notin \mathbb{Z}^-) \tag{8}$$

with

$$N^{-n} = \lim_{\nu \to -n} N^\nu \quad (n \in \mathbb{Z}^+). \tag{9}$$

We find it to be worthwhile to recall here the following useful lemmas and properties associated with the fractional differintegration which was defined earlier (cf. e.g., [6–10, 12]).

Lemma 2 (linearity property). *If the functions $f(z)$ and $g(z)$ are single-valued and analytic in some domain $\Omega \subseteq \mathbb{C}$, then*

$$(h_1 f(z) + h_2 g(z))_\nu = h_1 f_\nu(z) + h_2 g_\nu(z) \quad (\nu \in \mathbb{R}; z \in \Omega) \tag{10}$$

for any constants h_1 and h_2.

Lemma 3 (index law). *If the function $f(z)$ is single-valued and analytic in some domain $\Omega \subseteq \mathbb{C}$, then*

$$(f_\sigma(z))_\nu = f_{\sigma+\nu}(z) = (f_\nu(z))_\sigma$$
$$(f_\sigma(z) \ne 0; f_\nu(z) \ne 0; \sigma, \nu \in \mathbb{R}; z \in \Omega). \tag{11}$$

Lemma 4 (generalized Leibniz rule). *If the functions $f(z)$ and $g(z)$ are single-valued and analytic in some domain $\Omega \subseteq \mathbb{C}$, then*

$$(f(z) \cdot g(z))_\nu = \sum_{n=0}^{\infty} \binom{\nu}{n} f_{\nu-n}(z) \cdot g_n(z) \quad (\nu \in \mathbb{R}; z \in \Omega), \tag{12}$$

where $g_n(z)$ is the ordinary derivative of $g(z)$ of order n ($n \in \mathbb{N}_0 := \mathbb{N} \cup \{0\}$), being tacitly assumed (for simplicity) that $g(z)$ is the polynomial part (if any) of the product $f(z)g(z)$.

Property 1. For a constant λ,

$$(e^{\lambda z})_\nu = \lambda^\nu e^{\lambda z} \quad (\lambda \ne 0; \nu \in \mathbb{R}; z \in \mathbb{C}). \tag{13}$$

Property 2. For a constant λ,

$$(e^{-\lambda z})_\nu = e^{-i\pi\nu} \lambda^\nu e^{-\lambda z} \quad (\lambda \ne 0; \nu \in \mathbb{R}; z \in \mathbb{C}). \tag{14}$$

Property 3. For a constant λ,

$$(z^\lambda)_\nu = e^{-i\pi\nu} \frac{\Gamma(\nu-\lambda)}{\Gamma(-\lambda)} z^{\lambda-\nu}$$
$$\left(\nu \in \mathbb{R}; z \in \mathbb{C}; \left|\frac{\Gamma(\nu-\lambda)}{\Gamma(-\lambda)}\right| < \infty\right). \tag{15}$$

Now, let apply N-fractional method to nonhomogeneous Bessel equation.

2. The N^ν-Method Applied to Bessel Equation

Theorem 5. *Let $y \in \{y : 0 \ne |y_\nu| < \infty; \nu \in \mathbb{R}\}$ and $f \in \{f : 0 \ne |f_\nu| < \infty; \nu \in \mathbb{R}\}$. We consider the nonhomogeneous Bessel equation:*

$$L[y, x, \lambda, p] = y_2 + y\left[\lambda - \frac{p^2 - 1/4}{x^2}\right] = f \quad (0 < x \le 1), \tag{16}$$

and it has particular solutions of the forms

$$y^{I} = x^{p+1/2}e^{-i\sqrt{\lambda}x} \times \left\{\left[\left(fx^{1/2-p}e^{i\sqrt{\lambda}x}\right)_{-p-1/2} e^{-2i\sqrt{\lambda}x}x^{p-1/2}\right]_{-1} \times e^{2i\sqrt{\lambda}x}x^{-p-1/2}\right\}_{p-1/2}, \quad (17)$$

$$y^{II} = x^{p+1/2}e^{i\sqrt{\lambda}x} \times \left\{\left[\left(fx^{1/2-p}e^{-i\sqrt{\lambda}x}\right)_{-p-1/2} e^{2i\sqrt{\lambda}x}x^{p-1/2}\right]_{-1} \times e^{-2i\sqrt{\lambda}x}x^{-p-1/2}\right\}_{p-1/2}, \quad (18)$$

$$y^{III} = x^{-p+1/2}e^{-i\sqrt{\lambda}x} \times \left\{\left[\left(fx^{1/2+p}e^{i\sqrt{\lambda}x}\right)_{p-1/2} e^{-2i\sqrt{\lambda}x}x^{-p-1/2}\right]_{-1} \times e^{2i\sqrt{\lambda}x}x^{p-1/2}\right\}_{-p-1/2}, \quad (19)$$

$$y^{IV} = x^{-p+1/2}e^{i\sqrt{\lambda}x} \times \left\{\left[\left(fx^{1/2+p}e^{-i\sqrt{\lambda}x}\right)_{p-1/2} e^{2i\sqrt{\lambda}x}x^{-p-1/2}\right]_{-1} \times e^{-2i\sqrt{\lambda}x}x^{p-1/2}\right\}_{-p-1/2}, \quad (20)$$

where $y_2 = d^2y/dx^2$, $y = y(z)$ $(z \in \mathbb{C})$, $f = f(z)$ *(an arbitrary given function), and* p, λ *are given constants.*

Remark 6. The cases $p = 0$ of (19) and (20) coincide with those (17) and (18).

Proof. Set

$$y = x^{\eta}\psi, \quad \psi = \psi(x). \quad (21)$$

Thus

$$y_1 = \eta x^{\eta-1}\psi + x^{\eta}\psi_1, \quad y_2 = \eta(\eta-1)x^{\eta-2}\psi + 2\eta x^{\eta-1}\psi_1 + x^{\eta}\psi_2. \quad (22)$$

Putting (21) and (22) in (16), we obtained

$$\psi_2 x^{\eta} + \psi_1 2\eta x^{\eta-1} + \psi\eta(\eta-1)x^{\eta-2} + x^{\eta}\psi\left[\lambda - \frac{p^2 - 1/4}{x^2}\right] = f. \quad (23)$$

□

With some rearrangement of the terms in (23), we have

$$\psi_2 x + \psi_1 2\eta + \psi\left[x^{-1}\left(\eta^2 - \eta + \frac{1}{4} - p^2\right) + \lambda x\right] = fx^{1-\eta}. \quad (24)$$

Here, we choose η such that

$$\eta^2 - \eta + \frac{1}{4} - p^2 = 0. \quad (25)$$

That is,

$$\eta = \frac{1}{2} \pm p. \quad (26)$$

(I) Let $\eta = p + (1/2)$. From (21) and (24), we have

$$y = x^{p+(1/2)}\psi, \quad (27)$$

$$\psi_2 x + \psi_1(2p+1) + \psi\lambda x = fx^{(1/2)-p}. \quad (28)$$

Set

$$\psi = e^{\mu x}\phi, \quad \phi = \phi(x). \quad (29)$$

Rewrite (28) in the form

$$(e^{\mu x}\phi)_2 x + (e^{\mu x}\phi)_1(2p+1) + e^{\mu x}\phi\lambda x = fx^{(1/2)-p}. \quad (30)$$

At this point, differentiating $e^{\mu x}\phi$ two times,

$$(e^{\mu x}\phi)_1 = e^{\mu x}(\mu\phi + \phi_1), \quad (e^{\mu x}\phi)_2 = e^{\mu x}(\mu^2\phi + 2\mu\phi_1 + \phi_2), \quad (31)$$

and substituting from (29) and (31) in (30), we can express (30) as

$$\phi_2 x + \phi_1(2\mu x + 2p + 1) + \phi\left(2p\mu + \mu + x\left(\mu^2 + \lambda\right)\right) = fx^{(1/2)-p}e^{-\mu x}. \quad (32)$$

Choose μ such that

$$\mu^2 + \lambda = 0. \quad (33)$$

That is,

$$\mu = \pm\sqrt{\lambda}i. \quad (34)$$

(I) (i): For instance, taking $\mu = -\sqrt{\lambda}i$, we have

$$\psi = e^{-\sqrt{\lambda}ix}\phi, \quad (35)$$

$$\phi_2 x + \phi_1\left[-\sqrt{\lambda}i2x + 2p + 1\right] + \phi\left[-i(2p+1)\sqrt{\lambda}\right] = fx^{(1/2)-p}e^{-\mu x} \quad (36)$$

from (29) and (32).

Applying the operator N^{ν} to both members of (36), we find the following equality:

$$[\phi_2 x]_{\nu} + \left\{\phi_1\left[-\sqrt{\lambda}i2x + 2p + 1\right]\right\}_{\nu} + \left\{\phi\left[-i(2p+1)\sqrt{\lambda}\right]\right\}_{\nu} = \left[fx^{(1/2)-p}e^{-\mu x}\right]_{\nu}. \quad (37)$$

Using (3)–(12), we have

$$[\phi_2 x]_\nu = x\phi_{2+\nu} + \nu\phi_{1+\nu},$$
$$\{\phi_1[-\sqrt{\lambda}i2x + 2p + 1]\}_\nu \quad (38)$$
$$= \phi_{1+\nu}[-\sqrt{\lambda}i2x + 2p + 1] - \sqrt{\lambda}i2\nu\phi_\nu.$$

Making use of the relations (38), rewrite (37) in the following form:

$$\phi_{2+\nu}x + \phi_{1+\nu}[-\sqrt{\lambda}i2x + 2p + 1 + \nu] - \phi_\nu[\sqrt{\lambda}i2\nu + i(2p+1)\sqrt{\lambda}] = [fx^{(1/2)-p}e^{i\sqrt{\lambda}x}]_\nu. \quad (39)$$

Choose ν such that

$$\nu = -p - \frac{1}{2}. \quad (40)$$

We then have

$$\phi_{-p+3/2}x + \phi_{-p+1/2}\left[p + \frac{1}{2} - 2\sqrt{\lambda}ix\right] = [fx^{(1/2)-p}e^{i\sqrt{\lambda}x}]_{-p-1/2} \quad (41)$$

from (39).

Next, writing

$$\phi_{-p+1/2} = \omega = \omega(x), \quad (42)$$

we obtain the following equality from (41):

$$\omega_1 + \omega\left[x^{-1}\left(p + \frac{1}{2}\right) - 2\sqrt{\lambda}i\right] = [fx^{(1/2)-p}e^{i\sqrt{\lambda}x}]_{-p-1/2}x^{-1}. \quad (43)$$

This is an ordinary differential equation of the first order which has a particular solution,

$$\omega = \left[[fx^{(1/2)-p}e^{i\sqrt{\lambda}x}]_{-p-1/2}x^{-1}e^{-2\sqrt{\lambda}ix}x^{p+1/2}\right]_{-1} \times e^{2\sqrt{\lambda}ix}x^{-p-1/2}. \quad (44)$$

Making use of the reverse process to obtain y^I, we finally obtain the solution (17) from (44), (42), (35), and (27).

Inversely, (44) satisfies (43); then

$$\phi = \omega_{p-1/2} \quad (45)$$

satisfies (41). Therefore, (17) satisfies (16) because we have (27), (35), (44), and (45).

(I) (ii): In the case when $\mu = \sqrt{\lambda}i$, we have

$$\psi = e^{\sqrt{\lambda}ix}\phi, \quad (46)$$

$$\phi_2 x + \phi_1[2\sqrt{\lambda}ix + 2p + 1] + \phi[(2p+1)\sqrt{\lambda}i] = fx^{(1/2)-p}e^{-i\sqrt{\lambda}x} \quad (47)$$

from (29) and (32).

Applying the operator N^ν to both members of (47), we have

$$\phi_{2+\nu}x + \phi_{1+\nu}[2\sqrt{\lambda}ix + 2p + 1 + \nu] + \phi_\nu[\sqrt{\lambda}i(2p+1)i] = (fx^{(1/2)-p}e^{-i\sqrt{\lambda}x})_\nu. \quad (48)$$

Choosing ν such that

$$\nu = -p - \frac{1}{2} \quad (49)$$

and replacing

$$\phi_{-p+1/2} = \vartheta = \vartheta(x), \quad (50)$$

we then obtain

$$\vartheta_1 + \vartheta\left[2\sqrt{\lambda}i + \left(p + \frac{1}{2}\right)x^{-1}\right] = (fx^{(1/2)-p}e^{-i\sqrt{\lambda}x})_{-p-1/2}x^{-1} \quad (51)$$

from (48). A particular solution of (51) is given by

$$\vartheta = \left[(fx^{1/2-p}e^{-i\sqrt{\lambda}x})_{-p-1/2}x^{-1}e^{2\sqrt{\lambda}ix}x^{p+1/2}dx\right]_{-1} \times e^{-2\sqrt{\lambda}ix}x^{-p-1/2}. \quad (52)$$

Thus, we have (18) from (52), (50), (46), and (27).

(II) Let $\eta = -p + (1/2)$.

With the help of the similar method in (I), replacing p by $-p$ in (I) (i) and (I) (ii), we have other solutions (19) and (20) different from (17) and (18), respectively, if $p \neq 0$.

3. The Operator N^ν-Method to a Homogeneous Bessel Equation

Theorem 7. *If $y \in \overset{o}{\wp}$, just as in Theorem 5, then the homogeneous Bessel equation*

$$L[y, x, p] = y_2 + y\left[\lambda - \frac{p^2 - 1/4}{x^2}\right] = 0, \quad (0 < x \leq 1), \quad (53)$$

has solutions of the forms

$$y_I = kx^{p+1/2}e^{-i\sqrt{\lambda}x}\{e^{2i\sqrt{\lambda}x}x^{-p-1/2}\}_{p-1/2}, \quad (54)$$

$$y_{II} = kx^{p+1/2}e^{i\sqrt{\lambda}x}\{e^{-2i\sqrt{\lambda}x}x^{-p-1/2}\}_{p-1/2}, \quad (55)$$

$$y_{III} = kx^{-p+1/2}e^{-i\sqrt{\lambda}x}\{e^{2i\sqrt{\lambda}x}x^{p-1/2}\}_{-p-1/2}, \quad (56)$$

$$y_{IV} = kx^{-p+1/2}e^{i\sqrt{\lambda}x}\{e^{-2i\sqrt{\lambda}x}x^{p-1/2}\}_{-p-1/2}, \quad (57)$$

for $p \neq 0$, where k is an arbitrary constant.

Remark 8. In the case when $p = 0$, (56) and (57) coincide with (54) and (55).

Proof. When $f = 0$ in Section 2, we have

$$\omega_1 + \omega\left[\left(p + \frac{1}{2}\right)x^{-1} - 2\sqrt{\lambda}i\right] = 0, \tag{58}$$

$$\vartheta_1 + \vartheta\left[\left(p + \frac{1}{2}\right)x^{-1} + 2\sqrt{\lambda}i\right] = 0, \tag{59}$$

for $\mu = -i\sqrt{\lambda}$ and $\mu = i\sqrt{\lambda}$, instead of (43) and (51).

Therefore, we get (54) for (58) and (55) for (59).

And, for $\eta = -p + (1/2)$, replacing p by $-p$ in (58) and (59), we have (56) and (57). □

Theorem 9. *Let* $y \in \overset{o}{\wp}$ *and* $f \in \overset{o}{\wp}$ *just as in Theorem 5. Then the nonhomogeneous modified Sturm-Liouville equation* (16) *is satisfied by the fractional differintegrated functions*

$$y = y^{\iota} + y_{\iota}. \tag{60}$$

Proof. It is clear by Theorems 5 and 7. □

Application 1. If we substitute $p = 0$, $\lambda = 1/4$, and $f = ix^{-1/2}e^{-(i/2)x}$ in (16), then we obtain the following equation:

$$y_2 + y\left(\frac{1}{4} + \frac{1}{4x^2}\right) = ix^{-1/2}e^{-(i/2)x}, \tag{61}$$

and its solution is

$$y = x^{1/2}e^{(-i/2)x}\left\{\left[\left(ix^{-1/2}e^{-(i/2)x}x^{1/2}e^{(i/2)x}\right)_{-1/2} \times e^{-ix}x^{-1/2}\right]_{-1} e^{ix}x^{-1/2}\right\}_{-1/2}. \tag{62}$$

By performing the necessary operations in (62), we get

$$y = x^{1/2}e^{(-i/2)x}\left\{\left[\frac{2i\sqrt{x}}{\sqrt{\pi}}e^{-ix}x^{-1/2}\right]_{-1} e^{ix}x^{-1/2}\right\}_{-1/2}, \tag{63}$$

where Riemann Liouville operator is

$$[i]_{-1/2} = \frac{1}{\Gamma(1/2)}\int_0^x \frac{i}{\sqrt{x-t}}dt = \frac{2i\sqrt{x}}{\sqrt{\pi}},$$
$$y = x^{1/2}e^{(-i/2)x}\left(-\frac{2x^{-1/2}}{\sqrt{\pi}}\right)_{-1/2}, \tag{64}$$

and using the definitions of Riemann Liouville operator again, we obtain the following solution:

$$\left(-\frac{2x^{-1/2}}{\sqrt{\pi}}\right)_{-1/2} = -\frac{1}{\Gamma(1/2)}\int_0^x \frac{2t^{-1/2}}{\sqrt{\pi}\sqrt{x-t}}dt = -2, \tag{65}$$

$$y = -2x^{1/2}e^{-(i/2)x}. \tag{66}$$

Now, let us show that the last equality is the solution of (61):

$$y_2 = e^{-(i/2)x}\left[\frac{1}{2}x^{1/2} + \frac{1}{2}x^{-3/2} + ix^{-1/2}\right]. \tag{67}$$

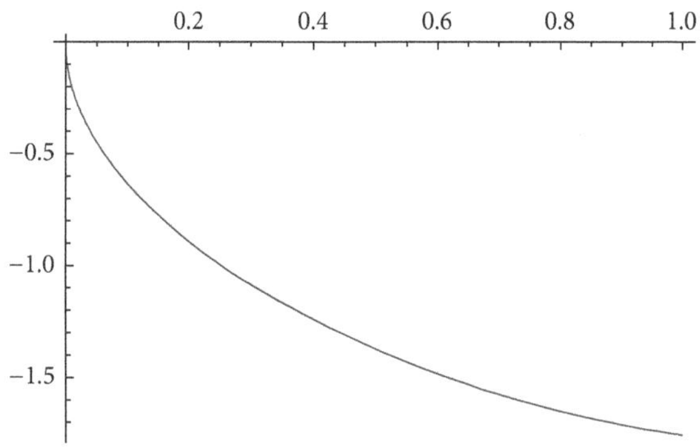

FIGURE 1: Rey.

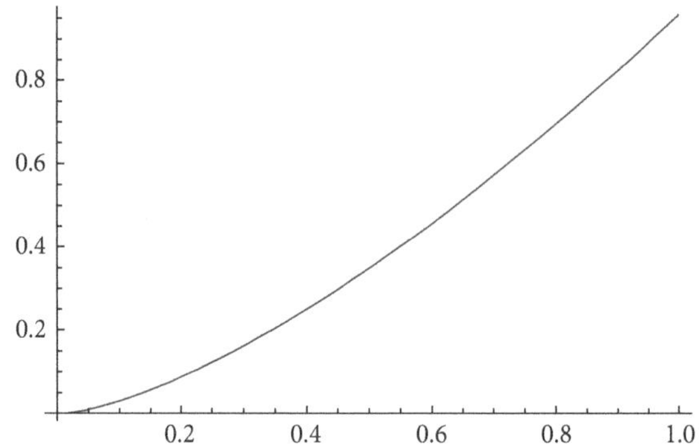

FIGURE 2: Imy.

Obviously, if (66) and (67) are put in (61), it is satisfied. The graph of the solution of (61) is given in Figures 1 and 2.

Application 2. If we substitute $p = -1$ and $\lambda = 0$ in (53), then we obtain the following equation:

$$y_2 - \frac{3}{4x^2}y = 0 \quad (0 < x \le 1), \tag{68}$$

and its solution is

$$y = kx^{-1/2}\left\{x^{1/2}\right\}_{-3/2}. \tag{69}$$

We prove that y_{ι} is the solution of (67). With the help of Riemann Liouville operator,

$$\left[x^{-1/2}\right]_{-1/2} = \frac{1}{\Gamma(3/2)}\int_0^x \frac{t^{1/2}}{\sqrt{x-t}}dt = \frac{\sqrt{\pi}x^2}{4}, \tag{70}$$

$$y = \frac{kx^{3/2}\sqrt{\pi}}{4}. \tag{71}$$

Now, let us show that the last equality is the solution of (67),

$$y_2 = \frac{3kx^{-1/2}\sqrt{\pi}}{16}. \tag{72}$$

Obviously, if (71) and (72) are put into (68), it is satisfied. The graph of the solution of (68) is given in Figure 3.

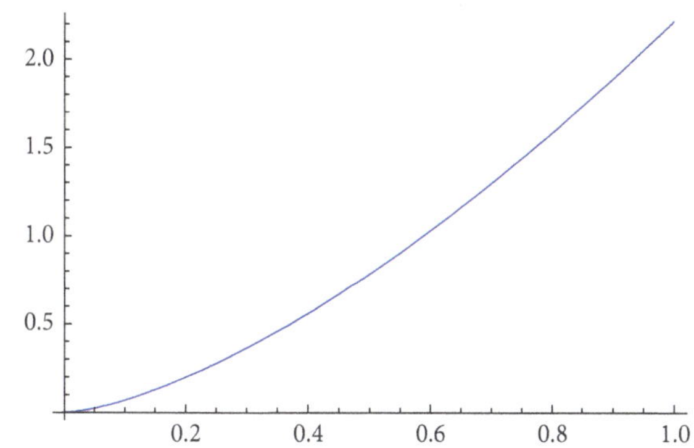

FIGURE 3

4. Two Further Cases of Modified Bessel Equation

Theorem 10. *In the similar way as in the previous sections, we can solve the following nonhomogeneous modified Bessel equation:*

$$
\begin{aligned}
&y_2 + y\left[\lambda + \frac{(1/4) + p^2}{x^2}\right] = f, \\
&y_2 + y\left[-\lambda + \frac{(1/4) + p^2}{x^2}\right] = f,
\end{aligned}
\tag{73}
$$

which are obtained by replacing p by ip ($-\lambda$ instead of λ) in (16); that is,

$$
y_2 + y\left[\lambda + \frac{(1/4) - (ip)^2}{x^2}\right] = f, \tag{74}
$$

$$
y_2 + y\left[-\lambda + \frac{(1/4) - (ip)^2}{x^2}\right] = f. \tag{75}
$$

(i) Therefore, the solutions for (74) are given by replacing p by ip in (17), (18), (19), and (20) as follows:

$$
\begin{aligned}
y^{(i)} &= x^{pi+1/2} e^{-i\sqrt{\lambda}x} \\
&\quad \times \left\{\left[\left(f x^{1/2-pi} e^{i\sqrt{\lambda}x}\right)_{-pi-1/2} e^{-2i\sqrt{\lambda}x} x^{pi-1/2}\right]_{-1} \right. \\
&\quad \left. \times e^{2i\sqrt{\lambda}x} x^{-pi-1/2}\right\}_{pi-1/2}, \\
y^{(ii)} &= x^{pi+1/2} e^{i\sqrt{\lambda}x} \\
&\quad \times \left\{\left[\left(f x^{1/2-pi} e^{-i\sqrt{\lambda}x}\right)_{-pi-1/2} e^{2i\sqrt{\lambda}x} x^{pi-1/2}\right]_{-1} \right. \\
&\quad \left. \times e^{-2i\sqrt{\lambda}x} x^{-pi-1/2}\right\}_{pi-1/2}, \\
y^{(iii)} &= x^{-pi+1/2} e^{-i\sqrt{\lambda}x} \\
&\quad \times \left\{\left[\left(f x^{1/2+pi} e^{i\sqrt{\lambda}x}\right)_{pi-1/2} e^{-2i\sqrt{\lambda}x} x^{-pi-1/2}\right]_{-1} \right. \\
&\quad \left. \times e^{2i\sqrt{\lambda}x} x^{pi-1/2}\right\}_{-pi-1/2}, \\
y^{(iv)} &= x^{-pi+1/2} e^{i\sqrt{\lambda}x} \\
&\quad \times \left\{\left[\left(f x^{1/2+pi} e^{-i\sqrt{\lambda}x}\right)_{pi-1/2} e^{2i\sqrt{\lambda}x} x^{-pi-1/2}\right]_{-1} \right. \\
&\quad \left. \times e^{-2i\sqrt{\lambda}x} x^{pi-1/2}\right\}_{-pi-1/2}.
\end{aligned}
\tag{76}
$$

(ii) In the same way, for the solutions for (75), substituting the relations (21), and (22) into (75), we have

$$
\phi_2 x + \phi_1 2\nu + \phi\left[\left(\nu^2 - \nu + \frac{1}{4} - (pi)^2\right)x^{-1} - \lambda x\right] = f x^{1-\nu}. \tag{77}
$$

Choose ν as follows:

$$
\nu^2 - \nu + \frac{1}{4} + p^2 = 0. \tag{78}
$$

That is

$$
\nu = \frac{1}{2} \pm pi. \tag{79}
$$

Let $\nu = pi + (1/2)$. From (21) and (77), we have

$$
y = x^{pi+(1/2)}\phi, \tag{80}
$$

$$
\phi_2 x + \phi_1 (2pi + 1) - \phi\lambda x = f x^{(1/2)-pi}. \tag{81}
$$

Next, set (29); then (81) is rewritten in the form

$$
\left(e^{\mu x}\psi\right)_2 x + \left(e^{\mu x}\psi\right)_1 (2ip + 1) - e^{\mu x}\psi\lambda x = f x^{(1/2)-pi}. \tag{82}
$$

Substituting the relations (29) and (31) into (82), we have

$$
\begin{aligned}
&\psi_2 x + \psi_1 (2\mu x + 2pi + 1) \\
&\quad + \psi\left[\left(\mu^2 - \lambda\right)x + (2pi + 1)\mu\right] = f x^{(1/2)-ip} e^{-\mu x}.
\end{aligned}
\tag{83}
$$

Choose μ as follows:

$$
\mu^2 - \lambda = 0. \tag{84}
$$

That is,

$$
\mu = \pm\sqrt{\lambda}. \tag{85}
$$

(ii. 1) In the case when $\mu = -\sqrt{\lambda}$, we have

$$
\phi = e^{-\sqrt{\lambda}x}\psi, \tag{86}
$$

$$
\begin{aligned}
&\psi_2 x + \psi_1 \left(-2\sqrt{\lambda}x + 2pi + 1\right) \\
&\quad - \psi\left[\sqrt{\lambda}(2pi + 1)\right] = f x^{(1/2)-pi} e^{\sqrt{\lambda}x}
\end{aligned}
\tag{87}
$$

from (29) and (83).

Applying the operator N^{ν} to both members of (87), we then obtain

$$\left(\psi_2 x\right)_{\nu} + \left[\psi_1\left(-2\sqrt{\lambda}x + 2pi + 1\right)\right]_{\nu} + \left\{\psi\left[-\sqrt{\lambda}\left(2pi + 1\right)\right]\right\}_{\nu} = \left(fx^{(1/2)-pi}e^{\sqrt{\lambda}x}\right)_{\nu}. \tag{88}$$

Using (4), (10), and (11), we have

$$\psi_{2+\nu}x + \psi_{1+\nu}\left(-2\sqrt{\lambda}x + 2pi + 1 + \nu\right) + \psi_{\nu}\left[-\sqrt{\lambda}\left(2pi + 1 + 2\nu\right)\right] = \left(fx^{(1/2)-pi}e^{\sqrt{\lambda}x}\right)_{\nu}. \tag{89}$$

Choose ν such that

$$\nu = -pi - \frac{1}{2}. \tag{90}$$

We then have

$$\psi_{2-pi-1/2}x + \psi_{1-pi-1/2}\left[-2\sqrt{\lambda}x + pi + \frac{1}{2}\right] = \left(fx^{(1/2)-pi}e^{\sqrt{\lambda}x}\right)_{-pi-1/2} \tag{91}$$

from (89).

Next, writing

$$\psi_{1/2-pi} = u = u\left(x\right), \tag{92}$$

we obtain

$$u_1 + u\left[-2\sqrt{\lambda} + \left(pi + \frac{1}{2}\right)x^{-1}\right] = \left(fx^{(1/2)-pi}e^{\sqrt{\lambda}x}\right)_{-pi-1/2}x^{-1} \tag{93}$$

from (91). This is an ordinary differential equation of the first order which has a particular solution

$$u = \left[\left(fx^{(1/2)-pi}e^{\sqrt{\lambda}x}\right)_{-pi-1/2}e^{-2\sqrt{\lambda}x}x^{pi-1/2}\right]_{-1} \times e^{2\sqrt{\lambda}x}x^{-pi-1/2}. \tag{94}$$

We finally obtain the solution

$$y^{(i)} = x^{pi+1/2}e^{-\sqrt{\lambda}x} \times \left\{\left[\left(fx^{1/2-pi}e^{\sqrt{\lambda}x}\right)_{-pi-1/2}e^{-2\sqrt{\lambda}x}x^{pi-1/2}\right]_{-1} \times e^{2\sqrt{\lambda}x}x^{-pi-1/2}\right\}_{pi-1/2}. \tag{95}$$

from (94), (92), (86) and (80).

(ii. 2) Similarly, in the case when $\mu = \sqrt{\lambda}$, we obtain

$$y^{(ii)} = r^{pi+1/2}e^{\sqrt{\lambda}x} \times \left\{\left[\left(fx^{1/2-pi}e^{-\sqrt{\lambda}x}\right)_{-pi-1/2}e^{2\sqrt{\lambda}x}x^{pi-1/2}\right]_{-1} \times e^{-2\sqrt{\lambda}x}x^{-pi-1/2}\right\}_{pi-1/2}. \tag{96}$$

Let $\nu = -ip + (1/2)$. In the same way as in the procedure in (ii), replacing ip by $-ip$ (ii. 1) and (ii. 2), we can obtain $y^{(iii)}$ and $y^{(iv)}$.

Theorem 11. *In the homogeneous case for* (74) *with* $f = 0$, *using the solutions* (54), (55), (56), *and* (57) *and replacing p by pi, we obtain*

$$\begin{aligned} y_{(i)} &= \alpha x^{pi+1/2}e^{-i\sqrt{\lambda}x}\left(e^{2i\sqrt{\lambda}x}x^{-pi-1/2}\right)_{pi-1/2}, \\ y_{(ii)} &= \alpha x^{pi+1/2}e^{i\sqrt{\lambda}x}\left(e^{-2i\sqrt{\lambda}x}x^{-pi-1/2}\right)_{pi-1/2}, \\ y_{(iii)} &= \alpha x^{-pi+1/2}e^{-i\sqrt{\lambda}x}\left(e^{2i\sqrt{\lambda}x}x^{pi-1/2}\right)_{-pi-1/2}, \\ y_{(iv)} &= \alpha x^{-pi+1/2}e^{i\sqrt{\lambda}x}\left(e^{-2i\sqrt{\lambda}x}x^{pi-1/2}\right)_{-pi-1/2}, \end{aligned} \tag{97}$$

for $p \neq 0$, *where* α *is an arbitrary constant.*

5. Conclusion

The N-fractional calculus operator N^{ν}-method is applied to the nonhomogeneous and homogeneous Bessel equation. Explicit fractional solutions of Bessel equations are obtained. Furthermore, similar solutions were obtained for the modified same equation by using the method.

Acknowledgment

The authors sincerely thank the reviewers for their valuable suggestions and useful comments.

References

[1] L. Adam, *Fractional Calculus: History, Definitions and Applications For the Engineer*, 2004.

[2] A. A. Kilbas, H. M. Srivastava, and J. J. Trujillo, *Theory and Applications of Fractional Differential Equations*, vol. 204 of *North-Holland Mathematical Studies*, Elsevier (North-Holland) Science Publishers, Amsterdam, The Netherlands, 2006.

[3] I. Podlubny, *Fractional Differential Equations: An Introduction To Fractional Derivatives, Fractional Differential Equations, Methods of Their Solution and Some of Their Applications, Mathematics in Science and Enginering*, vol. 198, Academic Press, New York, NY, USA, 1999.

[4] S. G. Samko, A. A. Kilbas, and O. I. Marichev, *Fractional Integrals and Derivatives: Theory and Applications Translated From the Russian: Integrals and Derivatives of Fractional Order and Some of Their Applications*, "Nauka i Tekhnika", Minsk, (1987), Gordon and Breach Science Publishers, Reading, UK, 1993.

[5] K. B. Oldham and J. Spanier, *The Fractional Calculus*, Akademic Press, New York, NY, USA, 1974.

[6] K. Nishimoto, *Fractional Calculus*, vol. I, Descartes Press, Koriyama, Japan, 1984.

[7] K. Nishimoto, *Fractional Calculus*, vol. II, Descartes Press, Koriyama, Japan, 1987.

[8] K. Nishimoto, *Fractional Calculus*, vol. III, Descartes Press, Koriyama, Japan, 1989.

[9] K. Nishimoto, *Fractional Calculus*, vol. IV, Descartes Press, Koriyama, Japan, 1991.

[10] K. Nishimoto, *Fractional Calculus*, vol. V, Descartes Press, Koriyama, Japan, 1996.

[11] K. Nishimoto and S. S. de Romero, "N-fractional calculus operator N^{ν} method to nonhomogeneous and homogeneous Whittaker Equations," *Journal of Fractional Calculus*, vol. 9, pp. 17–22, 1996.

[12] K. Nishimoto, *An Essence of Nishimoto's Fractional Calculus (Calculus of the 21st Century): Integrations and Differentiations of Arbitrary Order*, Descartes Press, Koriyama, Japan, 1991.

[13] M. Coz and P. Rochus, "Translation kernels for velocity dependent interactions," *Journal of Mathematical Physics*, vol. 18, no. 11, pp. 2232–2240, 1976.

[14] A. Erdely, W. Magnus, F. Oberhettinger, and F. G. Tricomi, *Tables of Integral Transform*, vol. II, McGraw-Hill, New York, NY, USA, 1954.

[15] A. Erdely, "Axially symmetric potentials and fractional integration," *Journal of the Society for Industrial and Applied Mathematics*, vol. 13, pp. 216–228, 1965.

[16] A. Weinstein, "Generalized axially symmetric potential theory," *Bulletin of the American Mathematical Society*, vol. 59, pp. 20–38, 1953.

[17] B. Ross, *Fractional Calculus and Its Applications*, Conference Proceedings held at the University of New Haven, June 1974, Springer, New York, NY, USA, 1975.

[18] S. Salinas De Romero and H. M. Srivastava, "An application of the N-fractional calculus operator method to a modified Whittaker equation," *Applied Mathematics and Computation*, vol. 115, no. 1, pp. 11–21, 2000.

[19] H. M. Srivastava and R. G. Buschman, *Theory and Applications of Convolution Integral Equations*, vol. 79 of *Kluwer Series on Mathematics and Its Applications*, Kluwer Academic Publishers, Dordrecht, The Netherlands, 1992.

[20] H. M. Srivastava, S. Owa, and K. Nishimoto, "Some fractional differintegral equations," *Journal of Mathematical Analysis and Applications*, vol. 106, no. 2, pp. 360–366, 1985.

[21] H. Koyunbakan and E. S. Panakhov, "Transformation operator for singular Sturm-Liouville equations," *International Journal of Applied Mathematics*, vol. 14, pp. 135–143, 2003.

A New Linearized Crank-Nicolson Mixed Element Scheme for the Extended Fisher-Kolmogorov Equation

Jinfeng Wang,[1] Hong Li,[2] Siriguleng He,[2] Wei Gao,[2] and Yang Liu[2]

[1] *School of Statistics and Mathematics, Inner Mongolia University of Finance and Economics, Hohhot 010070, China*
[2] *School of Mathematical Sciences, Inner Mongolia University, Hohhot 010021, China*

Correspondence should be addressed to Hong Li; smslh@imu.edu.cn and Yang Liu; mathliuyang@aliyun.com

Academic Editors: W.-S. Du and T. Ozawa

We present a new mixed finite element method for solving the extended Fisher-Kolmogorov (EFK) equation. We first decompose the EFK equation as the two second-order equations, then deal with a second-order equation employing finite element method, and handle the other second-order equation using a new mixed finite element method. In the new mixed finite element method, the gradient ∇u belongs to the weaker $(L^2(\Omega))^2$ space taking the place of the classical $\mathbf{H}(\text{div};\Omega)$ space. We prove some a priori bounds for the solution for semidiscrete scheme and derive a fully discrete mixed scheme based on a linearized Crank-Nicolson method. At the same time, we get the optimal a priori error estimates in L^2 and H^1-norm for both the scalar unknown u and the diffusion term $w = -\Delta u$ and a priori error estimates in $(L^2)^2$-norm for its gradient $\chi = \nabla u$ for both semi-discrete and fully discrete schemes.

1. Introduction

In recent years, a lot of numerical methods for solving fourth-order partial differential equations have been presented and analyzed by many researchers. In [1–6], authors studied some mixed finite element methods for fourth-order elliptic and parabolic equations. Shi et al. [7], Wang et al. [8, 9], and H. R. Chen and S. C. Chen [10] proposed some nonconforming finite element methods for fourth-order elliptic equation (or biharmonic equation). In [11], some rectangular finite element methods for fourth-order elliptic singular perturbation problems were considered. Hu and Shi [12] studied the best L^2 norm error estimate of lower order finite element methods for the fourth-order problem. Chen and Wang [13] discussed a mixed finite element method for thin film epitaxy. In [14], a conforming finite element approximation for the fourth-order Steklov eigenvalue problem is discussed. In [15], a Crank-Nicolson time-stepping was used to approximate the differential term and the product trapezoidal method was employed to treat the integral term, and the quasi-wavelets numerical method for space discretization. Some numerical methods were proposed and studied for Cahn-Hilliard equations, such as (mixed) finite element methods [16–18], time-stepping methods [19, 20], spectral Galerkin method [21], discontinuous Galerkin method [22], and a conservative nonlinear difference scheme [23]. Liu et al. [24] studied a C^1-conforming finite element method for a fourth-order nonlinear hyperbolic equation. He et al. [25], proposed some mixed element schemes for fourth-order linear wave equation. In [26–28], some (mixed) finite element methods were studied for the extended Fisher-Kolmogorov equations.

In [29, 30], authors proposed and discussed a new mixed finite element scheme based on new mixed spaces for second-order linear elliptic equations. In the new mixed method, a weaker $(L^2(\Omega))^2$ space was provided to replace the complex $\mathbf{H}(\text{div};\Omega)$ space. Considering the advantage of the new mixed method, some problems have been studied, such as second-order Sobolev equation [31], second-order parabolic equations [32, 33], and reaction-convection-diffusion problems [34, 35]. However, the new mixed element method for the extended Fisher-Kolmogorov equation has not been studied in the literatures.

In this paper, our aim is to apply the new mixed method [29, 30] to solve the extended Fisher-Kolmogorov (EFK) equation [28]

$$u_t + \gamma\Delta^2 u - \Delta u + f(u) = 0, \quad (\mathbf{x}, t) \in \Omega \times J, \tag{1}$$

with initial condition

$$u(\mathbf{x}, 0) = u_0(\mathbf{x}), \quad \mathbf{x} \in \Omega, \tag{2}$$

and Dirichlet boundary conditions

$$u(\mathbf{x}, t) = \Delta u(\mathbf{x}, t) = 0, \quad (\mathbf{x}, t) \in \partial\Omega \times \overline{J}, \tag{3}$$

where Ω is a bounded convex polygonal domain in $R^d, d \le 2$ with boundary $\partial\Omega$, and $J = (0, T]$ is the time interval with $0 < T < \infty$. $u_0(\mathbf{x})$ is given function, $f(u) = u^3 - u$ coefficient γ is a positive constant.

For formulating the new mixed scheme, we first introduce a diffusion term $w = -\Delta u$ to get the following two lower order equations:

$$u_t - \gamma\Delta w + w = -f(u), \quad (\mathbf{x}, t) \in \Omega \times J, \tag{4a}$$

$$-\Delta u = w, \quad (\mathbf{x}, t) \in \Omega \times J. \tag{4b}$$

Then, we introduce another auxiliary variable $\boldsymbol{\chi} = \nabla u$ in (4b) to obtain the following lower-order system:

$$u_t - \gamma\Delta w + w = -f(u), \tag{5a}$$

$$-\nabla\cdot\boldsymbol{\chi} = w, \tag{5b}$$

$$\boldsymbol{\chi} - \nabla u = 0. \tag{5c}$$

For the lower-order system (5a), (5b), and (5c) we will formulate a new mixed scheme based on [29, 30] and discuss some a priori bounds for the mixed element solution and a priori error estimates for semidiscrete scheme. What is more, we will get the fully discrete scheme based on linearized Crank-Nicolson method and some a priori error estimates. From the obtained results, we can find easily that a priori error estimates in L^2 and H^1-norm for the scalar unknown u and the diffusion term w are optimal. Moreover, we also prove a priori error estimate in $(L^2)^2$-norm for the gradient term $\boldsymbol{\chi}$.

The remainder of this paper is organized as follows. In Section 2, a new mixed weak formulation and semidiscrete mixed scheme are formulated for system (5a), (5b), and (5c), then, some a priori bounds for the new mixed element solution are derived. In Section 3, a new mixed elliptic projection operator $\mathcal{Q}_h$ and a projection operator $\mathcal{S}_h$ associated with the coupled systems are presented and some a priori error estimates for semidiscrete scheme are proved. In Section 4, some a priori error estimates for fully discrete linearized Crank-Nicolson scheme are analyzed. Finally in Section 5, some concluding remarks and extensions about the new mixed finite element method are given. In this paper, $C > 0$ is a generic constant which does not depend on the spatial mesh parameter h or time step parameter Δt. At the same time, we denote the natural inner product in $L^2(\Omega)$ or $(L^2(\Omega))^2$ by $(\cdot,\cdot)$ with the corresponding norm $\|\cdot\|$. The others notations and definitions of Sobolev spaces as in [36, 37] are used. In order to facilitate the expression, we denote $B = B(\Omega)$, $\mathbf{F} = (F(\Omega))^d$, such as $H_0^1 \triangleq H_0^1(\Omega)$, $\mathbf{L}^2 \triangleq (L^2(\Omega))^2$, $\mathbf{H}^1 \triangleq (H^1(\Omega))^2$.

2. New Mixed Weak Formulation and Semidiscrete Scheme

In the following analysis, we will discuss the new mixed formulation and some a priori error estimates for semidiscrete scheme. Based on the lower-order system (5a), (5b), and (5c) the new mixed weak formulation is to find $\{u, w, \boldsymbol{\chi}\} : [0, T] \mapsto H_0^1 \times H_0^1 \times \mathbf{L}^2$ such that

$$(u_t, v) + \gamma(\nabla w, \nabla v) + (w, v) + (f(u), v) = 0, \quad \forall v \in H_0^1, \tag{6a}$$

$$(\boldsymbol{\chi}, \nabla g) = (w, g), \quad \forall g \in H_0^1, \tag{6b}$$

$$(\boldsymbol{\chi}, \mathbf{z}) - (\nabla u, \mathbf{z}) = 0, \quad \forall \mathbf{z} \in \mathbf{L}^2. \tag{6c}$$

Let H_h and $\mathbf{L}_h$, respectively, be finite dimensional subspaces of H_0^1 and $\mathbf{L}^2$ defined by the finite element pair $P_1 - P_0^2$ [29, 30]

$$H_h = \left\{v_h \in C^0(\Omega) \cap H_0^1 \mid v_h \in P_1(K), \forall K \in \mathscr{K}_h\right\},$$
$$\mathbf{L}_h = \left\{\mathbf{w}_h = (w_{1h}, w_{2h}) \in \mathbf{L}^2 \mid w_{1h}, w_{2h} \in P_0(K), \forall K \in \mathscr{K}_h\right\}. \tag{7}$$

Then, the semidiscrete mixed scheme for (6a), (6b), and (6c) is to determine $\{u_h, w_h, \boldsymbol{\chi}_h\} : [0, T] \mapsto H_h \times H_h \times \mathbf{L}_h$ such that

$$(u_{ht}, v_h) + \gamma(\nabla w_h, \nabla v_h) + (w_h, v_h) + (f(u_h), v_h) = 0, \quad \forall v_h \in H_h, \tag{8a}$$

$$(\boldsymbol{\chi}_h, \nabla g_h) = (w_h, g_h), \quad \forall g_h \in H_h, \tag{8b}$$

$$(\boldsymbol{\chi}_h, \mathbf{z}_h) - (\nabla u_h, \mathbf{z}_h) = 0, \quad \forall \mathbf{z}_h \in \mathbf{L}_h. \tag{8c}$$

In the following discussion, we derive some a priori bounds for the finite element solution.

Theorem 1. *Let $\{u_h, w_h, \boldsymbol{\chi}_h\}$ be the solution for system* (8a), (8b), *and* (8c) *then there exists a constant $C > 0$ such that, for $1 \le p < +\infty$*

$$\|u_h\|_{L^p} + \|u_h\|_1 + \|w_h\| + \|\boldsymbol{\chi}_h\| + \left(\int_0^t \|u_{ht}\|^2 ds\right)^{1/2} \le C(\gamma, \|u_{0h}\|, \|w_{0h}\|). \tag{9}$$

Proof. Differentiating (8b) and (8c) with respect to t, we obtain

$$(u_{ht}, v_h) + \gamma (\nabla w_h, \nabla v_h) + (w_h, v_h) + (f(u_h), v_h) = 0, \quad \forall v_h \in H_h, \tag{10a}$$

$$-(\chi_{ht}, \nabla g_h) + (w_{ht}, g_h), \quad \forall g_h \in H_h, \tag{10b}$$

$$(\chi_{ht}, \mathbf{z}_h) - (\nabla u_{ht}, \mathbf{z}_h) = 0, \quad \forall \mathbf{z}_h \in \mathbf{L}_h. \tag{10c}$$

Take $v_h = u_{ht}$, $g_h = w_h$, $\mathbf{z}_h = \nabla w_h$ in (10a), (10b), and (10c) to obtain

$$\| u_{ht}\|^2 + \gamma (\nabla w_h, \nabla u_{ht}) + (w_h, u_{ht}) + (f(u_h), u_{ht}) = 0, \tag{11a}$$

$$-\gamma (\chi_{ht}, \nabla w_h) + \frac{\gamma}{2}\frac{d}{dt}\|w_h\|^2 = 0, \tag{11b}$$

$$\gamma (\chi_{ht}, \nabla w_h) - \gamma (\nabla u_{ht}, \nabla w_h) = 0. \tag{11c}$$

We define $F(\sigma) \triangleq (1/4)(1-\sigma^2)^2$ and add the three equations for system (11a), (11b), and (11c) to get

$$\begin{aligned} &\frac{d}{dt}\left(\frac{\gamma}{2}\|w_h\|^2 + (F(u_h), 1)\right) + \|u_{ht}\|^2 \\ &= -(w_h, u_{ht}) \le \frac{1}{2}\|w_h\|^2 + \frac{1}{2}\|u_{ht}\|^2. \end{aligned} \tag{12}$$

Integrate (12) with respect to time from 0 to t and use Gronwall lemma to have

$$\begin{aligned} &\frac{\gamma}{2}\|w_h\|^2 + (F(u_h), 1) + \int_0^t \|u_{ht}\|^2 ds \\ &\le C\left(\gamma\|w_h(0)\|^2 + (F(u_h(0)), 1)\right). \end{aligned} \tag{13}$$

We note that $F(u_h) \ge 0$ to obtain

$$\|u_h\| + \|w_h\| \le C(\gamma, \|u_{0h}\|, \|w_{0h}\|). \tag{14}$$

Take $g_h = u_h$ in (8b) and $\mathbf{z}_h = \nabla u_h$ in (8c) to obtain

$$\|\nabla u_h\| \le C(\|u_h\| + \|w_h\|). \tag{15}$$

Substitute (14) into (15) to obtain

$$\|\nabla u_h\| \le C(\gamma, \|u_{0h}\|, \|w_{0h}\|). \tag{16}$$

Choose $\mathbf{z}_h = \chi_h$ in (8c) and use (16) to obtain

$$\|\chi_h\| \le \|\nabla u_h\| \le C(\gamma, \|u_{0h}\|, \|w_{0h}\|). \tag{17}$$

For $\Omega \subset R^2$, we use Sobolev imbedding theorem to get

$$\|u_h\|_{L^p} \le \|u_h\|_1 \le C(\gamma, \|u_{0h}\|, \|w_{0h}\|). \tag{18}$$

By (14), (16)–(18), we accomplish the proof of Theorem 1. □

Remark 2. Based on the discussion in [29, 30], we can see easily that the discrete LBB condition for the mixed finite element space $(H_h, \mathbf{L}_h)$ is satisfied. In our weak formulation (6a), (6b), and (6c) the $\mathbf{H}(\text{div}; \Omega)$ space is replaced by the weaker $\mathbf{L}^2$ space. Compared to the $\mathbf{H}(\text{div}; \Omega)$ space, the regularity requirement for our $\mathbf{L}^2$ space is reduced.

3. A Priori Error Estimates for Semidiscrete Scheme

In order to analyze the convergence of the method, we first introduce the projection operator $\mathcal{S}_h$ and a new mixed elliptic projection operator $\mathcal{Q}_h$ associated with the coupled equations.

Lemma 3. *Let $\mathcal{S}_h : [0,T] \to H_0^1 \to H_h$ be satisfied by the following relation:*

$$\begin{aligned} &(\nabla (w - \mathcal{S}_h w), \nabla v_h) = 0, \quad \forall v_h \in H_h, \\ &\|w - \mathcal{S}_h w\|_{L^2(\Omega)} + h\|w - \mathcal{S}_h w\|_{H^1} \le Ch^2\|w\|_{H^2}. \end{aligned} \tag{19}$$

Lemma 4. *Let $(\mathcal{Q}_h u, \mathcal{Q}_h \chi) : [0,T] \to H_h \times \mathbf{L}_h$ be given by the following new mixed relations:*

$$-(\chi - \mathcal{Q}_h\chi, \nabla g_h) = 0, \quad \forall g_h \in H_h, \tag{20a}$$

$$(\chi - \mathcal{Q}_h\chi, \mathbf{z}_h) - (\nabla (u - \mathcal{Q}_h u), \mathbf{z}_h) = 0, \quad \forall \mathbf{z}_h \in \mathbf{L}_h; \tag{20b}$$

there exists a constant $C > 0$ independent of h such that

$$\begin{aligned} &\|u - \mathcal{Q}_h u\| + h(\|\chi - \mathcal{Q}_h\chi\| + \|u - \mathcal{Q}_h u\|_1) \\ &\le Ch^2(\|u\|_{H^2} + \|\chi\|_{\mathbf{H}^1}), \\ &\|u_t - \mathcal{Q}_h u_t\| \le Ch^2(\|u_t\|_{H^2} + \|\chi_t\|_{\mathbf{H}^1}), \end{aligned} \tag{21}$$

from [29, 30], we can obtain the proof for Lemma 4.

Further, assuming that the triangulation is quasi-uniform, the following error estimates [27] hold, for $j = 0, 1$ and $p = 2, \infty$:

$$\|\mathcal{Q}_h u\|_{W^{j,p}} \le \|\tau\|_{W^{j,p}} + \|u\|_{W^{j,p}} \le C_u. \tag{22}$$

Based on Lemmas 3 and 4, the following theorem for semidiscrete error estimates is obtained.

Theorem 5. *With $u_h(0) = \mathcal{Q}_h u_0$, assume that the solution's regular properties for system (6a), (6b), and (6c) satisfy $w, u, u_t \in L^\infty(H^2)$, $\chi, \chi_t \in L^\infty(\mathbf{H}^1)$. Then there exists a positive constant $C > 0$ independent of h such that*

$$\begin{aligned} &\|u - u_h\| + \|w - w_h\| + h(\|u - u_h\|_1 + \|\chi - \chi_h\|) \\ &\le Ch^2\Bigg[\|u\|_{L^\infty(H^2)} + \|\chi\|_{L^\infty(\mathbf{H}^1)} \\ &\quad + \left(\int_0^t \left(\|u\|_{H^2}^2 + \|u_t\|_{H^2}^2 + \|\chi\|_{\mathbf{H}^1}^2 \right.\right. \\ &\quad \left.\left. + \|\chi_t\|_{\mathbf{H}^1}^2 + \|w\|_{H^2}^2\right) ds\right)^{1/2}\Bigg]. \end{aligned} \tag{23}$$

Proof. To get a priori error estimates, we decompose the errors as

$$\begin{aligned} w - w_h &= w - \mathcal{S}_h w + \mathcal{S}_h w - w_h = \pi + \kappa; \\ u - u_h &= u - \mathcal{Q}_h u + \mathcal{Q}_h u - u_h = \tau + \vartheta; \\ \chi - \chi_h &= \chi - \mathcal{Q}_h\chi + \mathcal{Q}_h\chi - \chi_h = \lambda + \zeta. \end{aligned} \tag{24}$$

Combining (6a)–(8c), (19) and (20a) and (20b), we can get the error equations

$$(\vartheta_t, v_h) + \gamma(\nabla\kappa, \nabla v_h) + (\kappa, v_h) + (f(u) - f(u_h), v_h) = -(\tau_t, v_h) - (\pi, v_h), \quad \forall v_h \in H_h, \tag{25a}$$

$$-(\boldsymbol{\zeta}, \nabla g_h) + (\kappa, g_h) = -(\pi, g_h), \quad \forall g_h \in H_h, \tag{25b}$$

$$(\boldsymbol{\zeta}, \mathbf{z}_h) - (\nabla\vartheta, \mathbf{z}_h) = 0, \quad \forall \mathbf{z}_h \in \mathbf{L}_h. \tag{25c}$$

Differentiating (25b) and (25c) with respect to t, we get

$$(\vartheta_t, v_h) + \gamma(\nabla\kappa, \nabla v_h) + (\kappa, v_h) + (f(u) - f(u_h), v_h) = -(\tau_t, v_h) - (\pi, v_h), \quad \forall v_h \in H_h, \tag{26a}$$

$$-(\boldsymbol{\zeta}_t, \nabla g_h) + (\kappa_t, g_h) = -(\pi_t, g_h), \quad \forall g_h \in H_h, \tag{26b}$$

$$(\boldsymbol{\zeta}_t, \mathbf{z}_h) - (\nabla\vartheta_t, \mathbf{z}_h) = 0, \quad \forall \mathbf{z}_h \in \mathbf{L}_h. \tag{26c}$$

Choose $v_h = \vartheta_t$, $g_h = \kappa$, and $\mathbf{z}_h = \nabla\kappa$ in (26a), (26b), and (26c), respectively, to obtain

$$\|\vartheta_t\|^2 + \gamma(\nabla\kappa, \nabla\vartheta_t) + (\kappa, \vartheta_t) + (f(u) - f(u_h), \vartheta_t) = -(\tau_t, \vartheta_t) - (\pi, \vartheta_t), \tag{27a}$$

$$\frac{1}{2}\frac{d}{dt}\|\kappa\|^2 - (\boldsymbol{\zeta}_t, \nabla\kappa) = -(\pi_t, \kappa), \tag{27b}$$

$$(\boldsymbol{\zeta}_t, \nabla\kappa) - (\nabla\vartheta_t, \nabla\kappa) = 0. \tag{27c}$$

Using (27a), (27b), and (27c) and Cauchy-Schwarz inequality and Young inequality, we have

$$\begin{aligned}
&\|\vartheta_t\|^2 + (f(u) - f(u_h), \vartheta_t) \\
&= -\gamma(\nabla\kappa, \nabla\vartheta_t) - (\kappa, \vartheta_t) - (\tau_t, \vartheta_t) - (\pi, \vartheta_t) \\
&= -\gamma(\boldsymbol{\zeta}_t, \nabla\kappa) - (\kappa, \vartheta_t) - (\tau_t, \vartheta_t) - (\pi, \vartheta_t) \\
&= -\frac{\gamma}{2}\frac{d}{dt}\|\kappa\|^2 - (\pi_t, \kappa) - (\kappa, \vartheta_t) - (\tau_t, \vartheta_t) - (\pi, \vartheta_t) \\
&\le -\frac{\gamma}{2}\frac{d}{dt}\|\kappa\|^2 + C\left(\|\pi_t\|^2 + \|\tau_t\|^2 + \|\pi\|^2\right) \\
&\quad + \|\kappa\|^2 + \frac{1}{4}\|\vartheta_t\|^2.
\end{aligned} \tag{28}$$

Using Cauchy-Schwarz inequality, Young inequality, Sobolev imbedding theorem, and inequality (22), we use the similar method as the one in [27] to obtain

$$\begin{aligned}
&-(f(u) - f(u_h), \vartheta_t) \\
&= -(f(u) - f(Q_h u), \vartheta_t) - (f(Q_h u) - f(u_h), \vartheta_t) \\
&= -(f(u) - f(Q_h u), \vartheta_t) - (f(\vartheta), \vartheta_t) - 3(u_h Q_h u \vartheta, \vartheta_t) \\
&\le \left|\left(\tau\left(u^2 + u Q_h u + (Q_h u)^2 - 1\right), \vartheta_t\right)\right| - \frac{d}{dt}(F(\vartheta), 1) \\
&\quad - 3(u_h Q_h u \vartheta, \vartheta_t) \le C\left(\|\tau\|^2 + \|\vartheta\|^2\right) \\
&\quad - \frac{d}{dt}(F(\vartheta), 1) + \frac{1}{4}\|\vartheta_t\|^2.
\end{aligned} \tag{29}$$

Substituting (29) into (28), we have

$$\begin{aligned}
&\frac{d}{dt}\left[\frac{\gamma}{2}\|\kappa\|^2 + (F(\vartheta), 1)\right] + \frac{1}{2}\|\vartheta_t\|^2 \\
&\le C\left(\|\tau_t\|^2 + \|\tau\|^2 + \|\pi\|^2 + \|\pi_t\|^2 + \|\vartheta\|^2 + \|\kappa\|^2\right).
\end{aligned} \tag{30}$$

Integrate (30) with respect to time from 0 to t to obtain

$$\begin{aligned}
&\frac{\gamma}{2}\|\kappa\|^2 + (F(\vartheta), 1) + \frac{1}{2}\int_0^t \|\vartheta_t\|^2 ds \\
&\le C\int_0^t \left(\|\tau_t\|^2 + \|\tau\|^2 + \|\pi\|^2 + \|\pi_t\|^2 + \|\vartheta\|^2 + \|\kappa\|^2\right) ds.
\end{aligned} \tag{31}$$

Using Gronwall lemma, we have

$$\begin{aligned}
&\frac{\gamma}{2}\|\kappa\|^2 + \|\vartheta\|^2 + \frac{1}{2}\int_0^t \|\vartheta_t\|^2 ds \\
&\le C\int_0^t \left(\|\tau_t\|^2 + \|\tau\|^2 + \|\pi\|^2 + \|\pi_t\|^2\right) ds.
\end{aligned} \tag{32}$$

Taking $g_h = \vartheta$ in (25b) and $\mathbf{z}_h = \nabla\vartheta$ in (25c), we have

$$(\boldsymbol{\zeta}, \nabla\vartheta) - (\kappa, \vartheta) = (\pi, \vartheta), \tag{33a}$$

$$\|\nabla\vartheta\|^2 - (\boldsymbol{\zeta}, \nabla\vartheta) = 0. \tag{33b}$$

Add the two equations to get

$$\|\nabla\vartheta\|^2 \le C\left(\|\kappa\|^2 + \|\vartheta\|^2 + \|\pi\|^2\right). \tag{34}$$

Choose $\mathbf{z}_h = \boldsymbol{\zeta}$ in (25c) to obtain

$$\|\boldsymbol{\zeta}\|^2 \le \|\nabla\vartheta\|^2 \le C\left(\|\kappa\|^2 + \|\vartheta\|^2 + \|\pi\|^2\right). \tag{35}$$

Substitute (32) into (34) and (35) to get

$$\begin{aligned}
&\|\boldsymbol{\zeta}\|^2 + \|\nabla\vartheta\|^2 \\
&\le C\left(\|\pi\|^2 + \int_0^t \left(\|\tau_t\|^2 + \|\tau\|^2 + \|\pi\|^2 + \|\pi_t\|^2\right) ds\right).
\end{aligned} \tag{36}$$

Combining Lemmas 3 and 4, (32)–(36) with the triangle inequality, we obtain the error estimates for Theorem 5. □

Remark 6. The conclusion for Theorem 5 demonstrates that the optimal convergence order in L^2-norm for both the scalar unknown u and the diffusion term w is obtained. At the same time, the optimal convergence order in H^1-norm for the scalar unknown u is gotten, too.

Theorem 7. *Assume that the solution's regular properties for system (6a), (6b), and (6c) satisfy $u, w, u_t \in L^\infty(H^2)$ and $\boldsymbol{\chi}, \boldsymbol{\chi}_t \in L^\infty(\mathbf{H}^1)$. Then there exists a positive constant C independent of h such that*

$$\left\|w - w_h\right\|_1 \leq Ch\left[\|u\|_{L^\infty(H^2)} + \|w\|_{L^\infty(H^2)} + \|\boldsymbol{\chi}\|_{L^\infty(\mathbf{H}^1)} + \left(\int_0^t \left(\|u\|_{H^2}^2 + \|u_t\|_{H^2}^2 + \|\boldsymbol{\chi}\|_{\mathbf{H}^1}^2 + \|\boldsymbol{\chi}_t\|_{\mathbf{H}^1}^2 + \|w\|_{H^2}^2\right) ds\right)^{1/2}\right]. \tag{37}$$

Proof. Choose $v_h = \kappa_t$, $g_h = \vartheta_t$, and $\mathbf{z}_h = \boldsymbol{\zeta}_t$ in (26a), (26b), and (26c), respectively, to obtain

$$(\vartheta_t, \kappa_t) + \frac{\gamma}{2}\frac{d}{dt}\|\kappa\|_1^2 + (f(u) - f(u_h), \kappa_t) = -(\tau_t, \kappa_t) - (\pi, \kappa_t), \tag{38a}$$

$$(\boldsymbol{\zeta}_t, \nabla\vartheta_t) - (\kappa_t, \vartheta_t) = (\pi_t, \vartheta_t), \tag{38b}$$

$$\|\boldsymbol{\zeta}_t\|^2 - (\nabla\vartheta_t, \boldsymbol{\zeta}_t) = 0. \tag{38c}$$

Adding the three equations of system (38a), (38b), and (38c), we have

$$\begin{aligned}
&\frac{\gamma}{2}\frac{d}{dt}\|\kappa\|_1^2 + \|\boldsymbol{\zeta}_t\|^2 \\
&= -(f(u) - f(u_h), \kappa_t) - (\tau_t, \kappa_t) - (\pi, \kappa_t) + (\pi_t, \vartheta_t) \\
&= -\frac{d}{dt}(f(u) - f(u_h), \kappa) + (f'(u)u_t - f'(u_h)u_{ht}, \kappa) \\
&\quad - \frac{d}{dt}(\tau_t, \kappa) + (\tau_{tt}, \kappa) - \frac{d}{dt}(\pi, \kappa) + (\pi_t, \kappa) + (\pi_t, \vartheta_t).
\end{aligned} \tag{39}$$

Integrate (39) with respect to t and use Cauchy-Schwarz inequality and Young inequality to get

$$\begin{aligned}
&\frac{\gamma}{2}\|\kappa\|_1^2 + \int_0^t \|\boldsymbol{\zeta}_t\|^2 ds \\
&= -(f(u) - f(u_h), \kappa) \\
&\quad + \int_0^t (f'(u)u_t - f'(u_h)u_{ht}, \kappa)\, ds \\
&\quad - (\tau_t, \kappa) + \int_0^t (\tau_{tt}, \kappa)\, ds - (\pi, \kappa) + \int_0^t (\pi_t, \kappa)\, ds \\
&\quad + \int_0^t (\pi_t, \vartheta_t)\, ds \leq -(f(u) - f(u_h), \kappa) \\
&\quad + \int_0^t (f'(u)u_t - f'(u_h)u_{ht}, \kappa)\, ds, \\
&C\left[\|\tau_t\|^2 + \|\pi\|^2 + \int_0^t \left(\|\tau_{tt}\|^2 + \|\pi_t\|^2 + \|\kappa\|^2 + \|\vartheta_t\|^2\right) ds\right] \\
&\quad + \frac{\gamma}{8}\|\kappa\|^2.
\end{aligned} \tag{40}$$

Using the similar method to the estimate for (29), we obtain

$$\begin{aligned}
&-(f(u) - f(u_h), \kappa) \\
&= -(f(u) - f(R_h u), \kappa) - (f(R_h u) - f(u_h), \kappa) \\
&\leq \left|\left(\tau\left(u^2 + uR_h u + (R_h u)^2 - 1\right), \kappa\right)\right| \\
&\quad + \left(f'(\widehat{u}_h)(R_h u - u_h), \kappa\right) \\
&= \left|\left(\tau\left(u^2 + uR_h u + (R_h u)^2 - 1\right), \kappa\right)\right| \\
&\quad + \left(\left(3(\widehat{u}_h)^2 - 1\right)\vartheta, \kappa\right) \\
&\leq C\left(\|\tau\|^2 + \|\vartheta\|^2\right) + \frac{\gamma}{8}\|\kappa\|_1^2, \\
&\int_0^t (f'(u)u_t - f'(u_h)u_{ht}, \kappa)\, ds \\
&= \int_0^t \left(f'(u)(u_t - u_{ht}) + (f'(u) - f'(u_h))u_{ht}, \kappa\right) ds \\
&= \int_0^t \left(f'(u)(\vartheta_t + \tau_t) + f''(u_1)(\vartheta + \tau)u_{ht}, \kappa\right) ds \\
&\leq C\int_0^t \left(\|\tau\|^2 + \|\vartheta\|^2 + \|\tau_t\|^2 + \|\vartheta_t\|^2 + \|\kappa\|_1^2\right) ds.
\end{aligned} \tag{41}$$

Substitute (41) into (40) to get

$$\begin{aligned}&\frac{\gamma}{4}\|\kappa\|_1^2+\int_0^t\|\zeta_t\|^2ds\\&\le C\Big[\|\tau_t\|^2+\|\pi\|^2\\&\quad+\int_0^t\big(\|\tau\|^2+\|\tau_t\|^2+\|\tau_{tt}\|^2+\|\pi_t\|^2\\&\quad+\|\kappa\|_1^2+\|\vartheta\|^2+\|\vartheta_t\|^2\big)ds\Big].\end{aligned}\tag{42}$$

Substitute (32) into (42) and use Gronwall lemma to get

$$\begin{aligned}&\frac{\gamma}{4}\|\kappa\|_1^2+\int_0^t\|\zeta_t\|^2ds\\&\le C\Big[\|\tau_t\|^2+\|\pi\|^2\\&\quad+\int_0^t\big(\|\tau\|^2+\|\tau_t\|^2+\|\tau_{tt}\|^2+\|\pi\|^2+\|\pi_t\|^2\big)ds\Big].\end{aligned}\tag{43}$$

Combining Lemmas 3 and 4, (43) with the triangle inequality, we obtain the error estimates for Theorem 7. □

Remark 8. In Theorem 7, the optimal convergence order in H^1-norm for the diffusion term w is obtained.

4. A Priori Error Estimates Based on Linearized C-N Scheme

4.1. Linearized Crank-Nicolson Mixed Scheme. In the following discussion, we will derive the fully discrete a priori error estimates based on a linearized Crank-Nicolson method. Let $0=t_0<t_1<t_2<\cdots<t_M=T$ be a given partition of the time interval $[0,T]$ with step length $\Delta t=T/M$ and nodes $t_n=n\Delta t$, for some positive integer M. For a smooth function ϕ on $[0,T]$, define $\psi^n=\phi(t_n)$ and $\psi^{n-1/2}=(\psi^n+\psi^{n-1})/2$.

Equations (6a), (6b), and (6c) can be written as the following formulation at time $t=t_{n-1/2}$:

$$\begin{aligned}&(u_t(t_{n-1/2}),v)+\gamma(\nabla w(t_{n-1/2}),\nabla v)\\&\quad+(w(t_{n-1/2}),v)+(f(u(t_{n-1/2})),v)=0,\quad \forall v\in H_0^1,\end{aligned}\tag{44a}$$

$$(\chi(t_{n-1/2}),\nabla g)-(w(t_{n-1/2}),g)=0,\quad \forall g\in H_0^1m,\tag{44b}$$

$$(\chi(t_{n-1/2}),\mathbf{z})-(\nabla u(t_{n-1/2}),\mathbf{z})=0,\quad \forall\mathbf{z}\in\mathbf{L}^2.\tag{44c}$$

Then, the following equivalent formulation for (44a), (44b), and (44c) is as follows:

$$\begin{aligned}&\left(\frac{u^n-u^{n-1}}{\Delta t},v\right)+\gamma\left(\nabla w^{n-1/2},\nabla v\right)\\&\quad+\left(w^{n-1/2},v\right)+\left(f\left(\widehat{u}^{n-1/2}\right),v\right)\\&\quad=\left(R_1^{n-1/2},v\right)+\left(R_2^{n-1/2},v\right),\quad \forall v\in H_0^1,\end{aligned}\tag{45a}$$

$$\left(\chi^{n-1/2},\nabla g\right)-\left(w^{n-1/2},g\right)=0,\quad \forall g\in H_0^1,\tag{45b}$$

$$\left(\chi^{n-1/2},\mathbf{z}\right)-\left(\nabla u^{n-1/2},\mathbf{z}\right)=0,\quad \forall\mathbf{z}\in\mathbf{L}^2,\tag{45c}$$

where

$$\begin{aligned}\widehat{u}^{n-1/2}&=\frac{3}{2}u^{n-1}-\frac{1}{2}u^{n-2},\\R_1^{n-1/2}&=\frac{u^n-u^{n-1}}{\Delta t}-u_t(t_{n-1/2}),\\R_2^{n-1/2}&=f\left(\widehat{u}^{n-1/2}\right)-f(u(t_{n-1/2})).\end{aligned}\tag{46}$$

Now a linearized C-N fully discrete procedure is to find $(u_h^n,w_h^n,\chi_h^n)\in H_h\times H_h\times\mathbf{L}_h$, $(n=0,1,\ldots,M)$ such that

$$\begin{aligned}&\left(\frac{u_h^n-u_h^{n-1}}{\Delta t},v_h\right)+\gamma\left(\nabla w_h^{n-1/2},\nabla v_h\right)\\&\quad+\left(w_h^{n-1/2},v_h\right)+\left(f\left(\widehat{u}_h^{n-1/2}\right),v_h\right)=0,\quad \forall v_h\in H_h,\end{aligned}\tag{47a}$$

$$\left(\chi_h^{n-1/2},\nabla g_h\right)-\left(w_h^{n-1/2},g_h\right)=0,\quad \forall g_h\in H_h,\tag{47b}$$

$$\left(\chi_h^{n-1/2},\mathbf{z}_h\right)-\left(\nabla u_h^{n-1/2},\mathbf{z}_h\right)=0,\quad \forall\mathbf{z}_h\in\mathbf{L}_h.\tag{47c}$$

Remark 9. We can find that the system (47a), (47b), and (47c) is a linear scheme by a linearized term

$$f\left(\widehat{u}_h^{n-1/2}\right)=\left(\frac{3}{2}u_h^{n-1}-\frac{1}{2}u_h^{n-2}\right)^3-\left(\frac{3}{2}u_h^{n-1}-\frac{1}{2}u_h^{n-2}\right).\tag{48}$$

In the following subsection, we derive a priori error estimates based on a linearized Crank-Nicolson fully discrete scheme.

4.2. A Priori Error Estimates for Fully Discrete Scheme. In order to derive the linearized C-N fully discrete error estimates, we now write the errors as follows:

$$\begin{aligned}w(t_n)-w_h^n&=w(t_n)-\mathcal{S}_hw^n+\mathcal{S}_hw^n-w_h^n=\pi^n+\kappa^n;\\u(t_n)-u_h^n&=u(t_n)-\mathcal{Q}_hu^n+\mathcal{Q}_hu^n-u_h^n=\tau^n+\vartheta^n;\\\chi(t_n)-\chi_h^n&=\chi(t_n)-\mathcal{Q}_h\chi^n+\mathcal{Q}_h\chi^n-\chi_h^n=\lambda^n+\zeta^n.\end{aligned}\tag{49}$$

Subtracting (47a), (47b), and (47c) from (45a), (45b), and (45c) and using (19) and (20a) and (20b) at $t = t_{n-1/2}$, we have the following error equations

$$\left(\frac{\vartheta^n - \vartheta^{n-1}}{\Delta t}, v_h\right) + \gamma\left(\nabla\kappa^{n-1/2}, \nabla v_h\right) + \left(\kappa^{n-1/2}, v_h\right) + \left(f\left(\widehat{u}^{n-1/2}\right) - f\left(\widehat{u}_h^{n-1/2}\right), v_h\right) = -\left(\frac{\tau^n - \tau^{n-1}}{\Delta t}, v_h\right) - \left(\pi^{n-1/2}, v_h\right) + \left(R_1^{n-1/2}, v_h\right) + \left(R_2^{n-1/2}, v_h\right), \quad \forall v_h \in H_h, \tag{50a}$$

$$\left(\zeta^{n-1/2}, \nabla g_h\right) - \left(\kappa^{n-1/2}, g_h\right) = \left(\pi^{n-1/2}, g_h\right), \quad \forall g_h \in H_h, \tag{50b}$$

$$\left(\zeta^{n-1/2}, \mathbf{z}_h\right) - \left(\nabla\vartheta^{n-1/2}, \mathbf{z}_h\right) = 0, \quad \forall \mathbf{z}_h \in \mathbf{L}_h. \tag{50c}$$

In order to get the fully discrete error estimates, we introduce the following lemma.

Lemma 10. *For $R_1^{n-1/2}$ and $R_2^{n-1/2}$, the following estimates hold:*

$$\begin{aligned} \left\|R_1^{n-1/2}\right\| &\le C\Delta t^2 \|u_{ttt}\|_{L^\infty(L^2)}, \\ \left\|R_2^{n-1/2}\right\| &\le C(u)\,\Delta t^2 \|u_{ttt}\|_{L^\infty(L^2)}. \end{aligned} \tag{51}$$

Proof. Using the Taylor expansion, we can obtain easily the conclusion for Lemma 10. □

Remark 11. From the estimate for $\|R_2^{n-1/2}\|$ in Lemma 10, we know that the convergence rate in time direction for the system (47a), (47b), and (47c) is order 2, so the system (47a), (47b), and (47c) is called a linearized Crank-Nicolson scheme based on the linearized term $f(\widehat{u}_h^{n-1/2})$.

In the following discussion, we will derive some fully discrete a priori error estimates.

Theorem 12. *With $u_h^0 = \mathcal{Q}_h u^0$, assume that the solution's regular properties for system* (6a), (6b), *and* (6c) *satisfy $u_t \in L^2(H^2)$, $u_{ttt} \in L^\infty(L^2)$, $u, w \in L^\infty(H^2)$, $\boldsymbol{\chi} \in L^\infty(\mathbf{H}^1)$, $\boldsymbol{\chi}_t \in L^2(\mathbf{H}^1)$. Then there exists a positive constant C independent of h and Δt such that*

$$\left\|u(t_J) - u_h^J\right\| \le C\left[h^2 |||\cdot||| + \Delta t^2 \|u_{ttt}\|_{L^\infty(L^2)}\right], \tag{52}$$

$|||\cdot||| \triangleq \|u\|_{L^\infty(H^2)} + \|u_t\|_{L^2(H^2)} + \|w\|_{L^\infty(H^2)} + \|\boldsymbol{\chi}\|_{L^\infty(\mathbf{H}^1)} + \|\boldsymbol{\chi}_t\|_{L^2(\mathbf{H}^1)}$.

Proof. Setting $v_h = \vartheta^{n-1/2}$, $g_h = \kappa^{n-1/2}$, and $\mathbf{z}_h = \nabla\kappa^{n-1/2}$ in (50a), (50b), and (50c), respectively, we get

$$\frac{\|\vartheta^n\|^2 - \|\vartheta^{n-1}\|^2}{2\Delta t} + \gamma\left(\nabla\kappa^{n-1/2}, \nabla\vartheta^{n-1/2}\right) + \left(\kappa^{n-1/2}, \vartheta^{n-1/2}\right) + \left(f\left(\widehat{u}^{n-1/2}\right) - f\left(\widehat{u}_h^{n-1/2}\right), \vartheta^{n-1/2}\right) = -\left(\frac{\tau^n - \tau^{n-1}}{\Delta t}, \vartheta^{n-1/2}\right) - \left(\pi^{n-1/2}, \vartheta^{n-1/2}\right) + \left(R_1^{n-1/2}, \vartheta^{n-1/2}\right) + \left(R_2^{n-1/2}, \vartheta^{n-1/2}\right), \tag{53a}$$

$$\gamma\left\|\kappa^{n-1/2}\right\|^2 - \gamma\left(\zeta^{n-1/2}, \nabla\kappa^{n-1/2}\right) = -\gamma\left(\pi^{n-1/2}, \kappa^{n-1/2}\right), \tag{53b}$$

$$\gamma\left(\zeta^{n-1/2}, \nabla\kappa^{n-1/2}\right) - \gamma\left(\nabla\vartheta^{n-1/2}, \nabla\kappa^{n-1/2}\right) = 0. \tag{53c}$$

Adding the three equations of (53a), (53b), and (53c) and using Cauchy-Schwarz inequality and Young inequality, we have

$$\begin{aligned} &\gamma\left\|\kappa^{n-1/2}\right\|^2 + \frac{1}{2\Delta t}\left(\|\vartheta^n\|^2 - \|\vartheta^{n-1}\|^2\right) \\ &= -\left(\kappa^{n-1/2}, \vartheta^{n-1/2}\right) - \left(f\left(\widehat{u}^{n-1/2}\right) - f\left(\widehat{u}_h^{n-1/2}\right), \vartheta^{n-1/2}\right) \\ &\quad - \left(\frac{\tau^n - \tau^{n-1}}{\Delta t}, \vartheta^{n-1/2}\right) \\ &\quad - \left(\pi^{n-1/2}, \vartheta^{n-1/2}\right) + \left(R_1^{n-1/2}, \vartheta^{n-1/2}\right) + \left(R_2^{n-1/2}, \vartheta^{n-1/2}\right) \\ &\quad - \gamma\left(\pi^{n-1/2}, \kappa^{n-1/2}\right) \\ &\le \left(\left\|\frac{\tau^n - \tau^{n-1}}{\Delta t}\right\| + \left\|R_1^{n-1/2}\right\| + \left\|R_2^{n-1/2}\right\|\right)\left\|\vartheta^{n-1/2}\right\| \\ &\quad + \left\|\kappa^{n-1/2}\right\|\left\|\vartheta^{n-1/2}\right\| + \gamma\left\|\pi^{n-1/2}\right\|\left\|\kappa^{n-1/2}\right\| \\ &\quad + \left(\left\|f\left(\widehat{u}^{n-1/2}\right) - f\left(\widehat{u}_h^{n-1/2}\right)\right\| + \left\|\pi^{n-1/2}\right\|\right)\left\|\vartheta^{n-1/2}\right\| \\ &\le C\left(\left\|\frac{\tau^n - \tau^{n-1}}{\Delta t}\right\|^2 + \left\|R_1^{n-1/2}\right\|^2 + \left\|R_2^{n-1/2}\right\|^2 + \left\|\pi^{n-1/2}\right\|^2\right) \\ &\quad + \frac{\gamma}{2}\left\|\kappa^{n-1/2}\right\|^2 + C\left\|f\left(\widehat{u}^{n-1/2}\right) - f\left(\widehat{u}_h^{n-1/2}\right)\right\|^2 \\ &\quad + C\left\|\vartheta^{n-1/2}\right\|^2. \end{aligned} \tag{54}$$

In order to accomplish our process, we have to estimate the nonlinear term $\|f(\widehat{u}^{n-1/2}) - f(\widehat{u}_h^{n-1/2})\|^2$. Using Young inequality, we have

$$\begin{aligned}
&\left\| f\left(\widehat{u}^{n-1/2}\right) - f\left(\widehat{u}_h^{n-1/2}\right)\right\|^2 \\
&\leq 2\left\|\left(\widehat{u}^{n-1/2}\right)^3 - \left(\widehat{u}_h^{n-1/2}\right)^3\right\|^2 + 2\left\|\widehat{u}^{n-1/2} - \widehat{u}_h^{n-1/2}\right\|^2 \\
&= 2\left\|\left[\frac{3}{2}\left(u^{n-1} - u_h^{n-1}\right) - \frac{1}{2}\left(u^{n-2} - u_h^{n-2}\right)\right]\right. \\
&\quad \times \left[\left(\frac{3}{2}u^{n-1} - \frac{1}{2}u^{n-2}\right)^2 + \left(\frac{3}{2}u_h^{n-1} - \frac{1}{2}u_h^{n-2}\right)^2\right. \\
&\quad \left.\left. + \left(\frac{3}{2}u^{n-1} - \frac{1}{2}u^{n-2}\right)\left(\frac{3}{2}u_h^{n-1} - \frac{1}{2}u_h^{n-2}\right)\right]\right\|^2 \\
&\quad + 2\left\|\frac{3}{2}\left(u^{n-1} - u_h^{n-1}\right) - \frac{1}{2}\left(u^{n-2} - u_h^{n-2}\right)\right\|^2 \\
&\leq 2\left\|\frac{3}{2}\left(\vartheta^{n-1} + \tau^{n-1}\right) - \frac{1}{2}\left(\vartheta^{n-2} + \tau^{n-2}\right)\right\|^2 \\
&\quad \times \left[\left\|\left(\frac{3}{2}u^{n-1} - \frac{1}{2}u^{n-2}\right)^2 + \left(\frac{3}{2}u_h^{n-1} - \frac{1}{2}u_h^{n-2}\right)^2\right.\right. \\
&\quad \left.\left. + \left(\frac{3}{2}u^{n-1} - \frac{1}{2}u^{n-2}\right)\left(\frac{3}{2}u_h^{n-1} - \frac{1}{2}u_h^{n-2}\right)\right\|_{L^\infty}^2 + 1\right] \\
&= \left(24C_u^2 + 2\right)\left\|\frac{3}{2}\left(\vartheta^{n-1} + \tau^{n-1}\right) - \frac{1}{2}\left(\vartheta^{n-2} + \tau^{n-2}\right)\right\|^2 \\
&\leq C\left(\left\|\tau^{n-1}\right\|^2 + \left\|\tau^{n-2}\right\|^2 + \left\|\vartheta^{n-1}\right\|^2 + \left\|\vartheta^{n-2}\right\|^2\right).
\end{aligned} \tag{55}$$

Note that

$$\left\|\frac{\tau^n - \tau^{n-1}}{\Delta t}\right\|^2 \leq \frac{1}{\Delta t}\int_{t_{n-1}}^{t_n} \left\|\tau_t(s)\right\|^2 ds. \tag{56}$$

Substitute (55) and (56) into (54) to obtain

$$\begin{aligned}
&\frac{\gamma}{2}\left\|\kappa^{n-1/2}\right\|^2 + \frac{1}{2\Delta t}\left(\left\|\vartheta^n\right\|^2 - \left\|\vartheta^{n-1}\right\|^2\right) \\
&\leq C\left(\frac{1}{\Delta t}\int_{t_{n-1}}^{t_n} \left\|\tau_t(s)\right\|^2 ds + \left\|R_1^{n-1/2}\right\|^2\right. \\
&\quad \left. + \left\|R_2^{n-1/2}\right\|^2 + \left\|\pi^{n-1/2}\right\|^2\right) \\
&\quad + C\left(\left\|\tau^{n-1}\right\|^2 + \left\|\tau^{n-2}\right\|^2 + \left\|\vartheta^n\right\|^2 + \left\|\vartheta^{n-1}\right\|^2 + \left\|\vartheta^{n-2}\right\|^2\right).
\end{aligned} \tag{57}$$

Summing from $n = 2, \ldots, J$, the resulting inequality becomes

$$\begin{aligned}
&\gamma\Delta t\sum_{n=2}^{J}\left\|\kappa^{n-1/2}\right\|^2 + (1 - C\Delta t)\left\|\vartheta^J\right\|^2 \\
&\leq \left\|\vartheta^1\right\|^2 + C\Delta t\sum_{n=2}^{J}\left(\left\|R_1^{n-1/2}\right\|^2 + \left\|R_2^{n-1/2}\right\|^2 + \left\|\pi^{n-1/2}\right\|^2\right) \\
&\quad + C\Delta t\sum_{n=0}^{J-1}\left(\left\|\tau^n\right\|^2 + \left\|\vartheta^n\right\|^2\right) + C\int_0^{t_J}\left\|\tau_t\right\|^2 ds.
\end{aligned} \tag{58}$$

For sufficiently small Δt, using Gronwall lemma with Lemma 10, we get

$$\begin{aligned}
&\gamma\Delta t\sum_{n=2}^{J}\left\|\kappa^{n-1/2}\right\|^2 + \frac{1}{2}\left\|\vartheta^J\right\|^2 \\
&\leq \left\|\vartheta^1\right\|^2 + C\Delta t\sum_{n=0}^{J-1}\left\|\tau^n\right\|^2 \\
&\quad + C\Delta t\sum_{n=2}^{J}\left(\left\|R_1^{n-1/2}\right\|^2 + \left\|R_2^{n-1/2}\right\|^2 + \left\|\pi^{n-1/2}\right\|^2\right) \\
&\quad + C\int_0^{t_J}\left\|\tau_t\right\|^2 ds \leq \left\|\vartheta^1\right\|^2 \\
&\quad + C\left(\left\|\tau^n\right\|_{L^\infty(L^2)}^2 + \left\|\pi^{n-1/2}\right\|_{L^\infty(L^2)}^2\right. \\
&\quad \left. + \int_0^{t_J}\left\|\tau_t\right\|^2 ds\right) + C\Delta t^4\left\|u_{ttt}\right\|_{L^\infty(L^2)}^2.
\end{aligned} \tag{59}$$

Use Lemmas 3 and 4, (59) with the triangle inequality to accomplish the proof of Theorem 12. □

Theorem 13. *With $u_h^0 = \mathcal{Q}_h u^0$, assume that the solution's regular properties for system* (6a), (6b), *and* (6c) *satisfy $u_t \in L^2(H^2)$, $u_{ttt} \in L^\infty(L^2)$, $u, w \in L^\infty(H^2)$, $\boldsymbol{\chi} \in L^\infty(\mathbf{H}^1)$, $\boldsymbol{\chi}_t \in L^2(\mathbf{H}^1)$. Then there exists a positive constant C independent of h and Δt such that*

$$\left\|u(t_J) - u_h^J\right\|_1 \leq C\left[h|||\cdot||| + \Delta t^2\left\|u_{ttt}\right\|_{L^\infty(L^2)}\right], \tag{60a}$$

$$\left(\Delta t\sum_{n=1}^{J}\left\|w(t_{n-1/2}) - w_h^{n-1/2}\right\|_j^2\right)^{1/2} \leq C\left[h^{2-j}|||\cdot||| + \Delta t^2\left\|u_{ttt}\right\|_{L^\infty(L^2)}\right], \quad j = 0, 1, \tag{60b}$$

$$\left\|\boldsymbol{\chi}(t_{n-1/2}) - \boldsymbol{\chi}_h^{n-1/2}\right\| \leq C\left[h|||\cdot||| + \Delta t^2\left\|u_{ttt}\right\|_{L^\infty(L^2)}\right]. \tag{60c}$$

Proof. Choose $v_h = \kappa^{n-1/2}$, $g_h = (\vartheta^n - \vartheta^{n-1})/\Delta t$, and $\mathbf{z}_h = (\nabla\vartheta^n - \nabla\vartheta^{n-1})/\Delta t$ in (50a), (50b), and (50c) to obtain

$$\left(\frac{\vartheta^n - \vartheta^{n-1}}{\Delta t}, \kappa^{n-1/2}\right) + \gamma\left\|\nabla\kappa^{n-1/2}\right\|^2 + \left\|\kappa^{n-1/2}\right\|^2 + \left(f\left(\widehat{\widehat{u}}^{n-1/2}\right) - f\left(\widehat{\widehat{u}}_h^{n-1/2}\right), \kappa^{n-1/2}\right) = -\left(\frac{\tau^n - \tau^{n-1}}{\Delta t}, \kappa^{n-1/2}\right) - \left(\pi^{n-1/2}, \kappa^{n-1/2}\right) + \left(R_1^{n-1/2}, \kappa^{n-1/2}\right) + \left(R_2^{n-1/2}, \kappa^{n-1/2}\right), \tag{61a}$$

$$\left(\zeta^{n-1/2}, \frac{\nabla\vartheta^n - \nabla\vartheta^{n-1}}{\Delta t}\right) - \left(\kappa^{n-1/2}, \frac{\vartheta^n - \vartheta^{n-1}}{\Delta t}\right) = \left(\pi^{n-1/2}, \frac{\vartheta^n - \vartheta^{n-1}}{\Delta t}\right), \tag{61b}$$

$$-\left(\zeta^{n-1/2}, \frac{\nabla\vartheta^n - \nabla\vartheta^{n-1}}{\Delta t}\right) + \frac{\left\|\nabla\vartheta^n\right\|^2 - \left\|\nabla\vartheta^{n-1}\right\|^2}{2\Delta t} = 0. \tag{61c}$$

We add the above three equations of (61a), (61b), and (61c) to get

$$\begin{aligned}
&\frac{\left\|\nabla\vartheta^n\right\|^2 - \left\|\nabla\vartheta^{n-1}\right\|^2}{2\Delta t} + \gamma\left\|\nabla\kappa^{n-1/2}\right\|^2 + \left\|\kappa^{n-1/2}\right\|^2 \\
&= -\left(f\left(\widehat{\widehat{u}}^{n-1/2}\right) - f\left(\widehat{\widehat{u}}_h^{n-1/2}\right), \kappa^{n-1/2}\right) - \left(\frac{\tau^n - \tau^{n-1}}{\Delta t}, \kappa^{n-1/2}\right) - \left(\pi^{n-1/2}, \kappa^{n-1/2}\right) \\
&\quad + \left(R_1^{n-1/2}, \kappa^{n-1/2}\right) + \left(R_2^{n-1/2}, \kappa^{n-1/2}\right) + \left(\pi^{n-1/2}, \frac{\vartheta^n - \vartheta^{n-1}}{\Delta t}\right) \\
&= -\left(f\left(\widehat{\widehat{u}}^{n-1/2}\right) - f\left(\widehat{\widehat{u}}_h^{n-1/2}\right), \kappa^{n-1/2}\right) - \left(\frac{\tau^n - \tau^{n-1}}{\Delta t}, \kappa^{n-1/2}\right) - \left(\pi^{n-1/2}, \kappa^{n-1/2}\right) \\
&\quad + \left(R_1^{n-1/2}, \kappa^{n-1/2}\right) + \left(R_2^{n-1/2}, \kappa^{n-1/2}\right) + \frac{\left(\pi^n, \vartheta^n\right) - \left(\pi^{n-1}, \vartheta^{n-1}\right)}{\Delta t} - \left(\vartheta^{n-1/2}, \frac{\pi^n - \pi^{n-1}}{\Delta t}\right).
\end{aligned} \tag{62}$$

For (62), we employ Cauchy-Schwarz inequality as well as Young inequality to obtain

$$\begin{aligned}
&\frac{\left\|\nabla\vartheta^n\right\|^2 - \left\|\nabla\vartheta^{n-1}\right\|^2}{2\Delta t} + \gamma\left\|\nabla\kappa^{n-1/2}\right\|^2 + \frac{1}{2}\left\|\kappa^{n-1/2}\right\|^2 \\
&\le C\left(\left\|f\left(\widehat{\widehat{u}}^{n-1/2}\right) - f\left(\widehat{\widehat{u}}_h^{n-1/2}\right)\right\|^2 + \left\|\frac{\tau^n - \tau^{n-1}}{\Delta t}\right\|^2 + \left\|\pi^{n-1/2}\right\|^2 + \left\|R_1^{n-1/2}\right\|^2 + \left\|R_2^{n-1/2}\right\|^2 + \left\|\vartheta^{n-1/2}\right\|^2 + \left\|\frac{\pi^n - \pi^{n-1}}{\Delta t}\right\|^2\right) \\
&\quad + \frac{\left(\pi^n, \vartheta^n\right) - \left(\pi^{n-1}, \vartheta^{n-1}\right)}{\Delta t}.
\end{aligned} \tag{63}$$

Multiply (63) by $2\Delta t$ and then sum from $n = 2, \ldots, J$ to get

$$\begin{aligned}
&\left\|\nabla\vartheta^J\right\|^2 + 2\gamma\Delta t\sum_{n=2}^{J}\left\|\nabla\kappa^{n-1/2}\right\|^2 + \Delta t\sum_{n=2}^{J}\left\|\kappa^{n-1/2}\right\|^2 \le \left\|\nabla\vartheta^1\right\|^2 \\
&\quad + C\Delta t\sum_{n=2}^{J}\left(\left\|f\left(\widehat{\widehat{u}}^{n-1/2}\right) - f\left(\widehat{\widehat{u}}_h^{n-1/2}\right)\right\|^2 + \left\|\frac{\tau^n - \tau^{n-1}}{\Delta t}\right\|^2 + \left\|\pi^{n-1/2}\right\|^2 + \left\|R_1^{n-1/2}\right\|^2 + \left\|R_2^{n-1/2}\right\|^2 + \left\|\vartheta^{n-1/2}\right\|^2 + \left\|\frac{\pi^n - \pi^{n-1}}{\Delta t}\right\|^2\right) \\
&\quad + \left(\pi^J, \vartheta^J\right) - \left(\pi^1, \vartheta^1\right).
\end{aligned} \tag{64}$$

Substituting the following inequality

$$\left\|\frac{\pi^n - \pi^{n-1}}{\Delta t}\right\|^2 \le \frac{1}{\Delta t}\int_{t_{n-1}}^{t_n}\left\|\pi_t(s)\right\|^2 ds, \tag{65}$$

and (55) into (64), we obtain

$$\begin{aligned}
&\left\|\nabla\vartheta^J\right\|^2 + 2\gamma\Delta t\sum_{n=2}^{J}\left\|\nabla\kappa^{n-1/2}\right\|^2 + \Delta t\sum_{n=2}^{J}\left\|\kappa^{n-1/2}\right\|^2 \\
&\le \left\|\vartheta^1\right\|_1^2 + \left\|\pi^J\right\|^2 + C\Delta t\sum_{n=2}^{J}\left(\left\|\tau^{n-1}\right\|^2 + \left\|\tau^{n-2}\right\|^2 + \left\|\vartheta^{n-1}\right\|^2 + \left\|\vartheta^{n-2}\right\|^2 + \left\|\pi^{n-1/2}\right\|^2 + \left\|R_1^{n-1/2}\right\|^2 + \left\|R_2^{n-1/2}\right\|^2 + \left\|\vartheta^{n-1/2}\right\|^2\right) + \left\|\vartheta^J\right\|^2.
\end{aligned} \tag{66}$$

Substitute (59) into (66) to obtain

$$\begin{aligned}
&\left\|\nabla\vartheta^J\right\|^2 + 2\gamma\Delta t\sum_{n=2}^{J}\left\|\nabla\kappa^{n-1/2}\right\|^2 \\
&\quad + \Delta t\sum_{n=2}^{J}\left\|\kappa^{n-1/2}\right\|^2 \le \left\|\vartheta^1\right\|_1^2 \\
&\quad + C\Delta t\sum_{n=2}^{J}\left(\left\|\tau^{n-1}\right\|^2 + \left\|\tau^{n-2}\right\|^2 + \left\|\pi^{n-1/2}\right\|^2\right. \\
&\qquad \left. + \left\|R_1^{n-1/2}\right\|^2 + \left\|R_2^{n-1/2}\right\|^2\right) + \left\|\pi^J\right\|^2 \\
&\quad + C\left(\left\|\tau^n\right\|_{L^\infty(L^2)}^2 + \int_0^{t_J}\left\|\tau_t\right\|^2 ds\right. \\
&\qquad \left. + \Delta t^4\left\|u_{ttt}\right\|_{L^\infty(L^2)} + \left\|\pi^{n-1/2}\right\|_{L^\infty(L^2)}^2\right).
\end{aligned} \tag{67}$$

Taking $\mathbf{z}_h = \boldsymbol{\zeta}^{n-1/2}$ in (50c), we have

$$\left\|\zeta^{n-1/2}\right\|^2 \le C\left\|\nabla\vartheta^{n-1/2}\right\|^2 \le C\left(\left\|\vartheta^n\right\|^2 + \left\|\vartheta^{n-1}\right\|^2\right). \tag{68}$$

Combining Lemmas 3–10, (67), and (68) with the triangle inequality, we accomplish the proof. □

5. Some Concluding Remarks and Extensions

In [29, 30], authors proposed a new mixed finite element method, which has been applied to solve second-order evolution equations, such as Sobolev equations [31], parabolic equations [32, 33], and reaction-convection-diffusion problems [34, 35]. In this paper, we apply the new mixed finite element scheme [29, 30] to solve the extended Fisher-Kolmogorov (EFK) equation (fourth-order nonlinear reaction diffusion equation). Compared to the classical mixed methods, the weaker square integrable $\mathbf{L}^2$ space, which takes the place of the complex $\mathbf{H}(\text{div};\Omega)$, is used in our method. We derive some a priori bounds for the solution and a priori error estimates for semidiscrete scheme. What's more, we obtain a priori error estimates for fully discrete scheme by a linearized Crank-Nicolson method.

If we take $\gamma = 0$ in (1), we can get the following second-order nonlinear reaction-diffusion equation:

$$u_t - \Delta u + f(u) = 0, \quad (\mathbf{x}, t) \in \Omega \times J, \tag{69}$$

with initial condition

$$u(\mathbf{x}, 0) = u_0(\mathbf{x}), \quad \mathbf{x} \in \Omega, \tag{70}$$

and Dirichlet boundary conditions

$$u(\mathbf{x}, t) = 0, \quad (\mathbf{x}, t) \in \partial\Omega \times \bar{J}, \tag{71}$$

where Ω is a bounded convex polygonal domain in $R^d, d \le 2$ with boundary $\partial\Omega$, and $J = (0, T]$ is the time interval with $0 < T < \infty$. $u_0(\mathbf{x})$ is given function, $f(u) = u^3 - u$.

Using a similar method as the one in this paper, we can get the following weak formulation for system (69)–(71):

$$(u_t, v) + (\boldsymbol{\lambda}, \nabla v) + (f(u), v) = 0, \quad \forall v \in H_0^1, \tag{72a}$$

$$(\boldsymbol{\lambda}, \mathbf{z}) - (\nabla u, \mathbf{z}) = 0, \quad \forall \mathbf{z} \in \mathbf{L}^2, \tag{72b}$$

where $\boldsymbol{\lambda} = \nabla u$. Based on the mixed weak formulation (72a) and (72b), we can get the similar theoretical analysis as our method in this paper.

In the future work, we will apply the new mixed scheme to solve fourth-order linear/nonlinear wave equations [24, 25, 38]. At the same time, we will study the large time-stepping method based on the new mixed element scheme for the Cahn-Hilliard equation [19, 20].

Acknowledgments

This work was supported by the National Natural Science Fund of China (11061021), the Key Project of Chinese Ministry of Education (12024), the Scientific Research Projection of Higher Schools of Inner Mongolia (NJZZ12011, NJZY13199) and the Natural Science Fund of Inner Mongolia Province (2012MS0108, 2012MS0106, and 2011BS0102), the Program of Higher-level talents of Inner Mongolia University (125119, Z200901004, and 30105-125132), and the Scientific Research Projection of Inner Mongolia University of Finance and Economics (KY1101).

References

[1] Z. X. Chen, "Analysis of expanded mixed methods for fourth-order elliptic problems," *Numerical Methods for Partial Differential Equations*, vol. 13, no. 5, pp. 483–503, 1997.

[2] J. C. Li, "Optimal error estimates of mixed finite element methods for a fourth-order nonlinear elliptic problem," *Journal of Mathematical Analysis and Applications*, vol. 334, no. 1, pp. 183–195, 2007.

[3] J. C. Li, "Mixed methods for fourth-order elliptic and parabolic problems using radial basis functions," *Advances in Computational Mathematics*, vol. 23, no. 1-2, pp. 21–30, 2005.

[4] J. C. Li, "Optimal convergence analysis of mixed finite element methods for fourth-order elliptic and parabolic problems," *Numerical Methods for Partial Differential Equations*, vol. 22, no. 4, pp. 884–896, 2006.

[5] X. L. Cheng, W. M. Han, and H. C. Huang, "Some mixed finite element methods for biharmonic equation," *Journal of Computational and Applied Mathematics*, vol. 126, no. 1-2, pp. 91–109, 2000.

[6] Y. Liu, H. Li, S. He, W. Gao, and Z. C. Fang, "H^1-Galerkin mixed element method and numerical simulation for the fourth-order parabolic partial differential equations," *Mathematica Numerica Sinica*, vol. 34, no. 2, pp. 259–274, 2012.

[7] D. Y. Shi, S. P. Mao, and S. C. Chen, "On the anisotropic accuracy analysis of ACM's nonconforming finite element," *Journal of Computational Mathematics*, vol. 23, no. 6, pp. 635–646, 2005.

[8] M. Wang, Z. C. Shi, and J. C. Xu, "Some n-rectangle nonconforming elements for fourth order elliptic equations," *Journal of Computational Mathematics*, vol. 25, no. 4, pp. 408–420, 2007.

[9] W. Ming and J. C. Xu, "The Morley element for fourth order elliptic equations in any dimensions," *Numerische Mathematik*, vol. 103, no. 1, pp. 155–169, 2006.

[10] H. R. Chen and S. C. Chen, "C^0-nonconforming elements for fourth order elliptic problem," *Mathematica Numerica Sinica*, vol. 35, no. 1, pp. 21–30, 2012.

[11] L. Wang, Y. K. Wu, and X. P. Xie, "Uniformly stable rectangular elements for fourth order elliptic singular perturbation problems," *Numerical Methods for Partial Differential Equations*, vol. 29, no. 3, pp. 721–737, 2013.

[12] J. Hu and Z. C. Shi, "The best L^2 norm error estimate of lower order finite element methods for the fourth order problem," *Journal of Computational Mathematics*, vol. 30, no. 5, pp. 449–460, 2012.

[13] W. B. Chen and Y. Q. Wang, "A mixed finite element method for thin film epitaxy," *Numerische Mathematik*, vol. 122, no. 4, pp. 771–793, 2012.

[14] H. Bi, S. X. Ren, and Y. D. Yang, "Conforming finite element approximations for a fourth-order Steklov eigenvalue problem," *Mathematical Problems in Engineering*, vol. 2011, Article ID 873152, 13 pages, 2011.

[15] X. H. Yang, D. Xu, and H. X. Zhang, "Crank-Nicolson/quasi-wavelets method for solving fourth order partial integro-differential equation with a weakly singular kernel," *Journal of Computational Physics*, vol. 234, no. 1, pp. 317–329, 2013.

[16] T. Zhang, "Finite element analysis for Cahn-Hilliard equation," *Mathematica Numerica Sinica*, vol. 28, no. 3, pp. 281–292, 2006.

[17] X. B. Feng and A. Prohl, "Error analysis of a mixed finite element method for the Cahn-Hilliard equation," *Numerische Mathematik*, vol. 99, no. 1, pp. 47–84, 2004.

[18] Q. Du, L. Ju, and L. Tian, "Finite element approximation of the Cahn-Hilliard equation on surfaces," *Computer Methods in Applied Mechanics and Engineering*, vol. 200, no. 29–32, pp. 2458–2470, 2011.

[19] Y. N. He, Y. X. Liu, and T. Tang, "On large time-stepping methods for the Cahn-Hilliard equation," *Applied Numerical Mathematics*, vol. 57, no. 5–7, pp. 616–628, 2007.

[20] Y. F. Yang, X. L. Feng, and Y. N. He, "Large time-stepping method based on the finite element discretization for the Cahn-Hilliard equation," *Journal of Applied Mathematics & Informatics*, vol. 29, no. 5-6, pp. 1129–1141, 2011.

[21] Y. N. He and Y. X. Liu, "Stability and convergence of the spectral Galerkin method for the Cahn-Hilliard equation," *Numerical Methods for Partial Differential Equations*, vol. 24, no. 6, pp. 1485–1500, 2008.

[22] S. M. Choo and Y. J. Lee, "A discontinuous Galerkin method for the Cahn-Hilliard equation," *Journal of Applied Mathematics and Computing*, vol. 18, no. 1-2, pp. 113–126, 2005.

[23] S. M. Choo, S. K. Chung, and K. I. Kim, "Conservative nonlinear difference scheme for the Cahn-Hilliard equation," *Computers & Mathematics with Applications*, vol. 39, no. 1-2, pp. 229–243, 2000.

[24] Y. Liu, H. Li, S. He, W. Gao, and Z. C. Fang, "A C^1-conforming finite element method for nonlinear fourth-order hyperbolic equation," *World Academy of Science, Engineering and Technology*, vol. 80, pp. 1423–1427, 2011.

[25] S. He, H. Li, and Y. Liu, "Analysis of mixed finite element methods for fourth-order wave equations," *Computers & Mathematics with Applications*, vol. 65, no. 1, pp. 1–16, 2013.

[26] L. J. T. Doss and A. P. Nandini, "An H^1-Galerkin mixed finite element method for the extended Fisher-Kolmogorov equation," *International Journal of Numerical Analysis & Modeling B*, vol. 3, no. 4, pp. 460–485, 2012.

[27] P. Danumjaya and A. K. Pani, "Mixed finite element methods for a fourth order reaction diffusion equation," *Numerical Methods for Partial Differential Equations*, vol. 28, no. 4, pp. 1227–1251, 2012.

[28] P. Danumjaya and A. K. Pani, "Numerical methods for the extended Fisher-Kolmogorov (EFK) equation," *International Journal of Numerical Analysis & Modeling*, vol. 3, no. 2, pp. 186–210, 2006.

[29] S. C. Chen and H. R. Chen, "New mixed element schemes for second order elliptic problem," *Mathematica Numerica Sinica*, vol. 32, no. 2, pp. 213–218, 2010.

[30] F. Shi, J. P. Yu, and K. T. Li, "A new stabilized mixed finite-element method for Poisson equation based on two local Gauss integrations for linear element pair," *International Journal of Computer Mathematics*, vol. 88, no. 11, pp. 2293–2305, 2011.

[31] D. Y. Shi and Y. D. Zhang, "High accuracy analysis of a new nonconforming mixed finite element scheme for Sobolev equations," *Applied Mathematics and Computation*, vol. 218, no. 7, pp. 3176–3186, 2011.

[32] Z. Weng, X. Feng, and P. Huang, "A new mixed finite element method based on the Crank-Nicolson scheme for the parabolic problems," *Applied Mathematical Modelling*, vol. 36, no. 10, pp. 5068–5079, 2012.

[33] L. Li, P. Sun, and Z. D. Luo, "A new mixed finite element formulation and error estimates for parabolic problems," *Acta Mathematica Scientia*, vol. 32, no. 6, pp. 1158–1165, 2012.

[34] Y. Liu, H. Li, W. Gao, S. He, and Z. C. Fang, "A novel characteristic expanded mixed method for reaction-convection-diffusion problems," *Journal of Applied Mathematics*, vol. 2013, Article ID 683205, 11 pages, 2013.

[35] D. Y. Shi and Q. L. Tang, "A new characteristic nonconforming mixed finite element scheme for convection-dominated diffusion problem," *Journal of Applied Mathematics*, vol. 2013, Article ID 951692, 10 pages, 2013.

[36] P. G. Ciarlet, *Finite Element Method for Elliptic Problems*, North-Holland, Amsterdam, The Netherlands, 1978.

[37] Z. D. Luo, *Mixed Finite Element Methods and Applications*, Chinese Science Press, Beijing, China, 2006.

[38] S. M. Zheng and A. Milani, "Exponential attractors and inertial manifolds for singular perturbations of the Cahn-Hilliard equations," *Nonlinear Analysis: Theory, Methods and Applications*, vol. 57, no. 5-6, pp. 843–877, 2004.

21

On the Normed Space of Equivalence Classes of Fuzzy Numbers

Dong Qiu, Chongxia Lu, and Wei Zhang

College of Mathematics and Physics, Chongqing University of Posts and Telecommunications, Nanan, Chongqing 400065, China

Correspondence should be addressed to Dong Qiu; dongqiumath@163.com

Academic Editors: T. Allahviranloo and A. N. Yannacopoulos

We study the norm induced by the supremum metric on the space of fuzzy numbers. And then we propose a method for constructing a norm on the quotient space of fuzzy numbers. This norm is very natural and works well with the induced metric on the quotient space.

1. Introduction

Since the concept of fuzzy numbers was firstly introduced in the 1970s, it has been studied extensively from many different aspects of the theory and applications such as fuzzy topology, fuzzy analysis, fuzzy logic, and fuzzy decision making (see, e.g., [1–6]). The operations in the set of fuzzy numbers are usually obtained by the Zadeh extension principle [7–9]. In the study of algebraic structures and topological structures for fuzzy numbers, many results have been obtained (see, e.g., [10–19]).

In the classical mathematics, if X is a normed space with norm $\|\cdot\|$, it is readily checked that the formula $d(x, y) = \|x - y\|$, for $x, y \in X$, defines a metric d on X. Thus a normed space is naturally a metric space and all metric space concepts are meaningful. However, we will show that such proposition does not hold true for the well known supremum metric on the space of fuzzy numbers. To overcome this weakness, we will consider the quotient space of fuzzy numbers up to an equivalence relation which is introduced by Mareš [20, 21] and is studied extensively by many researchers [4, 12, 22–24]. We will propose a method for constructing a norm on the quotient space of fuzzy numbers. This norm is very natural and works well with the induced metric on the quotient space.

2. Preliminaries

A fuzzy set of $\mathbb{R}$ is a function $\mu : \mathbb{R} \to [0,1]$. For each such fuzzy set μ, we denote by $[\mu]^a = \{x \in \mathbb{R} : \mu(x) \geq a\}$ for any $a \in (0,1]$ its a-level set. We define the set $[\mu]^0$ by $[\mu]^0 = \overline{\{x \in \mathbb{R} : \mu(x) > 0\}}$, where $\overline{A}$ denotes the closure of a set A. A fuzzy number μ is a fuzzy set with nonempty bounded closed level sets $[\mu]^a = [\mu_L(a), \mu_R(a)]$ for all $a \in [0,1]$. We denote the class of fuzzy numbers by $\mathcal{F}$. Notice that the real numbers $\mathbb{R}$ can be imbedded in $\mathcal{F}$ by defining a fuzzy number as follows

$$\mathbf{a}(x) = \begin{cases} 1, & \text{if } x = a, \\ 0, & \text{otherwise,} \end{cases} \tag{1}$$

for each $a \in \mathbb{R}$.

For any $\mu, \nu \in \mathcal{F}$ and $a \in \mathbb{R}$, owing to Zadeh's extension principle [7–9], scalar multiplication and addition are defined for any $x \in \mathbb{R}$ by

$$(a \times \mu)(x) = (a\mu)(x) = \begin{cases} \mu\left(\dfrac{x}{a}\right), & \text{if } a \neq 0, \\ \mathbf{0}, & \text{if } a = 0, \end{cases} \tag{2}$$
$$(\mu + \nu)(x) = \sup_{x_1, x_2 : x_1 + x_2 = x} \min\{\mu(x_1), \nu(x_2)\}.$$

For any $\mu \in \mathcal{F}$, we define the fuzzy number $-\mu \in \mathcal{F}$ by $-\mu = (-1) \times \mu$, that is, $-\mu(x) = \mu(-x)$, for all $x \in \mathbb{R}$. Since the addition does not possess an inverse subtraction, $\mathcal{F}$ is not a real vector space. We say that a fuzzy number $s \in \mathcal{F}$ is symmetric [20], if

$$s(x) = s(-x), \tag{3}$$

for all $x \in \mathbb{R}$, that is, $s = -s$. The set of all symmetric fuzzy numbers will be denoted by $\mathcal{S}$.

Definition 1 (see [22]). Let $\mu, \nu \in \mathcal{F}$. We say that μ is equivalent to ν and write $\mu \sim \nu$ if and only if there exist symmetric fuzzy numbers $s_1, s_2 \in \mathcal{S}$ such that

$$\mu + s_1 = \nu + s_2. \tag{4}$$

The equivalence relation defined above is reflexive, symmetric, and transitive [20]. Let $\langle\mu\rangle$ denote the equivalence class containing the element μ and denote the set of equivalence classes by $\mathcal{F}/\mathcal{S}$. By the level set representations for fuzzy numbers, one can easily prove the following lemmas.

Lemma 2 (see [25]). *For any $\mu \in \mathcal{F}$, $\mu - \mu \in \mathcal{S}$.*

Lemma 3 (see [25]). *For any $\mu, \nu \in \mathcal{F}$, $\mu \sim \nu$ if and only if $\mu - \nu \in \mathcal{S}$.*

Definition 4 (see [26]). For a fuzzy number μ, we define a function $\mu_M : [0,1] \to \mathbb{R}$ by assigning the midpoint of each a-level set to $\mu_M(a)$ for all $a \in [0,1]$, that is,

$$\mu_M(a) = \frac{\mu_L(a) + \mu_R(a)}{2}. \tag{5}$$

Lemma 5 (see [25]). *For any $\mu, \nu \in \mathcal{F}$, $\mu \sim \nu$ if and only if $\mu_M = \nu_M$.*

Lemma 6 (see [26]). *$(\mathcal{F}/\mathcal{S}, +, \times)$ is a real vector space.*

3. Main Results

In this section, we will give a norm structure on $\mathcal{F}/\mathcal{S}$ which is compatible with a metric. For the family of fuzzy numbers $\mathcal{F}$, the d_∞-metric is induced by the Hausdorff metric as [17]

$$\begin{aligned} d_\infty(\mu,\nu) &= \sup_{0\le a\le 1} H\left([\mu]^a, [\nu]^a\right) \\ &= \sup_{0\le a\le 1} \max\left\{\left|\mu_L(a) - \nu_L(a)\right|, \left|\mu_R(a) - \nu_R(a)\right|\right\}, \end{aligned} \tag{6}$$

for all $\mu, \nu \in \mathcal{F}$.

Define a function $\|\cdot\| : \mathcal{F} \to \mathbb{R}$ as

$$\|\mu\| = d(\mu, \mathbf{0}) = \sup_{0\le a\le 1} \max\left\{\left|\mu_L(a)\right|, \left|\mu_R(a)\right|\right\}. \tag{7}$$

Theorem 7. *The function $\|\cdot\|$ in (7) is a norm on $\mathcal{F}$.*

Proof. (i) It is obvious that $\|\mu\| \ge 0$ for all $\mu \in \mathcal{F}$ and $\|\mu\| = 0$ if $\mu = \mathbf{0}$. Conversely, if $\mu \ne \mathbf{0}$, then we have that there exists $a_0 \in [0,1]$ such that $[\mu_L(a_0), \mu_R(a_0)] \ne \{0\}$. Thus we have $\|\mu\| > 0$.

(ii) For all $\mu \in \mathcal{F}$ and $b \in \mathbb{R}$, we have

$$\begin{aligned} \|b\mu\| &= \sup_{0\le a\le 1} \max\left\{|b|\left|\mu_L(a)\right|, |b|\left|\mu_R(a)\right|\right\} \\ &= |b|\,\|\mu\|. \end{aligned} \tag{8}$$

(iii) For all $\mu, \nu \in \mathcal{F}$, we have that

$$\begin{aligned} \|\mu+\nu\| &= \sup_{0\le a\le 1} \max\left\{\left|\mu_L(a) + \nu_L(a)\right|, \left|\mu_R(a) + \nu_R(a)\right|\right\} \\ &\le \sup_{0\le a\le 1} \max\left\{\left|\mu_L(a)\right| + \left|\nu_L(a)\right|, \left|\mu_R(a)\right| + \left|\nu_R(a)\right|\right\} \\ &\le \sup_{0\le a\le 1} \max\left\{\left|\mu_L(a)\right|, \left|\mu_R(a)\right|\right\} \\ &\quad + \sup_{0\le a\le 1} \max\left\{\left|\nu_L(a)\right|, \left|\nu_R(a)\right|\right\} \\ &= \|\mu\| + \|\nu\|. \end{aligned} \tag{9}$$

We conclude that $\|\cdot\|$ is a norm on $\mathcal{F}$. □

Although $\|\cdot\|$ is a norm on $\mathcal{F}$, the function $d : \mathcal{F}\times\mathcal{F} \to \mathbb{R}$ induced by $\|\cdot\|$ as $d(\mu,\nu) = \|\mu - \nu\|$ is not a metric on $\mathcal{F}$. Consequently, we get that $d_\infty(\mu,\nu) \ne \|\mu - \nu\|$.

Theorem 8. *The function d has the following properties:*

(i) *$d(\mu,\nu) \ge 0$, for any $\mu, \nu \in \mathcal{F}$;*

(ii) *$d(\mu,\nu) = d(\nu,\mu)$, for any $\mu, \nu \in \mathcal{F}$;*

(iii) *$d(\mu,\nu) \le d(\mu,\omega) + d(\omega,\nu)$, for any $\mu, \nu, \omega \in \mathcal{F}$;*

(iv) *$d(\mu,\mu) \ne 0$, for any fuzzy number $\mu \notin \mathbb{R}$.*

Proof. (i) By the definition of d, it is obvious that $d(\mu,\nu) \ge 0$, for any $\mu, \nu \in \mathcal{F}$.

(ii) For all $\mu, \nu \in \mathcal{F}$, we have that

$$\begin{aligned} d(\mu,\nu) &= \|\mu - \nu\| \\ &= \sup_{0\le a\le 1} \max\left\{\left|\mu_L(a) - \nu_R(a)\right|, \left|\mu_R(a) - \nu_L(a)\right|\right\} \\ &= \sup_{0\le a\le 1} \max\left\{\left|\nu_L(a) - \mu_R(a)\right|, \left|\nu_R(a) - \mu_L(a)\right|\right\} \\ &= \|\nu - \mu\| \\ &= d(\nu,\mu). \end{aligned} \tag{10}$$

(iii) In order to prove the triangle inequality, for any fixed $a \in [0,1]$, and for any $\mu, \nu, \omega \in \mathcal{F}$, we only proof the following six cases. Similarly, the others can be proved.

Case 1 ($\mu_L(a) \le \nu_L(a) \le \mu_R(a) \le \nu_R(a)$ and $\omega_L(a) \le \mu_L(a)$). In this case we have that

$$\begin{aligned} &\max\left\{\left|\mu_L(a) - \nu_R(a)\right|, \left|\mu_R(a) - \nu_L(a)\right|\right\} \\ &\quad = \left|\mu_L(a) - \nu_R(a)\right| \\ &\quad \le \left|\omega_L(a) - \nu_R(a)\right| \\ &\quad \le \max\left\{\left|\mu_L(a) - \omega_R(a)\right|, \left|\mu_R(a) - \omega_L(a)\right|\right\} \\ &\qquad + \max\left\{\left|\omega_L(a) - \nu_R(a)\right|, \left|\omega_R(a) - \nu_L(a)\right|\right\}. \end{aligned} \tag{11}$$

Case 2 ($\mu_L(a) \le \nu_L(a) \le \mu_R(a) \le \nu_R(a)$ and $\mu_L(a) \le \omega_L(a) \le \omega_R(a) \le \nu_R(a)$). In this case we have that

$$\begin{aligned}
&\max\{|\mu_L(a) - \nu_R(a)|, |\mu_R(a) - \nu_L(a)|\} \\
&\quad = |\mu_L(a) - \nu_R(a)| \\
&\quad \le |\mu_L(a) - \omega_R(a)| + |\omega_L(a) - \nu_R(a)| \\
&\quad \le \max\{|\mu_L(a) - \omega_R(a)|, |\mu_R(a) - \omega_L(a)|\} \\
&\qquad + \max\{|\omega_L(a) - \nu_R(a)|, |\omega_R(a) - \nu_L(a)|\}.
\end{aligned} \tag{12}$$

Case 3 ($\mu_L(a) \le \nu_L(a) \le \nu_R(a) \le \mu_R(a)$ and $\omega_L(a) \le \mu_L(a)$). In this case we have that

$$\begin{aligned}
&\max\{|\mu_L(a) - \nu_R(a)|, |\mu_R(a) - \nu_L(a)|\} \\
&\quad \le |\mu_L(a) - \nu_R(a)| \\
&\quad \le |\omega_L(a) - \mu_R(a)| \\
&\quad \le \max\{|\mu_L(a) - \omega_R(a)|, |\mu_R(a) - \omega_L(a)|\} \\
&\qquad + \max\{|\omega_L(a) - \nu_R(a)|, |\omega_R(a) - \nu_L(a)|\}.
\end{aligned} \tag{13}$$

Case 4 ($\mu_L(a) \le \omega_L(a) \le \omega_R(a) \le \nu_L(a) \le \nu_R(a) \le \mu_R(a)$). If $\max\{|\mu_L(a) - \nu_R(a)|, |\mu_R(a) - \nu_L(a)|\} = |\mu_L(a) - \nu_R(a)|$, then we have that

$$\begin{aligned}
&\max\{|\mu_L(a) - \nu_R(a)|, |\mu_R(a) - \nu_L(a)|\} \\
&\quad = |\mu_L(a) - \nu_R(a)| \\
&\quad = |\mu_L(a) - \omega_L(a)| + |\omega_L(a) - \nu_R(a)| \\
&\quad \le |\mu_L(a) - \omega_R(a)| + |\omega_L(a) - \nu_R(a)| \\
&\quad \le \max\{|\mu_L(a) - \omega_R(a)|, |\mu_R(a) - \omega_L(a)|\} \\
&\qquad + \max\{|\omega_L(a) - \nu_R(a)|, |\omega_R(a) - \nu_L(a)|\}.
\end{aligned} \tag{14}$$

Otherwise, we have that

$$\begin{aligned}
&\max\{|\mu_L(a) - \nu_R(a)|, |\mu_R(a) - \nu_L(a)|\} \\
&\quad = |\mu_R(a) - \nu_L(a)| \\
&\quad \le |\mu_R(a) - \omega_L(a)| \\
&\quad \le \max\{|\mu_L(a) - \omega_R(a)|, |\mu_R(a) - \omega_L(a)|\} \\
&\qquad + \max\{|\omega_L(a) - \nu_R(a)|, |\omega_R(a) - \nu_L(a)|\}.
\end{aligned} \tag{15}$$

Case 5 ($\mu_L(a) \le \nu_L(a) \le \omega_L(a) \le \omega_R(a) \le \nu_R(a) \le \mu_R(a)$). In this case we have that

$$\begin{aligned}
&|\mu_L(a) - \nu_R(a)| \le |\mu_L(a) - \omega_R(a)| + |\omega_L(a) - \nu_R(a)|, \\
&|\mu_R(a) - \nu_L(a)| \le |\omega_L(a) - \nu_R(a)| + |\omega_R(a) - \nu_L(a)|.
\end{aligned} \tag{16}$$

Consequently, we have that

$$\begin{aligned}
&\max\{|\mu_L(a) - \nu_R(a)|, |\mu_R(a) - \nu_L(a)|\} \\
&\quad \le \max\{|\mu_L(a) - \omega_R(a)|, |\mu_R(a) - \omega_L(a)|\} \\
&\qquad + \max\{|\omega_L(a) - \nu_R(a)|, |\omega_R(a) - \nu_L(a)|\}.
\end{aligned} \tag{17}$$

Case 6 ($\mu_L(a) \le \omega_L(a) \le \nu_L(a) \le \nu_R(a) \le \omega_R(a) \le \mu_R(a)$). In this case we have that

$$\begin{aligned}
&|\mu_L(a) - \nu_R(a)| \le |\mu_L(a) - \omega_R(a)|, \\
&|\mu_R(a) - \nu_L(a)| \le |\mu_R(a) - \omega_L(a)|.
\end{aligned} \tag{18}$$

Consequently, we have that

$$\begin{aligned}
&\max\{|\mu_L(a) - \nu_R(a)|, |\mu_R(a) - \nu_L(a)|\} \\
&\quad \le \max\{|\mu_L(a) - \omega_R(a)|, |\mu_R(a) - \omega_L(a)|\} \\
&\qquad + \max\{|\omega_L(a) - \nu_R(a)|, |\omega_R(a) - \nu_L(a)|\}.
\end{aligned} \tag{19}$$

From what is proved above, we can get that

$$\begin{aligned}
d(\mu, \nu) &= \|\mu - \nu\| \\
&= \sup_{0\le a\le 1} \max\{|\mu_L(a) - \nu_R(a)|, |\mu_R(a) - \nu_L(a)|\} \\
&\le \sup_{0\le a\le 1} \max\{|\mu_L(a) - \omega_R(a)|, |\mu_R(a) - \omega_L(a)|\} \\
&\quad + \sup_{0\le a\le 1} \max\{|\omega_L(a) - \nu_R(a)|, |\omega_R(a) - \nu_L(a)|\} \\
&= \|\mu - \omega\| + \|\omega - \nu\| \\
&= d(\mu, \omega) + d(\omega, \nu).
\end{aligned} \tag{20}$$

(iv) Since $\mu \notin \mathbb{R}$, there exists $a_0 \in [0, 1]$ such that $\mu_R(a_0) - \mu_L(a_0) > 0$. Thus we have that

$$\begin{aligned}
d(\mu, \mu) &= \|\mu - \mu\| \\
&= \sup_{0\le a\le 1} |\mu_L(a) - \mu_R(a)| \\
&> \mu_R(a_0) - \mu_L(a_0) \\
&> 0.
\end{aligned} \tag{21}$$

□

In order to induce a metric which is compatible with the norm, we consider the quotient space of fuzzy numbers. It is very natural to define a function $\|\cdot\| : \mathscr{F}/\mathscr{S} \to \mathbb{R}$ as

$$\|\langle\mu\rangle\| = \inf_{\nu\in\langle\mu\rangle} \|\nu\|, \tag{22}$$

for each $\langle\mu\rangle \in \mathscr{F}/\mathscr{S}$.

Theorem 9. *The function* $\|\cdot\|$ *in (22) is a norm on* $\mathscr{F}/\mathscr{S}$.

Proof. (i) It is obvious that $\|\langle\mu\rangle\| \geq 0$ for all $\mu \in \mathscr{F}$ and $\|\langle\mu\rangle\| = 0$ if $\langle\mu\rangle = \langle\mathbf{0}\rangle$. Conversely, if $\langle\mu\rangle \neq \langle\mathbf{0}\rangle$, from Lemma 5, it follows that the midpoint function $\mu_M \neq \mathbf{0}_M$. Thus we have

$$\begin{aligned}
\|\langle\mu\rangle\| &= \inf_{\nu\in\langle\mu\rangle} \|\nu\| \\
&= \inf_{\nu\in\langle\mu\rangle} \left(\sup_{0\le a\le 1} \max\left\{|\nu_L(a)|, |\nu_R(a)|\right\} \right) \\
&\geq \inf_{\nu\in\langle\mu\rangle} \left(\sup_{0\le a\le 1} \frac{|\nu_L(a) + \nu_R(a)|}{2} \right) \\
&= \sup_{0\le a\le 1} |\mu_M(a)| \\
&> 0.
\end{aligned} \tag{23}$$

(ii) For all $\langle\mu\rangle \in \mathscr{F}/\mathscr{S}$ and $b \in \mathbb{R}$, we have

$$\begin{aligned}
\|b\langle\mu\rangle\| &= \|\langle b\mu\rangle\| \\
&= \inf_{\nu\in\langle b\mu\rangle} \|\nu\| \\
&= \inf_{b\nu'\in\langle b\mu\rangle} \|b\nu'\| \\
&= \inf_{\nu'\in\langle\mu\rangle} \|b\nu'\| \\
&= |b|\,\|\langle\mu\rangle\|.
\end{aligned} \tag{24}$$

(iii) For all $\langle\mu\rangle, \langle\nu\rangle \in \mathscr{F}/\mathscr{S}$, we have that

$$\begin{aligned}
\|\langle\mu\rangle + \langle\nu\rangle\| &= \|\langle\mu + \nu\rangle\| \\
&= \inf_{\omega\in\langle\mu+\nu\rangle} \|\omega\| \\
&\leq \inf_{\mu'\in\langle\mu\rangle, \nu'\in\langle\nu\rangle} \|\mu + \nu\| \\
&\leq \inf_{\mu'\in\langle\mu\rangle, \nu'\in\langle\nu\rangle} (\|\mu\| + \|\nu\|) \\
&\leq \inf_{\mu'\in\langle\mu\rangle} \|\mu\| + \inf_{\nu'\in\langle\nu\rangle} \|\nu\| \\
&= \|\langle\mu\rangle\| + \|\langle\nu\rangle\|.
\end{aligned} \tag{25}$$

We conclude that $\|\cdot\|$ is a norm on $\mathscr{F}/\mathscr{S}$. □

Now we show that the function $\rho : \mathscr{F}/\mathscr{S} \times \mathscr{F}/\mathscr{S} \to \mathbb{R}$ induced by $\|\cdot\|$ as $\rho(\langle\mu\rangle, \langle\nu\rangle) = \|\langle\mu - \nu\rangle\|$ is exactly a metric on $\mathscr{F}/\mathscr{S}$.

Theorem 10. *The function* ρ *is a metric on* $\mathscr{F}/\mathscr{S}$.

Proof. (i) By the definition of ρ, it is obvious that $\rho(\langle\mu\rangle, \langle\nu\rangle) \geq 0$, for any $\langle\mu\rangle, \langle\nu\rangle \in \mathscr{F}/\mathscr{S}$. If $\rho(\langle\mu\rangle, \langle\nu\rangle) = 0$, then $\|\langle\mu - \nu\rangle\| = 0$. Thus by Theorem 9, we get $\langle\mu\rangle = \langle\nu\rangle$. In addition, by Lemma 2 we have $\rho(\langle\mu\rangle, \langle\mu\rangle) = \|\langle\mu - \mu\rangle\| = 0$.

(ii) For all $\langle\mu\rangle, \langle\nu\rangle \in \mathscr{F}/\mathscr{S}$, we have that

$$\begin{aligned}
\rho(\langle\mu\rangle, \langle\nu\rangle) &= \|\langle\mu - \nu\rangle\| = \|(-1)\langle\nu - \mu\rangle\| \\
&= \|\nu - \mu\| = \rho(\langle\nu\rangle, \langle\mu\rangle).
\end{aligned} \tag{26}$$

(iii) For any $\langle\mu\rangle, \langle\nu\rangle, \langle\omega\rangle \in \mathscr{F}/\mathscr{S}$, we have

$$\begin{aligned}
\rho(\langle\mu\rangle, \langle\nu\rangle) &= \|\langle\mu - \nu\rangle\| \\
&= \|\langle\mu - \omega + \omega - \nu\rangle\| \\
&\leq \|\langle\mu - \omega\rangle\| + \|\langle\omega - \nu\rangle\| \\
&= \rho(\langle\mu\rangle, \langle\omega\rangle) + \rho(\langle\omega\rangle, \langle\nu\rangle).
\end{aligned} \tag{27}$$

We conclude that ρ is a metric on $\mathscr{F}/\mathscr{S}$. □

4. Conclusions

In this present investigation, we studied the norm induced by the supremum metric d_∞ on the space of fuzzy numbers. And then we proposed a method for constructing a norm on the quotient space of fuzzy numbers. This norm is very natural and works well with the induced metric on $\mathscr{F}/\mathscr{S}$. The works in this paper enable us to study the fuzzy numbers in the new environment. We hope that our results in this paper may lead to significant, new, and innovative results in other related fields.

Acknowledgments

This work was supported by the Mathematical Tianyuan Foundation of China (Grant no. 11126087), the National Natural Science Foundation of China (Grant no. 11201512) and the Natural Science Foundation Project of CQ CSTC (cstc2012jjA00001).

References

[1] B. Bede and S. G. Gal, "Generalizations of the differentiability of fuzzy-number-valued functions with applications to fuzzy differential equations," *Fuzzy Sets and Systems*, vol. 151, no. 3, pp. 581–599, 2005.

[2] D. Dubois and H. Prade, "Operations on fuzzy numbers," *International Journal of Systems Science*, vol. 9, no. 6, pp. 613–626, 1978.

[3] D. Dubois and H. Prade, *Fuzzy Sets and Systems*, Academic Press, New York, NY, USA, 1980.

[4] D. H. Hong, "An additive decomposition of fuzzy numbers," *Kybernetika*, vol. 39, no. 3, pp. 289–294, 2003.

[5] M. Michta, "On set-valued stochastic integrals and fuzzy stochastic equations," *Fuzzy Sets and Systems*, vol. 177, no. 1, pp. 1–19, 2011.

[6] L. Stefanini, "A generalization of Hukuhara difference and division for interval and fuzzy arithmetic," *Fuzzy Sets and Systems*, vol. 161, no. 11, pp. 1564–1584, 2010.

[7] L. A. Zadeh, "The concept of alinguistic variable and its applications to approximate reasoning—I," *Information Sciences*, vol. 8, no. 3, pp. 199–249, 1975.

[8] L. A. Zadeh, "The concept of alinguistic variable and its applications to approximate reasoning,—II," *Information Sciences*, vol. 8, no. 4, pp. 301–357, 1975.

[9] L. A. Zadeh, "The concept of a linguistic variable and its application to approximate reasoning—III," *Information Sciences*, vol. 9, no. 1, pp. 43–80, 1975.

[10] A. M. Bica, "Algebraic structures for fuzzy numbers from categorial point of view," *Soft Computing*, vol. 11, no. 11, pp. 1099–1105, 2007.

[11] D. Dubois and H. Prade, "Additions of interactive fuzzy numbers," *IEEE Transactions on Automatic Control*, vol. 26, no. 4, pp. 926–936, 1981.

[12] D. H. Hong and S. Lee, "Some algebraic properties and a distance measure for interval-valued fuzzy numbers," *Information Sciences*, vol. 148, no. 1–4, pp. 1–10, 2002.

[13] A. Maturo, "On some structures of fuzzy numbers," *Iranian Journal of Fuzzy Systems*, vol. 6, no. 4, pp. 49–59, 2009.

[14] R. Mesiar, "A note to the T-sum of L-R fuzzy numbers," *Fuzzy Sets and Systems*, vol. 79, no. 2, pp. 259–261, 1996.

[15] S. G. Pushkov, "Fuzzy modules with respect to a t-norm and some of their properties," *Journal of Mathematical Sciences*, vol. 154, no. 3, pp. 374–378, 2008.

[16] D. Qiu and L. Shu, "Notes on 'On the restudy of fuzzy complex analysis: part I and part II'," *Fuzzy Sets and Systems*, vol. 159, no. 16, pp. 2185–2189, 2008.

[17] D. Qiu and L. Shu, "Supremum metric on the space of fuzzy sets and common fixed point theorems for fuzzy mappings," *Information Sciences*, vol. 178, no. 18, pp. 3595–3604, 2008.

[18] D. Qiu, L. Shu, and Z. Mo, "Notes on fuzzy complex analysis," *Fuzzy Sets and Systems*, vol. 160, no. 11, pp. 1578–1589, 2009.

[19] D. Qiu, L. Shu, and Z. Mo, "On starshaped fuzzy sets," *Fuzzy Sets and Systems*, vol. 160, no. 11, pp. 1563–1577, 2009.

[20] M. Mareš, "Addition of fuzzy quantities: disjunction-conjunction approach," *Kybernetika*, vol. 25, no. 2, pp. 104–116, 1989.

[21] M. Mareš, "Additive decomposition of fuzzy quantities with finite supports," *Fuzzy Sets and Systems*, vol. 47, no. 3, pp. 341–346, 1992.

[22] D. H. Hong and H. Y. Do, "Additive decomposition of fuzzy quantities," *Information Sciences*, vol. 88, no. 1–4, pp. 201–207, 1996.

[23] M. Mareš, "Algebraic equivalences over fuzzy quantities," *Kybernetika*, vol. 29, no. 2, pp. 122–132, 1993.

[24] M. Mareš, "Brief note on distributivity of triangular fuzzy numbers," *Kybernetika*, vol. 31, no. 5, pp. 451–457, 1995.

[25] D. Qiu and W. Zhang, "Symmetric fuzzy numbers and additive equivalence of fuzzy numbers," *Soft Computing*, vol. 17, no. 8, pp. 1471–1477, 2013.

[26] K. D. Jamison, "Possibilities as cumulative subjective probabilities and a norm on the space of congruence classes of fuzzy numbers motivated by an expected utility functional," *Fuzzy Sets and Systems*, vol. 111, no. 3, pp. 331–339, 2000.

Stacked Central Configurations for the Spatial Nine-Body Problem

Su Xia and Deng Chunhua

Faculty of Mathematics and Physics, Huaiyin Institute of Technology, Huai'an 223003, China

Correspondence should be addressed to Su Xia; dchsx2005@163.com

Academic Editors: Y.-S. Piao and M. Shibata

We show the existence of the twisted stacked central configurations for the 9-body problem. More precisely, the position vectors x_1, x_2, x_3, x_4, and x_5 are at the vertices of a square pyramid Σ; the position vectors x_6, x_7, x_8, and x_9 are at the vertices of a square Π.

1. Introduction and Main Results

The classical n-body problem [1, 2] concerns the motion of n mass points moving in space according to Newton's law:

$$m_i \ddot{x}_i = -\sum_{j=1, j\neq i}^{n} \frac{m_i m_j \left(x_i - x_j\right)}{r_{ij}^3}, \quad i = 1, 2, \ldots, n. \quad (1)$$

Here, $x_i \in \mathbb{R}^d$ is the position of mass $m_i > 0$, the gravitational constant is taken equal to 1, and $r_{ij} = |x_i - x_j|$ is the Euclidean distance between x_i and x_j.

The space of configuration is defined by

$$X = \left\{ \left(x_1, \ldots, x_n\right) \in \left(\mathbb{R}^d\right)^n : x_i \neq x_j \ \forall i \neq j \right\}, \quad (2)$$

while the center of mass is given by

$$c = \frac{m_1 x_1 + \cdots + m_n x_n}{M}, \quad (3)$$

where $M = m_1 + \cdots + m_n$ is the total mass.

A configuration $x = (x_1, \ldots, x_n) \in X$ is called a *central configuration* [2, 3] if there exists a constant λ, called the multiplier, such that

$$-\lambda \left(x_i - c\right) = \sum_{j=1, j\neq i}^{n} \frac{m_j \left(x_j - x_i\right)}{r_{ij}^3}, \quad i = 1, 2, \ldots, n. \quad (4)$$

It is easy to see that a central configuration remains a central configuration after a rotation in $\mathbb{R}^d$ and a scalar multiplication. More precisely, let $A \in SO(d)$ and $a > 0$, if $x = (x_1, \ldots, x_n)$ is a central configuration, so are $Ax = (Ax_1, \ldots, Ax_n)$ and $ax = (ax_1, \ldots, ax_n)$.

Two central configurations are said to be equivalent if one can be transformed to the other by a scalar multiplication and a rotation. In this paper, when we say a central configuration, we mean a class of central configurations as defined by the above equivalent relation.

Central configurations of the n-body problem are important because they allow the computation of homographic solutions; if the n bodies are heading for a simultaneous collision, then the bodies tend to a central configuration (see [3, 4]); there is a relation between central configurations and the bifurcations of the hypersurfaces of constant energy and angular momentum (see [5]).

In this paper, we are interested in spatial central configurations, that is, $d = 3$. In 2005, Hampton [6] provides a new family of planar central configurations for the 5-body problem with an interesting property: the central configuration has a subset of three bodies forming a central configuration of the 3-body problem. The authors [7] find new classes of central configurations of the 5-body problem which are the ones studied by Hampton [6] having three bodies in the vertices of an equilateral triangle, but the other two, instead of being located symmetrically with respect to a perpendicular bisector, are on the perpendicular bisector. The stacked central configurations studied by Hampton [6] were completed by Llibre et al. [8] (see also [9]).

Zhang and Zhou [10] showed the existence of double pyramidal central configurations of $N+2$-body problem. The authors [11–13] provided new examples of stacked central configurations for the spatial 7-body problem where four bodies are at the vertices of a regular tetrahedron and the other three bodies are located at the vertices of an equilateral triangle.

In this paper, we find new classes of stacked spatial central configurations for the 9-body problem which have five bodies at the vertices of a square pyramid, and the other four bodies are located at the vertices of a square. More precisely, the spatial central configurations considered here satisfy the following (see Figure 1): the position vectors x_1, x_2, x_3, x_4, and x_5 are at the vertices of a square pyramid Σ; the position vectors x_6, x_7, x_8, and x_9 are at the vertices of a square Π.

Without loss of generality, we can assume that

$$\begin{aligned} &x_1=(1,0,0), \quad x_2=(0,1,0), \quad x_3=(-1,0,0),\\ &x_4=(0,-1,0), \quad x_5=(0,0,h), \quad x_6=(x,0,y),\\ &x_7=(0,x,y), \quad x_8=(-x,0,y), \quad x_9=(0,-x,y), \end{aligned} \tag{5}$$

where $x>0$, $y\in\mathbb{R}$, and $y\neq 0$; the positive constant h satisfies the equation

$$\frac{2}{r_{15}^3}=\frac{1}{r_{12}^3}+\frac{1}{r_{13}^3}, \tag{6}$$

(see [10] and the references therein); that is, $h=1.26276522$.

The main results of this paper are the following.

Theorem 1. *Consider the spatial configurations according to Figure 1, in order that the nine mass points are in a central configuration, the following statements are necessary:*

(1) *the masses m_1, m_2, m_3, and m_4 must be equal;*

(2) *the masses m_6, m_7, m_8, and m_9 must be equal.*

Theorem 2. *There exist points $(x_0,y_0)\in T^{-1}(0)\cap D$ (see Figure 2) such that the nine bodies take the coordinates*

$$\begin{aligned} &x_1=(1,0,0), \quad x_2=(0,1,0),\\ &x_3=(-1,0,0), \quad x_4=(0,-1,0),\\ &x_5=(0,0,h), \quad x_6=(x_0,0,y_0),\\ &x_7=(0,x_0,y_0), \quad x_8=(-x_0,0,y_0),\\ &x_9=(0,-x_0,y_0). \end{aligned} \tag{7}$$

Then, there are positive solutions of m_1, m_5, m_6 such that these bodies form a spatial central configuration according to Figure 1.

The proofs of the theorems are given in the next sections.

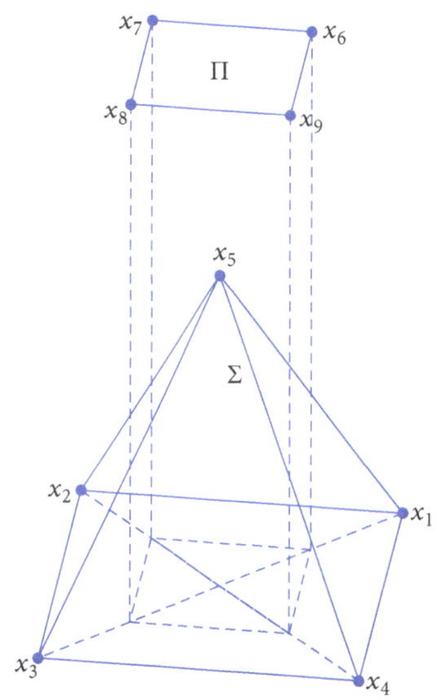

FIGURE 1: The configuration for the 9-body problem.

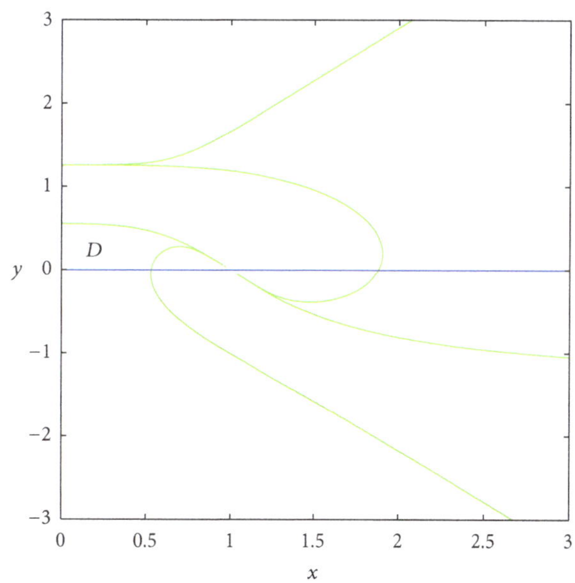

FIGURE 2: The region D.

2. Proof of Theorem 1

For the spatial central configurations, instead of working with (4), we consider the Dziobek-Laura-Andoyer equations (see [9, 11–13] and the references therein):

$$f_{ijk}=\sum_{l=1, l\neq i,j,k}^{n} m_l\left(d_{il}-d_{jl}\right)\Delta_{ijkl}=0 \tag{8}$$

for $1\le i<j\le n$, $k=1,\ldots,n$, $k\neq i,j$. Here, $d_{ij}=1/r_{ij}^3$ and $\Delta_{ijkl}=(x_i-x_j)\wedge(x_i-x_k)\cdot(x_i-x_l)$. Thus, Δ_{ijkl} gives six times the signed volume of the tetrahedron formed by the bodies with positions x_i, x_j, x_k, and x_l; (8) is a system of $n(n-1)(n-2)/2$ equations.

For the 9-body problem, (8) is a system of 252 equations. According to Figure 1, our class of configurations with nine bodies must satisfy

$$\begin{aligned}
& r_{12} = r_{23} = r_{34} = r_{14} = \sqrt{2}, \qquad r_{13} = r_{24} = 2, \\
& r_{67} = r_{78} = r_{89} = r_{69} = \sqrt{2}x, \qquad r_{68} = r_{79} = 2x, \\
& r_{16} = r_{27} = r_{38} = r_{49} = \sqrt{(x-1)^2 + y^2}, \\
& r_{17} = r_{19} = r_{26} = r_{28} = r_{37} = r_{39} = r_{46} \\
& \quad = r_{48} = \sqrt{x^2 + 1 + y^2}, \\
& r_{18} = r_{29} = r_{36} = r_{47} = \sqrt{(x+1)^2 + y^2}, \\
& r_{15} = r_{25} = r_{35} = r_{45} = \sqrt{1 + h^2}, \\
& r_{56} = r_{57} = r_{58} = r_{59} = \sqrt{x^2 + (y-h)^2}.
\end{aligned} \tag{9}$$

Due to assumption (5) and the definition of Δ_{ijkl}, we have several symmetries in the signed volumes.

By using the symmetries and the properties of Δ_{ijkl}, we obtain the following results.

Lemma 3. *In order to have a spatial central configuration according to Figure 1, a necessary condition is that the masses m_1, m_2, m_3, and m_4 must be equal.*

Proof. It is sufficient to consider the equations $f_{687} = 0$ and $f_{796} = 0$:

$$\begin{aligned}
f_{687} &= (m_1 - m_3)(d_{16} - d_{18})\Delta_{6871} = 0, \\
f_{796} &= (m_2 - m_4)(d_{16} - d_{18})\Delta_{7962} = 0.
\end{aligned} \tag{10}$$

For our class of central configurations, we have $d_{16} - d_{18} \neq 0$, $\Delta_{6871} \neq 0$, and $\Delta_{7962} \neq 0$. So the above equations hold if and only if $m_1 = m_3$, $m_2 = m_4$. Consider the expression of $f_{678} = 0$:

$$\begin{aligned}
f_{678} = {} & (m_1 - m_2)(d_{16} - d_{17})\Delta_{6781} \\
& + (m_3 - m_4)(d_{18} - d_{17})\Delta_{6783} = 0.
\end{aligned} \tag{11}$$

Substituting $m_1 = m_3$, $m_2 = m_4$ into the above equation, we have

$$f_{678} = (m_1 - m_2)(d_{16} + d_{18} - 2d_{17})\Delta_{6781} = 0. \tag{12}$$

For our class of central configurations, we have $d_{16} + d_{18} - 2d_{17} \neq 0$, since the function $g(x) = x^{-3/2}$ is convex for all $x > 0$, and $\Delta_{6781} \neq 0$. So the above equation holds if and only if $m_1 = m_2$. So statement 1 of Theorem 1 is proved. □

Lemma 4. *If the configuration, according to Figure 1, is a central configuration, a necessary condition is that the masses m_6, m_7, m_8, and m_9 must be equal.*

Proof. It is sufficient to consider the equations $f_{132} = 0$ and $f_{241} = 0$:

$$\begin{aligned}
f_{132} &= (m_6 - m_8)(d_{16} - d_{18})\Delta_{1326} = 0, \\
f_{241} &= (m_7 - m_9)(d_{16} - d_{18})\Delta_{2417} = 0.
\end{aligned} \tag{13}$$

For our class of central configurations, we have $d_{16} - d_{18} \neq 0$, $\Delta_{1326} \neq 0$, and $\Delta_{2417} \neq 0$. So the above equations hold if and only if $m_6 = m_8$, $m_7 = m_9$. Consider the expression of $f_{123} = 0$:

$$\begin{aligned}
f_{123} = {} & (m_6 - m_7)(d_{16} - d_{17})\Delta_{1236} \\
& + (m_8 - m_9)(d_{18} - d_{17})\Delta_{1238} = 0.
\end{aligned} \tag{14}$$

Substituting $m_6 = m_8$, $m_7 = m_9$ into the above equation, we have

$$f_{123} = (m_6 - m_7)(d_{16} + d_{18} - 2d_{17})\Delta_{1236} = 0. \tag{15}$$

For our class of central configurations, we have $d_{16} + d_{18} - 2d_{17} \neq 0$, and $\Delta_{1236} \neq 0$. So the above equation holds if and only if $m_6 = m_7$. Hence, statement 2 of Theorem 1 is proved.

The proof Theorem 1 is completed. □

We restrict the set of admissible masses to $m_1 = m_2 = m_3 = m_4 = \alpha$ and $m_6 = m_7 = m_8 = m_9 = \beta$. Substituting $m_1 = m_2 = m_3 = m_4 = \alpha$ and $m_6 = m_7 = m_8 = m_9 = \beta$ into (8), they reduce to the following 4 equations:

$$\begin{aligned}
f_{152} = {} & \beta((d_{16} + d_{17} - 2d_{56})\Delta_{1526} \\
& + (d_{17} + d_{18} - 2d_{56})\Delta_{1528}) = 0,
\end{aligned} \tag{16}$$

$$\begin{aligned}
f_{162} = {} & \alpha(d_{12} + d_{13} - d_{17} - d_{18})\Delta_{1623} \\
& + m_5(d_{15} - d_{56})\Delta_{1625} \\
& + \beta(d_{17} + d_{18} - d_{67} - d_{68})\Delta_{1628} = 0,
\end{aligned} \tag{17}$$

$$\begin{aligned}
f_{175} = {} & \alpha((d_{12} - d_{16})\Delta_{1752} + (d_{13} - d_{17})\Delta_{1753} \\
& + (d_{12} - d_{18})\Delta_{1754}) \\
& + \beta((d_{16} - d_{67})\Delta_{1756} + (d_{18} - d_{67})\Delta_{1758} \\
& + (d_{17} - d_{68})\Delta_{1759}) = 0,
\end{aligned} \tag{18}$$

$$\begin{aligned}
f_{562} = {} & \alpha((d_{15} - d_{16})\Delta_{5621} + (d_{15} - d_{18})\Delta_{5623} \\
& + (d_{15} - d_{17})\Delta_{5624}) \\
& + \beta((d_{56} - d_{67})\Delta_{5627} + (d_{56} - d_{68})\Delta_{5628} \\
& + (d_{56} - d_{67})\Delta_{5629}) = 0.
\end{aligned} \tag{19}$$

If we write $f_{152} = \beta T = \beta((d_{16} + d_{17} - 2d_{56})\Delta_{1526} + (d_{17} + d_{18} - 2d_{56})\Delta_{1528}) = 0$, it follows that $T = 0$ in order to have central configurations. So in the following, we restrict our central configurations to the set $T^{-1}(0)$.

Lemma 5. *According to one's assumptions and the set $T^{-1}(0)$, (8) is satisfied if (17) and (18) are satisfied.*

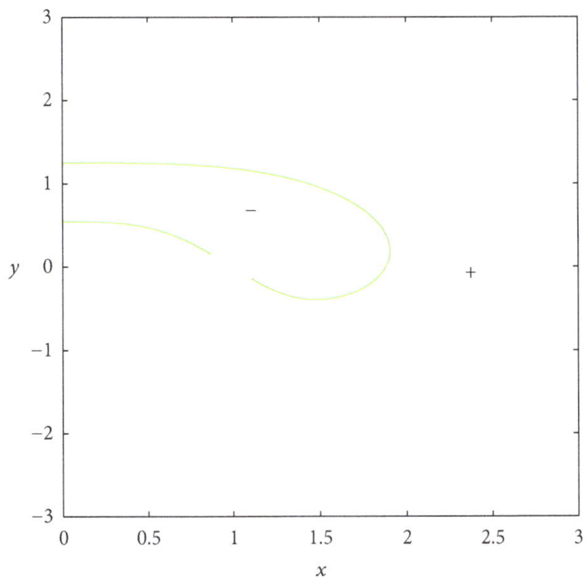

FIGURE 3: The curve $a_{21} = 0$.

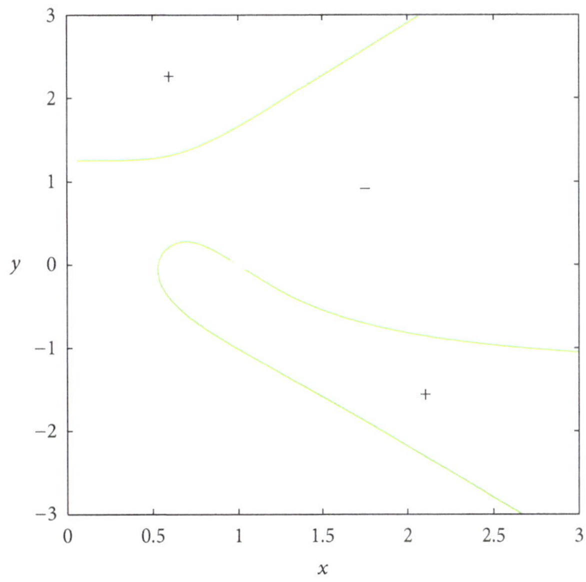

FIGURE 4: The curve $a_{23} = 0$.

Proof. Under the assumptions (5), we have

$$T = (d_{16} + 2d_{17} + d_{18} - 4d_{56})(y - h) + hx(d_{16} - d_{18}) = 0; \quad (20)$$

that is,

$$4(y - h)d_{56} = (y - h)(d_{16} + 2d_{17} + d_{18}) + hx(d_{16} - d_{18}). \quad (21)$$

Substituting (21) into (19), we obtain the equation $f_{175} = 0$.

Hence in the set $T^{-1}(0)$, $f_{175} = 0$ implies $f_{562} = 0$. This completes the proof. □

From Lemma 5, in order to study central configurations according to Figure 1 in the set $T^{-1}(0)$, it is sufficient to study the following 2 equations:

$$f_{162} = 0, \qquad f_{175} = 0. \quad (22)$$

Denote by $A = (a_{ij})$ the matrix of the coefficients of the homogeneous linear system in the variables α, m_4, β defined by (22). Thus,

$$\begin{aligned} a_{11} &= (d_{12} + d_{13} - d_{17} - d_{18})\Delta_{1623} \\ &= -2y\left(\frac{1}{2\sqrt{2}} + \frac{1}{8} - \frac{1}{(x^2 + 1 + y^2)^{3/2}} - \frac{1}{((x+1)^2 + y^2)^{3/2}}\right), \end{aligned}$$

$$\begin{aligned} a_{12} &= (d_{15} - d_{56})\Delta_{1625} \\ &= (-y - hx + h) \times \left(\frac{1}{(1 + h^2)^{3/2}} - \frac{1}{(x^2 + (y - h)^2)^{3/2}}\right), \end{aligned}$$

$$\begin{aligned} a_{13} &= (d_{17} + d_{18} - d_{67} - d_{68})\Delta_{1628} \\ &= -2xy\left(\frac{1}{(x^2 + 1 + y^2)^{3/2}} + \frac{1}{((x+1)^2 + y^2)^{3/2}} - \frac{1}{8x^3} - \frac{1}{2\sqrt{2}x^3}\right), \end{aligned}$$

$$\begin{aligned} a_{21} &= (d_{12} - d_{16})\Delta_{1752} + (d_{13} - d_{17})\Delta_{1753} + (d_{12} - d_{18})\Delta_{1754} \\ &= (y - h)\left(\frac{1}{((x+1)^2 + y^2)^{3/2}} - \frac{1}{((x-1)^2 + y^2)^{3/2}}\right) \\ &\quad + hx\left(\frac{1}{4} + \frac{1}{\sqrt{2}} - \frac{1}{((x-1)^2 + y^2)^{3/2}} - \frac{1}{((x+1)^2 + y^2)^{3/2}} - \frac{2}{(x^2 + 1 + y^2)^{3/2}}\right), \end{aligned}$$

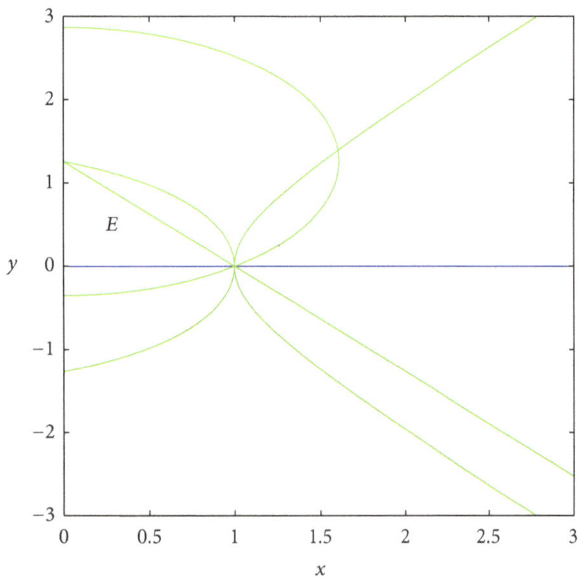

FIGURE 5: The region E.

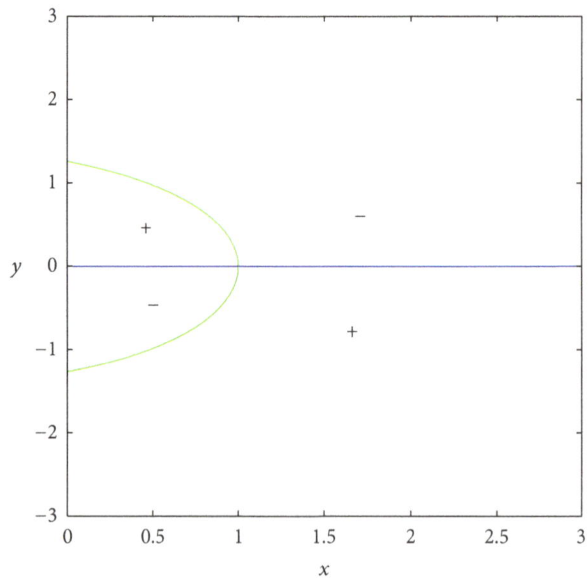

FIGURE 6: The curve $a_{11} = 0$.

$$
\begin{aligned}
a_{22} &= 0, \\
a_{23} &= (d_{16} - d_{67})\,\Delta_{1756} + (d_{18} - d_{67})\,\Delta_{1758} \\
&\quad + (d_{17} - d_{68})\,\Delta_{1759} \\
&= hx^2 \left(\frac{1}{\left((x+1)^2 + y^2\right)^{3/2}} - \frac{1}{\left((x+1)^2 + y^2\right)^{3/2}} \right) \\
&\quad - x\,(y-h) \\
&\quad \times \left(\frac{1}{\left((x+1)^2 + y^2\right)^{3/2}} + \frac{1}{\left((x+1)^2 + y^2\right)^{3/2}} \right. \\
&\quad \left. + \frac{2}{\left(x^2 + 1 + y^2\right)^{3/2}} - \frac{1}{4x^3} - \frac{1}{\sqrt{2}x^3} \right).
\end{aligned}
\tag{23}
$$

Let $x = \begin{pmatrix} \alpha \\ m_4 \\ \beta \end{pmatrix}$. Then in order to get the spatial central configuration as Figure 1, we need to find a positive solution α, m_4, β of the following system:

$$Ax = 0, \tag{24}$$

where $T = 0$.

3. The Existence of Spatial Central Configurations

In order to prove the existence of positive solutions of (24) in the set $T^{-1}(0)$, it is sufficient to prove that the entries in each row of A change the signs. So if the entries of some row of A have the same signs, there are no admissible masses such that the bodies are in a central configuration according to Figure 1.

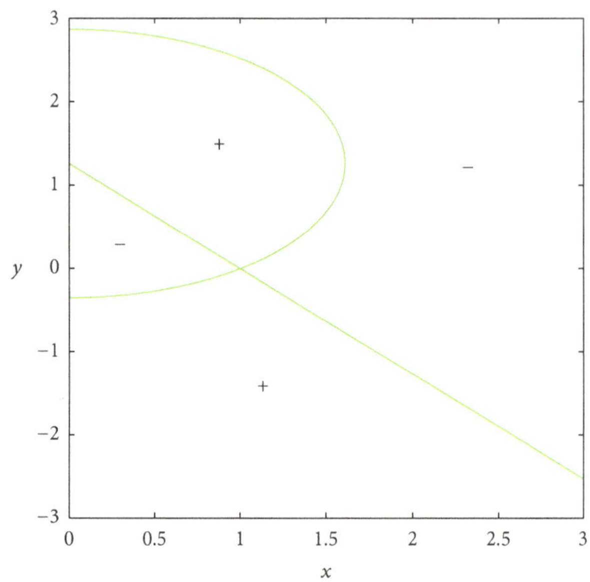

FIGURE 7: The curve $a_{12} = 0$.

Proof of Theorem 2. Since the rank of matrix A is two in the set $T^{-1}(0)$, there are nontrivial solutions of (24) in the set $T^{-1}(0)$.

Now we prove the existence of spatial central configurations according to Figure 1 for some points in the set D (see Figure 2). In order to prove the existence of positive solutions of (24) in the set $T^{-1}(0)$, the entries a_{21}, a_{23} of the second line in the matrix A should have opposite signs. Thus, we consider the following set D, where D is surrounded by curves $x = 0$, $y = 0$, $a_{21} = 0$, and $a_{23} = 0$.

In the set D, the entries of matrix A have the following signs: $a_{21} > 0$, $a_{23} < 0$ (see Figures 3 and 4); $a_{11} > 0$, $a_{12} < 0$, $a_{13} > 0$ because the set D is included in the set E, where E is

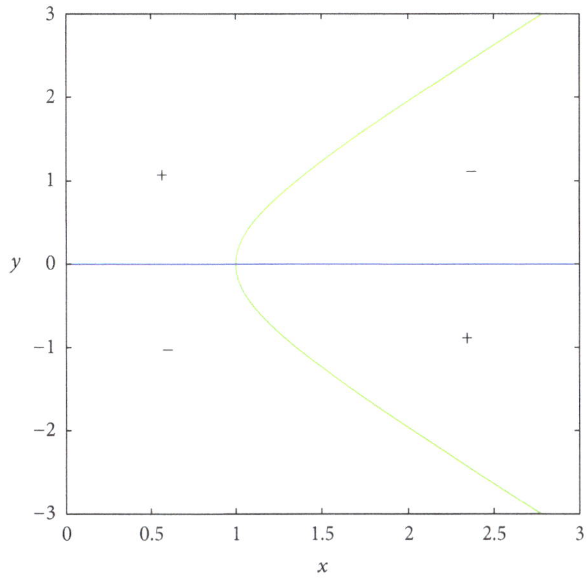

FIGURE 8: The curve $a_{13} = 0$.

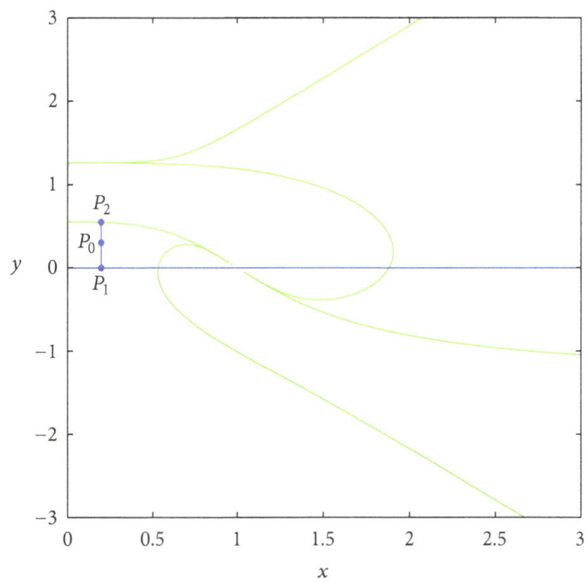

FIGURE 9: The existence of central configurations for the 9-body problem.

surrounded by curves $x = 0$, $y = 0$, and $y = h(1 - x)$ (see Figures 5, 6, 7, and 8). In short, the signs of the entries of the matrix A restricted to the set D are the following:

$$A = \begin{pmatrix} + & - & + \\ + & 0 & - \end{pmatrix}. \quad (25)$$

In the rest of the proof, we show that the set $T^{-1}(0)$ has intersection with the set D. We consider the subset of D:

$$L = \{(x, y) : x = x_1, 0 < y < y_1\}, \quad (26)$$

where $x_1 \in (0, 1)$. Obviously L is a segment with endpoints

$$P_1 = (x_1, 0), \qquad P_2 = (x_1, y_1), \quad (27)$$

(see Figure 9), and the point (x_1, y_1) satisfies the equation $a_{21} = 0$. Evaluating the function T at these points, we have

$$T(P_1) < 0, \qquad T(P_2) > 0. \quad (28)$$

Thus, there exists a point $P_0 = (x_0, y_0) \in L$, such that $T(P_0) = 0$. So at the point P_0 we have nontrivial positive solutions of (24), since the signs of the entries of the matrix A at this point are the following:

$$A(P_0) = \begin{pmatrix} + & - & + \\ + & 0 & - \end{pmatrix}. \quad (29)$$

Thus, the proof of Theorem 2 is completed. □

Acknowledgments

The authors are supported by the Natural Science Foundation of China (NFSC11201168) and the Scientific Research Foundation of Huaiyin Institute of Technology (HGA1102).

References

[1] V. Arnold, V. Kozlov, and A. Neishtadt, *Dynamical Systems, Mathematical Aspects of Classical and Celestial Machanics*, Springer, Berlin, Germany, 1988.

[2] A. Wintner, *Analytical Foundations of Celestial Mechanics*, Princeton University Press, Princeton, NJ, USA, 1941.

[3] R. Moeckel, "On central configurations," *Mathematische Zeitschrift*, vol. 205, no. 4, pp. 499–517, 1990.

[4] D. Sarri, "On the role and the properties of n-body central configurations," *Celestial Mechanics*, vol. 21, no. 1, pp. 9–20, 1980.

[5] S. Smale, "Topology and mechanics II: the planar n-body problem," *Inventiones Mathematicae*, vol. 11, no. 1, pp. 45–64, 1970.

[6] M. Hampton, "Stacked central configurations: new examples in the planar five-body problem," *Nonlinearity*, vol. 18, no. 5, pp. 2299–2304, 2005.

[7] J. Llibre and L. F. Mello, "New central configurations for the planar 5-body problem," *Celestial Mechanics and Dynamical Astronomy*, vol. 100, no. 2, pp. 141–149, 2008.

[8] J. Llibre, L. F. Mello, and E. Perez-Chavela, "New stacked central configurations for the planar 5-body problem," *Celestial Mechanics and Dynamical Astronomy*, vol. 110, no. 1, pp. 43–52, 2011.

[9] L. F. Mello, F. E. Chaves, A. C. Fernandes, and B. A. Garcia, "Stacked central configurations for the spatial six-body problem," *Journal of Geometry and Physics*, vol. 59, no. 9, pp. 1216–1226, 2009.

[10] S. Q. Zhang and Q. Zhou, "Double pyramidal central configurations," *Physics Letters A*, vol. 281, no. 4, pp. 240–248, 2001.

[11] M. Hampton and M. Santoprete, "Seven-body central configurations: a family of central configurations in the spatial seven-body problem," *Celestial Mechanics and Dynamical Astronomy*, vol. 99, no. 4, pp. 293–305, 2007.

[12] L. F. Mello and A. C. Fernandes, "Stacked central configurations for the spatial seven-body problem," *Qualitative Theory of Dynamical Systems*, vol. 12, no. 1, pp. 101–114, 2013.

[13] X. Su and T. Q. An, "Twisted stacked central configurations for the spatial seven-body problem," *Journal of Geometry and Physics*, vol. 70, pp. 164–171, 2013.

g-Weak Contraction in Ordered Cone Rectangular Metric Spaces

S. K. Malhotra,[1] J. B. Sharma,[2] and Satish Shukla[3]

[1] *Department of Mathematics, Govt. S.G.S.P.G. College Ganj Basoda, Vidisha 464221, India*
[2] *Department of Mathematics, Choithram College of Professional Studies, Dhar Road, Indore 453001, India*
[3] *Department of Applied Mathematics, Shri Vaishnav Institute of Technology & Science, Gram Baroli Sanwer Road, Indore 453331, India*

Correspondence should be addressed to Satish Shukla; satishmathematics@yahoo.co.in

Academic Editors: T. Ozawa and S. A. Tersian

We prove some common fixed-point theorems for the ordered g-weak contractions in cone rectangular metric spaces without assuming the normality of cone. Our results generalize some recent results from cone metric and cone rectangular metric spaces into ordered cone rectangular metric spaces. Examples are provided which illustrate the results.

1. Introduction and Preliminaries

There are a number of generalizations of metric spaces. One such generalization is obtained by replacing the real valued metric function with a vector valued metric function. In the mid-20th century (see [1]), the notions of K-metric and K-normed spaces were introduced, in such spaces an ordered Banach space instead of the real numbers was used as a codomain for metric function. Indeed, this idea of replacement of real numbers by an ordered "set" can be seen in [2, 3] (see also references therein). Huang and Zhang [4] reintroduced such spaces under the name of cone metric spaces, defining convergent and Cauchy sequence in terms of interior points of underlying cone. They proved the basic version of the fixed-point theorem with the assumption that the cone is normal. Subsequently several authors (see, e.g., [5–14]) generalized the results of Huang and Zhang. In [13], Rezapour and Hamlbarani removed the normality of cone and proved the results of Huang and Zhang in nonnormal cone metric spaces.

In [15], Branciari introduced a class of generalized metric spaces with replacing triangular inequality by similar ones which involve four or more points instead of three and improved Banach contraction principle. Azam and Arshad [16] proved fixed-point result for Kannan-type contraction in rectangular metric spaces. After the work of Huang and Zhang [4], Azam et al. [17] introduced the notion of cone rectangular metric spaces and proved fixed-point result for Banach-type contraction in cone rectangular space. Samet and Vetro [18] obtained the fixed-point results in c-chainable cone rectangular metric spaces.

Ordered normed spaces and cones have applications in applied mathematics, for instance, in using Newton's approximation method [19] and in optimization theory [20]. The existence of fixed point in partially ordered sets was investigated by Ran and Reurings [21] and then by Nieto and Rodríguez-López [22]. Fixed-point results in ordered cone metric spaces were obtained by several authors (see, e.g., [11, 23–25]). Very recently, Malhotra et al. [26] proved the fixed-point results in ordered cone rectangular metric spaces for Reich-type contractions.

The notion of g-weak contraction is introduced by Vetro (see [14]) in cone metric spaces. In this paper, we prove some common fixed point theorems for g-weak contractions in ordered cone rectangular metric spaces. Our results generalize and extend the results of Huang and Zhang [4], Azam et al. [17], Azam and Arshad [16], Malhotra et al. [26], and the result of Vetro [14] on ordered cone rectangular metric spaces.

We need the following definitions and results, consistent with [4, 20].

Definition 1 (see [4]). Let E be a real Banach space and P a subset of E. The set P is called a cone if

(i) P is closed, nonempty, and $P \neq \{\theta\}$; here θ is the zero vector of E;

(ii) $a, b \in \mathbb{R}$, $a, b \geq 0$, $x, y \in P \Rightarrow ax + by \in P$;

(iii) $x \in P$ and $-x \in P \Rightarrow x = \theta$.

Given a cone $P \subset E$, we define a partial ordering "$\preceq$" with respect to P by $x \preceq y$ if and only if $y - x \in P$. We write $x \prec y$ to indicate that $x \preceq y$ but $x \neq y$, while $x \ll y$ if and only if $y - x \in P^0$, where P^0 denotes the interior of P.

Let P be a cone in a real Banach space E; then P is called normal, if there exists a constant $K > 0$ such that for all $x, y \in E$,

$$\theta \preceq x \preceq y \text{ implies } \|x\| \leq K \|y\|. \tag{1}$$

The least positive number K satisfying the above inequality is called the normal constant of P.

Definition 2 (see [4]). Let X be a nonempty set and E a real Banach space. Suppose that the mapping $d : X \times X \to E$ satisfies

(i) $\theta \preceq d(x, y)$ for all $x, y \in X$ and $d(x, y) = \theta$ if and only if $x = y$;

(ii) $d(x, y) = d(y, x)$ for all $x, y \in X$;

(iii) $d(x, y) \preceq d(x, z) + d(y, z)$ for all $x, y, z \in X$.

Then, d is called a cone metric on X, and (X, d) is called a cone metric space. In the following, we always suppose that E is a real Banach space, and P is a solid cone in E; that is, $P^0 \neq \phi$ and "$\preceq$" is partial ordering with respect to P.

The concept of cone metric space is more general than that of a metric space because each metric space is a cone metric space with $E = \mathbb{R}$ and $P = [0, +\infty)$.

For examples and basic properties of normal and nonnormal cones and cone metric spaces, we refer to [4, 13].

The following remark will be useful in sequel.

Remark 3 (see [27]). Let P be a cone in a real Banach space E, and a, b, $c \in P$ we then have the following

(a) If $a \preceq b$ and $b \ll c$ then $a \ll c$.

(b) If $a \ll b$ and $b \ll c$ then $a \ll c$.

(c) If $\theta \preceq u \ll c$ for each $c \in P^0$, then $u = \theta$.

(d) If $c \in P^0$ and $a_n \to \theta$, then there exist $n_0 \in \mathbb{N}$ such that, for all $n > n_0$, we have $a_n \ll c$.

(e) If $\theta \preceq a_n \preceq b_n$ for each n and $a_n \to a$, $b_n \to b$, then $a \preceq b$.

(f) If $a \preceq \lambda a$, where $0 \leq \lambda < 1$, then $a = \theta$.

Definition 4 (see [17]). Let X be a nonempty set. Suppose that the mapping $d : X \times X \to E$ satisfies

(i) $\theta \preceq d(x, y)$ for all $x, y \in X$ and $d(x, y) = \theta$ if and only if $x = y$;

(ii) $d(x, y) = d(y, x)$ for all $x, y \in X$;

(iii) $d(x, y) \preceq d(x, w) + d(w, z) + d(z, y)$ for all $x, y \in X$ and for all distinct points $w, z \in X - \{x, y\}$ (rectangular property).

Then, d is called a cone rectangular metric on X, and (X, d) is called a cone rectangular metric space. Let $\{x_n\}$ be a sequence in (X, d) and $x \in X$. If for every $c \in E$, with $\theta \ll c$ there is $n_0 \in \mathbb{N}$ such that for all $n > n_0$, $d(x_n, x) \ll c$, then $\{x_n\}$ is said to be convergent, $\{x_n\}$ converges to x, and x is the limit of $\{x_n\}$. We denote this by $\lim_n x_n = x$ or $x_n \to x$, as $n \to \infty$. If for every $c \in E$ with $\theta \ll c$ there is $n_0 \in \mathbb{N}$ such that for all $n > n_0$ and $m \in \mathbb{N}$ we have $d(x_n, x_{n+m}) \ll c$, then $\{x_n\}$ is called a Cauchy sequence in (X, d). If every Cauchy sequence is convergent in (X, d), then (X, d) is called a complete cone rectangular metric space. If the underlying cone is normal, then (X, d) is called normal cone rectangular metric space.

Example 5. Let $X = \mathbb{N}$, $E = \mathbb{R}^2$, and $P = \{(x, y) : x, y \geq 0\}$.

Define $d : X \times X \to E$ as follows:

$$d(x, y) = \begin{cases} (0,0) & \text{if } x = y, \\ (3,9) & \text{if } x \text{ and } y \text{ are in } \{1,2\},\ x \neq y, \\ (1,3) & \text{if } x \text{ and } y \text{ cannot both at} \\ & \quad \text{a time in } \{1,2\},\ x \neq y. \end{cases} \tag{2}$$

Now (X, d) is a cone rectangular metric space but (X, d) is not a cone metric space because it lacks the triangular property:

$$(3,9) = d(1,2) > d(1,3) + d(3,2) = (1,3) + (1,3) = (2,6), \tag{3}$$

as $(3, 9) - (2, 6) = (1, 3) \in P$.

Note that in the above example (X, d) is a normal cone rectangular metric space. The following is an example of non-normal cone rectangular metric space.

Example 6. Let $X = \mathbb{N}$, $E = C^1_{\mathbb{R}}[0, 1]$ with $\|x\| = \|x\|_\infty + \|x'\|_\infty$, and $P = \{x \in E : x(t) \geq 0 \text{ for } t \in [0, 1]\}$. Then, this cone is not normal (see [13]).

Define $d : X \times X \to E$ as follows:

$$d(x, y) = \begin{cases} 0 & \text{if } x = y, \\ 3e^t & \text{if } x \text{ and } y \text{ are in } \{1,2\},\ x \neq y, \\ e^t & \text{if } x \text{ and } y \text{ cannot both at} \\ & \quad \text{a time in } \{1,2\},\ x \neq y. \end{cases} \tag{4}$$

Then (X, d) is nonnormal cone rectangular metric space but (X, d) is not a cone metric space because it lacks the triangular property.

Definition 7 (see [5]). Let f and g be self-mappings of a nonempty set X and $C(f, g) = \{x \in X : fx = gx\}$. The pair (f, g) is called weakly compatible if $fgx = gfx$ for all $x \in C(f, g)$. If $w = fx = gx$ for some x in X, then x is called a coincidence point of f and g, and w is called a point of coincidence of f and g.

Definition 8. If a nonempty set X is equipped with a partial order "$\sqsubseteq$" and mapping $d : X \times X \to E$ such that (X, d)

is a cone rectangular metric space, then $(X, \sqsubseteq, d)$ is called an ordered cone rectangular metric space. Let $f, g : X \to X$ be two mappings. The mapping f is called nondecreasing with respect to "$\sqsubseteq$", if for each $x, y \in X$, $x \sqsubseteq y$ implies $fx \sqsubseteq fy$. The mapping f is called g-nondecreasing if for each $x, y \in X$, $gx \sqsubseteq gy$ implies $fx \sqsubseteq fy$. A subset $\mathscr{A}$ of X is called well ordered if for all the elements of $\mathscr{A}$ are comparable; that is, for all $x, y \in \mathscr{A}$ either $x \sqsubseteq y$ or $y \sqsubseteq x$. $\mathscr{A}$ is called g-well ordered if all the elements of $\mathscr{A}$ are g-comparable; that is, for all $x, y \in \mathscr{A}$ either $gx \sqsubseteq gy$ or $gy \sqsubseteq gx$.

In the trivial case, that is, for $g = I_X$ (the identity mapping of X), the g-well orderedness reduces into well orderedness. But, for nontrivial cases, that is, when $g \neq I_X$ the concepts of g-well orderedness and well orderedness are independent.

Example 9. Let $X = \{0, 1, 2, 3, 4\}$, let "$\sqsubseteq$" be a partial order relation on X defined by $\sqsubseteq = \{(0,0), (1,1), (2,2), (3,3), (4,4), (1,2), (2,3), (1,3), (1,4)\}$. Let $\mathscr{A} = \{0, 1, 3\}$, $\mathscr{B} = \{1, 4\}$ and $g : X \to X$ be defined by $g0 = 1$, $g1 = 2$, $g2 = 3$, $g3 = 3$, $g4 = 0$. Then it is clear that $\mathscr{A}$ is not well ordered but it is g-well ordered, while $\mathscr{B}$ is not g-well ordered but it is well ordered.

Let $(X, \sqsubseteq, d)$ be an ordered cone rectangular metric space $f, g : X \to X$ two mappings. The mapping f is called ordered Reich-type contraction if for all $x, y \in X$ with $x \sqsubseteq y$, λ, μ, $\delta \in [0, 1)$ such that $\lambda + \mu + \delta < 1$ and

$$d(fx, fy) \preceq \lambda d(x, y) + \mu d(x, fx) + \delta d(y, fy). \quad (5)$$

If (5) is satisfied for all $x, y \in X$, then f is called Reich contraction.

The mapping f is called an ordered g-weak contraction if

$$d(fx, fy) \preceq \lambda d(gx, gy) + \mu d(gx, fx) + \delta d(gy, fy) \quad (6)$$

for all $x, y \in X$ with $gx \sqsubseteq gy$, where λ, μ, and δ are nonnegative constants such that $\lambda + \mu + \delta < 1$. If inequality (6) is satisfied for all $x, y \in X$, then f is called a g-weak contraction.

Note that for $g = I_X$ (the identity mapping of X) the ordered g-weak contraction reduces into the ordered Reich contraction.

Now, we can state our main results.

2. Main Results

Theorem 10. *Let $(X, \sqsubseteq, d)$ be an ordered cone rectangular metric space $f, g : X \to X$ two mappings such that $f(X) \subset g(X)$ and $g(X)$ is complete. Suppose that the following conditions are satisfied:*

(i) *f is an ordered g-weak contraction, that is, satisfies* (6)*;*

(ii) *f is g-nondecreasing;*

(iii) *there exists $x_0 \in X$ such that $gx_0 \sqsubseteq fx_0$;*

(iv) *if $\{gx_n\}$ were any nondecreasing sequence in X converging to some gz, then $gx_n \sqsubseteq gz$ for all n and $gz \sqsubseteq ggz$.*

Then, f and g have a coincidence point. Furthermore, if f and g are weakly compatible then they have a common fixed point. In addition, the set of common fixed points of f and g is g-well ordered if and only if the common fixed point of f and g is unique.

Proof. Starting with given $x_0 \in X$, we define a sequence $\{y_n\}$ as follows: let $fx_0 = gx_1 = y_1$ (which is possible as $f(X) \subset g(X)$). As $gx_0 \sqsubseteq fx_0$, we have $gx_0 \sqsubseteq gx_1$, and as f is g-nondecreasing, we obtain $fx_0 \sqsubseteq fx_1$. Again, $f(X) \subset g(X)$ therefore let $fx_1 = gx_2 = y_2$. Since $gx_1 \sqsubseteq gx_2$ and f is g-nondecreasing, we obtain $gx_1 \sqsubseteq gx_2$. On repeating this process, we obtain

$$\begin{gathered} fx_0 \sqsubseteq fx_1 \sqsubseteq \cdots \sqsubseteq fx_n \sqsubseteq fx_{n+1} \sqsubseteq \cdots, \\ gx_1 \sqsubseteq gx_2 \sqsubseteq \cdots \sqsubseteq gx_n \sqsubseteq gx_{n+1} \sqsubseteq \cdots, \\ \forall n \in \mathbb{N}. \\ fx_{n-1} = gx_n = y_n \end{gathered} \quad (7)$$

Thus, $\{y_n\} = \{gx_n\}$ is a nondecreasing sequence with respect to $\sqsubseteq$.

We will show that f and g have a point of coincidence. If, $y_n = y_{n+1}$ for any $n \in \mathbb{N}$, we have $y_n = gx_n = fx_n$; therefore, y_n is a point of coincidence of f and g with coincidence point x_n. Therefore, we assume that $y_n \neq y_{n+1}$ for all $n \in \mathbb{N}$.

As, $gx_n \sqsubseteq gx_{n+1}$ for all $n \in \mathbb{N}$, it follows from (6) that

$$\begin{aligned} d(y_n, y_{n+1}) &= d(fx_{n-1}, fx_n) \\ &\preceq \lambda d(gx_{n-1}, gx_n) + \mu d(gx_{n-1}, fx_{n-1}) \\ &\quad + \delta d(gx_n, fx_n) \\ &= \lambda d(y_{n-1}, y_n) + \mu d(y_{n-1}, y_n) + \delta d(y_n, y_{n+1}) \\ &= (\lambda + \mu) d(y_{n-1}, y_n) + \delta d(y_n, y_{n+1}). \end{aligned} \quad (8)$$

For simplicity, set $d_n = d(y_n, y_{n+1})$ for all $n \in \mathbb{N}$; then it follows from above inequality that

$$d_n \preceq \frac{\lambda + \mu}{1 - \delta} d_{n-1} = \alpha d_{n-1}, \quad \forall n \in \mathbb{N}, \quad (9)$$

where $\alpha = (\lambda + \mu)/(1 - \delta) < 1$ (as $\lambda + \mu + \delta < 1$). By repeating this process, we obtain

$$d_n \preceq \alpha^n d_0, \quad \forall n \in \mathbb{N}. \quad (10)$$

If $y_n = y_{n+p}$ for any $n \in \mathbb{N}$ and positive integer $p > 1$, then as $y_n \sqsubseteq y_{n+2}$, it follows from (6) that

$$\begin{aligned} d(y_n, y_{n+1}) &= d(y_{n+p}, y_{n+1}) = d(fx_{n+p-1}, fx_n) \\ &\preceq \lambda d(gx_{n+p-1}, gx_n) + \mu d(gx_{n+p-1}, fx_{n+p-1}) \\ &\quad + \delta d(gx_n, fx_n) \\ &= \lambda d(y_{n+p-1}, y_n) + \mu d(y_{n+p-1}, y_{n+p}) \\ &\quad + \delta d(y_n, y_{n+1}) \\ &= \lambda d(y_{n+p-1}, y_{n+p}) + \mu d(y_{n+p-1}, y_{n+p}) \\ &\quad + \delta d(y_n, y_{n+1}); \end{aligned} \quad (11)$$

that is,

$$d_n \preceq \lambda d_{n+p-1} + \mu d_{n+p-1} + \delta d_n,$$
$$d_n \preceq \alpha d_{n+p-1}. \quad (12)$$

Repeating this process p times, we obtain

$$d_n \preceq \alpha^p d_n < d_n \quad \left(\text{as } \alpha = \frac{\lambda+\mu}{1-\delta} < 1\right), \quad (13)$$

a contradiction. Therefore, we can assume that $y_n \neq y_m$ for all distinct $n, m \in \mathbb{N}$.

Again, as $y_n \sqsubseteq y_{n+2}$, we obtain from (6) and (10) that

$$\begin{aligned}
d(y_n, y_{n+2}) &= d(fx_{n-1}, fx_{n+1}) \\
&\preceq \lambda d(gx_{n-1}, gx_{n+1}) + \mu d(gx_{n-1}, fx_{n-1}) \\
&\quad + \delta d(gx_{n+1}, fx_{n+1}) \\
&= \lambda d(y_{n-1}, y_{n+1}) + \mu d(y_{n-1}, y_n) \\
&\quad + \delta d(y_{n+1}, y_{n+2}) \\
&\preceq \lambda [d(y_{n-1}, y_n) + d(y_n, y_{n+2}) + d(y_{n+2}, y_{n+1})] \\
&\quad + \mu d(y_{n-1}, y_n) + \delta d(y_{n+1}, y_{n+2}) \\
&= \lambda [d_{n-1} + d(y_n, y_{n+2}) + d_{n+1}] \\
&\quad + \mu d_{n-1} + \delta d_{n+1};
\end{aligned} \quad (14)$$

that is,

$$\begin{aligned}
d(y_n, y_{n+2}) &\preceq \frac{\lambda+\mu}{1-\lambda} d_{n-1} + \frac{\lambda+\delta}{1-\lambda} d_{n+1} \\
&\preceq \frac{\lambda+\mu}{1-\lambda}\alpha^{n-1} d_0 + \frac{\lambda+\delta}{1-\lambda}\alpha^{n+1} d_0 \\
&\preceq \frac{\lambda+\mu+[\lambda+\delta]\alpha^2}{1-\lambda}\alpha^{n-1} d_0.
\end{aligned} \quad (15)$$

As $\alpha < 1$, we have

$$d(y_n, y_{n+2}) \preceq \frac{2\lambda+\mu+\delta}{1-\lambda}\alpha^{n-1} d_0,$$
$$d(y_n, y_{n+2}) \preceq \beta\alpha^{n-1} d_0, \quad \forall n \in \mathbb{N}, \quad (16)$$

where $\beta = (2\lambda+\mu+\delta)/(1-\lambda) \geq 0$.

For the sequence $\{y_n\}$, we consider $d(y_n, y_{n+p})$ in two cases.

If p is odd say $2m+1$, then using rectangular inequality and (10), we obtain

$$\begin{aligned}
d(y_n, y_{n+2m+1}) &\preceq d(y_{n+2m}, y_{n+2m+1}) + d(y_{n+2m-1}, y_{n+2m}) \\
&\quad + d(y_n, y_{n+2m-1}) \\
&= d_{n+2m} + d_{n+2m-1} + d(y_n, y_{n+2m-1}) \\
&\preceq d_{n+2m} + d_{n+2m-1} + d_{n+2m-2} + d_{n+2m-3} \\
&\quad + \cdots + d_n \\
&\preceq \alpha^{n+2m} d_0 + \alpha^{n+2m-1} d_0 + \alpha^{n+2m-2} d_0 \\
&\quad + \cdots + \alpha^n d_0 \\
&= [\alpha^{2m} + \alpha^{2m-1} + \cdots + 1]\alpha^n d_0 \\
&\preceq \frac{\alpha^n}{1-\alpha} d_0.
\end{aligned} \quad (17)$$

Therefore,

$$d(y_n, y_{n+2m+1}) \preceq \frac{\alpha^n}{1-\alpha} d_0. \quad (18)$$

If p is even, say $2m$, then using rectangular inequality and (10) and (16), we obtain

$$\begin{aligned}
d(y_n, y_{n+2m}) &\preceq d(y_{n+2m-1}, y_{n+2m}) + d(y_{n+2m-1}, y_{n+2m-2}) \\
&\quad + d(y_{n+2m-2}, y_n) \\
&= d_{n+2m-1} + d_{n+2m-2} + d(y_{n+2m-2}, y_n) \\
&\preceq d_{n+2m-1} + d_{n+2m-2} + d_{n+2m-3} + d_{n+2m-4} \\
&\quad + \cdots + d_{n+2} + d(y_n, y_{n+2}) \\
&\preceq \alpha^{n+2m-1} d_0 + \alpha^{n+2m-2} d_0 + \cdots + \alpha^{n+2} d_0 \\
&\quad + \beta\alpha^{n-1} d_0 \\
&= [\alpha^{2m-1} + \alpha^{2m-2} + \cdots + \alpha^2]\alpha^n d_0 + \beta\alpha^{n-1} d_0 \\
&\preceq \frac{\alpha^n}{1-\alpha} d_0 + \beta\alpha^{n-1} d_0.
\end{aligned} \quad (19)$$

Therefore,

$$d(y_n, y_{n+2m}) \preceq \frac{\alpha^n}{1-\alpha} d_0 + \beta\alpha^{n-1} d_0. \quad (20)$$

As $\beta \geq 0$ and $0 \leq \alpha < 1$, we have $(\alpha^n/(1-\alpha))d_0 \to \theta$ and $\beta\alpha^{n-1}d_0 \to \theta$. So, it follows from (18), (20), and (a), (d) of Remark 3 that for every $c \in E$ with $\theta \ll c$ there exists $n_0 \in \mathbb{N}$ such that $d(y_n, y_{n+2m+1}) \ll c$ and $d(y_n, y_{n+2m}) \ll c$ for all $n > n_0$. Thus, $\{y_n\} = \{gx_n\}$ is a Cauchy sequence in $g(X)$. As $g(X)$ is complete, there exist $z, u \in X$ such that

$$\lim_{n\to\infty} y_n = \lim_{n\to\infty} gx_n = \lim_{n\to\infty} fx_{n-1} = gz = u. \quad (21)$$

We will show that $fz = gz = u$.

Now,

$$\begin{aligned} d(u, fz) &\preceq d(u, y_n) + d(y_n, y_{n+1}) + d(y_{n+1}, fz) \\ &= d(u, y_n) + d_n + d(y_{n+1}, fz). \end{aligned} \tag{22}$$

From (iv), we have $gx_n \sqsubseteq gz$; that is, $y_n \sqsubseteq u$; therefore it follows from (6) that

$$\begin{aligned} d(y_{n+1}, fz) &= d(fx_n, fz) \\ &\preceq \lambda d(gx_n, gz) + \mu d(gx_n, fx_n) + \delta d(gz, fz) \\ &= \lambda d(y_n, u) + \mu d(y_n, y_{n+1}) + \delta d(u, fz) \\ &\preceq \lambda d(y_n, u) + \mu d_n \\ &\quad + \delta [d(u, y_n) + d(y_n, y_{n+1}) + d(y_{n+1}, fz)] \\ &= [\lambda + \delta] d(y_n, u) + [\mu + \delta] d_n + \delta d(y_{n+1}, fz); \end{aligned} \tag{23}$$

that is,

$$d(y_{n+1}, fz) \preceq \frac{\lambda + \delta}{1 - \delta} d(y_n, u) + \frac{\mu + \delta}{1 - \delta} d_n. \tag{24}$$

In view of (10), (21), and (a), (d) of Remark 3, for every $c \in E$ with $\theta \ll c$, there exists $n_1 \in \mathbb{N}$ such that $d(y_n, u) \ll c(1 - \delta)/2(\lambda + \delta)$, $d_n \ll c(1 - \delta)/2(\mu + \delta)$ for all $n > n_1$. Therefore, it follows from above inequality that

$$d(y_{n+1}, fz) \ll c \quad \forall n > n_1 \text{ and every } c \in E \text{ with } \theta \ll c. \tag{25}$$

Therefore, again with same arguments, from (22) and (c) of Remark 3, we obtain that $d(u, fz) = \theta$; that is, $fz = gz = u$. Thus, z is a coincidence point and u is point of coincidence of f and g.

Now suppose that f and g are weakly compatible; then we have $fu = fgz = gfz = gu$. As $gz \sqsubseteq ggz$, therefore using (6) we obtain

$$\begin{aligned} d(u, fu) &= d(fz, ffz) \\ &\preceq \lambda d(gz, gfz) + \mu d(gz, fz) + \delta d(gfz, ffz) \\ &= \lambda d(gz, fgz) + \mu d(gz, gz) + \delta d(fgz, fgz) \\ &= \lambda d(u, fu). \end{aligned} \tag{26}$$

As $\lambda \in [0, 1)$, it follows from (f) of Remark 3 and the above inequality that $d(u, fu) = \theta$; that is, $fu = u = gu$. Thus, u is a common fixed point of f and g.

Suppose the set of common fixed points of f and g, that is, $\mathcal{F}$, is g-well ordered and $u, v \in \mathcal{F}$. As $\mathcal{F}$ is g-well ordered, let, for example, $gu \sqsubseteq gv$. Then, it follows from (6) that

$$\begin{aligned} d(u, v) &= d(fu, fv) \\ &\preceq \lambda d(gu, gv) + \mu d(gu, fu) + \delta d(gv, fv) \\ &= \lambda d(u, v) + \mu d(u, u) + \delta d(v, v) \\ &= \lambda d(u, v). \end{aligned} \tag{27}$$

As $\lambda \in [0, 1)$, it follows from (f) of Remark 3 that $d(u, v) = \theta$; that is, $u = v$. Therefore, the common fixed point of f and g is unique. For converse, let common fixed point be unique, then $\mathcal{F}$ will be singleton and therefore g-well ordered. □

Remark 11. For the existence of common fixed point of f and g, Vetro [14] used the condition

$$fgx = ggx \quad \text{whenever } fx = gx. \tag{28}$$

Here, we have used the weak compatibility of mappings f and g, and it is obvious that the condition used by Vetro implies the weak compatibility of mappings f and g.

Taking $g = I_X$ (identity mapping of X), we obtain the main result of [26].

Corollary 12. *Let $(X, \sqsubseteq, d)$ be an ordered complete cone rectangular metric space and $f : X \to X$ a mapping such that the following conditions are satisfied:*

(i) *f is an ordered Reich-type contraction, that is, satisfies (5);*

(ii) *f is nondecreasing with respect to "$\sqsubseteq$";*

(iii) *there exists $x_0 \in X$ such that $x_0 \sqsubseteq fx_0$;*

(iv) *if $\{x_n\}$ is any nondecreasing sequence in X converging to some z then $x_n \sqsubseteq z$, for all n.*

Then, f has a fixed point. In addition, the set of fixed points of f is well ordered if and only if the fixed point of f is unique.

With suitable values of control constants λ, μ, and δ, we obtain the following generalizations of Theorems 2.1 and 2.3 of Abbas and Jungck [5] on ordered cone rectangular metric spaces.

Corollary 13. *Let $(X, \sqsubseteq, d)$ be an ordered cone rectangular metric space $f, g : X \to X$ two mappings such that $f(X) \subset g(X)$ and $g(X)$ is complete. Suppose that the following conditions are satisfied:*

(i) *$d(fx, fy) \preceq \lambda d(gx, gy)$ for all $x, y \in X$ with $gx \sqsubseteq gy$, where $\lambda \in [0, 1)$;*

(ii) *f is g-nondecreasing;*

(iii) *there exists $x_0 \in X$ such that $gx_0 \sqsubseteq fx_0$;*

(iv) *if $\{gx_n\}$ were any nondecreasing sequence in X converging to some gz then $gx_n \sqsubseteq gz$, for all n and $gz \sqsubseteq ggz$.*

Then, f and g have a coincidence point. Furthermore, if f and g are weakly compatible then they have a common fixed point. In addition, the set of common fixed points of f and g is g-well ordered if and only if the common fixed point of f and g is unique.

Corollary 14. *Let $(X, \sqsubseteq, d)$ be an ordered cone rectangular metric space and $f, g : X \to X$ two mappings such that $f(X) \subset g(X)$ and $g(X)$ is complete. Suppose that the following conditions are satisfied:*

(i) *$d(fx, fy) \preceq \lambda[d(gx, fy) + d(gy, fy)]$ for all $x, y \in X$ with $gx \sqsubseteq gy$, where $\lambda \in [0, 1/2)$;*

(ii) *f is g-nondecreasing;*

(iii) *there exists $x_0 \in X$ such that $gx_0 \sqsubseteq fx_0$;*

(iv) *if $\{gx_n\}$ were any nondecreasing sequence in X converging to some gz, then $gx_n \sqsubseteq gz$ for all n and $gz \sqsubseteq ggz$.*

Then, f and g have a coincidence point. Furthermore, if f and g are weakly compatible then they have a common fixed point. In addition, the set of common fixed points of f and g is g-well ordered if and only if the common fixed point of f and g is unique.

The following example shows that the results of this paper are a proper generalization of the results of Malhotra et al. [26] and Vetro [14].

Example 15. Let $X = \{1, 2, 3, 4\}$ and $E = C^1_{\mathbb{R}}[0, 1]$ with $\|x\| = \|x\|_\infty + \|x'\|_\infty$, $P = \{x(t) : x(t) \geq 0 \text{ for } t \in [0, 1]\}$. Define $d : X \times X \to E$ as follows:

$$\begin{aligned} d(1,2) &= d(2,1) = 3e^t, \\ d(2,3) &= d(3,2) = d(1,3) = d(3,1) = e^t, \\ d(1,4) &= d(4,1) = d(2,4) = d(4,2) = d(3,4) \\ &= d(4,3) = 4e^t, \\ d(x,y) &= \theta \quad \text{if } x = y. \end{aligned} \tag{29}$$

Then, (X, d) is a complete nonnormal cone rectangular metric space but not cone metric space. Define mappings $f, g : X \to X$ and partial order "$\sqsubseteq$" on X as follows:

$$\begin{aligned} f &= \begin{pmatrix} 1 & 2 & 3 & 4 \\ 1 & 3 & 3 & 1 \end{pmatrix}, \\ g &= \begin{pmatrix} 1 & 2 & 3 & 4 \\ 1 & 2 & 2 & 3 \end{pmatrix}, \\ \sqsubseteq &= \{(1,1), (2,2), (3,3), (4,4), (1,2), (1,3)\}. \end{aligned} \tag{30}$$

Then it is easy to verify that f is an ordered g-weak contraction in $(X, \sqsubseteq, d)$ with $\lambda \in [1/3, 1), \mu = \delta = 0$. Indeed, we have to check the validity of (6) only for $(x, y) = (1, 2), (1, 4)$. Then,

$$\begin{aligned} d(f1, f2) &= d(1,3) = e^t, \\ \lambda d(g1, g2) + \mu d(g1, f1) + \delta d(g2, f2) &= e^t [3\lambda + \delta]; \end{aligned} \tag{31}$$

therefore, (6) holds for $\lambda \in [1/3, 1)$, $\mu = \delta = 0$. Again,

$$d(f1, f4) = d(1,1) = \theta; \tag{32}$$

therefore, (6) holds for arbitrary λ, δ, and μ such that $\lambda + \mu + \delta < 1$.

All other conditions of Theorem 10 are satisfied and 1 is the unique common fixed point of f and g. Note that f is not an ordered Reich-type contraction. Indeed, for point $(x, y) = (1, 3)$ there are no $\lambda, \mu, \delta \in [0, 1)$ such that condition (5) is satisfied. Therefore, the results of Malhotra et al. [26] are not applicable here.

The following example illustrates the crucial role of weak compatibility of mappings for the existence of common fixed point in Theorem 10.

Example 16. Let (X, d) be the cone rectangular metric space as in Example 15. Then, (X, d) is a complete nonnormal cone rectangular metric space but not cone metric space. Define mappings $f, g : X \to X$ and partial order "$\sqsubseteq$" on X as follows:

$$\begin{aligned} f &= \begin{pmatrix} 1 & 2 & 3 & 4 \\ 2 & 2 & 2 & 3 \end{pmatrix}, \\ g &= \begin{pmatrix} 1 & 2 & 3 & 4 \\ 1 & 3 & 2 & 4 \end{pmatrix}, \\ \sqsubseteq &= \{(1,1), (2,2), (3,3), (4,4), (2,4), (2,3), (1,3)\}. \end{aligned} \tag{33}$$

Then, it is easy to verify that f is an ordered g-weak contraction in $(X, \sqsubseteq, d)$ with $\lambda = \delta \in [1/8, 1/2)$, $\mu = 0$. Indeed, we have to check the validity of (6) only for $(x, y) = (3, 4), (3, 2), (1, 2)$. Then,

$$\begin{aligned} d(f3, f4) &= d(2,3) = e^t, \\ \lambda d(g3, g4) + \mu d(g3, f3) + \delta d(g4, f4) &= 4e^t [\lambda + \delta]; \end{aligned} \tag{34}$$

Therefore, (6) holds for $\lambda = \delta \in [1/8, 1/2)$, $\mu = 0$. Again,

$$d(f3, f2) = d(2,2) = \theta; \tag{35}$$

Therefore, (6) holds for arbitrary λ, δ, and μ such that $\lambda + \mu + \delta < 1$.

Similarly, for $(x, y) = (1, 2)$, and condition (6) holds for arbitrary λ, δ, and μ such that $\lambda + \mu + \delta < 1$.

All other conditions of Theorem 10 (except f and g are weakly compatible) are satisfied and 3 is a coincidence point of f and g. Note that $f3 = g3 = 2$; that is, 3 is a coincidence point of f and g but $fg3 \neq gf3$; therefore, f and g are not weakly compatible and have no common fixed point.

In the following theorem, the conditions on f, "nondecreasing" and "completeness of space," are replaced by another condition.

Theorem 17. *Let $(X, \sqsubseteq, d)$ be an ordered cone rectangular metric space and $f, g : X \to X$ is two mappings such that $f(X) \subset g(X)$. Suppose that the following conditions are satisfied:*

(A) *f is an ordered g-weak contraction that satisfies (6);*

(B) *there exists $u \in X$ such that $gu \sqsubseteq fu$ and $d(gu, fu) \preceq d(gx, fx)$ for all $x \in X$.*

Then, f and g have a coincidence point. Furthermore, if f and g are weakly compatible, then they have a common fixed point. In addition, the set of common fixed points of f and g is g-well ordered if and only if the common fixed point of f and g is unique.

Proof. Let $F(x) = d(gx, fx)$ for all $x \in X$ and $gz = fu$ for some $z \in X$ (which is possible, since $f(X) \subset g(X)$); then $F(u) \preceq F(x)$ for all $x \in X$. If $F(u) = \theta$, then $gu = fu$; that is, u is a coincidence point of f and g. If $\theta \prec F(u)$, then by assumption (B) $gu \sqsubseteq fu$, so $gu \sqsubseteq gz$, and by (A), we obtain

$$\begin{aligned} F(z) &= d(gz, fz) = d(fu, fz) \\ &\preceq \lambda d(gu, gz) + \mu d(gu, fu) + \delta d(gz, fz) \\ &= \lambda d(gu, fu) + \mu d(gu, fu) + \delta d(gz, fz) \\ &= \lambda F(u) + \mu F(u) + \delta F(z), \end{aligned} \quad (36)$$

$$F(z) \preceq \frac{\lambda + \mu}{1 - \delta} F(u) \prec F(u) \quad (\text{as } \lambda + \mu + \delta < 1),$$

a contradiction. Therefore, we must have $F(u) = \theta$; that is, $gu = fu$, and so u is a coincidence point of f and g.

The existence, necessary and sufficient condition for uniqueness of common fixed point follows from a similar process as used in Theorem 10. □

References

[1] P. P. Zabreĭko, "K-metric and K-normed spaces: survey," *Collectanea Mathematica*, vol. 48, no. 4–6, pp. 825–859, 1997.

[2] D. R. Kurepa, "Free power or width of some kinds of mathematical structure," *Publications de L'Institute Mathématique Nouvelle Série Tone*, vol. 42, pp. 3–12, 1987.

[3] D. R. Kurepa, "Tableaux ramiés d'ensembles. Espaces pseudodistanciés," *Comptes Rendus de L'Académie des Sciences*, vol. 198, pp. 1563–1565, 1934.

[4] L. G. Huang and X. Zhang, "Cone metric spaces and fixed point theorems of contractive mappings," *Journal of Mathematical Analysis and Applications*, vol. 332, no. 2, pp. 1468–1476, 2007.

[5] M. Abbas and G. Jungck, "Common fixed point results for noncommuting mappings without continuity in cone metric spaces," *Journal of Mathematical Analysis and Applications*, vol. 341, no. 1, pp. 416–420, 2008.

[6] C. di Bari and P. Vetro, "Weakly φ-pairs and common fixed points in cone metric spaces," *Rendiconti del Circolo Matematico di Palermo*, vol. 58, pp. 125–132, 2009.

[7] W.-S. Du, "A note on cone metric fixed point theory and its equivalence," *Nonlinear Analysis, Theory, Methods and Applications*, vol. 72, no. 5, pp. 2259–2261, 2010.

[8] D. Ilic and V. Rakocevic, "Quasi-contraction on a cone metric space," *Applied Mathematics Letters*, vol. 22, no. 5, pp. 728–731, 2009.

[9] S. Jankovi, Z. Kadelburg, and S. Radenović, "On cone metric spaces: a survey," *Nonlinear Analysis, Theory, Methods and Applications*, vol. 74, no. 7, pp. 2591–2601, 2011.

[10] Z. Kadelburg, S. Radenović, and V. Rakočević, "A note on the equivalence of some metric and cone metric fixed point results," *Applied Mathematics Letters*, vol. 24, no. 3, pp. 370–374, 2011.

[11] S. K. Malhotra, S. Shukla, and R. Sen, "A generalization of Banach contraction principle in ordered cone metric spaces," *Journal of Advanced Mathematical Studies*, vol. 5, no. 2, pp. 59–67, 2012.

[12] S. K. Malhotra, S. Shukla, and R. Sen, "Some coincidence and common fixed point theorems in cone metric spaces," *Bulletin of Mathematical Analysis and Applications*, vol. 4, no. 2, pp. 64–71, 2012.

[13] S. Rezapour and R. Hamlbarani, "Some notes on the paper 'Cone metric spaces and fixed point theorems of contractive mappings,'" *Journal of Mathematical Analysis and Applications*, vol. 345, no. 2, pp. 719–724, 2008.

[14] P. Vetro, "Common fixed points in cone metric spaces," *Rendiconti del Circolo Matematico di Palermo*, vol. 56, no. 3, pp. 464–468, 2007.

[15] A. Branciari, "A fixed point theorem of Banach-Caccioppoli type on a class of generalized metric spaces," *Publicationes Mathematicae*, vol. 57, no. 1-2, pp. 31–37, 2000.

[16] A. Azam and M. Arshad, "Kannan fixed point theorem on generalized metric spaces," *The Journal of Nonlinear Science and Its Applications*, no. 1, pp. 45–48, 2008.

[17] A. Azam, M. Arshad, and I. Beg, "Banach contraction principle on cone rectangular metric spaces," *Applicable Analysis and Discrete Mathematics*, vol. 3, no. 2, pp. 236–241, 2009.

[18] B. Samet and C. Vetro, "A fixed point theorem for uniformly locally contractive mappings in a c-chainable cone rectangular metric space," *Surveys in Mathematics and Its Applications*, vol. 6, pp. 107–116, 2011.

[19] J. S. Vandergraft, "Newton method for convex operators in partially ordered spaces," *SIAM Journal on Numerical Analysis*, vol. 4, no. 3, pp. 406–432, 1967.

[20] K. Deimling, *Nonlinear Functional Analysis*, Springer, New York, NY, USA, 1985.

[21] A. C. M. Ran and M. C. B. Reurings, "A fixed point theorem in partially ordered sets and some applications to matrix equations," *Proceedings of the American Mathematical Society*, vol. 132, no. 5, pp. 1435–1443, 2004.

[22] J. J. Nieto and R. Rodríguez-López, "Contractive mapping theorems in partially ordered sets and applications to ordinary differential equations," *Order*, vol. 22, no. 3, pp. 223–239, 2005.

[23] I. Altun, B. Damnjanović, and D. Djorić, "Fixed point and common fixed point theorems on ordered cone metric spaces," *Applied Mathematics Letters*, vol. 23, no. 3, pp. 310–316, 2009.

[24] I. Altun and G. Durmaz, "Some fixed point theorems on ordered cone metric spaces," *Rendiconti del Circolo Matematico di Palermo*, vol. 58, no. 2, pp. 319–325, 2009.

[25] Z. Kadelburg, M. Pavlović, and S. Radenović, "Common fixed point theorems for ordered contractions and quasicontractions in ordered cone metric spaces," *Computers and Mathematics with Applications*, vol. 59, no. 9, pp. 3148–3159, 2010.

[26] S. K. Malhotra, S. Shukla, and R. Sen, "Some fixed point theorems for ordered Reich type contractions in cone rectangular metric spaces," *Acta Mathematica Universitatis Comenianae*. In press.

[27] G. Jungck, S. Radenović, S. Radojevíc, and V. Rakočevíc, "Common fixed point theorems for weakly compatible pairs on cone metric spaces," *Fixed Point Theory and Applications*, vol. 2009, Article ID 643840, pp. 1–13, 2009.

Dynamic Analysis of an SEIR Model with Distinct Incidence for Exposed and Infectives

Junhong Li and Ning Cui

Department of Mathematics and Sciences, Hebei Institute of Architecture and Civil Engineering, Zhangjiakou, Hebei 075000, China

Correspondence should be addressed to Junhong Li; jhli2011@163.com

Academic Editors: J. Banaś and M. M. Cavalcanti

An SEIR model with vaccination strategy that incorporates distinct incidence rates for the exposed and the infected populations is studied. By means of Lyapunov function and LaSalle's invariant set theorem, we proved the global asymptotical stable results of the disease-free equilibrium. The sufficient conditions for the global stability of the endemic equilibrium are obtained using the compound matrix theory. Furthermore, the method of direct numerical simulation of the system shows that there is a periodic solution, when the system has three equilibrium points.

1. Introduction

Mathematical models have become important tools in analyzing the spread and the control of infectious diseases. Many infectious diseases in nature, such as measles, HIV/AIDS, SARS, and tuberculosis (see [1–6]), incubate inside the hosts for a period of time before the hosts become infectious. Li and Fang (see [7]) studied the global stability of an age-structured SEIR model with infectivity in latent period. Yi et al. (see [8]) discussed the dynamical behaviors of an SEIR epidemic system with nonlinear transmission rate. Li and Zhou (see [9]) considered the global stability of an SEIR model with vertical transmission and saturating contact rate.

In this paper, we will consider an SEIR model that the diseases can be infected in the latent period and the infected period. The population size N is divided into four homogeneous classes: the susceptible $S(t)$, the exposed (in the latent period) $E(t)$, the infective $I(t)$, and the recovered $R(t)$. It is assumed that all the offsprings at birth are susceptible to the disease. The inflow rate (including birth and immigration) and outflow rate (including natural death and emigration) are denoted by b and d, respectively. The rate of disease-caused death is taken as α. We assume that susceptible individuals are vaccinated at a constant per capita rate a. Due to the partial efficiency of the vaccine, only σ fraction of the vaccinated susceptibles goes to the recovered class. The remained $1-\sigma$ fraction of the vaccinated susceptibles has no immunity at all and goes to the exposed class after infected by contact with the infectives. If $\sigma = 0$, it means that the vaccine has no effect at all, and if $\sigma = 1$, the vaccine is perfectly effective. The positive parameter ε is the rate at which the exposed individuals become infectious. γ is the constant rate, at which the infectious individuals recover with acquiring permanent immunity. The transfer mechanism from the class $S(t)$ to the class $E(t)$ is guided by the function $\beta(I + qE)/N$, where β is the force of infection. q denotes the relative measure of infectiousness for the asymptomatic class $E(t)$.

Based on these considerations, and with reference to [10–12], the SEIR model is given by the following system of differential equations:

$$
\begin{aligned}
S' &= bN - F(S, E, I) - (\sigma a + d) S, \\
E' &= F(S, E, I) - (d + \varepsilon) E, \\
I' &= \varepsilon E - (\alpha + \gamma + d) I, \\
R' &= \sigma a S + \gamma I - dR, \\
0 &= S + E + I + R - N,
\end{aligned}
\tag{1}
$$

where the derivative d/dt is denoted by $'$ and $F(S, E, I) = \beta S(1-a)(I + qE)/N + \beta a S(1-\sigma)(I + qE)/N$.

Thus, the total population size N implies $N' = (b-d)N - \alpha I$. Let $s = S/N$, $e = E/N$, $i = I/N$, and $r = R/N$.

Because the variable R does not appear in the equations of S, E, and I, we only need to consider the following subsystem:

$$\begin{aligned} s' &= b - \beta s (1-\sigma a)(i+qe) - (\sigma a + b)s + s\alpha i, \\ e' &= \beta s (1-\sigma a)(i+qe) - (b+\varepsilon)e + e\alpha i, \\ i' &= \varepsilon e - (\alpha+\gamma+b)i + \alpha i^2, \\ 0 &= s+e+i+r-1. \end{aligned} \tag{2}$$

The system (2) is equivalent to (1). From biological considerations, we study (2) in the following closed set:

$$T = \left\{(s,e,i) \in R_+^3 \mid 0 \le s+e+i \le 1\right\}, \tag{3}$$

where R_+^3 denotes the nonnegative cone of R^3 including its lower dimensional faces.

2. Equilibria and Global Stability

It is easy to visualize that (2) always has a disease-free equilibrium $P_0(b/(\sigma a + b), 0, 0)$. The Jacobian matrix of (2) at an arbitrary point $P(s,e,i)$ takes the following form:

$$J(P) = \begin{bmatrix} J_{11} & -q\beta(1-\sigma a)s & J_{13} \\ J_{21} & J_{22} & J_{23} \\ 0 & \varepsilon & J_{33} \end{bmatrix}, \tag{4}$$

where

$$\begin{aligned} J_{11} &= -\beta(1-\sigma a)(i+qe) - (\sigma a+b) + \alpha i, \\ J_{13} &= \alpha s - \beta(1-\sigma a)s, \\ J_{23} &= \alpha e + \beta(1-\sigma a)s, \\ J_{33} &= 2\alpha i - (\alpha+\gamma+b), \\ J_{22} &= \beta q(1-\sigma a)s - (b+\varepsilon) + \alpha i, \\ J_{21} &= \beta(1-\sigma a)(i+qe). \end{aligned} \tag{5}$$

Theorem 1. *If $R_0 < 1$, the disease-free equilibrium P_0 is locally asymptotically stable, where*

$$R_0 = \frac{b\beta(1-\sigma a)(q\alpha+q\gamma+qb+\varepsilon)}{(b+\varepsilon)(\alpha+b+\gamma)(\sigma a+b)}. \tag{6}$$

Proof. Let

$$\lambda_1 = \frac{(\varepsilon+q\alpha+q\gamma+qb)(\alpha+\gamma+2b+\varepsilon)}{(\alpha+\gamma+b)(b+\varepsilon)qR_0} - 1. \tag{7}$$

We calculate the characteristic equation of $J(P_0)$ as follows:

$$(\lambda+\sigma a+b) \times \left(\lambda^2 + \frac{bq\beta(1-\sigma a)\lambda_1\lambda}{(\sigma a+b)} + (b+\varepsilon)(\alpha+\gamma+b)(1-R_0)\right) = 0. \tag{8}$$

The stability of P_0 is equivalent to all eigenvalues of (8) being with negative real parts, which can be guaranteed by $R_0 < 1$. Consequently, the disease-free equilibrium is local asymptotical stability. This proves the theorem. □

Theorem 2. *If $R_{01} \le 1$, the disease-free equilibrium P_0 is globally asymptotically stable, where*

$$R_{01} = \frac{\beta(1-\sigma a)(q\alpha+q\gamma+qb+\varepsilon)}{(b+\varepsilon)(\alpha+b+\gamma)}. \tag{9}$$

Proof. Consider the following function:

$$L = (\alpha+\gamma+b)e + \beta(1-\sigma a)i. \tag{10}$$

Its derivative along the solutions to the system (2) is as follows:

$$\begin{aligned} L' &= i\left[(\alpha+\gamma+b)\beta(1-\sigma a)s - (1-\sigma a)(\alpha+\gamma+b-\alpha i)\right] \\ &\quad + e\left[(\alpha+\gamma+b)q\beta(1-\sigma a)s \right. \\ &\quad \left. - (\alpha+\gamma+b)(b+\varepsilon+\alpha i) + (1-\sigma a)\varepsilon\right] \\ &\le e(\alpha+\gamma+b)(b+\varepsilon)(R_{01}-1) \\ &\le 0. \end{aligned} \tag{11}$$

Furthermore, $L' = 0$ only if $e = 0$. The maximum invariant set in $\{(s,e,i) \in T : L' = 0\}$ is the singleton $\{P_0\}$. When $R_{01} \le 1$, the global stability of P_0 follows from LaSalle's invariance principle (see [13]). This completes the proof. □

Theorem 3. *Equation (2) has a unique endemic equilibrium $P^*(s^*, e^*, i^*)$ if $b \ge \alpha$ and $\Delta > 1$, where*

$$\Delta = (b+\varepsilon-\alpha)\left(\frac{\gamma}{b}+1\right) \times \left(\frac{1}{\varepsilon} - \frac{(\sigma a+b-\alpha)}{\beta(1-\sigma a)(qb+q\gamma+\varepsilon)}\right). \tag{12}$$

Proof. Let the right side of each of the first three differential equations equal to zero in (2); we obtain the following:

$$\begin{aligned} \beta s^*(1-\sigma a)(i^*+qe^*) - (b+\varepsilon)e^* + \alpha e^* i^* &= 0, \\ \varepsilon e^* - (\alpha+\gamma+b)i^* + \alpha i^{*2} &= 0, \\ b - \beta s^*(1-\sigma a)(i^*+qe^*) - (\sigma a+b)s^* + \alpha s^* i^* &= 0, \end{aligned} \tag{13}$$

with $s^* > 0$, $e^* > 0$ and $i^* > 0$. So we get

$$\begin{aligned} \frac{b}{s^*} &= \beta(1-\sigma a)(i^*+qe^*) - (\sigma a+b) + \alpha i^*, \\ \frac{\beta s^*(1-\sigma a)(i^*+qe^*)}{e^*} &= (b+\varepsilon) + \alpha i^*, \\ \varepsilon e^* &= (\alpha+\gamma+b)i^* - \alpha i^{*2}. \end{aligned} \tag{14}$$

When the three equations of (14) are multiplied together, we obtain the following:

$$\begin{aligned} b\beta\varepsilon(1-\sigma a)&\left[\frac{(\alpha+\gamma+b)q}{\varepsilon}+1-\frac{\alpha q i^*}{\varepsilon}\right] \\ &= (b+\varepsilon-\alpha i^*)(\alpha+\gamma+b-\alpha i^*) \\ &\times\left[\frac{-\alpha q\beta(1-\sigma a)i^{*2}}{\varepsilon}\right. \\ &+\left(\frac{\beta(1-\sigma a)(\alpha q+\gamma q+bq+\varepsilon)}{\varepsilon}+\alpha\right)i^* \\ &\left.-(\sigma a+b)\right]. \end{aligned} \tag{15}$$

Define the following:

$$\begin{aligned} f(i) &= f_1(i)(b+\varepsilon-\alpha i)(\alpha+\gamma+b-\alpha i), \\ g(i) &= b\beta\varepsilon(1-\sigma a)\left[\frac{(\alpha+\gamma+b)q}{\varepsilon}+1-\frac{\alpha q i}{\varepsilon}\right], \end{aligned} \tag{16}$$

where

$$\begin{aligned} f_1(i) = &\left[\frac{-\alpha q\beta(1-\sigma a)i^2}{\varepsilon}\right. \\ &+\left(\frac{\beta(1-\sigma a)(\alpha q+\gamma q+bq+\varepsilon)}{\varepsilon}+\alpha\right)i \\ &\left.-(\sigma a+b)\right], \end{aligned} \tag{17}$$

and the roots of $f(i)$ are $i_1 = (b+\varepsilon)/\alpha > 1$, $i_2 = (\alpha+\gamma+b)/\alpha > 1$, and the other two are i_3, i_4 which satisfy $f_1(i) = 0$. $i_0 = (\alpha+\gamma+b)/\alpha+\varepsilon/(\alpha q)$ is the root of $g(i)$. Direct calculations show the following:

$$\begin{aligned} &g(1) = b\beta\varepsilon(1-\sigma a)(qb+q\gamma+\varepsilon) > 0, \\ &f(0) = -(\sigma a+b)(b+\varepsilon)(\alpha+\gamma+b) < 0, \\ &f(1) = \Delta g(1), \qquad i_0 > i_1, \qquad i_0 > i_2, \\ &i_3 i_4 = \frac{\varepsilon(\sigma a+b)}{\alpha q\beta(1-\sigma a)} > 0, \\ &i_3+i_4 = 1+\frac{b+\gamma}{\alpha}+\varepsilon\left(\frac{1}{\alpha q}+\frac{1}{q\beta(1-\sigma a)}\right) > 2. \end{aligned} \tag{18}$$

Since $\Delta > 1$, the linear function $g(i)$ has exactly one intersection with the function $f(i)$ where i lies in the interval $(0,1)$. Furthermore, s^* and e^* can be uniquely determined from i^* by the following:

$$\begin{aligned} e^* &= \frac{i^*(\alpha+\gamma+b-\alpha i^*)}{\varepsilon}, \\ s^* &= \frac{b}{[\beta(1-\sigma a)(i^*+e^*)+(\sigma a+b)+\alpha i^*]}. \end{aligned} \tag{19}$$

From this, we can easily see that (2) has a unique endemic equilibrium. This completes the proof. □

Denote the interior of T by $\mathring{T}$. In this paper, we obtain sufficient conditions that the equilibrium is globally asymptotically stable using the geometrical approach of Li and Muldowney in [14].

Theorem 4. *The unique endemic equilibrium P^* is globally asymptotically stable in $\mathring{T}$, when*

$$b \geq \alpha, \qquad \Delta > 1, \qquad R_{01} > 1, \qquad \alpha < \min\{\sigma a, b, \varepsilon\}. \tag{20}$$

Proof. Since $R_{01} > 1$, namely, $\Delta < R_0$ and P_0 is unstable, we can easy see that (2) satisfies the assumptions (H_1) and (H_2) (see [14]) in the interior of its feasible region T. The unique equilibrium is locally asymptotically stable using simple calculation.

Let $x = (s,e,i)$ and $f(x)$ denotes the vector field of (2) and

$$J^{[2]}(P) = \begin{pmatrix} a_{11} & q\beta s(1-\sigma a)+\alpha e & -\alpha s+\beta s(1-\sigma a) \\ \varepsilon & a_{22} & -q\beta s(1-\sigma a) \\ 0 & \beta(1-\sigma a)(i+qe) & a_{33} \end{pmatrix}, \tag{21}$$

where

$$\begin{aligned} a_{11} &= -\beta(1-\sigma a)(i+qe-qs)-(\sigma a+2b+\varepsilon)+2\alpha i, \\ a_{22} &= -\beta(1-\sigma a)(i+qe)-(2b-d+\alpha+\beta)+3\alpha i, \\ a_{33} &= -q\beta s(1-\sigma a)-(2b+\varepsilon+\alpha+\beta)+3\alpha i. \end{aligned} \tag{22}$$

Set the the following function:

$$P(x) = P(s,e,i) = \operatorname{diag}\left(1,\frac{e}{i},\frac{e}{i}\right). \tag{23}$$

Then the matrix $B = P_f P^{-1} + PJ^{[2]}P^{-1}$ can be written in block form as follow:

$$B = \begin{pmatrix} B_{11} & B_{12} \\ B_{21} & B_{22} \end{pmatrix}, \tag{24}$$

where

$$\begin{aligned} &B_{11} = a_{11}, \qquad B_{21} = \left(\frac{\varepsilon e}{i} \;\; 0\right)', \\ &B_{12} = \left(\frac{\beta(1-\sigma a)si}{e}+\alpha i\frac{(\beta(1-\sigma a)-\alpha)si}{e}\right), \\ &B_{221} = \frac{e'}{e}-\frac{i'}{i}+a_{22}, \qquad B_{222} = \frac{e'}{e}-\frac{i'}{i}+a_{33}, \\ &B_{22} = \begin{pmatrix} B_{221} & -q\beta s(1-\sigma a) \\ \beta(1-\sigma a)(i+qe) & B_{222} \end{pmatrix}. \end{aligned} \tag{25}$$

Let (u,v,w) denote the vectors in $R^3 \cong R^{\binom{3}{2}}$, we select a norm in R^3 as $|(u,v,w)| = \max\{|u|, |v|+|w|\}$ and let ρ denote

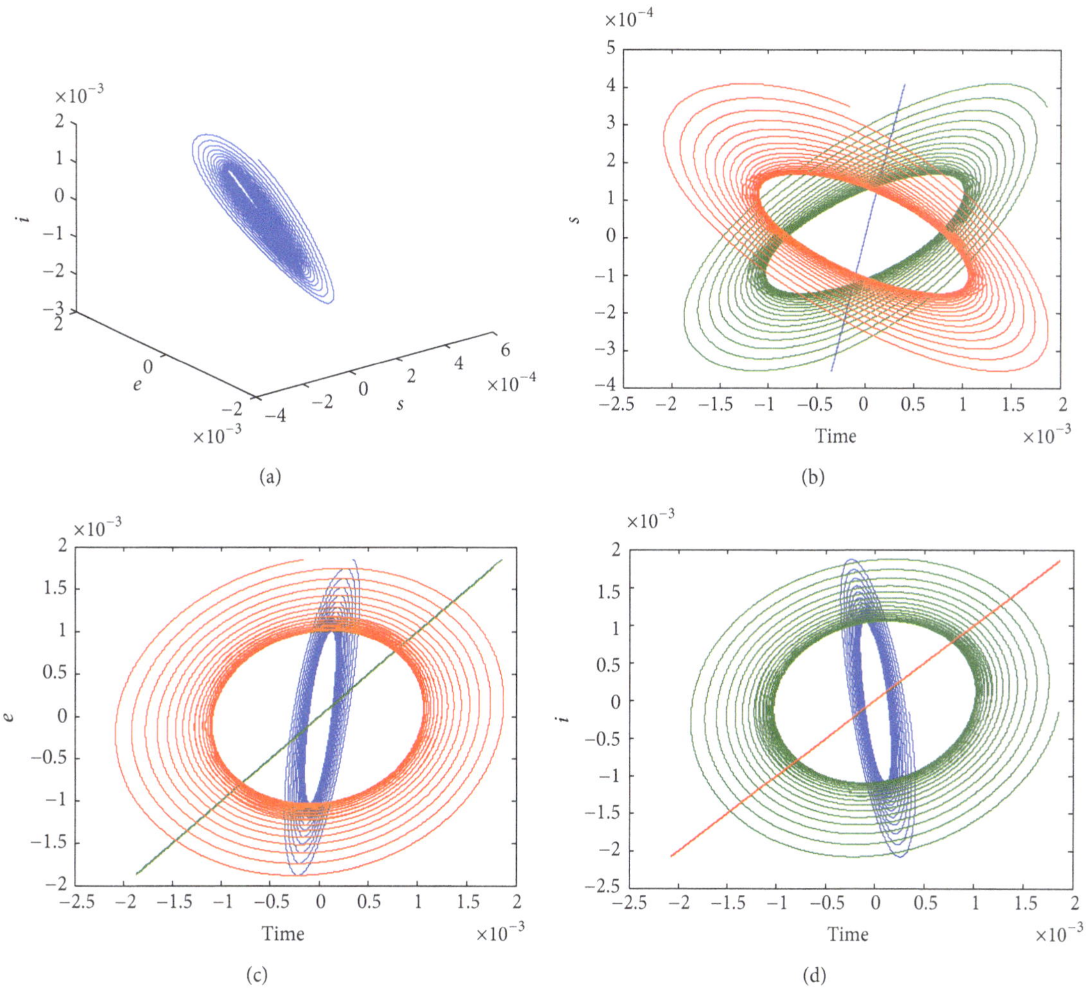

FIGURE 1: (a) Phase portraits of (2). (b) Time series of s. (c) Time series of e. (d) Time series of i.

the Lozinskii measure with respect to this norm. Using the method of estimating ρ in [15], we have $\rho(B) \le \sup(g_1, g_2)$, where

$$g_1 = \rho_1(B_{11}) + |B_{12}|, \qquad g_2 = |B_{21}| + \rho_1(B_{22}), \tag{26}$$

$|B_{12}|$, $|B_{21}|$ are matrix norms with respect to the l_1 vector norm, and ρ_1 denotes the Lozinskii measure with respect to l_1 norm. More specifically, $\rho_1(B_{11}) = B_{11}$, $|B_{21}| = \varepsilon e/i$, and $|B_{12}| = \beta(1-\sigma a)si/e + \alpha i$.

Rewriting the system (2), we have the following:

$$\frac{\varepsilon e}{i} = \frac{i'}{i} + (\alpha + \gamma + b) - \alpha i. \tag{27}$$

Therefore,

$$\begin{aligned} g_1 &= \frac{e'}{e} - \beta(1-\sigma a)(i + qe) - (\sigma a + b) + 2\alpha i, \\ g_2 &= \frac{e'}{e} + 2\alpha i - b - \min\{\sigma a, \varepsilon\}. \end{aligned} \tag{28}$$

Since $b \ge \alpha$ and $\alpha < \min\{\sigma a, b, \varepsilon\}$, there is $m = \max\{\sigma a + b - 2\alpha, b + \varepsilon - 2\alpha\} > 0$ and

$$\rho(B) \le \sup(g_1, g_2) \le \frac{e'}{e} - m \quad \text{for } t > \bar{t}. \tag{29}$$

Along each solution $x(t, x_0)$ to (2) such that $x_0 \in K$, where K is the compact absorbing set, we thus have the following:

$$\frac{1}{t}\int_0^t \rho(B)\,ds \le \frac{1}{t}\log\frac{e(t)}{e(\bar{t})} + \frac{1}{t}\int_0^{\bar{t}} \rho(B)\,ds - m\frac{t-\bar{t}}{t}, \tag{30}$$

which implies

$$\bar{q}_2 \le -\frac{m}{2} < 0. \tag{31}$$

This completes the proof. □

3. Conclusion

In this paper, we discuss an SEIR model that the diseases can be infected in the latent period and the infected period.

The vaccine effectiveness is also taken into account. We investigate the global dynamics of the reduced proportional system. If $R_{01} \le 1$, the disease-free equilibrium P_0 is globally asymptotically stable. The unique equilibrium P^* of the system (2) is globally asymptotically stable in $\overset{\circ}{T}$, when $b \ge \alpha$, $\Delta > 1$, $R_{01} > 1$, and $\alpha < \min\{\sigma a, b, \varepsilon\}$. When $q = a = \gamma = 0$, (2) becomes the SEIR model without infectivity in latent and disease-caused death (see [8]). When $q = 0$, (2) becomes the SEIR model without infectious in latent (see [12]).

The parameters are considered in the following cases:

$$\varepsilon = 0.05, \qquad \gamma = 0.003, \qquad \alpha = 0.002, \qquad \beta = 0.05, \tag{32}$$

(see [16]), and

$$q = 1, \qquad b = 0.00001, \qquad \sigma a = 0.000001. \tag{33}$$

At this case, there are three fixed points:

$$E_1\,(0.9091, 0, 0)\,, \qquad E_2\,(0.0911, 0.00018673, 0.0019)\,, \qquad E_3\,(0.0000083926, 0.025, 24.7973)\,, \tag{34}$$

and E_1, E_3 are unstable, E_2 stable. The method of direct numerical simulation of (2) shows that there is a periodic solution. The phase portraits of the system (2) in Figure 1(a), and time series of s, e, i are given in Figures 1(b), 1(c), and 1(d).

Acknowledgments

The authors are very grateful to the reviewers for their valuable comments and suggestions. This work was supported by the Youth Science Foundations of Education Department of Hebei Province (no. 2010233, 2011236).

References

[1] S. Bowong and J. Kurths, "Parameter estimation based synchronization for an epidemic model with application to tuberculosis in Cameroon," *Physics Letters A*, vol. 374, no. 44, pp. 4496–4505, 2010.

[2] B. Bokler, "Chaos and complexity in measles models: a comparative numerical study," *Mathematical Medicine and Biology*, vol. 10, no. 2, pp. 83–95, 1993.

[3] C. T. Bauch, E. Szusz, and L. P. Garrison, "Scheduling of measles vaccination in low-income countries: projections of a dynamic model," *Vaccine*, vol. 27, no. 31, pp. 4090–4098, 2009.

[4] Z. Mukandavire, C. Chiyaka, W. Garira, and G. Musuka, "Mathematical analysis of a sex-structured HIV/AIDS model with a discrete time delay," *Nonlinear Analysis: Theory, Methods and Applications*, vol. 71, no. 3-4, pp. 1082–1093, 2009.

[5] H. Huo and L. Feng, "Global stability for an HIV/AIDS epidemic model with different latent stages and treatment," *Applied Mathematical Modelling*, vol. 37, no. 3, pp. 1480–1489, 2013.

[6] Z. Yang and H. Jia, "Epidemic dynamics model with delay and impulsive vaccination control base on variable population," *Mathematical Methods in the Applied Sciences*, vol. 34, no. 15, pp. 1822–1832, 2011.

[7] X. Li and B. Fang, "Stability of an age-structured SEIR epidemic model with infectivity in latent period," *Applications and Applied Mathematics*, vol. 1, pp. 218–236, 2009.

[8] N. Yi, Q. Zhang, K. Mao, D. Yang, and Q. Li, "Analysis and control of an SEIR epidemic system with nonlinear transmission rate," *Mathematical and Computer Modelling*, vol. 50, no. 9-10, pp. 1498–1513, 2009.

[9] X. Z. Li and L. L. Zhou, "Global stability of an SEIR epidemic model with vertical transmission and saturating contact rate," *Chaos, Solitons and Fractals*, vol. 40, no. 2, pp. 874–884, 2009.

[10] B. Mukhopadhyay and R. Bhattacharyya, "Analysis of a spatially extended nonlinear SEIS epidemic model with distinct incidence for exposed and infectives," *Nonlinear Analysis: Real World Applications*, vol. 9, no. 2, pp. 585–598, 2008.

[11] O. Diekmann and J. A. P. Heesterbeek, *Mathematical Epidemiology of Infectious Disease*, John Wiley and Sons, Chichester, UK, 2000.

[12] C. Sun and Y. H. Hsieh, "Global analysis of an SEIR model with varying population size and vaccination," *Applied Mathematical Modelling*, vol. 34, no. 10, pp. 2685–2697, 2010.

[13] J. K. Hale, *Ordinary Differential Equations*, Wiley-Interscience, New York, NY, USA, 1969.

[14] M. Y. Li and J. S. Muldowney, "A geometric approach to global-stability problems," *SIAM Journal on Mathematical Analysis*, vol. 27, no. 4, pp. 1070–1083, 1996.

[15] R. H. Martin Jr., "Logarithmic norms and projections applied to linear differential systems," *Journal of Mathematical Analysis and Applications*, vol. 45, no. 2, pp. 432–454, 1974.

[16] E. Demirci, A. Unal, and N. Ozalp, "A fractional order SEIR model with density dependent death rate," *Hacettepe Journal of Mathematics and Statistics*, vol. 40, no. 2, pp. 287–295, 2011.

Exceptional Values of Meromorphic Function on Annulus

Hong-Yan Xu

Department of Informatics and Engineering, Jingdezhen Ceramic Institute, Jingdezhen, Jiangxi 333403, China

Correspondence should be addressed to Hong-Yan Xu; xhyhhh@126.com

Academic Editors: A. Agouzal, A. Barbagallo, and A. Fiorenza

The main purpose of this paper is to study the exceptional values of meromorphic function and its derivative on annulus. We also give some theorems and corollaries about exceptional values of meromorphic function on the annulus, which are the improvement of the previous results given by Chen and Wu.

1. Introduction

It is a very interesting problem on the exceptional values of meromorphic functions in the value distribution theory and argument distributed theory such as the Picard exceptional value and the Borel exceptional value. It is well known that every class of exceptional value is always responding to a singular direction, such as, the Picard exceptional value relating with Julia's direction and the Borel exceptional value relating with Borel's direction. In the 1970s, Gopalakrishna and Bhoosnurmath [1, 2] and Singh and Gopalakrishna [3] investigated Borel exceptional values of meromorphic function and its derivative on the whole complex plane. In 2011, Peng and Sun [4] gave some examples on T exceptional value which is an exceptional value relating with T direction.

In fact, Peng, Sun, Singh, Gopalakrishna, and so forth only studied exceptional values of meromorphic functions in the whole complex plane—single-connected region. For meromorphic function on the double-connected domain and several-connected region, there were few papers about its exceptional values. In 2005, Khrystiyanyn and Kondratyuk [5, 6] proposed the Nevanlinna theory for meromorphic functions on annuli (see also [7]). In 2010, Fernández [8] further investigated the value distribution of meromorphic functions on annuli. In 2012, Xu and Xuan [9] studied the uniqueness of meromorphic functions sharing some values on annuli. At the same year, Chen and Wu [10] firstly studied the Borel exceptional values of meromorphic function and its derivative on annulus. In this paper, we will further investigate the exceptional value of meromorphic function and its derivative on annulus and obtain a series of results which are improvement of previous theorems given by Chen and Wu [10].

The structure of this paper is as follows. In Section 2, we introduce the basic notations and fundamental theorems of meromorphic functions on annulus. Sections 3 and 4 are devoted to study the Borel exceptional values of meromorphic function on annulus and give some consequences of our theorems. Section 5 are devoted to study the Borel exceptional values of meromorphic function and its derivative on annulus.

2. Basic Notions in the Nevanlinna Theory on Annuli

Let f be a meromorphic function on the annulus $\mathbb{A} = \{z : 1/R_0 < |z| < R_0\}$, where $1 < R < R_0 \leq +\infty$. The notations of the Nevanlinna theory on annuli will be introduced as follows. Let

$$N_1(R,f) = \int_{1/R}^{1} \frac{n_1(t,f)}{t}dt, \qquad N_2(R,f) = \int_{1}^{R} \frac{n_2(t,f)}{t}dt,$$
$$m_0(R,f) = m(R,f) + m\left(\frac{1}{R}, f\right),$$
$$N_0(R,f) = N_1(R,f) + N_2(R,f), \tag{1}$$

where $n_1(t, f)$ and $n_2(t, f)$ are the counting functions of poles of the function f in $\{z : t < |z| \leq 1\}$ and $\{z : 1 < |z| \leq t\}$, respectively. The Nevanlinna characteristic of f on the annulus $\mathbb{A}$ is defined by

$$T_0(R, f) = m_0(R, f) - 2m(1, f) + N_0(R, f). \quad (2)$$

Similarly, for $a \in \overline{\mathbb{C}}$, we have

$$\begin{aligned}\overline{N}_0\left(r, \frac{1}{f-a}\right) &= \overline{N}_1\left(R, \frac{1}{f-a}\right) + \overline{N}_2\left(R, \frac{1}{f-a}\right) \\ &= \int_{1/R}^{1} \frac{\overline{n}_1(t, 1/(f-a))}{t} dt \quad (3) \\ &\quad + \int_{1}^{R} \frac{\overline{n}_2(t, 1/(f-a))}{t} dt\end{aligned}$$

in which each zero of the function $f - a$ is counted only once. In addition, we use $\overline{n}_1^{k)}(t, 1/(f-a))$ (or $\overline{n}_1^{(k}(t, 1/(f-a))$) to denote the counting function of poles of the function $1/(f-a)$ with multiplicities $\leq k$ (or $> k$) in $\{z : t < |z| \leq 1\}$, each point counted only once. Similarly, we have the notations $\overline{N}_1^{k)}(t, f)$, $\overline{N}_1^{(k}(t, f)$, $\overline{N}_2^{k)}(t, f)$, $\overline{N}_2^{(k}(t, f)$, and $\overline{N}_0^{k)}(t, f)$, $\overline{N}_0^{(k}(t, f)$.

For a nonconstant meromorphic function f on the annulus $\mathbb{A} = \{z : 1/R_0 < |z| < R_0\}$, where $1 < R < R_0 \leq +\infty$, the following properties will be used in this paper (see [5]):

(i) $T_0(R, f) = T_0(R, 1/f)$,

(ii) $\max\{T_0(R, f_1 \cdot f_2), T_0(R, f_1/f_2), T_0(R, f_1 + f_2)\} \leq T_0(R, f_1) + T_0(R, f_2) + O(1)$,

(iii) $T_0(R, 1/(f-a)) = T_0(R, f) + O(1)$, for every fixed $a \in \mathbb{C}$.

Khrystiyanyn and Kondratyuk [6] also obtained the lemma on the logarithmic derivative on the annulus $\mathbb{A}$.

Theorem 1 (see [6] lemma on the logarithmic derivative). *Let f be a nonconstant meromorphic function on the annulus $\mathbb{A} = \{z : 1/R_0 < |z| < R_0\}$, where $R_0 \leq +\infty$, and let $\lambda > 0$. Then*

(i) *in the case $R_0 = +\infty$,*

$$m_0\left(R, \frac{f'}{f}\right) = O(\log(RT_0(R, f))) \quad (4)$$

for $R \in (1, +\infty)$ except for the set Δ_R such that $\int_{\Delta_R} R^{\lambda-1} dR < +\infty$,

(ii) *if $R_0 < +\infty$, then*

$$m_0\left(R, \frac{f'}{f}\right) = O\left(\log\left(\frac{T_0(R, f)}{R_0 - R}\right)\right) \quad (5)$$

for $R \in (1, R_0)$ except for the set Δ'_R such that $\int_{\Delta'_R} (dR/(R_0 - R)^{\lambda-1}) < +\infty$.

In 2005, The second fundamental theorem on the annulus $\mathbb{A}$ was first obtained by Khrystiyanyn and Kondratyuk [6]. Cao et al. [11] introduce other forms of the second fundamental theorem on annuli as follows.

Theorem 2 ([11, Theorem 2.3] The second fundamental theorem). *Let f be a nonconstant meromorphic function on the annulus $\mathbb{A} = \{z : (1/R_0) < |z| < R_0\}$, where $1 < R_0 \leq +\infty$. Let $a_1, a_2, \ldots, a_q$ be q distinct complex numbers in the extended complex plane $\overline{\mathbb{C}}$. Let $k_1, k_2, \ldots, k_q$ be q positive integers, and let $\lambda \geq 0$. Then*

$$(q-2)T_0(R, f) < \sum_{j=1}^{q} \overline{N}_0\left(R, \frac{1}{f - a_j}\right) + S(R, f), \quad (6)$$

where (i) in the case $R_0 = +\infty$,

$$S(R, f) = O(\log(RT_0(R, f))) \quad (7)$$

for $R \in (1, +\infty)$ except for the set Δ_R such that $\int_{\Delta_R} R^{\lambda-1} dR < +\infty$,

(ii) *if $R_0 < +\infty$, then*

$$S(R, f) = O\left(\log\left(\frac{T_0(R, f)}{R_0 - R}\right)\right) \quad (8)$$

for $R \in (1, R_0)$ except for the set Δ'_R such that $\int_{\Delta'_R} (dR/(R_0 - R)^{\lambda-1}) < +\infty$.

Definition 3. Let $f(z)$ be a nonconstant meromorphic function on the annulus $\mathbb{A} = \{z : 1/R_0 < |z| < R_0\}$, where $1 < R_0 \leq +\infty$. The function f is called a transcendental or admissible meromorphic function on the annulus $\mathbb{A}$ which provided that

$$\limsup_{R\to\infty} \frac{T_0(R, f)}{\log R} = \infty, \quad 1 < R < R_0 = +\infty \quad (9)$$

or

$$\limsup_{R\to R_0} \frac{T_0(R, f)}{-\log(R_0 - R)} = \infty, \quad 1 < R < R_0 < +\infty, \quad (10)$$

respectively.

Thus for a transcendental or admissible meromorphic function on the annulus $\mathbb{A}$, $S(R, f) = o(T_0(R, f))$ holds for all $1 < R < R_0$ except for the set Δ_R or the set Δ'_R mentioned in Theorem 1, respectively.

Definition 4 (see [10], Definition 2.1). Let $f(z)$ be a nonconstant meromorphic function on the annulus $\mathbb{A} = \{z : 1/R_0 < |z| < R_0\}$, where $R_0 = +\infty$. Then the order of $f(z)$ is defined by

$$\rho(f) = \limsup_{R\to+\infty} \frac{T_0(R, f)}{\log R}. \quad (11)$$

Definition 5 (see [10], Definition 3.1). Let $f(z)$ be a nonconstant meromorphic function of order ρ $(0 < \rho < \infty)$ on the annulus $\mathbb{A} = \{z : 1/R_0 < |z| < R_0\}$, where $1 < R_0 \le +\infty$, and let $a \in \widehat{\mathbb{C}}$. Then we say that a is

(i) an evB (exceptional value in the sense of Borel) for f on $\mathbb{A}$ for distinct zeros of order $\le k$ if $\overline{\rho}_k(a, f) < \rho$;

(ii) an evB (exceptional value in the sense of Borel) for f on $\mathbb{A}$ for distinct zeros if $\overline{\rho}(a, f) < \rho$;

(iii) an evB (the Borel exceptional value) for f on $\mathbb{A}$ if $\rho(a, f) < \rho$, where

$$\begin{aligned} \overline{\rho}_k(a,f) &= \limsup_{R\to+\infty} \frac{\overline{N}_0^{k)}(R, 1/(f-a))}{\log R}, \\ \overline{\rho}(a,f) &= \limsup_{R\to+\infty} \frac{\overline{N}_0(R, 1/(f-a))}{\log R}, \\ \rho(a,f) &= \limsup_{R\to+\infty} \frac{N_0(R, 1/(f-a))}{\log R}. \end{aligned} \tag{12}$$

In particular, we say that a is an evB for f on $\mathbb{A}$ for simple zeros if $k = 1$ and a is an evB for f on $\mathbb{A}$ for simple and double zeros if $k = 2$.

3. The Main Results

Now, the main theorems of this paper are listed as follows.

Theorem 6. *Let f be transcendental meromorphic function of order ρ $(0 < \rho < \infty)$ on the annulus $\mathbb{A} := \{z : 1/R_0 < |z| < R_0\}$, where $R_0 = +\infty$. If there exist $a_1^1, \ldots, a_{p_1}^1, a_1^2, \ldots, a_{p_2}^2, \ldots, a_1^\nu, \ldots, a_{p_\nu}^\nu \in \widehat{\mathbb{C}}$ such that $a_1^i, a_2^i, \ldots, a_{p_i}^i$ are evBs for f on $\mathbb{A}$ for distinct zeros of order $\le k_i$, $i = 1, 2, \ldots, \nu$, where $p_1, \ldots, p_\nu, \nu \in \mathbb{N}_+$, and $k_1, k_2, \ldots, k_\nu$ are positive integers or infinity. Then*

$$\Xi := \sum_{i=1}^{\nu} \frac{p_i k_i}{1+k_i} + \sum_{i=1}^{\nu} \frac{1}{1+k_i} \sum_{j=1}^{p_i} \delta_0^{k_i}(a_j^i, f) \le 2. \tag{13}$$

Remark 7. Under the assumptions of Theorem 6, if $\nu = 3$, $p_1 = r$, $p_2 = t$, $p_3 = s$ and $k_1 = k$, $k_2 = l$, $k_3 = m$, since $\delta_0^{k_i}(a_j^i, f) \ge 0$, then we can get

$$\begin{aligned} \frac{kr}{k+1} + \frac{tl}{l+1} + \frac{sm}{m+1} &\le \sum_{i=1}^{\nu} \frac{p_i k_i}{1+k_i} \\ &+ \sum_{i=1}^{\nu} \frac{1}{1+k_i} \sum_{j=1}^{p_i} \delta_0^{k_i}(a_j^i, f) \le 2. \end{aligned} \tag{14}$$

This shows that Theorem 6 is an improvement of results given by Chen and Wu [10].

Definition 8. For positive integers k, m, we define that

$$\begin{aligned} \delta_0^k(a,f) &= 1 - \limsup_{R\to+\infty} \frac{N_0^k(R, 1/(f-a))}{T_0(R,f)}, \\ \Theta_0(a,f) &= 1 - \limsup_{R\to+\infty} \frac{\overline{N}_0(R, 1/(f-a))}{T_0(R,f)}, \end{aligned} \tag{15}$$

where $N_0^k(R, 1/(f-a))$ is the counting function of a-points of f on $\mathbb{A}$ where an a-point of multiplicity m is counted m times if $m \le k$ and $1+k$ times if $m > k$. In particular, if $k = \infty$, then

$$\begin{aligned} \delta_0^\infty(a,f) = \delta_0(a,f) &= \liminf_{R\to+\infty} \frac{m_0(R, 1/(f-a))}{T_0(R,f)} \\ &= 1 - \limsup_{R\to+\infty} \frac{N_0(R, 1/(f-a))}{T_0(R,f)}. \end{aligned} \tag{16}$$

Theorem 9. *Let f be transcendental meromorphic function of order ρ $(0 < \rho < \infty)$ on the annulus $\mathbb{A} := \{z : 1/R_0 < |z| < R_0\}$, where $R_0 = +\infty$. If there exist $a \in \widehat{\mathbb{C}}$ and two positive integers k and p such that*

$$(1+k)\Theta_0(a,f) + \sum_{b\ne a} \delta_0(b,f) > 2 - k(p-1), \tag{17}$$

then there exist at most p elements $\widehat{\mathbb{C}} \setminus \{a\}$ which are evBs for f on $\mathbb{A}$ for distinct zeros of order not exceeding k.

3.1. Proof of Theorem 6

Proof. For any positive integer k or ∞ and $a \in \widehat{\mathbb{C}}$, we have

$$\begin{aligned} \overline{N}_0\left(R, \frac{1}{f-a}\right) &\le \frac{k}{1+k} \overline{N}_0^{k)}\left(R, \frac{1}{f-a}\right) \\ &+ \frac{1}{1+k} N_0^k\left(R, \frac{1}{f-a}\right), \end{aligned} \tag{18}$$

where $k/(k+1) = 1$ and $1/(k+1) = 0$ if $k = \infty$. Then, from (18) and Theorem 2, we have

$$\begin{aligned} &\left(\sum_{i=1}^{\nu} p_i - 2\right) T_0(R,f) \\ &\quad \le \sum_{i=1}^{\nu} \frac{k_i}{1+k_i} \sum_{j=1}^{p_i} \overline{N}_0^{k_i)}\left(R, \frac{1}{f-a_j^i}\right) \\ &\quad + \sum_{i=1}^{\nu} \frac{1}{1+k_i} \sum_{j=1}^{p_i} N_0^{k_i}\left(R, \frac{1}{f-a_j^i}\right) + S(R,f). \end{aligned} \tag{19}$$

From Definition 5 and the assumptions of Theorem 6, there exists a constant η $(0 < \eta < \rho)$ such that for sufficiently large R,

$$\overline{N}_0^{k_i)}\left(R, \frac{1}{f-a_j^i}\right) < R^\eta, \quad j = 1, 2, \ldots, p_i, i = 1, 2, \ldots, \nu. \tag{20}$$

Hence, from (19) and (20) and for sufficiently large R, we have

$$\left(\sum_{i=1}^{\nu} p_i - 2\right) T_0(R,f) \leq \sum_{i=1}^{\nu} \frac{1}{1+k_i} \sum_{j=1}^{p_i} N_0^{k_i}\left(R, \frac{1}{f-a_j^i}\right) + O(R^{\eta}) + S(R,f). \quad (21)$$

Thus, for sufficiently large R and taking arbitrary $\varepsilon (> 0)$, we can get from (21) and the definition of $\delta_k(a,f)$ that

$$\left(\sum_{i=1}^{\nu} p_i - 2\right) T_0(R,f) \leq \sum_{i=1}^{\nu} \frac{1}{1+k_i} \sum_{j=1}^{p_i} \left(1 - \delta_{k_i}^{\theta}\left(a_j^i, f\right) + \varepsilon\right) T_0(R,f) + O(R^{\eta}) + S(R,f). \quad (22)$$

Since $R = +\infty$, it follows that

$$(\Xi - 2 - \varepsilon) T_0(R,f) \leq O(R^{\eta}) + O(\log R T_0(R,f)), \quad (23)$$

for $R \in (1, +\infty)$ except for the set Δ_R such that $\int_{\Delta_R} R^{\lambda-1} dR < +\infty$. If $I = [R, R') \in \Delta_R$, we have

$$R'^{\lambda} - R^{\lambda} = \lambda \int_R^{R'} x^{\lambda-1} dx \leq \lambda \int_{\Delta_R} R^{\lambda-1} dR < M, \quad (24)$$

where M is a constant. Thus,

$$(\Xi - 2 - \varepsilon) T_0(R,f) \leq O(R^{\eta}) + O(\log R T_0(R,f)) \quad (25)$$

holds for all R. If $\Xi > 2$, we can choose an arbitrary ε $(0 < \varepsilon < \Xi - 2)$ satisfying $\eta + \varepsilon < \rho$. Thus, from (25) and for sufficiently large R, we can get a contradiction to (13) easily.

Therefore, we can get the conclusion of Theorem 6. □

3.2. The Proof of Theorem 9

Lemma 10. *Let f be transcendental meromorphic function of order ρ $(0 < \rho < \infty)$ on the annulus $\mathbb{A} := \{z : 1/R_0 < |z| < R_0\}$, where $1 < R_0 \leq +\infty$. Then*

$$\limsup_{R \to +\infty} \frac{m_0(R, 1/f')}{T_0(R,f)} \leq 2 - \Theta_0(\infty, f) - \sum_{b \in \mathbb{C}} \delta_0(b,f). \quad (26)$$

Proof. Suppose that $b_1, b_2, \ldots, b_t \in \mathbb{C}$ are t distinct complex constants. From the definitions of $m_0(R,f), T_0(R,f)$, we have

$$\sum_{i=1}^{t} m_0\left(r, \frac{1}{f-b_i}\right) \leq m_0\left(r, \frac{1}{f'}\right) + S(R,f) \leq T_0(R,f') - N_0\left(R, \frac{1}{f'}\right) + S(R,f). \quad (27)$$

And since $T_0(R,f') \leq T_0(R,f) + \overline{N}_0(R,f) + S(R,f)$, then we have

$$\sum_{i=1}^{t} m_0\left(R, \frac{1}{f-b_i}\right) + N_0\left(R, \frac{1}{f'}\right) \leq T_0(R,f) + \overline{N}_0(R,f) + S(R,f). \quad (28)$$

It follows from the definitions of $\delta_0(a,f), \Theta_0(a,f)$ that

$$\sum_{i=1}^{t} \delta_0(b_i, f) + \limsup_{R \to +\infty} \frac{N_0(R, 1/f')}{T_0(R,f)} \leq 2 - \Theta_0(\infty, f). \quad (29)$$

From (29) and t is arbitrary, the proof of Lemma 10 is completed easily. □

Proof of Theorem 9. Without loss of generality, we assume that $a = \infty$. Next, we will employ reductio ad absurdum to prove the conclusion of Theorem 9. Suppose that there exist $p+1$ elements $a_1, a_2, \ldots, a_{p+1} \in \mathbb{C}$ which are evBs for f on $\mathbb{A}$ for distinct zeros of multiplicity $\leq k$. We know that if z_0 is a zero of $f - b$ on $\mathbb{A}$ of multiplicity s (> 1) for $b \in \mathbb{C}$, then z_0 is a zero of f' on $\mathbb{A}$ of multiplicity $s - 1$. It follows that

$$\sum_{i=1}^{p+1} \overline{N}_0\left(R, \frac{1}{f-a_i}\right) \leq \sum_{i=1}^{p+1} \overline{N}_0^{k)}\left(R, \frac{1}{f-a_i}\right) + \frac{1}{k} N_0\left(R, \frac{1}{f'}\right). \quad (30)$$

Therefore, by Theorem 2 we have

$$\begin{aligned} pT_0(R,f) &\leq \sum_{i=1}^{p+1} \overline{N}_0\left(R, \frac{1}{f-a_i}\right) + \overline{N}_0(R,f) + S(R,f) \\ &\leq \sum_{i=1}^{p+1} \overline{N}_0^{k)}\left(R, \frac{1}{f-a_i}\right) + \frac{1}{k} N_0\left(R, \frac{1}{f'}\right) \\ &\quad + \overline{N}_0(R,f) + S(R,f). \end{aligned} \quad (31)$$

From the definitions of order and evB of f on $\mathbb{A}$, there exists a constant η $(0 < \eta < \rho)$ such that for sufficiently large R, we have

$$\overline{N}_0^{k)}\left(R, \frac{1}{f-a_i}\right) < R^{\eta}, \quad i = 1, 2, \ldots, p+1. \quad (32)$$

From (31), (32), and Lemma 10, for sufficiently large R, it follows that

$$\left(p - 1 - \frac{1}{k}\left(2 - \Theta_0(\infty, f) - \sum_{b \neq \infty} \delta_0(b,f)\right) + \Theta_0(\infty, f)\right) \times T_0(R,f) \leq R^{\eta} + S(R,f), \quad (33)$$

for $R \in (1, +\infty)$ except for the set Δ_R such that $\int_{\Delta_R} R^{\lambda-1} dR < +\infty$. Since $R_0 = +\infty$, similar to the argument as in Theorem 6, from (33) we find that

$$\left(p - 1 - \frac{2 - (k+1)\Theta_0(\infty, f) - \sum_{b \neq \infty} \delta_0(b, f)}{k}\right) T_0(R, f) \leq R^{\eta} + O(\log(RT_0(R, f))) \tag{34}$$

holds for all R. Since $\eta < \rho$, from (34) and for sufficiently large R, we can get

$$(1+k)\Theta_0(\infty, f) + \sum_{b \neq \infty} \delta_0(b, f) > 2 - k(p-1), \tag{35}$$

which is contradiction with the assumption of Theorem 9.

Thus, this completes the proof of Theorem 9. □

4. Some Consequences of Theorems 6 and 9

In this section, we will give some consequences of Theorems 6 and 9. Before we give these results, some definitions will be introduced below.

Definition 11. Let f be meromorphic function on the annulus $\mathbb{A} := \{z : 1/R_0 < |z| < R_0\}$, where $R_0 = +\infty$. For $a \in \widehat{\mathbb{C}}$, then we say that

(i) a is called an exceptional value in the sense of Nevanlinna (evN for short) for f on $\mathbb{A}$, if $\delta_0(a, f) > 0$;

(ii) a is called a normal value in the sense of Nevanlinna (nvN for short) for f on $\mathbb{A}$, if $\delta_0(a, f) = 0$.

In addition, similar to the Picard exceptional value in the whole complex plane, we define that a is called an exceptional value in the sense of Picard (evP for short) for f on $\mathbb{A}$, if f has at most a finite number of a-points on $\mathbb{A}$.

Consequence 1. Under the assumptions of Theorem 6, if $k_1 = 1$, from Theorem 6, we get

$$p_1 + 2\sum_{i=2}^{\nu} \frac{p_i k_i}{1+k_i} + \sum_{j=1}^{p_1} \delta_0^1(a_j^1, f) + \sum_{i=2}^{\nu} \frac{2}{1+k_i} \sum_{j=1}^{p_i} \delta_0^1(a_j^i, f) \leq 4. \tag{36}$$

Since $p_i \geq 0$ and $\delta_0^1(a_j^i, f) \geq 0$ for $i = 1, \ldots, \nu$, it follows that

(i) if f has an evB for simple zeros which is also an evN for f on $\mathbb{A}$, then f has at most three evBs for simple zeros on $\mathbb{A}$;

(ii) if a_1 and a_2 are two evPs for f on $\mathbb{A}$ then no other element is an evB for f for simple zeros on $\mathbb{A}$;

(iii) there exist at most four elements which are evBs for f simple zeros on $\mathbb{A}$ since $\delta_1(a_j^1, f) \geq 0$. Moreover, all these four values are nvNs for f on $\mathbb{A}$.

Consequence 2. Under the assumptions of Theorem 6, let $k_1 = 1, p_1 = 1$, then

(a) if $k_2 = 2$, we have

$$\frac{2p_2}{3} + \sum_{i=3}^{\nu} \frac{p_i k_i}{1+k_i} + \frac{1}{2}\delta_0^1(a_1^1, f) + \frac{1}{3}\sum_{j=1}^{p_2} \delta_0^2(a_j^2, f) + \sum_{i=3}^{\nu} \frac{1}{1+k_i} \sum_{j=1}^{p_i} \delta_0^{k_i}(a_j^i, f) \leq \frac{3}{2}. \tag{37}$$

Since $p_i \geq 0$ and $\delta_0^{k_i}(a_j^i, f) \geq 0$, it follows from (37) that $p_2 < 3$. Thus, if a_1^1 is an evB for f on $\mathbb{A}$ for simple zeros, that is, $\delta_0^1(a_1^1, f) > 0$, then there exist at most two other elements which are evBs for f on $\mathbb{A}$ for distinct simple zeros and double zeros. Furthermore,

(i) if $\delta_0^1(a_1^1, f) = 1/3$, then two other elements evBs are also evNs for f on $\mathbb{A}$;

(ii) if any of the two other elements a_1^2 and a_2^2, say a_1^2, satisfies $\delta_0^1(a_1^2, f) = 1/2$, then a_1^1, a_2^2 are also evNs for f on $\mathbb{A}$;

(b) if $k_2 = 3$, we have

$$\frac{3p_2}{4} + \sum_{i=3}^{\nu} \frac{p_i k_i}{1+k_i} + \frac{1}{2}\delta_0^1(a_1, f) + \frac{1}{4}\sum_{j=1}^{p_2} \delta_0^3(a_j^2, f) + \sum_{i=3}^{\nu} \frac{1}{1+k_i} \sum_{j=1}^{p_i} \delta_0^{k_i}(a_j^i, f) \leq \frac{3}{2}. \tag{38}$$

From the above inequality and $\delta_0(a, f) \geq 0$ for $a \in \widehat{\mathbb{C}}$, we get that $p_1 = 1$, $p_2 \leq 2$ and $p_1 = 1$, $p_2 = 2$, $p_i = 0$ $(i = 3, \ldots, \nu)$. Thus, if f has an evB for simple zeros on $\mathbb{A}$, then there exist at most two other elements which are evBs for f distinct zeros of multiplicity ≤ 3 on $\mathbb{A}$. Moreover, all these exceptional values are nvNs for f on $\mathbb{A}$.

Consequence 3. Under the assumptions of Theorem 6, if $k_1 = 2$, then we have

$$\frac{2p_1}{3} + \sum_{i=2}^{\nu} \frac{p_i k_i}{1+k_i} + \frac{1}{3}\sum_{j=1}^{p_1} \delta_0^2(a_j^1, f) + \sum_{i=2}^{\nu} \frac{1}{1+k_i} \sum_{j=1}^{p_i} \delta_0^{k_i}(a_j^i, f) \leq 2. \tag{39}$$

Since $\sum_{i=2}^{\nu}(p_i k_i/(1+k_i)) + \sum_{i=2}^{\nu}(1/(1+k_i)) \sum_{j=1}^{p_i} \delta_0^{k_i}(a_j^i, f) \geq 0$, then from (39) we have

$$\frac{2p_1}{3} + \frac{1}{3}\sum_{j=1}^{p_1} \delta_0^2(a_j^1, f) \leq 2. \tag{40}$$

Thus, it follows that $p_1 \leq 3$ and $p_1 = 3$, $p_i = 0$, $i = 2, \ldots, \nu$. Hence, we have that f has at most three evBs for distinct simple and double zeros on $\mathbb{A}$. Moreover, all three evBs for distinct simple and double zeros are nvNs for f on $\mathbb{A}$.

Consequence 4. Under the assumptions of Theorem 6, let $k_1 = 2, p_1 = 1$.

(a) If $k_2 = 1$, then we have

$$p_2 + 2\sum_{i=3}^{\nu} \frac{p_i k_i}{1+k_i} + \frac{2}{3}\delta_0^2\left(a_1^1, f\right) + \sum_{j=1}^{p_2} \delta_0^1\left(a_j^2, f\right) + \sum_{i=3}^{\nu} \frac{2}{1+k_i} \sum_{j=1}^{p_i} \delta_0^{k_i}\left(a_j^i, f\right) \le \frac{8}{3}. \quad (41)$$

Thus, it follows that $p_2 \le 2$. So, if there exists an evB for f on $\mathbb{A}$ for distinct and double zeros, say a_1^1, then there exist at most two other evBs for f on $\mathbb{A}$ for simple zeros, say a_1^2, a_2^2. Furthermore, if $p_1 = 1, p_2 = 2$, it follows that

$$2\delta_0^2\left(a_1^1, f\right) + 3\delta_0^1\left(a_1^2, f\right) + \delta_0^1\left(a_2^2, f\right) \le 2. \quad (42)$$

Thus, we can see that any one of a_1^2, a_2^2 may not be an evP for f on $\mathbb{A}$. Furthermore, if a_1^1 is an evP for f on $\mathbb{A}$. Then a_1^2, a_2^2 are nvNs for f on $\mathbb{A}$;

(b) if $p_2 = 1, k_2 = 3, k_3 = 1$, then we have

$$p_3 + \sum_{i=4}^{\nu} \frac{2p_i k_i}{1+k_i} + \frac{2}{3}\delta_0^2\left(a_1^1, f\right) + \frac{1}{2}\delta_0^3\left(a_1^2, f\right) + \sum_{j=1}^{p_3} \delta_0^1\left(a_j^3, f\right) + \sum_{i=4}^{\nu} \frac{2}{1+k_i} \sum_{j=1}^{p_i} \delta_0^{k_i}\left(a_j^i, f\right) \le \frac{7}{6}. \quad (43)$$

Thus, we can get that $p_3 \le 1$. That is, if f have two evBs on $\mathbb{A}$, say a_1^1 and a_1^2 and a_1^1 is for distinct simple and double zeros, a_1^2 for distinct zeros of order ≤ 3, then f has at most one evB for simple zeros on $\mathbb{A}$. Furthermore, if $p_3 = 1$, then

$$\frac{2}{3}\delta_0^2\left(a_1^1, f\right) + \frac{1}{2}\delta_0^3\left(a_1^2, f\right) + \delta_0^1\left(a_1^3, f\right) \le \frac{1}{6}. \quad (44)$$

It follows that $\delta_0^2(a_1^1, f) \le 1/4, \delta_0^3(a_1^2, f) \le 1/3$ and $\delta_0^1(a_1^3, f) \le 1/6$. Thus, if any equality of these three inequalities holds, then the other two are nvNs for f on $\mathbb{A}$.

Now, we will give some consequences of Theorem 9 as follows.

Consequence 5. Under the assumptions of Theorem 9, if $k = 1$ and

$$2\Theta_0(a, f) + \sum_{b \ne a} \delta_0(b, f) > 3 - p, \quad (45)$$

we have the following:

(i) if $p = 1$ and $2\Theta_0(a, f) + \sum_{b \ne a} \delta_0(b, f) > 2$, then there exists at most one element $b \ne a$ which is an evB for f on $\mathbb{A}$ for simple zeros; in particular, this holds if there exists an $a \in \widehat{\mathbb{C}}$ satisfying $\Theta_0(a, f) = 1$;

(ii) if $p = 2$ and $2\Theta_0(a, f) + \sum_{b \ne a} \delta_0(b, f) > 1$, then there exists at most two elements $b_1, b_2 \ne a$ which are evBs for f on $\mathbb{A}$ for simple zeros; in particular, this holds if there exists an $a \in \widehat{\mathbb{C}}$ satisfying $\Theta_0(a, f) > 1/2$;

(iii) if $p = 3$ and $2\Theta_0(a, f) + \sum_{b \ne a} \delta_0(b, f) > 0$, then there exists at most three elements $b_1, b_2, b_3 \ne a$ which are evBs for f on $\mathbb{A}$ for simple zeros; in particular, this holds if there exists an $a \in \widehat{\mathbb{C}}$ satisfying $\Theta_0(a, f) > 0$.

Remark 12. Under the assumptions of Theorem 9, from Consequence 5, we have that if there exist four distinct elements $b_1, b_2, b_3, b_4 \in \widehat{\mathbb{C}}$ which are evBs for f on $\mathbb{A}$ for simple zeros, then $\Theta_0(b_i, f) \le 1/2$ and $\Theta_0(a, f) = 0$ for $a \ne b_i$ and $i = 1, 2, 3, 4$.

Consequence 6. Under the assumptions of Theorem 9, if $k = 2$ and

$$3\Theta_0(a, f) + \sum_{b \ne a} \delta_0(b, f) > 4 - 2p, \quad (46)$$

we have the following:

(i) if $p = 1$ and $3\Theta_0(a, f) + \sum_{b \ne a} \delta_0(b, f) > 2$, then there exists at most one element $b \ne a$ which is an evB for f on $\mathbb{A}$ for distinct simple and double zeros; in particular, this holds if there exists an $a \in \widehat{\mathbb{C}}$ satisfying $\Theta_0(a, f) > 2/3$;

(ii) if $p = 2$ and $3\Theta_0(a, f) + \sum_{b \ne a} \delta_0(b, f) > 0$, then there exists at most two elements $b_1, b_2 \ne a$ which are evBs for f on $\mathbb{A}$ for distinct simple and double zeros; in particular, this holds if there exists an $a \in \widehat{\mathbb{C}}$ satisfying $\Theta_0(a, f) > 0$.

Remark 13. Under the assumptions of Theorem 9, from Consequence 6, we have that if there exist four distinct elements $b_1, b_2, b_3 \in \widehat{\mathbb{C}}$ which are evBs for f on $\mathbb{A}$ for distinct simple and double zeros, then $\Theta_0(b_i, f) \le 2/3$ and $\Theta_0(a, f) = 0$ for $a \ne b_i$ and $i = 1, 2, 3$.

5. Exceptional Value of Meromorphic Function and Its Derivative

In this section, we will study the exceptional value of meromorphic function f on $\mathbb{A}$ and its differential polynomial of the form

$$M[f] := M\left[f, f', \ldots, f^{(k)}\right] = af(z)^{i_0}\left(f'\right)^{i_1} \cdots \left(f^{(k)}\right)^{i_k}, \quad (47)$$

where $i_0, i_1, \ldots, i_k \in \mathbb{N}$, $k \in \mathbb{N}_+$, and $a(\ne 0) \in \mathbb{C}$ is a complex constant.

Theorem 14. *Let f be transcendental meromorphic function of order ρ $(0 < \rho < \infty)$ on the annulus $\mathbb{A} := \{z : 1/R_0 < |z| < R_0\}$, where $R_0 = +\infty$. If 0 and ∞ are evBs for f on $\mathbb{A}$ for distinct zeros, then $M[f]$ has no evB for simple zeros in $\widehat{\mathbb{C}} \setminus \{0, \infty\}$.*

Proof. From (21), set $\sigma = i_0 + i_1 + \cdots + i_k$. Since f is a transcendental meromorphic function on $\mathbb{A}$, similar to the method of [12, Theorem 1.21], we can get that f and $M[f]$

have the same order ρ on $\mathbb{A}$. By applying Theorem 1 for $M[f]$, for $b(\neq 0) \in \mathbb{C}$ we have

$$\begin{aligned} T_0(R, M[f]) &\leq \overline{N}_0\left(R, \frac{1}{M[f]}\right) + \overline{N}_0(R, M[f]) \\ &\quad + \overline{N}_0\left(R, \frac{1}{M[f]-b}\right) + S(R, M[f]) \\ &\leq \overline{N}_0\left(R, \frac{1}{M[f]}\right) + \overline{N}_0(R, f) \\ &\quad + \frac{1}{2}\overline{N}_0^{1)}\left(R, \frac{1}{M[f]-b}\right) \\ &\quad + \frac{1}{2}\overline{N}_0^{1}\left(R, \frac{1}{M[f]-b}\right) + S(R, f) \\ &\leq \overline{N}_0\left(R, \frac{1}{M[f]}\right) + \overline{N}_0(R, f) \\ &\quad + \frac{1}{2}\overline{N}_0^{1)}\left(R, \frac{1}{M[f]-b}\right) \\ &\quad + \frac{1}{2}T_0(R, f) + S(R, f). \end{aligned} \tag{48}$$

Since

$$\begin{aligned} \overline{N}_0\left(R, \frac{1}{M[f]}\right) &\leq N_0\left(R, \frac{f^\sigma}{M[f]}\right) + \overline{N}_0\left(R, \frac{1}{f^\sigma}\right) \\ &\leq T_0\left(R, \frac{M[f]}{f^\sigma}\right) + \overline{N}_0\left(R, \frac{1}{f}\right) + S(R, f) \\ &\leq N_0\left(R, \frac{M[f]}{f^\sigma}\right) + \overline{N}_0\left(R, \frac{1}{f}\right) + S(R, f) \\ &\leq O\left(\overline{N}_0\left(R, \frac{1}{f}\right) + \overline{N}_0(R, f)\right) + S(R, f). \end{aligned} \tag{49}$$

and 0 and ∞ are evBs for f on $\mathbb{A}$ for distinct zeros, then there exists a constant η $(0 < \eta < \rho)$ such that for sufficiently large R, we have

$$\overline{N}_0\left(R, \frac{1}{f}\right) < R^\eta, \qquad \overline{N}_0(R, f) < R^\eta. \tag{50}$$

Suppose that b is an evB for $M[f]$ on $\mathbb{A}$ for simple zeros, since $M[f]$ is of order ρ. Then for sufficiently large R, there exists a constant $\beta < \rho$ such that

$$\overline{N}_0^{1)}\left(R, \frac{1}{M[f]-b}\right) < R^\beta. \tag{51}$$

Thus, from (48)–(50), it follows that

$$\frac{1}{2}T_0(R, M[f]) \leq O\left(r^{\beta'}\right) + O(\log RT_0(R, f)), \tag{52}$$

where $\beta' = \max\{\eta, \beta\}$ and for $R \in (1, +\infty)$ except for the set Δ_R such that $\int_{\Delta_R} R^{\lambda-1} dR < +\infty$. Since $M[f]$ is of order ρ, then there exists a sequence $\{R_n\}$ tending to ∞ such that

$$T_0(R_n, M[f]) > R_n^\alpha, \tag{53}$$

where α $(\beta' < \alpha < \rho)$ is a constant. Thus, we can get a contradiction by using the same argument as in Theorem 6.

Therefore, $M[f]$ has no evB for simple zeros in $\widehat{\mathbb{C}}\setminus\{0, \infty\}$. □

Acknowledgments

The author was supported by the NNSF of China (no. 61202313) and the Natural Science Foundation of Jiang-Xi Province in China (2010GQS0119 and 20132BAB211001).

References

[1] H. S. Gopalakrishna and S. S. Bhoosnurmath, "Exceptional values of meromorphic functions," *Annales Polonici Mathematici*, vol. 32, pp. 83–93, 1976.

[2] H. S. Gopalakrishna and S. Subhas Bhoosnurmath, "Exceptional values of meromorphic functions and its derivatives," *Annales Polonici Mathematici*, vol. 35, pp. 99–105, 1977-1978.

[3] S. K. Singh and H. S. Gopalakrishna, "Exceptional values of entire and meromorphic functions," *Mathematische Annalen*, vol. 191, no. 2, pp. 121–142, 1971.

[4] Y. H. Peng and D. C. Sun, "Examples on exceptional values of meromorphic functions," *Acta Mathematica Scientia*, vol. 31, no. 4, pp. 1327–1336, 2011.

[5] A. Ya. Khrystiyanyn and A. A. Kondratyuk, "On the Nevanlinna theory for meromorphic functions on annuli. I," *Matematychni Studii*, vol. 23, pp. 19–30, 2005.

[6] A. Ya. Khrystiyanyn and A. A. Kondratyuk, "On the Nevanlinna theory for meromorphic functions on annuli. II," *Matematychni Studii*, vol. 24, pp. 57–68, 2005.

[7] A. A. Kondratyuk and I. Laine, "Meromorphic functions in multiply connected domains," in *Fourier Series Methods in Complex Analysis. Proceedings of the Workshop*, I. Laine, Ed., pp. 24–29, University of Joensuu, Department of Mathematics, Mekrijärvi, Finland, July 2005, Report Series. Department of Mathematics, University of Joensuu, vol. 10, 9–111, 2006.

[8] A. Fernández, "On the value distribution of meromorphic function in the punctured plane," *Matematychni Studii*, vol. 34, pp. 136–144, 2010.

[9] H. Y. Xu and Z. X. Xuan, "The uniqueness of analyticfunctions on annuli sharing some values," *Abstract and Applied Analysis*, vol. 2012, Article ID 896596, 13 pages, 2012.

[10] Y. X. Chen and Z. J. Wu, "Exceptional values of a meromorphic function and its derivatives on annuli," *Annales Polonici Mathematici*, vol. 105, pp. 167–177, 2012.

[11] T.-B. Cao, H.-X. Yi, and H.-Y. Xu, "On the multiple values and uniqueness of meromorphic functions on annuli," *Computers and Mathematics with Applications*, vol. 58, no. 7, pp. 1457–1465, 2009.

[12] H. X. Yi and C. C. Yang, *Uniqueness Theory of Meromorphic Functions*, Kluwer Academic, Dordrecht, The Netherlands, 2003, Chinese Original: Science Press, Beijing, China, 1995.

26

On kth-Order Slant Weighted Toeplitz Operator

S. C. Arora[1] and Ritu Kathuria[2]

[1] *Department of Mathematics, University of Delhi, Delhi 110007, India*
[2] *Department of Mathematics, Motilal Nehru College, University of Delhi, Delhi 110007, India*

Correspondence should be addressed to Ritu Kathuria; ritu.kathuria@yahoo.co.in

Academic Editors: A. Agouzal and A. Ibeas

Let $\beta = \{\beta_n\}_{n\in\mathbb{Z}}$ be a sequence of positive numbers with $\beta_0 = 1$, $0 < \beta_n/\beta_{n+1} \le 1$ when $n \ge 0$ and $0 < \beta_n/\beta_{n-1} \le 1$ when $n \le 0$. A kth-order slant weighted Toeplitz operator on $L^2(\beta)$ is given by $U_\phi = W_k M_\phi$, where M_ϕ is the multiplication on $L^2(\beta)$ and W_k is an operator on $L^2(\beta)$ given by $W_k e_{nk}(z) = (\beta_n/\beta_{nk})e_n(z)$, $\{e_n(z) = z^k/\beta_k\}_{k\in\mathbb{Z}}$ being the orthonormal basis for $L^2(\beta)$. In this paper, we define a kth-order slant weighted Toeplitz matrix and characterise U_ϕ in terms of this matrix. We further prove some properties of U_ϕ using this characterisation.

1. Introduction

Toeplitz operators and slant Toeplitz operators [1] have been found immensely useful, especially in the study of prediction theory [2], wavelet analysis [3], and solution of differential equations [4]. Originally, these operators were defined and studied on the usual H^2 and L^2 spaces. During the past few decades, different generalisations of these spaces, like the weighted function spaces H_w^P and L_w^P and the weighted sequence spaces $H^2(\beta)$ and $L^2(\beta)$ have developed. Shields [5] has made a systematic study of the multiplication operator on $L^2(\beta)$. Lauric [6] has discussed the Toeplitz operator on $H^2(\beta)$. Motivated by these studies, we introduced and studied the notion of a slant weighted Toeplitz operator [7] on $L^2(\beta)$. In this paper, we study a kth-order slant weighted Toeplitz operator on the space $L^2(\beta)$. We begin with the following preliminaries.

Let $\beta = \{\beta_n\}_{n\in\mathbb{Z}}$ be a sequence of positive numbers with $\beta_0 = 1$, $0 < (\beta_n/\beta_{n+1}) \le 1$ when $n \ge 0$ and $0 < (\beta_n/\beta_{n-1}) \le 1$ when $n \le 0$.

Throughout the paper, we assume that for a fixed integer $k \ge 2$, $\beta_{k_n}/\beta_n \le M < \infty$. Consider the following spaces [5, 6]:

$$L^2(\beta) = \left\{ f(z) = \sum_{n=-\infty}^{\infty} a_n z^n \,\middle|\, a_n \in \mathbb{C}, \|f\|_\beta^2 = \sum_{n=-\infty}^{\infty} |a_n|^2 \beta_n^2 < \infty \right\},$$
$$H^2(\beta) = \left\{ f(z) = \sum_{n=0}^{\infty} a_n z^n \,\middle|\, a_n \in \mathbb{C}, \|f\|_\beta^2 = \sum_{n=0}^{\infty} |a_n|^2 \beta_n^2 < \infty \right\}. \tag{1}$$

Then, $(L^2(\beta), \|\cdot\|_\beta)$ is a Hilbert space [6] with an orthonormal basis given by $\{e_k(z) = z^k/\beta_k\}_{k\in\mathbb{Z}}$ and with an inner product defined by

$$\left\langle \sum_{n=-\infty}^{\infty} a_n z^n, \sum_{n=-\infty}^{\infty} b_n z^n \right\rangle = \sum_{n=-\infty}^{\infty} a_n \bar{b}_n \beta_n^2. \tag{2}$$

Also, $H^2(\beta)$ is a subspace of $L^2(\beta)$. Further, the space

$$L^{\infty}(\beta) = \left\{ \phi(z) = \sum_{n=-\infty}^{\infty} a_n z^n \;\middle|\; \phi L^2(\beta) \subseteq L^2(\beta),\ \exists c \in R \right. \\ \left. \text{such that } \|\phi f\|_{\beta} \le c \|f\|_{\beta}\ \forall f \in L^2(\beta) \right\} \tag{3}$$

is a Banach space with respect to the norm defined by

$$\|\phi\|_{\infty} = \inf\left\{ c \mid \|\phi f\|_{\beta} \le c \|f\|_{\beta}\ \forall f \in L^2(\beta) \right\}. \tag{4}$$

The mapping $P : L^2(\beta) \to H^2(\beta)$ is the orthogonal projection of $L^2(\beta)$ onto $H^2(\beta)$. For a given $\phi \in L^{\infty}(\beta)$, the induced weighted multiplication operator [5] is denoted by M_{ϕ} and is given by $M_{\phi} : L^2(\beta) \to L^2(\beta)$ such that

$$M_{\phi} e_k(z) = \frac{1}{\beta_k} \sum_{n=-\infty}^{\infty} a_n \beta_{n+k} e_{n+k}(z) = \frac{1}{\beta_k} \sum_{n=-\infty}^{\infty} a_{n-k} \beta_n e_n(z). \tag{5}$$

If we put $\phi_1(z) = z$, then $M_{\phi_1} = M_z$ is the operator defined as $M_z e_k(z) = w_k e_{k+1}(z)$, where $w_k = \beta_{k+1}/\beta_k$ for all $k \in \mathbb{Z}$, and it is known as a weighted shift [5].

The slant weighted Toeplitz operator [7] A_{ϕ} is an operator on $L^2(\beta)$ defined as $A_{\phi} : L^2(\beta) \to L^2(\beta)$ such that $A_{\phi} e_k(z) = (1/\beta_k) \sum_{n=-\infty}^{\infty} a_{2n-k} \beta_n e_n(z)$.

If $W : L^2(\beta) \to L^2(\beta)$ is such that

$$W e_{2n}(z) = \frac{\beta_n}{\beta_{2n}} e_n(z), \quad W e_{2n-1}(z) = 0 \quad \forall n \in \mathbb{Z}, \tag{6}$$

then A_{ϕ} can be alternately defined by

$$A_{\phi}(f) = W M_{\phi}(f) = W(\phi f) \quad \forall f \in L^2(\beta). \tag{7}$$

2. *k*th-Order Slant Weighted Toeplitz Operator

Suppose that the operator $W_k : L^2(\beta) \to L^2(\beta)$ is such that

$$W_k e_n(z) = \begin{cases} \dfrac{\beta_{n/k}}{\beta_n} e_{n/k}(z) & \text{if } n \text{ is divisible by } k, \\ 0 & \text{otherwise,} \end{cases} \tag{8}$$

Then the matrix of W_k is

$$\left[\begin{array}{c|cccccccccc} \cdots & \cdots & \cdots & \cdots & \cdots & \cdots & \cdots & \cdots & \cdots & \cdots & \cdots \\ \hline \cdots & \frac{\beta_0}{\beta_0} & 0 & 0 & 0 & 0 & 0 & 0 & 0 & \cdots & \\ \cdots & \underbrace{0 \quad 0 \quad 0 \quad 0}_{k\text{-times}} & & & & \frac{\beta_1}{\beta_k} & 0 & \cdots & \cdots & \cdots & \\ \cdots & 0 & 0 & 0 & 0 & \cdots & \cdots & 0 & \frac{\beta_2}{\beta_{2k}} & \cdots & \\ \cdots & 0 & 0 & \cdots & \cdots & \cdots & \cdots & \cdots & \cdots & \cdots & \\ \cdots & \cdots & \cdots & \cdots & \cdots & \cdots & \cdots & \cdots & \cdots & \cdots & \end{array}\right]. \tag{9}$$

The adjoint of W_k is given by

$$W_k^{*} e_n(z) = \frac{\beta_n}{\beta_{kn}} e_{kn}(z) \quad \forall n \in \mathbb{Z}. \tag{10}$$

Definition 1 (see [8]). For an integer $k \ge 2$, the kth order slant weighted Toeplitz operator $U_{\phi} : L^2(\beta) \to L^2(\beta)$ is such that $U_{\phi}(f) = W_k M_{\phi}(f)$ for all $f \in L^2(\beta)$. Thus, $U_{\phi} e_j(z) = (1/\beta_j) \sum_{n=-\infty}^{\infty} a_{nk-j} \beta_n e_n(z)$.

The (i, j)th entry of the matrix of U_{ϕ} is given by

$$\begin{aligned} \left\langle U_{\phi} e_j, e_i \right\rangle &= \left\langle \frac{1}{\beta_j} \sum_{n=-\infty}^{\infty} a_{nk-j} \beta_n e_n(z), e_i(z) \right\rangle \\ &= a_{ik-j} \frac{\beta_i}{\beta_j}. \end{aligned} \tag{11}$$

Hence, the matrix of U_{ϕ} with respect to this basis is as follows:

$$\left[\begin{array}{cc|cccc} \cdots & \cdots & \cdots & \cdots & \cdots & \cdots \\ \cdots & a_{-k+1}\frac{\beta_{-1}}{\beta_{-1}} & a_{-k}\frac{\beta_1}{\beta_0} & a_{-k-1}\frac{\beta_{-1}}{\beta_1} & a_{-k-2}\frac{\beta_{-1}}{\beta_2} & \cdots \\ \hline \cdots & a_1\frac{\beta_0}{\beta_{-1}} & a_0\frac{\beta_0}{\beta_0} & a_{-1}\frac{\beta_0}{\beta_1} & a_{-2}\frac{\beta_0}{\beta_2} & \cdots \\ \cdots & a_{k+1}\frac{\beta_1}{\beta_{-1}} & a_k\frac{\beta_1}{\beta_0} & a_{k-1}\frac{\beta_1}{\beta_1} & a_{k-2}\frac{\beta_1}{\beta_2} & \cdots \\ \cdots & a_{2k+1}\frac{\beta_2}{\beta_{-1}} & a_{2k}\frac{\beta_2}{\beta_0} & a_{2k-1}\frac{\beta_2}{\beta_1} & a_{2k-2}\frac{\beta_2}{\beta_2} & \cdots \\ \cdots & \cdots & \cdots & \cdots & \cdots & \cdots \end{array}\right]. \tag{12}$$

Theorem 2. *The mapping* $\phi \to U_{\phi}$ *is linear and one to one.*

Proof. For linearity, consider

$$\begin{aligned} U_{\alpha\phi+\beta\psi} &= W_k M_{\alpha\phi+\beta\psi} \\ &= W_k \left(M_{\alpha\phi} + M_{\beta\psi} \right) \\ &= W_k M_{\alpha\phi} + W_k M_{\beta\psi} \\ &= \alpha W_k M_{\phi} + \beta W_k M_{\psi} \\ &= \alpha U_{\phi} + \beta U_{\psi}. \end{aligned} \tag{13}$$

For checking one-to-one, let $U_{\phi} = U_{\psi}$.

Then, $U_{\phi} - U_{\psi} = 0$. By linearity, we get $U_{\phi-\psi} = 0$.

Hence,

$$\begin{aligned} & U_{\phi-\psi} e_n(z) = 0, \quad n \in \mathbb{Z}, \\ \Longrightarrow\ & W_k M_{\phi-\psi} e_n(z) = 0, \quad n \in \mathbb{Z}, \\ \Longrightarrow\ & W_k (\phi - \psi) e_n(z) = 0, \quad n \in \mathbb{Z}. \end{aligned} \tag{14}$$

This implies that either $\phi - \psi = 0$ or the degree of $(\phi - \psi) e_n(z)$ is not divisible by k. But since this is true for all $n \in \mathbb{Z}$, the second possibility is ruled out. Hence, $\phi - \psi = 0$ or $\phi = \psi$. □

Theorem 3. *Consider the following: (i)* $M_z W_k = W_k M_{z^k}$; *(ii)* $M_{z^m} W_k = W_k M_{z^{km}}$, $m \in \mathbb{Z}$.

Proof. (i) It is sufficient to prove that

$$M_z W_k e_n(z) = W_k M_{z^k} e_n(z) \quad \forall n \in \mathbb{Z}. \tag{15}$$

Suppose that n is not a multiple of k. Then,

$$M_z W_k e_n(z) = M_z 0 = 0 = W_k M_{z^k} e_n(z). \tag{16}$$

Now, when $n = pk$ (multiple of k),

$$\begin{aligned} M_z W_k e_n(z) &= M_z \frac{\beta_p}{\beta_{pk}} e_p(z) \\ &= \frac{\beta_{p+1}}{\beta_{pk}} e_{p+1}(z). \end{aligned} \tag{17}$$

On the other hand,

$$\begin{aligned} W_k M_{z^k} e_n(z) &= W_k M_{z^k} e_{pk}(z) \\ &= W_k \frac{\beta_{(p+1)k}}{\beta_{pk}} e_{(p+1)k}(z) \\ &= \frac{\beta_{p+1}}{\beta_{pk}} e_{p+1}(z). \end{aligned} \tag{18}$$

Hence from (17) and (18), we get that

$$M_z W_k e_n(z) = W_k M_{z^k} e_n(z) \quad \text{whenever } n \text{ is a multiple of } k. \tag{19}$$

We therefore conclude that for all $n \in \mathbb{Z}$,

$$\begin{aligned} & M_z W_k e_n(z) = W_k M_{z^k} e_n(z) \\ & \Longrightarrow M_z W_k = W_k M_{z^k}. \end{aligned} \tag{20}$$

(ii) We prove the result by induction on m.
For $m = 1$, the result is true by part (i).
Suppose that the result is true for $m = l$. Then,

$$M_{z^l} W_k = W_k M_{z^{lk}}. \tag{21}$$

Now

$$\begin{aligned} M_{z^{(l+1)k}} W_k &= M_z M_{z^l} W_k \\ &= M_z W_k M_{z^{lk}} \\ &= W_k M_{z^k} M_{z^{lk}} \\ &= W_k M_{z^{(l+1)k}}. \end{aligned} \tag{22}$$

Hence by induction, the result is true for all $m \in Z^+$.
For $m = -1$ and $1 \le j \le k-1$,

$$M_{z^m} W_k e_{nk+j}(z) = M_{z^{-1}} \quad 0 = 0 = W_k M_{z^{-k}} e_{nk+j}(z). \tag{23}$$

For $m = -1$ and $j = nk$ (a multiple of k),

$$\begin{aligned} M_{z^m} W_k e_j(z) &= M_{z^{-1}} W_k e_{nk}(z) \\ &= M_{z^{-1}} \frac{\beta_n}{\beta_{nk}} e_n(z) \\ &= \frac{\beta_n}{\beta_{nk}} M_{z^{-1}} e_n(z) \\ &= \frac{\beta_n}{\beta_{nk}} \frac{\beta_{n-1}}{\beta_n} e_{n-1}(z) \\ &= \frac{\beta_{n-1}}{\beta_{nk}} e_{n-1}(z). \end{aligned} \tag{24}$$

On the other hand,

$$\begin{aligned} W_k M_{z^{-1}} e_{nk}(z) &= W_k \left(z^{-1} \frac{z^{nk}}{\beta_{nk}} \right) \\ &= W_k \frac{z^{(n-1)k}}{\beta_{nk}} \frac{\beta_{(n-1)k}}{\beta_{(n-1)k}} \\ &= \frac{\beta_{(n-1)k}}{\beta_{nk}} W_k e_{(n-1)k}(z) \\ &= \frac{\beta_{(n-1)k}}{\beta_{nk}} \frac{\beta_{n-1}}{\beta_{(n-1)k}} e_{n-1}(z) \\ &= \frac{\beta_{n-1}}{\beta_{nk}} e_{n-1}(z). \end{aligned} \tag{25}$$

From (23), (24), and (25), we conclude that

$$M_{z^{-1}} W_k = W_k M_{z^{-k}}. \tag{26}$$

Hence, the result is true for $m = -1$ also. Therefore, by using induction, we can prove it for all negative integers m. The case when $m = 0$ is trivially true.

Hence the theorem is true. □

Corollary 4. *Let* $\phi(z) = \sum_{n=-\infty}^{\infty} a_n z^n \in L^\infty(\beta)$. *Then,*

$$M_{\phi(z)} W_k = W_k M_{\phi(z^k)}. \tag{27}$$

Theorem 5. *Consider the following:* $M_{\phi(z)} U_{\psi(z)} = U_{\phi(z^k)\psi(z)}$.

Proof. The proof is as follows:

$$\begin{aligned} M_{\phi(z)} U_{\psi(z)} &= M_{\phi(z)} W_k M_{\psi(z)} \\ &= W_k M_{\phi(z^k)} M_{\psi(z)} \\ &= W_k M_{\phi(z^k)\psi(z)} \\ &= U_{\phi(z^k)\psi(z)}. \end{aligned} \tag{28}$$

□

3. Matrix Characterisation of U_ϕ

Definition 6. Let $w_n = \beta_{n+1}/\beta_n$ for all $n \in \mathbb{Z}$. Then, the kth order slant weighted Toeplitz matrix corresponding to the weight sequence $\langle w_n \rangle$ is a bilaterally infinite matrix $\langle \lambda_{ij} \rangle$ such that

$$\lambda_{i+1,j+k} = \frac{w_i}{w_j w_{j+1} \cdots w_{j+k-1}} \lambda_{i,j}, \quad i, j = 0, \pm 1, \pm 2, \ldots. \quad (29)$$

We have proved earlier [8] that U is a kth order slant weighted Toeplitz operator if and only if $M_z U = U M_{z^k}$.

Next, we prove a characterisation of the kth order slant weighted Toeplitz operator in terms of the matrix previously defined.

Theorem 7. *A necessary and sufficient condition that an operator U on $L^2(\beta)$ is a kth order slant weighted Toeplitz operator is that its matrix with respect to the orthonormal basis $\{e_k(z) = z^k/\beta_k\}_{k\in\mathbb{Z}}$ is a kth order slant weighted Toeplitz matrix.*

Proof. Suppose that U is a kth order slant weighted Toeplitz operator. Then, the corresponding matrix $\langle \lambda_{ij} \rangle$ is given by

$$\lambda_{ij} = \left\langle U e_j, e_i \right\rangle = a_{ki-j} \frac{\beta_i}{\beta_j}. \quad (30)$$

Further,

$$\begin{aligned} \lambda_{i+1,j+k} &= \left\langle U e_{j+k}, e_{i+1} \right\rangle \\ &= a_{ki+k-j-k} \frac{\beta_{i+1}}{\beta_{j+k}} \\ &= a_{ki-j} \frac{\beta_{i+1}}{\beta_{j+k}} \\ &= \frac{\beta_{i+1}}{\beta_{j+k}} \frac{\beta_j}{\beta_i} \lambda_{i,j} \\ &= \frac{w_i}{w_j w_{j+1} \cdots w_{j+k-1}} \lambda_{i,j}, \end{aligned} \quad (31)$$

where $w_n = \beta_{n+1}/\beta_n$ for all $n \in \mathbb{Z}$. Thus, the matrix of U is a kth order slant weighted Toeplitz matrix. Conversely, we assume that U is an operator on $L^2(\beta)$ whose matrix is a kth order slant weighted Toeplitz matrix. This means that for all $i, j \in \mathbb{Z}$, we have

$$\left\langle U e_j, e_i \right\rangle = \frac{w_j w_{j+1} \cdots w_{j+k-1}}{w_i} \left\langle U e_{j+k}, e_{i+1} \right\rangle. \quad (32)$$

Now,

$$\begin{aligned} \left\langle M_z U e_j, e_i \right\rangle &= \left\langle U e_j, M_z^* e_i \right\rangle \\ &= \left\langle U e_j, w_{i-1} e_{i-1} \right\rangle \\ &= w_{i-1} \left\langle U e_j, e_{i-1} \right\rangle \\ &= w_{i-1} \frac{w_j w_{j+1} \cdots w_{j+k-1}}{w_{i-1}} \\ &\quad \times \left\langle U e_{j+k}, e_i \right\rangle \quad (\text{using } (32)) \\ &= \left\langle U M_{z^k} e_j, e_i \right\rangle, \quad \forall i, j \in \mathbb{Z}. \end{aligned} \quad (33)$$

Hence $M_z U = U M_{z^k}$. Therefore, we conclude that U is a kth order slant weighted Toeplitz operator. □

Next, consider the operator $S : L^2(\beta) \to L^2(\beta)$ given by $S e_j = (1/w_j) e_{j+1}$. Then, $S^* e_j = (1/w_{j-1}) e_{j-1}$.

Now, S is bounded as $\langle w_n \rangle$ is always positive and bounded.

Lemma 8. *Consider the following: $S^* = M_z^{-1}$.*

Proof. The proof is as follows:

$$\begin{aligned} S^* M_z e_j &= S^* w_j e_{j+1} = \frac{w_j}{w_j} e_j = e_j, \\ M_z S^* e_j &= M_z \frac{1}{w_{j-1}} e_{j-1} = \frac{w_{j-1}}{w_{j-1}} e_j = e_j. \end{aligned} \quad (34)$$

□

Theorem 9. *A bounded operator U on $L^2(\beta)$ is a kth order slant weighted Toeplitz operator if and only if $U = M_z^{-1} U M_{z^k}$, where M_z and M_{z^k}, are the multiplication operators induced by z and z^k, respectively.*

Proof. Let U be a kth order slant weighted Toeplitz operator on $L^2(\beta)$. Then, from (32) we get that

$$\begin{aligned} \left\langle U e_j, e_i \right\rangle &= \frac{w_j w_{j+1} \cdots w_{j+k-1}}{w_i} \left\langle U e_{j+k}, e_{i+1} \right\rangle \\ &= \left\langle U M_{z^k} e_j, S e_i \right\rangle \\ &= \left\langle S^* U M_{z^k} e_j, e_i \right\rangle \\ &= \left\langle M_z^{-1} U M_{z^k} e_j, e_i \right\rangle, \quad i, j = 0, \pm 1, \pm 2, \ldots. \end{aligned} \quad (35)$$

Thus, $U = M_z^{-1} U M_{z^k}$.

For the converse part, we assume that U is a bounded operator on $L^2(\beta)$ satisfying $U = M_z^{-1} U M_{z^k}$ for some fixed integer $k \geq 2$. Then, for all $i, j \in \mathbb{Z}$,

$$\begin{aligned} \left\langle U e_j, e_i \right\rangle &= \left\langle M_z^{-1} U M_{z^k} e_j, e_i \right\rangle \\ &= \left\langle S^* U M_{z^k} e_j, e_i \right\rangle \\ &= \left\langle U M_{z^k} e_j, S e_i \right\rangle \frac{w_j w_{j+1} \cdots w_{j+k-1}}{w_i} \left\langle U e_{j+k}, e_{i+1} \right\rangle. \end{aligned} \quad (36)$$

The previous equation shows that the matrix corresponding to U is a kth order slant weighted Toeplitz matrix. Hence, by Theroem 7, U is a kth order slant weighted Toeplitz operator. □

Corollary 10. *For a fixed integer $k \geq 2$, the set of all kth order slant weighted Toeplitz operators is weakly closed and hence strongly closed.*

Proof. We assume that for each positive integer n, U_n is a kth order slant weighted Toeplitz operator and $U_n \to U$ weakly. Then, for $f, g \in L^2(\beta)$, we get that $\langle U_n f, g \rangle \to \langle U f, g \rangle$.

From the previous theorem, this implies that

$$\begin{aligned}\left\langle M_z^{-1}U_nM_{z^k}f,g\right\rangle &= \left\langle S^*U_nM_{z^k}f,g\right\rangle \\ &= \left\langle U_nM_{z^k}f,Sg\right\rangle \\ &= \left\langle U_nz^k\cdot f,Sg\right\rangle \longrightarrow \left\langle Uz^k\cdot f,Sg\right\rangle \\ &= \left\langle M_z^{-1}UM_{z^k}f,g\right\rangle.\end{aligned} \tag{37}$$

But, for each n, $U_n = M_z^{-1}U_nM_{z^k}$.

Hence, $U = M_z^{-1}UM_{z^k}$. Therefore, U is a kth order slant weighted Toeplitz operator. □

4. Compression to $H^2(\beta)$

Definition 11. The compression of a kth order slant weighted Toeplitz operator on $H^2(\beta)$ is denoted by V_ϕ and defined as $V_\phi e_j(z) = (1/\beta_j)\sum_{n=0}^{\infty} a_{nk-j}\beta_n e_n(z)$. $j = 0,1,2,\ldots$.

Alternatively,

$$\begin{aligned}V_\phi &= PU_{\phi|_{H^2(\beta)}} \\ &= PW_kM_{\phi|_{H^2(\beta)}} \\ &= W_kPM_{\phi|_{H^2(\beta)}} \quad (\text{since } P \text{ reduces } W_k \text{ [8]}) \\ &= W_kT_\phi,\end{aligned} \tag{38}$$

where $T_\phi : H^2(\beta) \rightarrow H^2(\beta)$ is the Toeplitz operator on $H^2(\beta)$ induced by $\phi \in L^\infty(\beta)$. The matrix of V_ϕ is unilaterally infinite and has the form

$$\begin{bmatrix} a_0\dfrac{\beta_0}{\beta_0} & a_{-1}\dfrac{\beta_0}{\beta_1} & a_{-2}\dfrac{\beta_0}{\beta_2} & \cdots \\ a_k\dfrac{\beta_1}{\beta_0} & a_{k-1}\dfrac{\beta_1}{\beta_1} & a_{k-2}\dfrac{\beta_1}{\beta_2} & \cdots \\ a_{2k}\dfrac{\beta_2}{\beta_0} & a_{2k-1}\dfrac{\beta_2}{\beta_1} & a_{2k-2}\dfrac{\beta_2}{\beta_2} & \cdots \\ \cdots & \cdots & \cdots & \cdots \end{bmatrix} \tag{39}$$

for a given $\phi = \sum_{n=-\infty}^{\infty} a_n z^n$.

Further, we observe the following.

(i) The mapping $\phi \rightarrow V_\phi$ is linear and one to one.

(ii) $T_\phi W_k = PM_\phi W_k = PW_kM_\phi(z^k) = PU_{\phi(z^k)} = V_\phi(z^k)$.

References

[1] M. C. Ho, "Properties of slant Toeplitz operators," *Indiana University Mathematics Journal*, vol. 45, no. 3, pp. 843–862, 1996.

[2] H. Helson and G. Szegö, "A problem in prediction theory," *Annali di Matematica Pura ed Applicata*, vol. 51, no. 1, pp. 107–138, 1960.

[3] L. Villemoes, "Wavelet analysis of refinement equations," *SIAM Journal on Mathematical Analysis*, vol. 25, pp. 1433–1460, 1994.

[4] T. Goodman, C. Micchelli, and J. Ward, "Spectral radius formula for subdivision operators," in *Recent Advances in Wavelet AnalysisAcademic Press*, L. Schumaker and G. Webb, Eds., pp. 335–360, 1994.

[5] A. L. Shields, *Weighted Shift Operators and Analytic Function Theory, Topics in Operator Theory*, vol. 13 of *Mathematical Survey*, American Mathematical Society, Providence, RI, New England, 1974.

[6] V. Lauric, "On a weighted Toeplitz operator and its commutant," *International Journal of Mathematics and Mathematical Sciences*, vol. 2005, no. 6, pp. 823–835, 2005.

[7] S. C. Arora and R. Kathuria, "On weighted Toeplitz operators," *Australian Journal of Mathematical Analysis and Applications*, vol. 8, no. 1, article 11, 2011.

[8] S. C. Arora and R. Kathuria, "Generalised slant weighted Toeplitz operators," *World Academy of Science, Engineering and Technology*, vol. 75, pp. 1103–1106, 2011.

Permissions

The contributors of this book come from diverse backgrounds, making this book a truly international effort. This book will bring forth new frontiers with its revolutionizing research information and detailed analysis of the nascent developments around the world.

We would like to thank all the contributing authors for lending their expertise to make the book truly unique. They have played a crucial role in the development of this book. Without their invaluable contributions this book wouldn't have been possible. They have made vital efforts to compile up to date information on the varied aspects of this subject to make this book a valuable addition to the collection of many professionals and students.

This book was conceptualized with the vision of imparting up-to-date information and advanced data in this field. To ensure the same, a matchless editorial board was set up. Every individual on the board went through rigorous rounds of assessment to prove their worth. After which they invested a large part of their time researching and compiling the most relevant data for our readers. Conferences and sessions were held from time to time between the editorial board and the contributing authors to present the data in the most comprehensible form. The editorial team has worked tirelessly to provide valuable and valid information to help people across the globe.

Every chapter published in this book has been scrutinized by our experts. Their significance has been extensively debated. The topics covered herein carry significant findings which will fuel the growth of the discipline. They may even be implemented as practical applications or may be referred to as a beginning point for another development. Chapters in this book were first published by Hindawi Publishing Corporation; hereby published with permission under the Creative Commons Attribution License or equivalent.

The editorial board has been involved in producing this book since its inception. They have spent rigorous hours researching and exploring the diverse topics which have resulted in the successful publishing of this book. They have passed on their knowledge of decades through this book. To expedite this challenging task, the publisher supported the team at every step. A small team of assistant editors was also appointed to further simplify the editing procedure and attain best results for the readers.

Our editorial team has been hand-picked from every corner of the world. Their multi-ethnicity adds dynamic inputs to the discussions which result in innovative outcomes. These outcomes are then further discussed with the researchers and contributors who give their valuable feedback and opinion regarding the same. The feedback is then collaborated with the researches and they are edited in a comprehensive manner to aid the understanding of the subject.

Apart from the editorial board, the designing team has also invested a significant amount of their time in understanding the subject and creating the most relevant covers. They scrutinized every image to scout for the most suitable representation of the subject and create an appropriate cover for the book.

The publishing team has been involved in this book since its early stages. They were actively engaged in every process, be it collecting the data, connecting with the contributors or procuring relevant information. The team has been an ardent support to the editorial, designing and production team. Their endless efforts to recruit the best for this project, has resulted in the accomplishment of this book. They are a veteran in the field of academics and their pool of knowledge is as vast as their experience in printing. Their expertise and guidance has proved useful at every step. Their uncompromising quality standards have made this book an exceptional effort. Their encouragement from time to time has been an inspiration for everyone.

The publisher and the editorial board hope that this book will prove to be a valuable piece of knowledge for researchers, students, practitioners and scholars across the globe.

List of Contributors

Capsts Adela Elisabeta
Technical University of Cluj-Napoca, Department of Mathematics, Cluj-Napoca, G. Baritiu 25-28, 400027 JudCluj, Romania

Rabia Aktaş
Department of Mathematics, Faculty of Science, Ankara University, Tandogan, 06100 Ankara, Turkey

Esra Erkuş-Duman
Department of Mathematics, Faculty of Science, Gazi University, Teknikokullar, 06500 Ankara, Turkey

Abdullah Shoaib and Muhammad Arshad
Department of Mathematics, International Islamic University, H-10, Islamabad 44000, Pakistan

Jamshaid Ahma
Department of Mathematics, COMSATS Institute of Information Technology, Chak Shahzad, Islamabad 44000, Pakistan

Chengjie Yu
Department of Mathematics, Shantou University, Shantou, Guangdong 515063, China

Jong-Wuu Wu, Sheau-Chiann Chen and Heng-Yi Lai
Department of Applied Mathematics, National Chiayi University, Chiayi, Taiwan

Wen-Chuan Lee
Department of International Business, Chang Jung Christian University, Tainan 71101, Taiwan

Tian Yue, Xiao-qiu Song and Dong-qing Li
College of Science, China University of Mining and Technology, Xuzhou, Jiangsu 221008, China

Shyam Lal and Manoj Kumar
Department of Mathematics, Faculty of Science, Banaras Hindu University, Varanasi 221005, India

M. Z. Ahmad
Institute of Engineering Mathematics, Universiti Malaysia Perlis, Kampus Tetap Pauh Putra, 02600 Arau, Perlis, Malaysia

M. K. Hasan
School of Information Technology, Faculty of Information Science and Technology, Universiti Kebangsaan Malaysia, 43600 Bangi, Selangor, Malaysia

S. Abbasbandy
Department of Mathematics, Imam Khomeini International University, Ghazvin 34149-16818, Iran

Bin Zheng
School of Science, Shandong University of Technology, Zibo, Shandong 255049, China

Muhammad Arshad
Department of Mathematics, International Islamic University, H-10, Islamabad 44000, Pakistan

Jamshaid Ahmad
Department of Mathematics, COMSATS Institute of Information Technology, Chak Shahzad, Islamabad 44000, Pakistan

Josef Diblík
Department of Mathematics and Descriptive Geometry, Faculty of Civil Engineering, Brno University of Technology, Veverı 331/95, 602 00 Brno, Czech Republic
Department of Mathematics, Faculty of Electrical Engineering and Communication, Brno University of Technology, Technicka 3058/10,61600 Brno, Czech Republic

Mária Kúdelliková
Department of Mathematics, University of Zˇilina, Univerzitna´ 8215/1, 01026 Zˇilina, Slovakia

Bo-Yan Xi
College of Mathematics, Inner Mongolia University for Nationalities, Tongliao City, Inner Mongolia 028043, China

Paul Barry
School of Science, Waterford Institute of Technology, Waterford, Ireland

Andrew N. Guarendi and Abhilash J. Chandy
Department of Mechanical Engineering,The University of Akron, Akron, OH 44325-3903, USA

Jun-Fang Zhang
Department of Basic Science, Beijing University of Agriculture, Beijing 102206, China

Huizeng Qin and Bin Zheng
School of Science, Shandong University of Technology, Zibo, Shandong 255049, China

Erdal Bas, Resat Yilmazer and Etibar Panakhov
Department of Mathematics, Firat University, 23119 Elazig, Turkey

Jinfeng Wang
School of Statistics and Mathematics, Inner Mongolia University of Finance and Economics, Hohhot 010070, China

Hong Li, Siriguleng He, Wei Gao and Yang Liu
School of Mathematical Sciences, Inner Mongolia University, Hohhot 010021, China

Dong Qiu, Chongxia Lu and Wei Zhang
College of Mathematics and Physics, Chongqing University of Posts and Telecommunications, Nanan, Chongqing 400065, China

Su Xia and Deng Chunhua
Faculty of Mathematics and Physics, Huaiyin Institute of Technology, Huai'an 223003, China

S. K. Malhotra
Department of Mathematics, Govt. S.G.S.P.G. College Ganj Basoda, Vidisha 464221, India

J. B. Sharma
Department of Mathematics, Choithram College of Professional Studies, Dhar Road, Indore 453001, India

Satish Shukla
Department of Applied Mathematics, Shri Vaishnav Institute of Technology & Science, Gram Baroli Sanwer Road, Indore 453331, India

Junhong Li and Ning Cui
Department of Mathematics and Sciences, Hebei Institute of Architecture and Civil Engineering, Zhangjiakou, Hebei 075000, China

Hong-Yan Xu
Department of Informatics and Engineering, Jingdezhen Ceramic Institute, Jingdezhen, Jiangxi 333403, China

S. C. Arora
Department of Mathematics, University of Delhi, Delhi 110007, India

Ritu Kathuria
Department of Mathematics, Motilal Nehru College, University of Delhi, Delhi 110007, India